Edited by
Michael Hirscher

Handbook of Hydrogen Storage

Further Reading

Barton, P. I., Mitsos, A. (eds.)

Microfabricated Power Generation Devices

Design and Technology

2009

ISBN: 978-3-527-32081-3

Züttel, A., Borgschulte, A., Schlapbach, L. (eds.)

Hydrogen as a Future Energy Carrier

2008

ISBN: 978-3-527-30817-0

Sundmacher, K., Kienle, A., Pesch, H. J., Berndt, J. F., Huppmann, G. (eds.)

Molten Carbonate Fuel Cells

Modeling, Analysis, Simulation, and Control

2007

ISBN: 978-3-527-31474-4

Hynes, J. T., Klinman, J. T., Limbach, H.-H., Schowen, R. L. (eds.)

Hydrogen-Transfer Reactions

4 Volumes

2007

ISBN: 978-3-527-30777-7

Malanowski, N., Heimer, T., Luther, W., Werner, M. (eds.)

Growth Market Nanotechnology

An Analysis of Technology and Innovation

2006

ISBN: 978-3-527-31457-7

Olah, G. A., Goeppert, A., Prakash, G. K. S.

Beyond Oil and Gas: The Methanol Economy

2006

ISBN: 978-3-527-31275-7

Kockmann, N. (ed.)

Micro Process Engineering

Fundamentals, Devices, Fabrication, and Applications

2006

ISBN: 978-3-527-31246-7

Edited by
Michael Hirscher

Handbook of Hydrogen Storage

New Materials for Future Energy Storage

WILEY-VCH Verlag GmbH & Co. KGaA

The Editor

Dr. Michael Hirscher
Max-Planck-Institut für
Metallforschung
Heisenbergstr. 3
70569 Stuttgart

Cover
Crystal structure of metal-organic framework MOF-5 constructed of ZnO_4 tetrahedra and carboxylate ligands. Background showing a scanning electron microscope picture of cubic MOF-5 crystals.

Library of Congress Card No.: applied for

British Library Cataloguing-in-Publication Data
A catalogue record for this book is available from the British Library.

Bibliographic information published by the Deutsche Nationalbibliothek
The Deutsche Nationalbibliothek lists this publication in the Deutsche Nationalbibliografie; detailed bibliographic data are available on the Internet at http://dnb.d-nb.de.

Typesetting Thomson Digital, Noida, India
Printing and Binding betz-druck GmbH, Darmstadt
Cover Design Adam-Design, Weinheim

Printed in the Federal Republic of Germany
Printed on acid-free paper

ISBN: 978-3-527-32273-2

Foreword

Our children need your great ideas and effort for hydrogen storage technologies innovations.
Sustainable/Hydrogen society construction rely on your brain and hand of research.
Katsuhiko Hirose, Project General Manager, Strategic Planning Department, Fuel Cell System Engineering Division, Toyota Motor Corporation

The human race has a long history of desiring mobility. I believe that is why even in the Stone Age and Bronze Age people traveled long distances to exchange information and goods.

In the last century this mobility expanded dramatically through the invention of the automobile and the use of oil as its fuel. This combination of oil and automobile seemed perfect until it became an environmental problem. Local environmental issues were partly solved by the introduction of catalysts and precise emission control through the tremendous efforts of catalyst material scientists and engineers. However, automobiles are again facing very high hurdles such as global warming and energy security issues.

In 2008 the world experienced the great shock of high oil prices. Not only the automobile industry but also everyday lives are massively affected by high oil prices. The cause was not the previously predicted scenario (peak oil), the problem was brought about without the collapse of any oil fields, just high oil prices created in the market. We have recognized that we now need to accelerate efforts to move away from oil as an automobile fuel and, at the same time, meet future requirements for reduced carbon dioxide emissions. There are only a few technologies able to achieve these targets while meeting user requirements such as low cost, long range travel and quick, easy refueling capabilities.

Hydrogen is the most promising technology and the automobile industry has spent billions of dollars to bring it onto the roads. As a result, vehicle technologies have reached very high levels, out-performing the current internal combustion engines two- or three-fold in terms of thermal efficiency, and at the same time reaching current vehicle levels of cold start capability down to $-30\ ^\circ C$ and other performance targets. However, the biggest and most difficult issue remaining is cost reduction of the technologies to current vehicle levels, which is essential for hydro-

Handbook of Hydrogen Storage. Edited by Michael Hirscher

ISBN: 978-3-527-32273-2

gen to be popular enough to solve both the energy security and carbon dioxide issues. There had seemed to be a similar pattern of cost reduction for fuel cell stacks and components to that for current technologies, since they both include steps such as reducing the number of parts and improving performance to reduce size and materials usage. However, hydrogen storage is a little bit different. Currently, the only available technology is a carbon fiber reinforced plastic composite high-pressure tank system, but this technology leaves several problems for future cost reduction. The quantities of expensive materials used constrain potential cost reduction and the large cylindrical shape limits installation of the tank in conventional vehicles.

The world's major automotive companies have announced fuel cell vehicle commercialization from 2015. The automotive industry recognizes that hydrogen storage still needs innovative ideas and materials in order to transform it into a popular technology for the future. The important requirements for hydrogen storage technologies are energy efficiency, size and weight, as well as easy adaptation to supply infrastructures and cost.

This problem cannot be solved by either academia or industry alone, so scientists and engineers must work together to achieve this difficult but indispensable task. I believe the current good collaboration will bring this about and we will be able to leave a clean, green globe for our children.

December 2009 *Katsuhiko Hirose*

Contents

Handbook of Hydrogen Storage. Edited by Michael Hirscher

ISBN: 978-3-527-32273-2

Preface

The limited fossil fuel resources and the environmental impact of their use require a change to renewable energy sources in the near future. For mobile application an efficient energy carrier is needed that can be produced and used in a closed cycle. Presently, hydrogen is the only energy carrier that can be produced easily in large amounts and in an appropriate time scale. Electric energy, either from renewable energies, for example, solar and wind, or future fusion reactors, can be used to produce hydrogen from water by electrolysis. The combustion of hydrogen leads again only to water and the cycle is closed. A comprehensive overview of the hydrogen cycle was given recently in the book "Hydrogen as a Future Energy Carrier" edited by Züttel *et al.* [1].

For individual motor car traffic, currently, three concepts are discussed by major auto manufacturers, fuel-cell vehicles, extended-range electric vehicles and battery-electric vehicles, for example, see the GM road map [2, 3] in Figure 1. The use of these three technologies depends on the application field and is related to the different energy densities of these energy carriers (see Figure 2). To achieve a driving range of 500 km for a conventional vehicle with today's Diesel technology requires a tank system weighing approximately 43 kg with a volume of 46 L. A zero-emission vehicle driven by a fuel cell with hydrogen will need a 700 bar high-pressure tank system of about 125 kg and 260 L to achieve the same driving range. A battery-electric vehicle will require an energy storage system (comprising battery cells, heat management and power electronics) which, using the most advanced Li-ion-battery technology (energy density 120 Wh kg^{-1}), would weigh almost 1000 kg with a volume of 670 L. This value assumes that the total electric energy of the battery is used, which would significantly reduce the cycle life of the device. More realistic is a maximum usage of 80% of the stored energy. The loading of a battery will take between several hours at a high-voltage high-current station and about one day at a conventional 230 V power socket with 16 A. Fast charging stations (about 50 kW, 30 to 90 min reloading), on the other hand, would require a very considerable infrastructure investment of several billion euros for a European nation such as France or Germany, comparable to a hydrogen infrastructure. The refueling of the high-pressure hydrogen vessel will take about 3 min, which is comparable to the Diesel tank. Furthermore, a hydrogen storage system for a driving range of 500 km will cost about 3000 US$ if produced in

Handbook of Hydrogen Storage. Edited by Michael Hirscher

ISBN: 978-3-527-32273-2

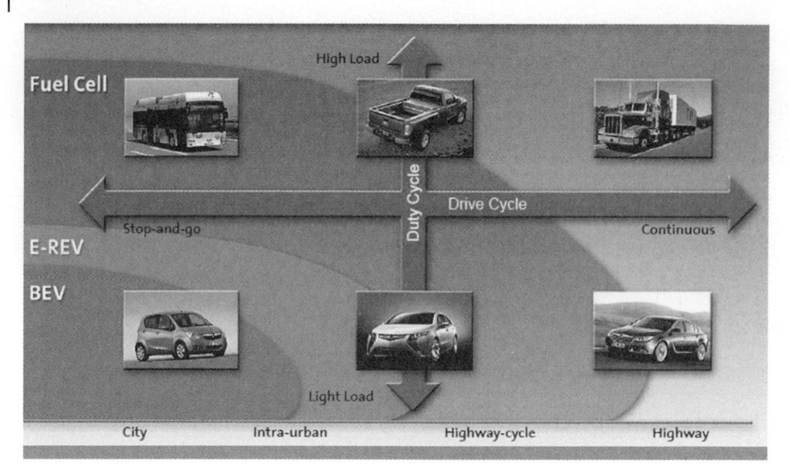

Figure 1 Application map for various electric vehicle technologies, from battery-electric vehicles (BEV), extended-range electric vehicles (E-REV) to fuel-cell vehicles, considering driving distance and load [2, 3].

high volumes. A comparable battery system would be in the price range of 50 000 US$.

According to these constraints, a battery-electric vehicle (BEV) would be the choice for light-weight vehicles and driving ranges up to 150 km. Extended-range electric vehicles (E-REV) could be the solution for users who only occasionally need longer driving ranges, up to 500 km. However, for long driving ranges or high load the fuel-cell vehicle possesses clear advantages due to zero emission in all operating

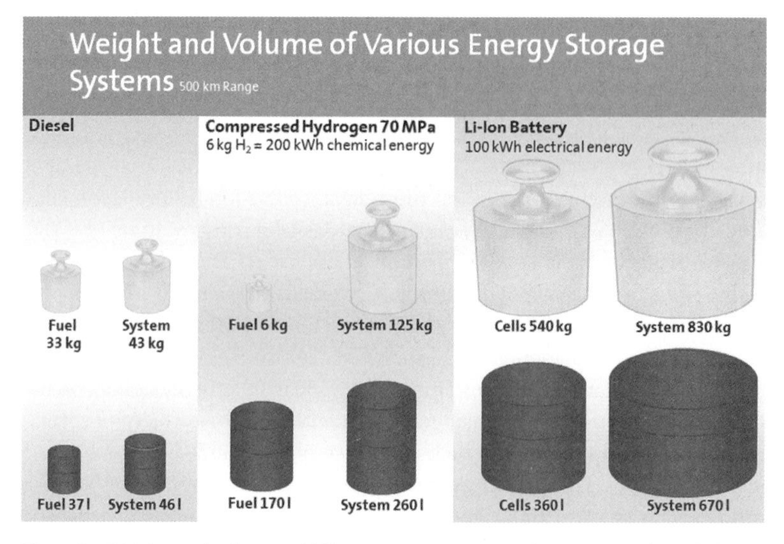

Figure 2 Weights and volumes of different energy carriers and systems used in vehicles to achieve a total driving range of 500 km [2, 3].

conditions and fast refueling times. Figure 1 shows an application map for the various technologies.

Nevertheless, current technologies, such as compressed gas or liquefied hydrogen, have severe disadvantages (especially in volumetric terms) compared to fossil fuels and the storage of hydrogen in light-weight solids could be the solution to further enhance the energy density of hydrogen tanks. The benchmark for all these solid-state systems needs to be an improvement over a 70 MPa tank system. In particular, fundamental knowledge of the atomistic processes is required to design optimized novel materials with new physical properties.

For automotive application of solid state absorbers, there are, more or less, two operating regimes envisaged to achieve such a technology breakthrough:

1) Room temperature storage at pressures up to 35 MPa
2) Cryogenic storage at pressures up to 5 MPa.

This handbook gives a comprehensive overview of novel solid hydrogen storage materials, highlighting their main advantages and drawbacks. The first chapter is devoted to the storage of hydrogen in a pure form giving the state-of-the-art for compressed and liquid hydrogen. In Chapter 2 adsorption materials for hydrogen storage by physisorption of hydrogen molecules are analyzed for possible cryo-adsorption systems. Chapter 3 describes the potential of clathrate hydrates as storage materials. Chapter 4 gives a review of conventional hydrides, including techniques to improve them by nanostructuring. Chapters 5–9 are devoted to novel light-weight materials, ranging from complex hydrides; amides, imides and mixtures; tailoring of reaction enthalpies; ammonia, borane and related compounds; and aluminum hydride. Finally, Chapter 10 concentrates on nanoparticles, mainly in systems confined by scaffold materials.

December 2009 *Michael Hirscher*

References

1 Züttel, A., Borgschulte, A., and Schlapbach L.,(eds) (2008) *Hydrogen as a Future Energy Carrier*, Wiley-VCH, Weinheim.

2 von Helmolt, R. and Eberle, U. (2007) *J. Power Sources*, **165**, 833–843.

3 Johnen F T., von Helmolt, R., and Eberle, U. Conference Proceedings "Technical Congress 2009" of the German Vehicle Manufacturers' Association VDA.

List of Contributors

Philipp Adelhelm
Utrecht University
Inorganic Chemistry and Catalysis
Debye Institute for Nanomaterials Science
Sorbonnelaan 16
3584 CA Utrecht
The Netherlands

Felix Baitalow
TU Bergakademie Freiberg
Department of Energy Process Engineering and Chemical Engineering
Fuchsmuehlenweg 9
Building 1
09596 Freiberg
Germany

Martin Dornheim
GKSS Research Centre Geesthacht
Institute of Materials Research
Max-Planck-Straße 1
21502 Geesthacht
Germany

Michael Felderhoff
Max-Planck-Institut für Kohlenforschung
Kaiser-Wilhelm-Platz 1
45470 Mülheim/Ruhr
Germany

Michael Hirscher
Max-Planck-Institut für Metallforschung
Heisenbergstr. 3
70569 Stuttgart
Germany

Jacques Huot
Université du Québec à Trois-Rivières
Institut de Recherche sur l'Hydrogène
3351 des Forges
PO Box 500, Trois-Rivières (Qc) G9A 5H7
Canada

Takayuki Ichikawa
Hiroshima University
Institute for Advanced Materials Research
1-3-1 Kagamiyama
Higashi-Hiroshima, 739-8530
Japan

Petra E. de Jongh
Utrecht University
Inorganic Chemistry and Catalysis,
Debye Institute for Nanomaterials Science
Sorbonnelaan 16
3584 CA Utrecht
The Netherlands

Handbook of Hydrogen Storage. Edited by Michael Hirscher

ISBN: 978-3-527-32273-2

Manfred Klell
Hydrogen Center Austria
Inffeldgasse 15
A-8010 Graz
Austria

Florian Mertens
TU Bergakademie Freiberg
Department of Physical Chemistry
Leipziger Str. 29
09596 Freiberg
Germany

Barbara Panella
ETH Zurich
Institute for Chemical and Bioengineering, Department of Chemistry and Applied Biosciences, Hönggerberg
8093 Zürich
Switzerland

and

Max-Planck-Institut für Metallforschung
Heisenbergstr. 3
70569 Stuttgart
Germany

Cor J. Peters
Delft University of Technology
Faculty of Mechanical, Maritime and Materials Engineering
Department of Process and Energy
Laboratory of Process Equipment
Leeghwaterstraat 44
2628 CA Delft
The Netherlands

and

The Petroleum Institute
Department of Chemical Engineering
P.O. Box 2533
Abu Dhabi
United Arab Emirates

Sona Raeissi
Shiraz University
Chemical and Petroleum Engineering Department
Shiraz 71345
Iran

Alireza Shariati
Shiraz University
Chemical and Petroleum Engineering Department
Shiraz 71345
Iran

Claudia Weidenthaler
Max-Planck-Institut für Kohlenforschung
Kaiser-Wilhelm-Platz 1
45470 Mülheim/Ruhr
Germany

Gert Wolf
TU Bergakademie Freiberg
Department of Physical Chemistry
Leipziger Str. 29
09596 Freiberg
Germany

Ragaiy Zidan
Savannah River National Laboratory
Energy Security Directorate
Aiken
SC 29803
USA

1 Storage of Hydrogen in the Pure Form

Manfred Klell

1.1 Introduction

Due to its low density, the storage of hydrogen at reasonable energy densities poses a technical and economic challenge. This chapter is dedicated to the storage of hydrogen in the pure form, which is defined by the thermodynamic variables of state and thus can best be analyzed on the basis of a depiction of these variables like the *T–s*-diagram.

Conventionally hydrogen is stored as compressed gas or as a cryogenic liquid. Apart from gravimetric and volumetric energy densities, the energies required for compression and liquefaction are evaluated. A short thermodynamic analysis of the storage infrastructure, including storage vessels, distribution, dispensary and refueling is given.

Hybrid storage of hydrogen, where a combination of technologies is applied, such as the storage of hydrogen as slush or as a supercritical fluid, is briefly mentioned.

A comparison of the energy densities of storage technologies for hydrogen and other energy carriers and a conclusion round off this chapter.

1.2 Thermodynamic State and Properties

Hydrogen is the most abundant element in the universe; more than 90% of all atoms are hydrogen. Hydrogen is the simplest atom and element number 1, consisting of one proton and one electron only. Apart from this ordinary isotope called protium, a small fraction of hydrogen atoms exist as deuterium (1 proton, 1 neutron, 1 electron) and an even smaller fraction as unstable tritium (1 proton, 2 neutrons, 1 electron).

An atomic property with relevance for the liquefaction of hydrogen molecules is its spin, the quantum analogy to the rotation of an elementary particle about its axis. If the spins of two hydrogen protons are parallel, the molecule is called ortho-hydrogen,

Handbook of Hydrogen Storage. Edited by Michael Hirscher

ISBN: 978-3-527-32273-2

if the spins are opposed, the molecule is called para-hydrogen. Ortho- and para-hydrogen have slightly different properties [29]. At standard (normal) conditions, molecular hydrogen is a mixture of about 75 vol% ortho- and 25 vol% para-hydrogen, which is called normal hydrogen. With a reduction in temperature, the content of para-hydrogen increases and reaches 100 vol% below −200 °C. The mixture of ortho- and para-hydrogen at thermodynamic equilibrium at a certain temperature is called equilibrium hydrogen. Para-hydrogen has a lower energy level than ortho-hydrogen, so during the liquefaction of hydrogen, additional energy has to be dissipated to convert ortho- to para-hydrogen.

Due to its single valence electron, hydrogen is very reactive and usually combines to yield the molecule H_2. On Earth hydrogen is rarely found in the pure form, but usually in a wide variety of inorganic and organic chemical compounds, the most common being water H_2O. Hydrogen forms chemical compounds (hydrides) with nearly all other elements. Due to their ability to form long chains and complex molecules, combinations with carbon play a key role for organic life (hydrocarbons, carbohydrate). Hydrogen is of crucial importance as an energy carrier in the metabolism of plants, animals, and humans. It is found in sugar, fat, proteins, alcohols, oils, and so on.

For the storage of pure hydrogen, H_2 has first to be separated from its compounds. A number of technologies exist for the production of hydrogen from different sources [3].

1.2.1 Variables of State

As the pure substance, hydrogen H_2 may exist in various physical phases as vapor, liquid and solid. As with any pure substance, the thermodynamic state of hydrogen is completely defined by specifying two independent intensive state variables. All other variables of state can then be determined by using one of the three relationships of state.

The *equation of state* is the mathematical relationship between the following three intensive thermodynamic properties: pressure p, temperature T, and specific volume v (or density $\varrho = 1/v$)

$$f_1(p, T, v) = 0 \tag{1.1}$$

The *calorific equation of state* relates the internal energy u to T and v or enthalpy h to T and p

$$f_2(u, T, v) = 0 \quad \text{or} \quad f_3(h, T, p) = 0 \tag{1.2}$$

The third state relationship relates entropy s to T and v or T and p

$$f_4(s, T, v) = 0 \quad \text{or} \quad f_5(s, T, p) = 0 \tag{1.3}$$

These relationships of state have to be determined experimentally and are available in the form of equations, tables or diagrams [21].

At ambient conditions, hydrogen fulfils well the relationship f_1 in the following simple form

$$p\,v = R\,T \quad \text{or} \quad p\,V_m = R_m\,T \tag{1.4}$$

In extensive form with mass

$$p\,V = n\,R_m\,T = m\,R\,T \tag{1.5}$$

With:

V [m^3]: volume, $V_m = V/n$ [m^3 mol^{-1}]: molar volume,
$n = N/N_A$ [mol]: number of mols, N: number of entities,
$N_A = 6.22 \cdot 10^{23}$ 1 mol^{-1}: Avogadro constant,
$M = m/n$ [kg kmol^{-1}, g mol^{-1}]: molar mass, m [kg]: mass,
$R_m = 8314.72$ J kmol^{-1} K^{-1}: universal gas constant,
$R = R_m/M$ [J kg^{-1} K^{-1}]: particular gas constant

A gas fulfilling Eq. (1.4) is called an ideal gas. Kinetic gas theory shows, that a gas behaves like an ideal gas, first if the he gas molecules are infinitesimally small, round, hard spheres occupying negligible volume and, secondly, if no forces exist amongst these molecules except during collisions. This holds true for most gases at low pressure and temperatures well above the critical temperature. This is the highest possible temperature at which a substance can condense.

An improvement of Eq. (1.4) was introduced by van der Waals. By replacing pressure p with $(p + a/V_m^2)$, intermolecular forces are accounted for, and by replacing volume V_m with $V_m - b$, the molecular volume is accounted for.

$$\left(p + \frac{a}{V_m^2}\right)(V_m - b) = R_m T \tag{1.6}$$

In the *van der Waals* Eq. (1.6) the substance specific parameters for hydrogen are $a = 0.025$ m^6 Pa mol^{-2}, $b = 2.66 \cdot 10^{-5}$ m^3 mol^{-1}.

A convenient approach to account for real gas behavior is to use the dimensionless *compressibility factor* Z in the equation of state.

$$\frac{p V_m}{R_m T} = \frac{p v}{R T} = Z \tag{1.7}$$

The deviation of Z from the value 1 is a measure of the deviation from ideal gas behavior, especially for the deviation of real mass from ideal mass.

$$\frac{pV}{n_{real} R_m T} = \frac{pV}{m_{real} RT} = Z = \frac{n_{ideal}}{n_{real}} = \frac{m_{ideal}}{m_{real}} \tag{1.8}$$

The compressibility factor Z has to be determined experimentally and can be found in the literature as a function of pressure p and temperature T for a number of gases [21]. By reducing the pressure p to the critical pressure p_{cr} and the temperature T to the critical temperature T_{cr}, a generalized compressibility factor for all gases can be drawn as a function of $p_R = p/p_{cr}$ and $T_R = T/T_{cr}$ [28].

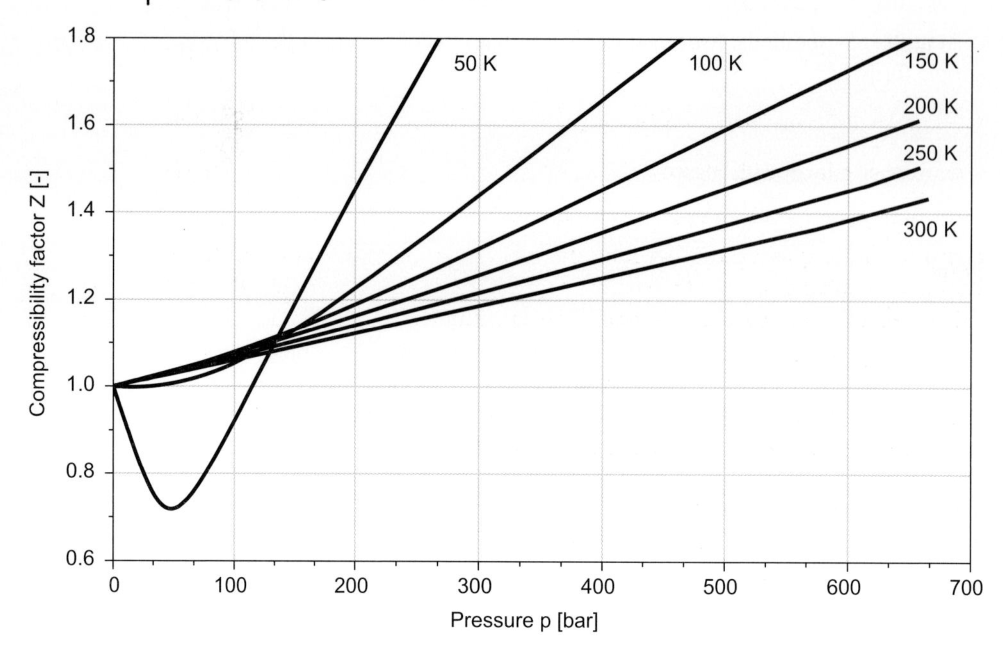

Figure 1.1 Compressibility factor Z of hydrogen.

The value of the compressibility factor Z for hydrogen at high pressures and low temperatures in Figure 1.1 shows that, at ambient temperature, a value of 1.2 is reached at 300 bar, and at low temperatures even earlier. This means that a calculation of the hydrogen mass in a container from a measurement of temperature and pressure using the ideal gas equation will result in a mass 20% greater than in reality.

The equation of state (1.3) can be depicted in a three-dimensional diagram with the three thermodynamic properties pressure p, temperature T, and specific volume v as axes. Two-dimensional projections with the third variable as a parameter are widely used to explain thermodynamic processes.

1.2.2
T–s-Diagram

For thermodynamic analysis, the *T–s*-diagram with temperature T versus entropy s as axes and lines of constant pressure, density, and enthalpy has proven very helpful. Changes of states and heat or work released or absorbed can be illustrated clearly using the *T–s*-diagram.

From the definition of entropy

$$ds = dq_{rev}/T \tag{1.9}$$

it follows that the reversible heat corresponds to the area below the curve of the process in the *T–s*-diagram:

$$q_{\text{rev}} = \int T\mathrm{d}s \tag{1.10}$$

With the definition of the reversible heat

$$\mathrm{d}q_{\text{rev}} = \mathrm{d}u + p\mathrm{d}v = \mathrm{d}h - v\mathrm{d}p \tag{1.11}$$

it follows that the heat, and thus the area below the curve of the process in the *T–s*-diagram corresponds to the change in internal energy u or enthalpy h given an isochoric or isobaric process. Moreover, the specific heat capacity corresponds to the subtangent of the curve of the process in the *T–s*-diagram:

$$c = \mathrm{d}q_{\text{rev}}/\mathrm{d}T = T\,\mathrm{d}s/\mathrm{d}T \tag{1.12}$$

The *T–s*-diagram for equality hydrogen for temperatures from 15 to 85 K is shown in Figure 1.2, and for temperatures from 85 to 330 K in Figure 1.3.

1.2.2.1 **Joule–Thomson Coefficient**

The Joule–Thomson coefficient μ_{JT} describes the extent and direction of the temperature change for an isenthalpic change of state (constant enthalpy h):

$$\mu_{\text{JT}} = \left(\frac{\partial T}{\partial p}\right)_h \tag{1.13}$$

A positive Joule–Thomson coefficient means that a decrease in temperature takes place along an isenthalpic pressure decrease. In the *T–s*-diagram, this is reflected as a falling isenthalpic line with a pressure decrease (cooling during expansion in a restriction). This effect is used in the liquefaction of hydrogen when cooling the fluid using a nozzle.

A negative Joule–Thomson coefficient means that an increase in temperature takes place along an isenthalpic pressure decrease. In the *T–s*-diagram, this is reflected as a rising isenthalpic line with pressure decrease (heating during relaxation in a restriction). This effect has to be accounted for when filling a high-pressure vessel with hydrogen.

The Joule–Thomson effect occurs when a gas or gas mixture experiences a change in temperature during an isenthalpic pressure change. An ideal gas does not show any Joule–Thomson effect. With ideal gases, the internal energy u and thus also the enthalpy h are only a function of temperature T. Thus ideal gases do not experience a change in temperature while the enthalpy remains constant, for example, in a flow through a restriction. This means that the Joule–Thomson coefficient is zero. Thus it can easily be judged from the *T–s*-diagram that a gas can be regarded as an ideal gas in an area where the isenthalpic lines are horizontal.

1.2.3 **Properties**

At normal temperature and pressure, hydrogen is a colorless, odorless gas with no toxic effects. It is the lowest density element, whilst also having a high diffusion

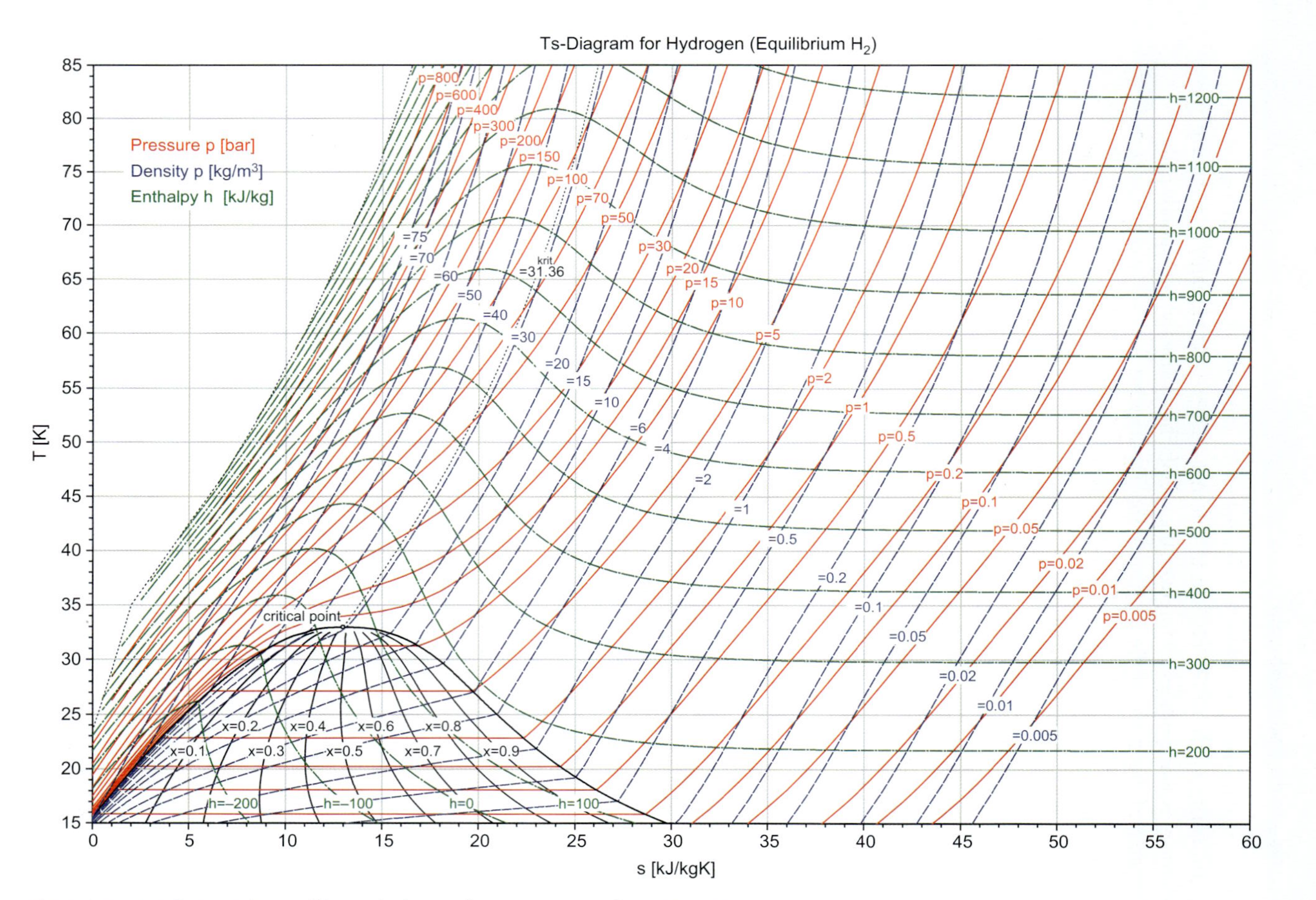

Figure 1.2 *T–s*-diagram for equilibrium hydrogen for temperatures from 15 to 85 K [8].

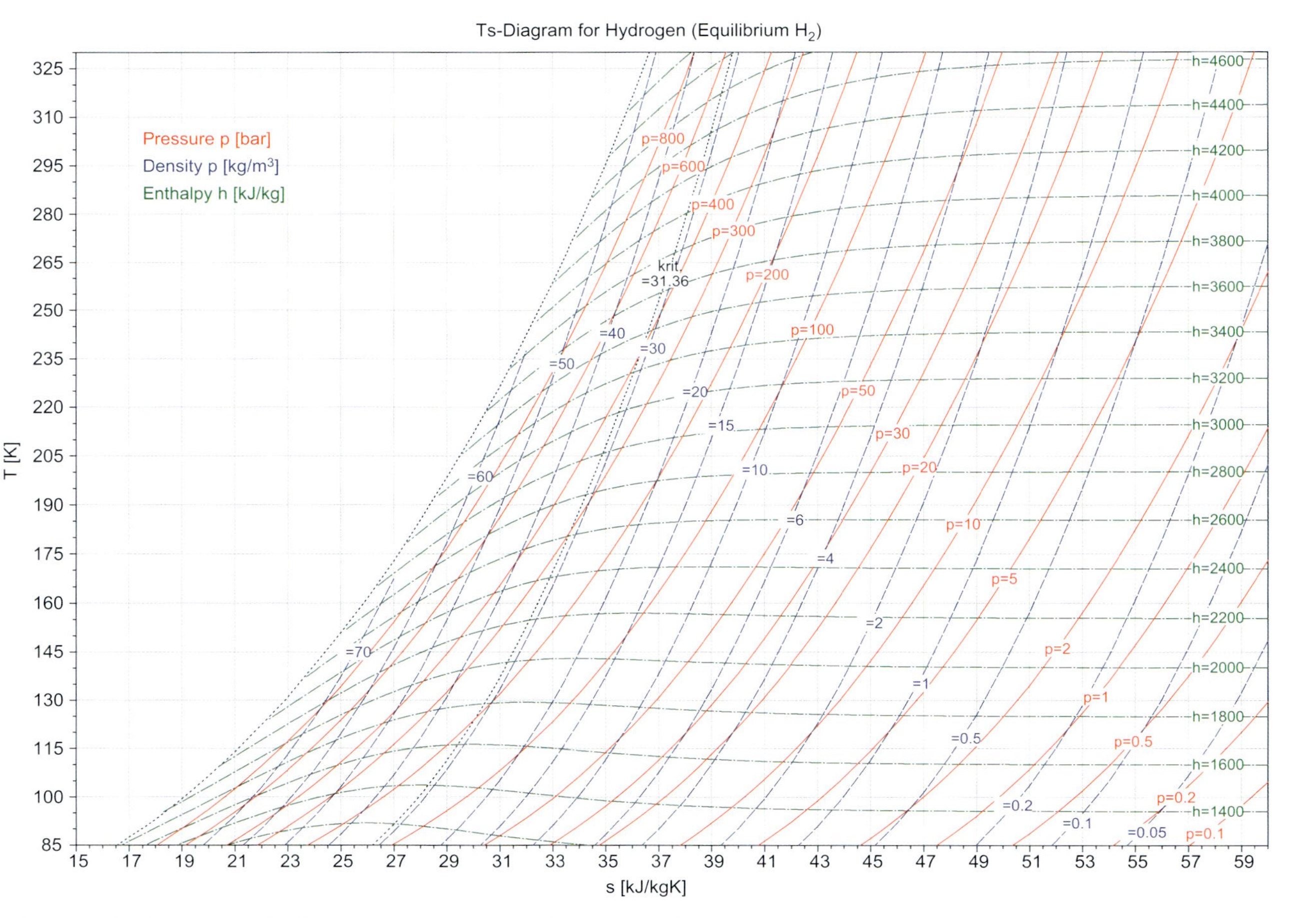

Figure 1.3 T–s-diagram for equilibrium hydrogen for temperatures from 85 to 330 K [8].

coefficient and a high specific heat capacity. After helium, hydrogen has the lowest melting and boiling points. Hydrogen is highly inflammable (EU rating F + and R12) with broadly spaced ignition limits in air (lower explosion limit 4% by volume, upper explosion limit 75.6% by volume) and low ignition energy (0.017 mJ for a stoichiometric air mixture).

As with all fuels, the use of hydrogen requires compliance with safety regulations, with EU safety sheets specifying:

- S9: keep containers in a well-aired location
- S16: keep away from ignition sources – do not smoke (explosion areas)
- S33: take precautions against electrostatic charge.

Because hydrogen is a very light and diffusive gas, explosive concentrations can normally be prevented easily through adequate ventilation. In the case of compressed gas CGH_2 storage, compliance with pressure vessel regulations is required. Direct contact with cryogenic liquids and gases can cause serious frostbite or freeze-burns. Furthermore, exposure to hydrogen can cause embrittlement and diffusion with a variety of materials, including most plastics and mild steel, which can in turn lead to fracture and leakage.

The properties of equilibrium hydrogen are summarized in Table 1.1. More extensive explanations about hydrogen and its properties can be found in [19, 21, 25, 29].

An important application of hydrogen is its combustion. Hydrogen can be burnt in internal combustion engines [3] producing low levels of pollutants, and in fuel cells free of pollutants [15]. Ideal combustion of hydrocarbon fuels takes place according to Eq. (1.14) [23].

$$C_xH_yO_z + \left(x + \frac{y}{4} - \frac{z}{2}\right)O_2 \rightarrow xCO_2 + \frac{y}{2}H_2O \tag{1.14}$$

With hydrogen containing no carbon, Eq. (1.14) becomes

$$H_2 + \tfrac{1}{2}O_2 \rightarrow H_2O \tag{1.15}$$

Thus hydrogen can be burnt without releasing CO_2, producing water only. A reaction enthalpy $\Delta_R H = -242$ MJ kmol^{-1} is released if the water remains in gaseous form. Hydrogen has the highest gravimetric calorific value of all fuels with $H_u = 120$ MJ kg^{-1} = 33.33 kWh kg^{-1}. If the water condenses, the enthalpy of condensation adds to a total reaction enthalpy of $\Delta_R H = -285$ MJ kmol^{-1} giving the gross calorific value (upper heating value) of hydrogen of $H_o = 142$ MJ kg^{-1} = 39.44 kWh kg^{-1}.

1.3 Gaseous Storage

The storage of gases in pressure vessels is a proven and tested technology. Most gases are available in containers at pressures up to 300 bar (30 MPa).

Table 1.1 Properties of hydrogen [3].

	Property	Value and Unit
	molar mass	$2.016\,kg\,kmol^{-1}$
	particular gas constant	$4124\,J\,kg^{-1}\,K^{-1}$
	(gravimetric) calorific value H_u	$120\,MJ\,kg^{-1} = 33.33\,kWh\,kg^{-1}$
at triple point:	temperature	$-259.35\,°C$ (13.80 K)
	pressure	0.07 bar
	density gaseous	$0.125\,kg\,m^{-3}$
	density liquid	$77\,kg\,m^{-3}$
	heat of fusion	$58.5\,kJ\,kg^{-1} = 16.25\,kWh\,kg^{-1}$
at boiling point	boiling temperature	$-252.85\,°C$ (20.30 K)
at 1.01325 bar:	heat of vaporization	$445.4\,kJ\,kg^{-1} = 123.7\,kWh\,kg^{-1}$
liquid phase:	density	$70.8\,kg\,m^{-3}$
	(volumetric) calorific value	$8.5\,MJ\,dm^{-3} = 2.36\,kWh\,kg^{-1}$
	specific heat capacity c_p	$9.8\,kJ\,kg^{-1}\,K^{-1}$
	specific heat capacity c_v	$5.8\,kJ\,kg^{-1}\,K^{-1}$
	thermal conductivity	$0.099\,W\,m^{-1}\,K^{-1}$
	dynamic viscosity	$11.9 \times 10^{-6}\,N\,s\,m^{-2}$
	speed of sound	$1089\,m\,s^{-1}$
gaseous phase:	density	$1.34\,kg\,m^{-3}$
	(volumetric) calorific value	$0.16\,MJ\,dm^{-3} = 0.044\,kWh\,dm^{-3}$
	specific heat capacity c_p	$12.2\,kJ\,kg^{-1}\,K^{-1}$
	specific heat capacity c_v	$6.6\,kJ\,kg^{-1}\,K^{-1}$
	thermal conductivity	$0.017\,W\,m^{-1}\,K^{-1}$
	dynamic viscosity	$1.11 \times 10^{-6}\,Ns\,m^{-2}$
	speed of sound	$355\,m\,s^{-1}$
at critical point:	temperature	$-239.95\,°C$ (33.20 K)
	pressure	13.1 bar
	density	$31.4\,kg\,m^{-3}$
at standard conditions:	density	$0.09\,kg\,m^{-3}$
(0 °C and 1.01325 bar)	(volumetric) calorific value	$0.01\,MJ\,dm^{-3} = 2.8\,Wh\,dm^{-3}$
	specific heat capacity c_p	$14.32\,kJ\,kg^{-1}\,K^{-1}$
	specific heat capacity c_v	$10.17\,kJ\,kg^{-1}\,K^{-1}$
	thermal conductivity	$0.184\,W\,m^{-1}\,K^{-1}$
	coefficient of diffusion	$0.61\,cm^2\,s^{-1}$
	dynamic viscosity	$8.91 \times 10^{-6}\,N\,s\,m^{-2}$
	speed of sound	$1246\,m\,s^{-1}$
mixtures with air:	lower explosion limit	4 Vol% H_2 ($\lambda = 10.1$)
	lower detonation limit	18 Vol% H_2 ($\lambda = 1.9$)
	stoichiometric mixture	29.6 Vol% H_2 ($\lambda = 1$)
	upper detonation limit	58.9 Vol% H_2 ($\lambda = 0.29$)
	upper explosion limit	75.6 Vol% H_2 ($\lambda = 0.13$)
	ignition temperature	585 °C (858 K)
	minimal ignition energy	0.017 mJ
	max. laminar flame speed	about $3\,m\,s^{-1}$
	adiabatic combustion temperature	about 2100 °C

Of thermodynamic interest are, on the one hand, the compression and the work necessary for it, and, on the other hand, the expansion during the filling process of a pressure vessel from a reservoir.

High demands are placed on pressure vessels for hydrogen from the materials side, from safety and component dimensioning. The infrastructure used includes pipelines for distribution and dispensers for refueling.

1.3.1
Compression and Expansion

From the first law of thermodynamics, the internal work w_i needed for the compression of a gas can be calculated from the enthalpies h_1 and h_2 before and after the compression and the cooling energy q_K:

$$w_i = h_2 - h_1 + q_K \tag{1.16}$$

The minimum required ideal compression work is the work for isothermal compression w_{is}, where the temperature stays constant during compression through cooling. The ideal isothermal compression work w_{is} can be found using the T–s-diagram. In the absence of internal friction, the cooling energy q_K equals the reversible heat and can be visualized in the T–s-diagram as the area below the change of states, see Figure 1.4a. The value can be calculated from

$$q_K = T \cdot \Delta s \tag{1.17}$$

For a compression from 1 bar to 900 bar at ambient temperature one finds the following value for equilibrium hydrogen: $w_{is} = 4383 - 3787 + 8181 = 8777\,\text{kJ kg}^{-1}$. This means that the compression needs an energy input of about 7.5% of the calorific value of hydrogen. Taking into account the efficiency of compressors at around 50%, one finds that the energy needed for the compression of hydrogen to 900 bar increases to about 15% of its calorific value. An overview of densities and energy densities at pressures relevant for containers is given in Tables 1.2 and 1.3, a comparison of densities and compressor work with liquid storage is given in Figure 1.22 (see later).

Substituting the definition of the reversible heat

$$q_{rev} = q_A + q_R = dh - v\,dp \tag{1.18}$$

in Eq. (1.16), we get, for compression without considering friction

$$w_i = \int_1^2 v\,dp \tag{1.19}$$

With the simplification of hydrogen as an ideal gas, we can substitute v from the ideal gas Eq. (1.4) and get:

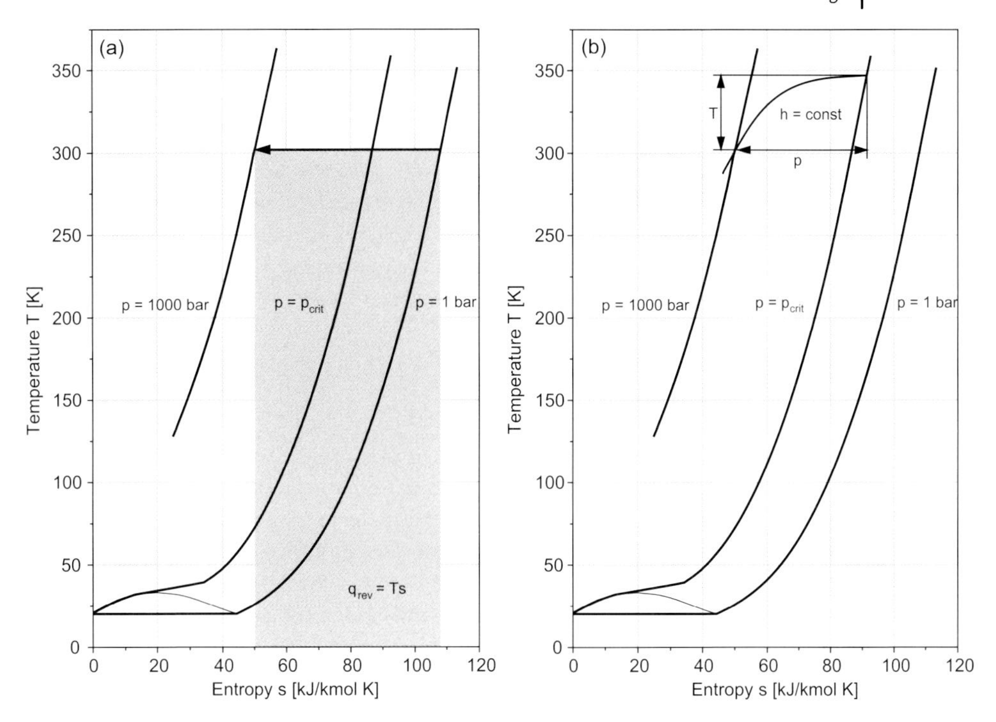

Figure 1.4 *T–s*-diagram: (a) isothermal compression from 1 to 1000 bar; (b) isenthalpic expansion from 1000 to 13 bar.

$$w_{\mathrm{i}} = RT \int_{1}^{2} \frac{\mathrm{d}p}{p} = RT \ln \frac{p_2}{p_1} \tag{1.20}$$

In our case, the calculation gives $w_{\mathrm{i}} = 8220\,\mathrm{kJ\,kg^{-1}}$. In comparison with the work calculated above this is 6.2% less, which is because hydrogen deviates from ideal gas behavior in the pressure range under consideration.

With flowing gases, compression is generally linked with a temperature increase, while expansion leads to a decrease in temperature. During the filling of a pressure vessel, gas flows from a reservoir of even higher pressure into the tank. Thermodynamically the process can be approximated by an adiabatic flow through a restriction where the total enthalpy remains constant. The associated change in temperature is described by the Joule–Thomson coefficient. In the relevant range of pressures, hydrogen has a negative Joule–Thomson coefficient, which means the temperature increases with a pressure decrease. The expansion of hydrogen from 1000 bar to 13 bar produces a rise in temperature of about 50 K, see Figure 1.4b.

The compression of the gas in the tank to be filled also causes its temperature to rise. This effect even outweighs the Joule–Thomson effect. A simulation can be done based on the first law of thermodynamics. Regarding the reservoir and the tank as a closed adiabatic system, the internal energies of the two containers before and after the filling process are equal. Regarding the tank to be filled alone as an open adiabatic system, the enthalpy flow into the tank equals its increase in internal energy.

The temperature of the gas in the tank will increase considerably during the filling process causing the container to warm up. As the filled container cools to ambient temperature, the pressure of the gas will decrease accordingly, thus leaving the container at ambient temperature below nominal filling pressure. This loss of filling mass can be avoided by a so-called cold-fill device, where the hydrogen is cooled during the filling process in a heat exchanger, for example, with liquid nitrogen.

1.3.2 Tank Systems

For compressed storage of hydrogen, the gas is usually compressed to pressures between 200 and 350 bar though, more recently, storage pressures of 700 bar and even higher have been under trial. Such enormous pressures require consideration of questions regarding material choice, component dimensioning and safety.

Hydrogen has a tendency to adsorb and dissociate at material surfaces, the atomic hydrogen then diffuses into the material and causes embrittlement and diffusion. Materials suitable for hydrogen applications are mainly austenitic stainless steel and aluminum alloys [12, 29].

Apart from the container itself, valves for reducing the pressure, pipelines, and sensors to control pressure, temperature and tightness are applied. For certification, tank systems have to undergo extensive tests, there are a number of regulations for pressure vessels and tank systems [16]. Electronic safety systems for monitoring pressure, temperature and tank filling level with leak monitoring and emergency stop measures have been developed especially for automotive applications [2].

Commercially available pressure containers are conventionally made of steel. These so called type I containers offer good properties concerning safety and strength, but at a high weight. They are available with net volumes from 2.5 to 50 l. Characteristics in Table 1.2 show that the pressures vary from 200 to 300 bar, the systems are quite heavy, the energy densities reach around 0.4 kWh kg^{-1}. To reduce the weight, steel containers have been replaced by composite containers. A thin inner liner of metal (steel or aluminum) ensures the gas tightness. Stability is given by a mesh partially (type II) or completely (type III) provided by carbon fibers. With type IV containers, the liners are also made of synthetic material. Composite containers are lighter but also expensive, especially with a growing demand for carbon fibers. For automotive applications, a number of type III and type IV tank systems are available. Their characteristics in Table 1.3 show that energy densities are considerably higher and reach gravimetrically 0.055 kg H_2 kg^{-1} or 1.833 kW h kg^{-1} and volumetrically 0.026 kg H_2 dm^{-3} or 0.867 kW h dm^{-3}. The costs of available tank systems vary from about 40 € per kWh of stored hydrogen energy for type III tanks for 350 bar up to about 150 € per kW h for type IV tanks for 700 bar. An example of a compressed hydrogen tank for automotive application is shown in Figure 1.5.

Table 1.2 Characteristics of commercially obtainable type I pressure containers.

Net volume [dm^3]	2.5	10	20	33	40	50
Nominal pressure [bar]	200	300	200	200	200	200/300
Testing pressure [bar]	300	450	300	300	300	300/450
Tank weight [kg]	3.5	21	31.6	41	58.5	58/94
Tank volume [dm^3]	3.6	14.3	27	41.8	49.8	60.1/64.7
H_2 density [$kg\,m^{-3}$] at 25 °C	14.5	20.6	14.5	14.5	14.5	14.5/20.6
H_2 content [Nm^3]	0.4	2.29	3.22	5.32	6.44	8.05/11.43
H_2 content [kg]	0.04	0.21	0.29	0.48	0.58	0.72/1.03
Grav. H_2 content [$kgH_2\,kg^{-1}$]	0.01	0.009	0.009	0.012	0.011	0.012/0.011
Vol. H_2 content [$kgH_2\,dm^{-3}$]	0.009	0.014	0.011	0.011	0.012	0.012/0.016
Grav. energy density [$kWh\,kg^{-1}$]	0.333	0.300	0.305	0.400	0.367	0.400/0.367
Vol. energy density [$kWh\,dm^{-3}$]	0.300	0.477	0.367	0.367	0.400	0.400/0.533

With adequate material and dimensioning, gaseous hydrogen storage takes place in a closed system and thus hydrogen can be stored without loss for extended periods of time.

Apart from storage in vessels, underground storage of large quantities of hydrogen in natural caverns has also been investigated [16].

1.3.3 High Pressure Infrastructure

The distribution of gaseous hydrogen from centralized production facilities is usually effected by transportation in large pressure containers by road or rail. To supply large amounts of hydrogen, distribution via pipelines is the most economical way [16].

Table 1.3 Characteristics of commercially obtainable automotive pressure containers.

Net volume [dm^3]	34	100	50	100	36	65	30	120
Type	III	III	III	III	IV	IV	IV	IV
Nominal pressure [bar]	350	350	700	700	350	350	700	700
Test pressure [bar]	525	525	1050	1050	525	525	1050	1050
Tank system weight [kg]	18	48	55	95	18	33	26	84
Tank system volume [dm^3]	50	150	80	150	60	100	60	200
H_2 density [$kg\,m^{-3}$] at 25 °C	23.3	23.3	39.3	39.3	23.3	23.3	39.3	39.3
H_2 content [Nm^3]	8.83	26	21.84	43.69	9.35	16.96	13.5	51.7
H_2 content [kg]	0.79	2.33	1.96	3.83	0.84	1.52	1.21	4.65
Grav. H_2 content [$kgH_2\,kg^{-1}$]	0.044	0.049	0.036	0.041	0.047	0.047	0.047	0.055
Vol. H_2 content [$kgH_2\,dm^{-3}$]	0.016	0.016	0.025	0.026	0.014	0.015	0.021	0.023
Grav. energy density [$kWh\,kg^{-1}$]	1.467	1.633	1.200	1.367	1.567	1.567	1.567	1.833
Vol. energy density [$kWh\,dm^{-3}$]	0.533	0.533	0.833	0.867	0.467	0.500	0.700	0.767

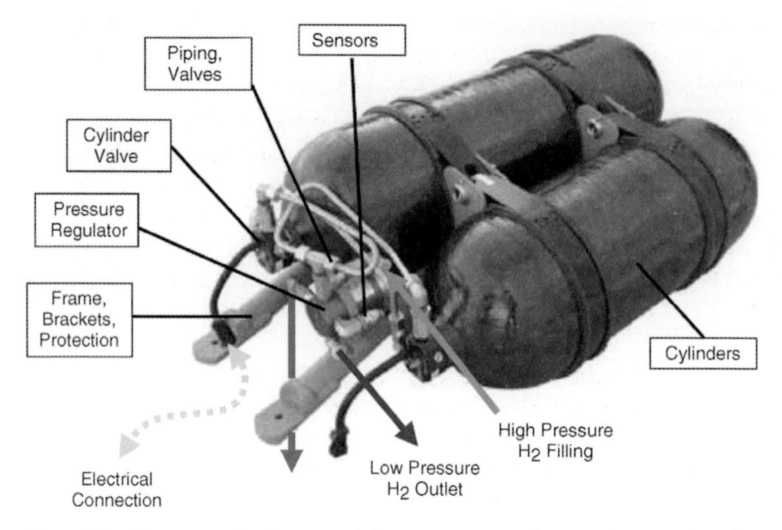

Figure 1.5 Compressed hydrogen tank for automotive application. Source Dynetek [1].

Hydrogen filling stations for vehicles resemble the filling stations for natural gas. Regulations and certification are usually derived from natural gas applications, whilst hydrogen-specific regulations are being developed [13]. As the gas is filled by a pressure gradient, the reservoir pressure has to be considerably higher than the nominal tank pressure to provide a short filling time.

Compressors for hydrogen have to fulfill high technical demands. High pressures of up to 900 bar can be achieved by multi-stage piston compressors. To assure the necessary purity of the compressed hydrogen, the piston rings have to be self-lubricating. Due to safety considerations, these compressors are often operated by hydraulic oil. For lower pressures, diaphragm compressors are also used. In newer concepts and for special applications, mechanical compressors and metal pistons are replaced by ionic liquids, metal hydrides or electrochemical compressors [16].

From the high pressure reservoir, hydrogen is filled into the tank system by a pressure gradient. Overfilling and overheating of the tank systems have to be avoided by adequate safety measures. The dispensers to refuel automotive tank systems resemble the usual dispensers for natural gas. Certificated couplings for 350 and 700 bar are available. After connection, a pressure pulse checks for tightness of the system. If successful, the valves for filling the tank open.

As mentioned before, the temperature of hydrogen increases and causes the tank to warm up during the filling process. After cooling to ambient temperature, the tank pressure would be below the nominal pressure, thus causing a loss of contained mass. This effect has to be accounted for by cooled filling or slow filling with heat dissipation into the external environment. In the second case, a significant increase in the filling time would have to be accepted, something that otherwise only requires a few minutes. Therefore, the gas filling infrastructure for high pressure requires a cold fill device, where hydrogen is cooled by a heat exchanger with liquid nitrogen. If the temperature of the tank system thus decreases below ambient temperature, it is necessary to ensure that, after heating to ambient temperature, the pressure of the tank system does not exceed the allowed maximum pressure, 25% above nominal.

Another important feature of the infrastructure is the filling time. Refueling of an automotive hydrogen gas tank takes a few minutes. The resulting effective energy flow can be estimated as follows: 10 kg of hydrogen contain energy of 1200 MJ. If a vessel is filled within 5 min or 300 s, the filling process corresponds to a power of 1200 MJ/300 s = 4 MW. This comes close to the values for refueling of gasoline or diesel and clearly exceeds the potential of solid storage or battery recharging.

Being gaseous, hydrogen has a number of features in common with natural gas and can be mixed with it in any ratio. The use of mixtures of hydrogen and natural gas H_2NG offers several advantages in terms of infrastructure, storage, and use, and is the subject of worldwide research. Synergies in infrastructure and customer acceptance can be achieved, for example, by running internal combustion engines on mixtures of hydrogen and natural gas. Alongside the bridging effect between natural gas and hydrogen, such mixing offers advantages in terms of reduced emissions and improvements to the combustion process. The wide ignition limits and high flame speed of hydrogen have as positive an impact on the combustion of H_2NG mixture as does the higher energy density of natural gas on range [2].

1.4 Liquid Storage

The storage of liquefied cryogenic gases is a proven and tested technology. Hydrogen was first liquefied by J. Dewar in 1898. Compared to compressed gases, the density of liquefied gases is considerably higher. On the other hand, liquefaction, storage and handling of the cryogenic boiling liquids require energy input as well as complex tank systems and infrastructure.

After an overview of the liquefaction process, a short thermodynamic analysis of a liquid hydrogen infrastructure with storage tanks and filling processes is presented. The tank systems are sophisticated containers with vacuum insulation and pressure regulation.

1.4.1 Liquefaction

Basic thermodynamic principles of the liquefaction of hydrogen are explained using the *T–s*-diagram, see Figure 1.6. In the hypothetical case of isobaric cooling at ambient pressure p_0 down to condensation at CP and the boiling point BP, the cooling energy would be:

$$q_K = h_2 - h_1 \quad (1.21)$$

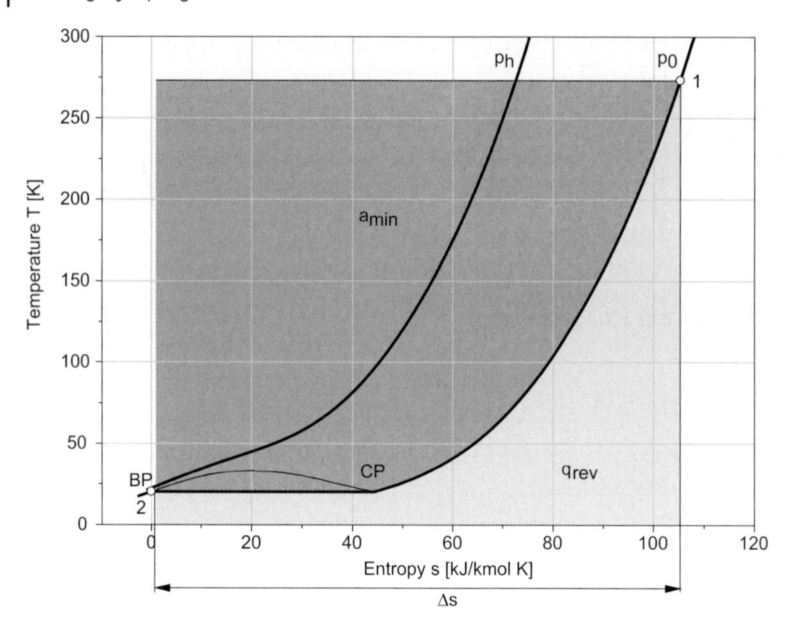

Figure 1.6 Cooling load and work of the cyclic process of liquefaction.

The cooling load for liquefying hydrogen at ambient pressure would correspond with the light gray area in Figure 1.6 and amount to

$$q_K = -0.888 - 3959 = -3960\,\text{kJ kg}^{-1}$$

This is about 3.3% of the calorific value of hydrogen. The only possible cooling medium for this case would be liquid helium, but such a process is technically and economically not viable. Therefore the liquefaction is treated as an ideal cyclic process. The minimum inner work for such a process can be calculated from:

$$a_{\min} = q_{zu} - q_{ab} = T_0 \Delta s - q_K \qquad (1.22)$$

The work for this case is depicted in dark gray in Figure 1.6 and will amount to

$$a_{\min} = 16092 - 3960 = 12132\,\text{kJ kg}^{-1}$$

This corresponds to about 10% of the calorific value of hydrogen.

Besides this low pressure process, there is also an alternative of a high pressure process at p_h above the critical pressure, see Figure 1.6. In this case the phase transition of the condensation can be avoided. This brings advantages in heat exchanger dimensioning, however, the plant design gets more complex.

Real-world efficiencies for liquefaction lie around 30%, which means, about 20% to 30% of the energy content of hydrogen is needed for its liquefaction. In the process, hydrogen is first compressed to about 30 bar. The gas is then cooled with liquid nitrogen to about 80 K. Between 80 and 30 K hydrogen is cooled by expansion turbines, where hydrogen is compressed, cooled and expanded. In this stage also the

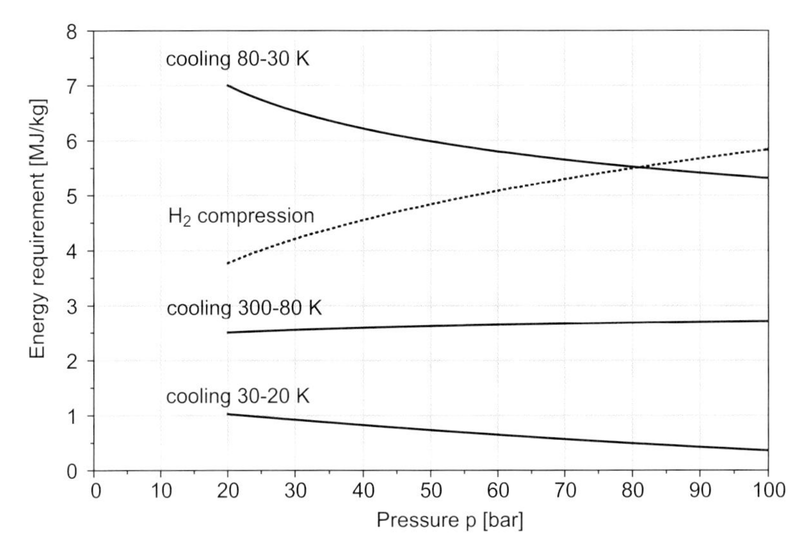

Figure 1.7 Energy required for different stages of liquefaction.

transition from ortho-hydrogen to para-hydrogen is effected, which means additional energy has to be dissipated. The last part of the cooling process from 30 to 20 K is done by Joule–Thomson valves. The positive Joule–Thomson effect is used here to cool during an expansion. Figure 1.7 with the energy required for the different stages of the liquefaction shows that the compression and the cooling from 80 to 30 K need most energy [24].

1.4.2 Thermodynamic Analysis

A thermodynamic analysis of a liquid hydrogen infrastructure will be given on the basis of the equipment at HyCentA Research GmbH in Graz [14]. The main storage tank, see Figure 1.8a, has a volume of 17 600 l and can store 1000 kg of hydrogen at a temperature of −253 °C at a pressure between 6 and 9 bar. The conditioning container, see Figure 1.8b, serves to adjust the pressure of the hydrogen to 2 to 4 bar to be used in the test stands or for filling tank systems.

The main tank is filled with liquid hydrogen from a trailer. Despite the sophisticated heat insulation in any container for cryogenic liquids, the small amount of remaining heat input will trigger off a warming process in the tank which causes the liquid in the container to evaporate and the pressure to rise. After a certain *pressure build-up* time the maximum operating pressure of the tank is reached. The pressure relief valve has to be opened. From this point onwards, gas must be released (*boil-off*). The container now acts as an open system with gas usually being lost to the environment.

Pressure build-up and boil-off of the main tank will be analyzed on the basis of the measurement of pressure and filling level over time, see Figure 1.9. From the

(a)

Figure 1.8 (a) Reservoir tank and (b) conditioning container at the HyCentA Graz.

(b)

Figure 1.8 (*Continued*)

measurements, heat input can be calculated. Simple thermodynamic models will be developed to simulate pressure build-up, boil-off and filling processes, including the cooling down of components.

The following characteristics are of interest:

Filling: Before filling, the pressure in the tank is reduced to enable liquid hydrogen to flow into the container from the trailer along a pressure gradient. As the blow-off valve remains open during the process, pressure drops till the end of the filling after about 50 min, see Figure 1.9b.

Pressure build-up: in spite of the insulation, due to the unavoidable heat input, hydrogen evaporates in the tank. Thus the pressure rises in the vessel. Pressure rise can be regarded as being linear over time, see Figure 1.9c. Pressure build-up continues until the maximum pressure p_{tank} in the vessel is reached and a blow-off valve has to be opened. The pressure build-up time in this case is nearly 30 h.

Boil-off: From reaching maximum system pressure of 6.7 bar in our case, the tank is an open system and hydrogen is blown out. From then, the pressure remains constant, the filling level decreases constantly. The vaporization rate and/or the rate of the effusing hydrogen is about 0.5 kg H_2 per hour.

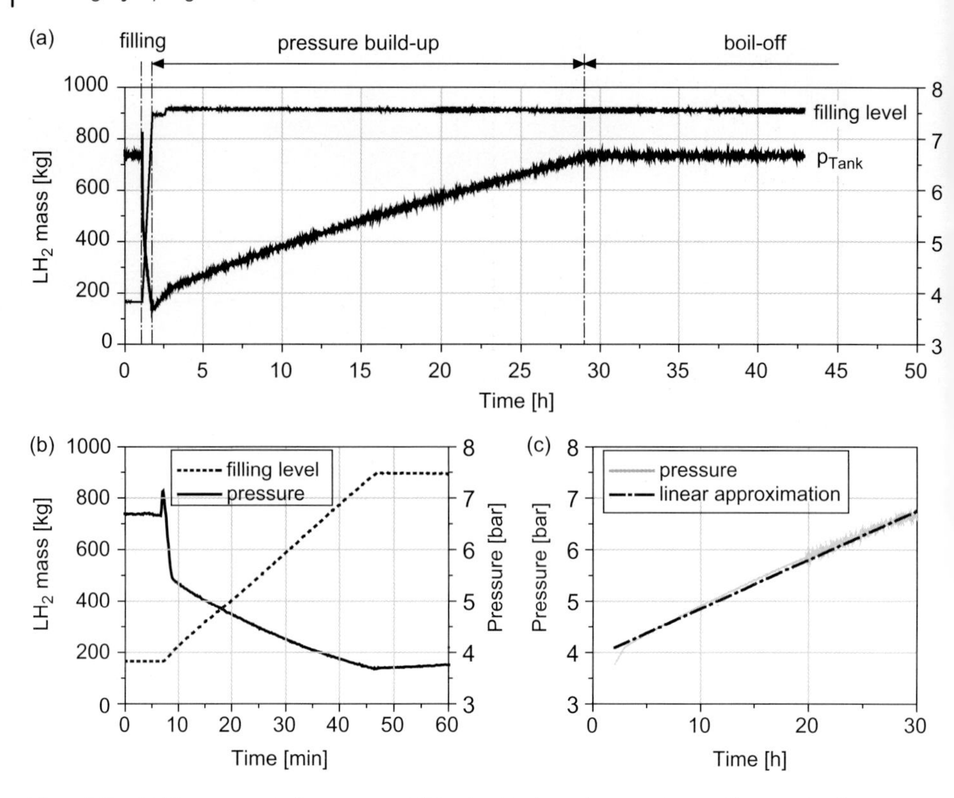

Figure 1.9 (a) Measurement of pressure and filling level in the tank versus time and in more detail: (b) during filling process, (c) during pressure build-up.

For a general illustration, we shall look first at a cryo-container with an in-flowing mass flow rate m_e and an out-flowing mass flow rate m_a as an open system, see Figure 1.10. It is assumed that this system is in thermodynamic equilibrium, that is, all state variables are equally distributed within the system. In particular, we find the same pressure throughout the system and the same temperature of the boiling liquid hydrogen and the saturated hydrogen vapor. This system can be described by applying the first law of thermodynamics and the law of conservation of mass. Despite the simplifying assumptions underlying the model, it can describe the principles of the relevant processes in the tank system, that is, the pressure build-up resulting from heat input, the evaporation resulting from heat input (boil-off), the effusion in order to decrease pressure, and the refueling process, see, for instance, [3, 9, 11, 12, 17, 27].

The first law of thermodynamics for open systems applies: the transport of work dW, heat dQ_a and mass dm_i with its enthalpy h_i and external energy e_{ai} (kinetic and potential energy) across the system boundaries equals the change in internal energy dU and external energy dE_a in the system.

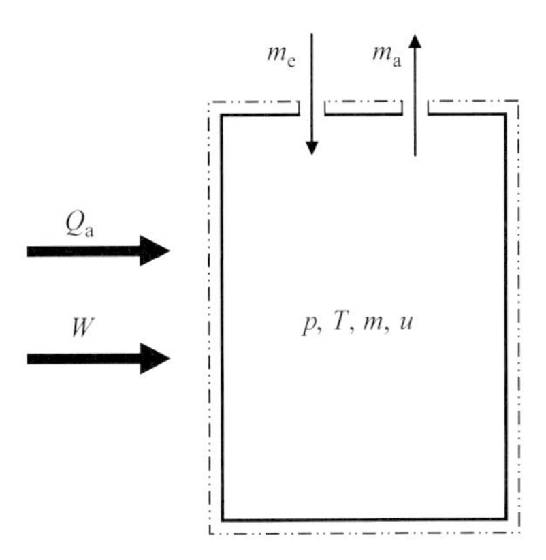

Figure 1.10 System of a cryo-container.

$$dW + dQ_a + \sum dm_i.(h_i + e_{ai}) = dU + dE_a \tag{1.23}$$

There is no work transferred. Neglecting kinetic and potential energy yields:

$$dQ_a + \sum dm_i \cdot (h_i) = dU \tag{1.24}$$

The conservation of mass states that added mass minus relieved mass equals the increase in mass of the system:

$$dm_e - dm_a = dm \tag{1.25}$$

1.4.2.1 **Pressure Build-Up**

In order to simulate the pressure increase and pressure build-up time in the storage system, we shall first assume a closed system in thermodynamic equilibrium with a constant heat input Q_a, see Figure 1.11. We assume that the pressure and temperature of the boiling liquid hydrogen and the gaseous saturated hydrogen vapor are the same throughout the whole system.

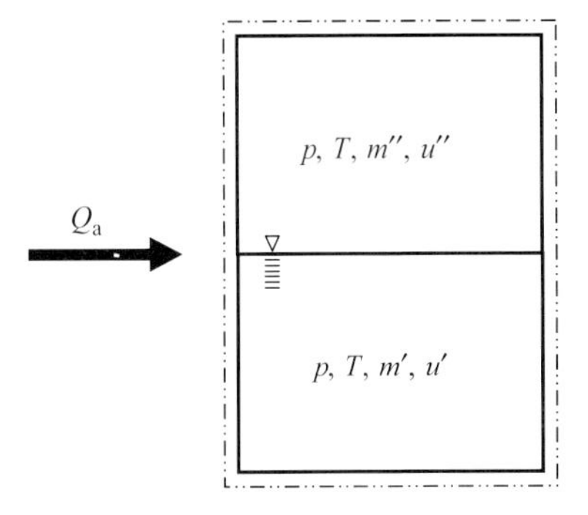

Figure 1.11 Cryo-storage system in a thermodynamic equilibrium system.

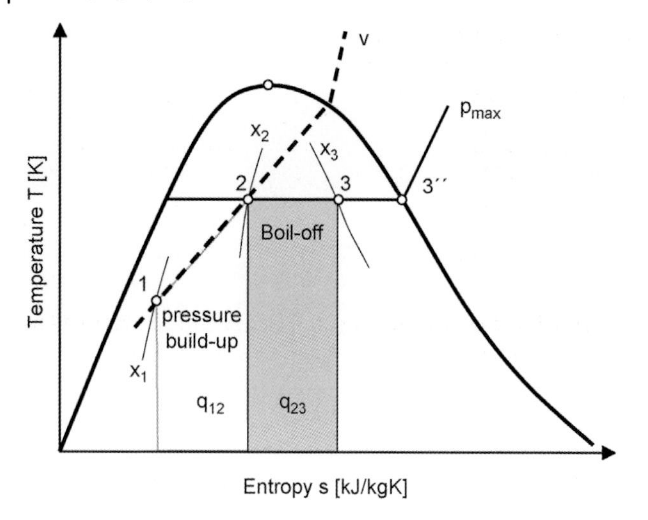

Figure 1.12 T–s-diagram for pressure build-up and boil-off due to heat input.

With a given container volume of 17 600 l and a given total hydrogen mass m, specific volume v (or density ϱ) is determined in the total system. All possible states move along an isochoric line in the T–s-diagram from starting point 1 to the highest permissible pressure in point 2, see Figure 1.12. If a pressure between points 1 and 2 is specified, the state of the system is clearly determined.

The constant container volume V consists of the variable shares for the liquid V' and gaseous hydrogen volume V'', in which the following applies:

$$V = \frac{m}{\varrho} = mv = V' + V'' = m'v' + m''v'' \tag{1.26}$$

The distribution of the mass in the boiling liquid m' and saturated vapor m'' is described by the vapor fraction x:

$$x = \frac{m''}{m} = \frac{m''}{m' + m''} \qquad m'' = x \cdot m \quad m' = (1-x) \cdot m \tag{1.27}$$

Thermophysical property tables give the specific volumes for liquid v' and saturated vapor v'' for each pressure so that the vapor fraction and the mass distribution can be determined:

$$x = \frac{v - v'}{v'' - v'} \tag{1.28}$$

With the vapor fractions of two states defined by a pressure increase Δp, the evaporated hydrogen mass $\Delta m''$ is also determined.

$$\Delta m'' = \Delta x \cdot m \tag{1.29}$$

The reason for the pressure build-up is the heat input. The specific heat quantity can be read from the T–s-diagram as the area below the constant-volume change in

Table 1.4 Pressure rise $\Delta p/\Delta t$ and resulting heat input Q_a.

FL [%]	p_1 [bar]	p_2 [bar]	Δp [bar]	Δt [h]	$\Delta p/\Delta t$ [mbar h^{-1}]	Q_a [kJ]	$\dot{Q}_a$ [W]
15.5	8.83	9.69	0.85	23.51	36.4	3271.89	38.65
38.0	9.10	9.72	0.62	16.14	38.5	5206.66	89.62
78.0	5.45	6.58	1.14	17.63	64.4	16146.41	254.42

state, see Figure 1.12. During the pressure build-up, the system can be described by means of the first law of thermodynamics for closed systems:

$$dQ_a = dU = d(mu) = m du \tag{1.30}$$

The internal energy u of the system consists of the internal energy of the liquid u' and of the vapor u'' and is determined at two points 1 and 2 with the help of the vapor fractions as

$$\begin{aligned} u_1 &= u_1' + x_1 \cdot (u_1'' - u_1') \\ u_2 &= u_2' + x_2 \cdot (u_2'' - u_2') \end{aligned} \tag{1.31}$$

With the filling level FL given through the measurement of the liquid hydrogen mass in the tank, the hydrogen mass is fixed. From the volume of the container, the constant specific volume is given. With this variable and a second property like pressure, all state variables of the hydrogen are given and can be read from the *T–s*-diagram or a property data base [21]. From two pressure measurements at two times, the vapor fractions and thus the internal energies of the hydrogen can be found, thus giving the heat input Q_a. For our example of the main tank, pressure rise and heat input for a number of cases are shown in Table 1.4.

1.4.2.2 **Boil-Off**

When the maximum operating pressure has been reached after pressure build-up to point 2, hydrogen is released by a valve in order to keep the pressure constant. From now on, saturated hydrogen vapor is released. The thermodynamic state in the tank is determined by the pressure and the vapor fraction x and moves from point 2 to point 3 in the *T–s*-diagram Figure 1.13 shows the relevant details from the measurements in Figure 1.9, the constant pressure in the tank and the liquid hydrogen mass in the tank, which decreases due to the boil-off without hydrogen being taken from the tank deliberately.

The first law of thermodynamics (1.24) for the now open system becomes:

$$dQ_a = dU + h'' dm \tag{1.32}$$

Integration between the two states 2 and 3 gives:

$$Q_a = \int_2^3 dU + \int_2^3 h'' dm = u_3 m_3 - u_2 m_2 + h''(m_2 - m_3) \tag{1.33}$$

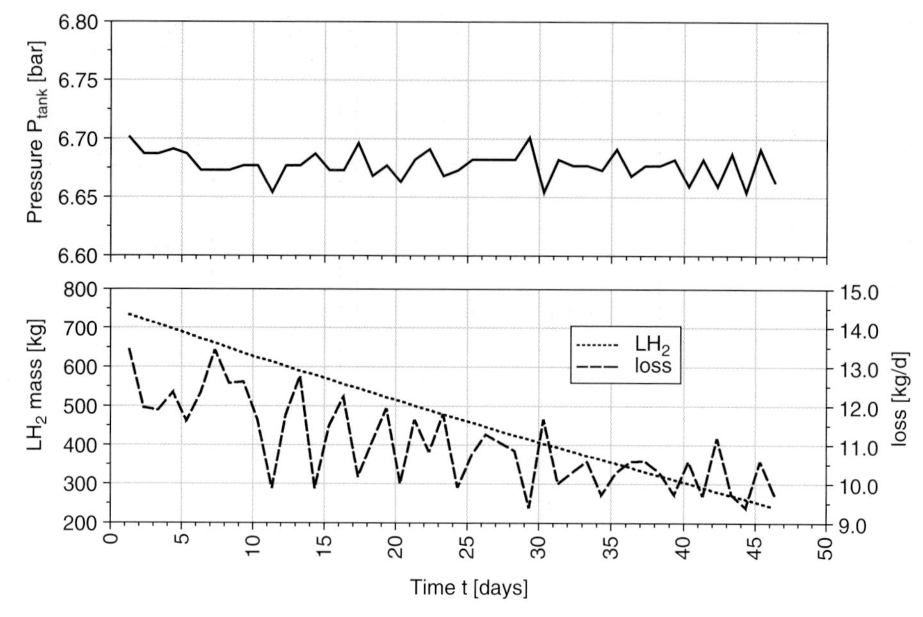

Figure 1.13 Tank pressure and filling level during boil-off.

The evaporated mass is the difference between the masses at two times:

$$\Delta m = m_3 - m_2 \tag{1.34}$$

With the fixed boil-off pressure and the measurement of the filling level, that is, of the hydrogen mass, all state variables of the hydrogen (u_2, u_3, h'') are known from the *T–s*-diagram or a data base. From two measurements of the filling level at two times, the evaporating mass can be calculated and thence the heat input Q_a. For our example of the main tank, evaporating mass and heat input for two cases are shown in Table 1.5. Evaporation losses of the tank here are between 1 and 1.2% per day.

1.4.2.3 Cooling and Filling

After the pressure build-up and boil-off behavior of a hydrogen container, the cooling and filling of a container will be analyzed. Generally, containers are filled using a pressure gradient. First, liquid hydrogen from the main container is filled into the conditioning vessel by a pressure gradient to adjust the hydrogen pressure there. All connecting pipes, valves and fittings have to be vacuum insulated. For each filling process, all pipelines, valves and fitting also have to be cooled to −253 °C.

Table 1.5 Evaporating mass $\Delta m/\Delta t$ and resulting heat input Q_a.

Pressure [bar]	m_1 [kg]	m_2 [kg]	Δm [kg]	Δt [h]	$\Delta m/\Delta t$ [kg h^{-1}]	Q_a [kJ]	$\dot{Q}_a$ [W]
6.67	689.02	603.05	85.97	169.8	0.51	43809.53	71.67
6.67	831.92.	523.15	308.77	709.16	0.44	38253.62	67.42

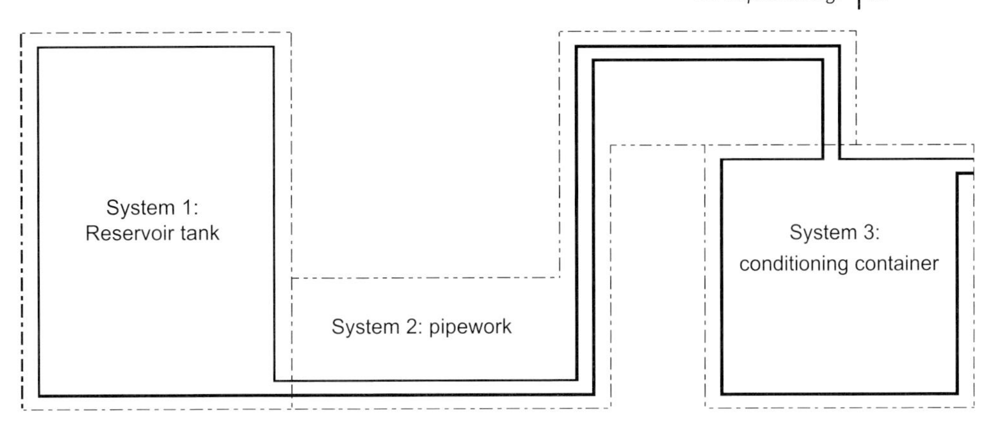

Figure 1.14 Filling system.

Figure 1.14 shows the complete system, which is divided into three subsystems.

For the main tank and the conditioning container subsystems 1 and 3, the relations of the general homogeneous model described above apply. It is assumed that both containers contain liquid hydrogen, that is, they are at low temperature. As the heat input into the reservoir tank can be neglected for the time the filling takes, the change in the internal energy dU_1 depends only on the effusing mass dm and its specific enthalpy h_1. The mass flow $\dot{m}$ coming from the reservoir tank eventually flows through system 2. If the temperature of system 2 (pipelines) is higher than the boiling temperature according to pressure – which is usually the case – the pipeline system must first be cooled before liquid hydrogen can be transported. The liquid hydrogen coming from the reservoir tank evaporates in the pipelines and chills them through enthalpy of evaporation.

The liquid mass inflow $\dot{m}$ leaves the system at the beginning of the filling process as gaseous mass flow $\dot{m}_g$. If the thermal mass of the pipelines reaches a temperature below the respective boiling temperature according to pressure, a liquid mass flow $\dot{m}_l$ manifests itself. A segment of the pipeline shows the chilling process in greater detail, see Figure 1.15 [4].

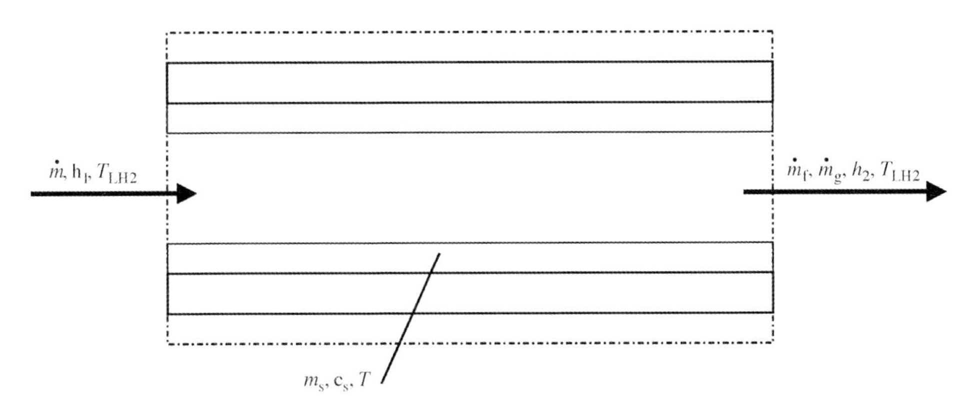

Figure 1.15 Chilling of the pipeline.

Energy balance of the chilling process:

$$m_S \cdot c_S \cdot \frac{dT}{dt} = -\dot{m}_{LH_2} \cdot r_{H_2} + \alpha \cdot A \cdot (T_0 - T) \tag{1.35}$$

The term $\alpha \cdot A \cdot (T_0 - T)$ corresponds with the transferred heat flow from radiation and convection and is much smaller than the term of the evaporation enthalpy $\dot{m}_{LH_2} \cdot r_{H_2}$ and can thus be disregarded. The following differential equation results:

$$T(t) = -\frac{\dot{m}_{LH_2} \cdot r_{H_2}}{m_S \cdot c_S} \cdot t + C \tag{1.36}$$

Boundary condition: $t = 0$, $T = T_0$

This results in the following temperature progression:

$$T(t) = T_U - \frac{\dot{m}_{LH_2} \cdot r_{H_2}}{m_S \cdot c_S} \cdot t \tag{1.37}$$

Energy balance for the pipelines:

$$dm_{12} \cdot h_1 - dm_{23} \cdot h_2 = dU_2 \tag{1.38}$$

The system of the conditioning container is represented in Figure 1.16. The inflowing mass consists either of gaseous or liquid hydrogen, depending on the temperature of the pipes.

Energy balance for the conditioning container:

$$dm_{23} \cdot h_2 = dU_3 \tag{1.39}$$

The chilling process usually requires several minutes and the evaporating hydrogen must generally be considered as a loss. It is disposed of into the environment via a

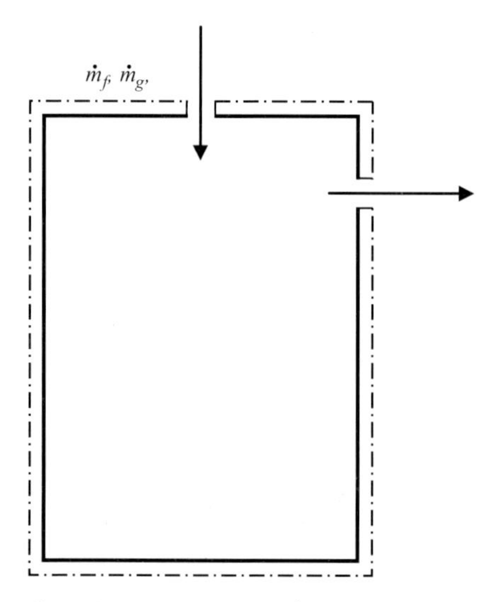

Figure 1.16 System 3: conditioning container.

flue at the HyCentA. There are further losses during the filling process that will be dealt with now.

1.4.2.4 **Back-Gas**

To fill a tank system in the test stand or in a car, liquid hydrogen is first transferred from the main tank to the conditioning container. There the pressure can be adjusted. From the conditioning container, the test tank is filled again by a pressure gradient. The whole system is shown in Figure 1.17.

Before liquid hydrogen can be transported and filled, all parts have to be cooled to −253 °C. During the filling process, a certain amount of gaseous hydrogen has to be transferred back from the test tank system, the so called back-gas m_{bg}. This will be analyzed in more detail, compare [5].

The back-gas is determined indirectly as the difference between the mass taken from the conditioner Δm_{condi} and the mass filled into the test tank Δm_{tank}. The relative amount of back-gas is defined as the rate of back-gas $\beta := \frac{m_{\text{bg}}}{\Delta m_{\text{tank}}} = \frac{\Delta m_{\text{condi}} - \Delta m_{\text{tank}}}{\Delta m_{\text{tank}}}$.

The back-gas consists of three kinds of losses:

1) Loss of heat input: The largest share of hydrogen is lost by evaporation to cool down the pipelines as described above. In spite of the vacuum insulation heat input cannot be avoided completely and causes hydrogen to evaporate also after the system has been cooled down.
2) Loss of filling volume: in order to fill a liquid hydrogen mass m′ into the tank, a volume V of hydrogen gas is replaced by liquid hydrogen $V = v''\, m'' = v'\, m'$. The gaseous mass m'' has to be removed from the tank, the increase of mass in the test tank is $m' - m''$. The rate of back-gas thus becomes

$$\beta_v = \frac{m''}{m' - m''} = \frac{v'}{v'' - v'} \tag{1.40}$$

Even though the ratio of the densities between the gaseous and liquid phase is much smaller with hydrogen than with other gases, this loss is relatively small.

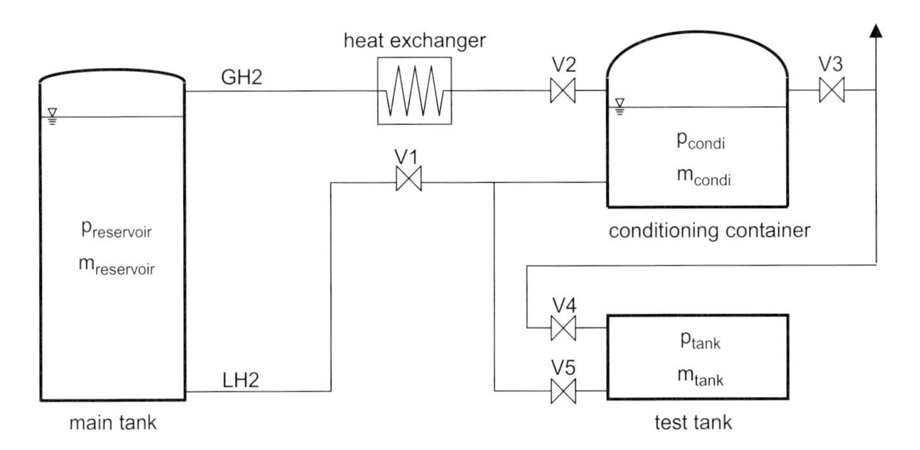

Figure 1.17 Experimental set-up at HyCentA.

Table 1.6 Back-gas losses for different filling processes [5].

Test	A	B	C	D	E
βtot	0.63	0.46	0.25	1.00	0.20
β_{heat1}	0.40	0.17	0.10	0.80	0.08
β_{heat2}	0.04	0.13	0.04	0.11	0.05
β_v	0.05	0.06	0.06	0.04	0.04
β_p	0.14	0.10	0.05	0.05	0.03

3) Loss of pressure drop: due to restrictions and friction, the pressure drops in the pipelines. This causes the boiling temperature to decrease and thus a fraction of hydrogen to evaporate. This part can be calculated in dependency of the vapor fraction x:

$$\beta_p = \frac{1}{1-x} \tag{1.41}$$

With a sophisticated experimental set-up, the back-gas was analyzed at HyCentA for different conditions. The results are shown in Table 1.6.

Total losses vary from 20% to 100%. Losses of filling volume and losses of pressure drop are relatively small, the largest contribution is the loss through heat input. This part decreases with repeated filling processes. To minimize losses, the filling process should meet the following requirements:

- Pipelines and system should be short and well isolated
- Pressure in the conditioner should be as high as possible, which shortens the filling time
- The higher the filling amount, the smaller the (relative) losses
- Undercooling the liquid in the conditioner will reduce losses.

The pressure in the conditioner can be regulated by valves V2 and V3, see Figure 1.17. After some time, thermodynamic equilibrium can be assumed in the container, so that the boiling temperature corresponds to the pressure according to the saturated lines in the *T–s*-diagram. If the pressure is varied quickly in the conditioner and not enough time is allowed for the temperature to reach equilibrium, the fluid will be in thermodynamic non-equilibrium in an undercooled or overheated state. If pressure is decreased and liquid hydrogen is undercooled, it will consume some additional heat to reach thermodynamic equilibrium without evaporating and in that way losses are minimized.

1.4.3 Tank Systems

Liquid storage requires highly sophisticated tank systems. Heat transfer into the tank through conduction, convection and radiation has to be minimized. Therefore, the specially insulated vessels consist of an inner tank and an outer container with an

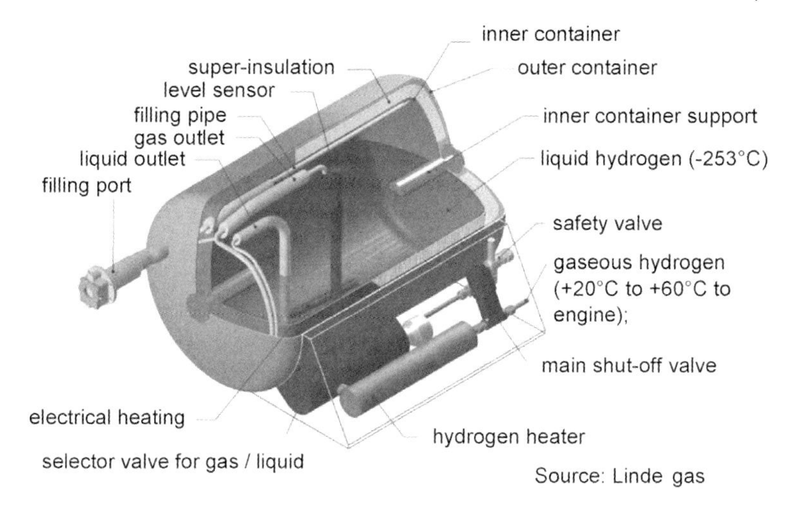

Figure 1.18 Liquid hydrogen tank [18].

insulating vacuum layer between them, see Figure 1.18. The vacuum layer should inhibit heat conduction and heat convection. The evacuated space between the nested containers is filled with multilayer insulation (MLI) having several layers of aluminum foil alternating with glass fiber matting to avoid heat radiation.

Nevertheless, due to the inevitable inward heat leakage, hydrogen evaporates in the container leading to an increase in pressure. Liquid hydrogen containers must therefore always be equipped with a suitable pressure relief system and safety valve. Liquid storage thus takes place in an open system in which released hydrogen has to be dealt with by means of catalytic combustion, dilution or alternative consumption. Evaporation losses on today's tank installations are somewhere between 0.3% and 3% per day, though larger tank installations have an advantage as a result of their lower surface area to volume ratio [16].

Active cooling with liquid nitrogen is used for large vessels for transportation. To avoid embrittlement and diffusion, appropriate materials like stainless steel and aluminum alloys have to be chosen. The austenitic stainless steel most commonly used for such tanks retains its excellent plasticity even at very low temperature and does not embrittle. Apart from cylindrical tank systems, flat shapes help to save volume in automotive applications. To release the appropriate amounts of hydrogen for acceleration, a hydrogen heater has to be installed. With this adjustable heat input, the hydrogen is evaporated and brought to the energy converter.

Liquid hydrogen tank systems reach today's highest energy densities with 0.06 kg $H_2\,kg^{-1}$ or $2\,kW\,h\,kg^{-1}$ and 0.04 kg $H_2\,dm^{-3}$ or $1.2\,kW\,h\,dm^{-3}$. The liquid storage system for the first small series hydrogen vehicle with internal combustion engine, BMW Hydrogen 7, was built by MAGNA STEYR in Graz [20]. Each system was tested for functionality, insulation, pressure build-up and so on, at HyCentA, see Figure 1.19. The tank system for about 9 kg of hydrogen has a volume of about $170\,dm^3$ and a weight of about 150 kg, which allows a maximum driving range of about 250 km. Costs for the system are not published [6].

Figure 1.19 Testing of a LH_2-tank system at HyCentA.

1.4.4 Distribution Facilities

As was explained in the thermodynamic analysis, the infrastructure for liquid hydrogen is sophisticated. It has to meet high safety requirements. All valves, pipelines, fittings and couplings have to be vacuum isolated and are made of appropriate materials. Backgas from cooling and filling causes loss of hydrogen to the environment. Filling of tank systems is usually effected by a pressure gradient, whilst cryo-pumps for liquid hydrogen are under investigation [16].

1.5 Hybrid Storage

By applying the term *hybrid* for a system combining two technologies, a number of hybrid storage systems for hydrogen are currently discussed. By cooling pure hydrogen below the freezing point at $-259\,^{\circ}$C, a mixture of solid and liquid hydrogen, called slush, can be produced. This promises higher energy densities, but needs more effort for production. Also cryo-compressed storage of hydrogen as a supercritical fluid in cryogenic tank systems for high pressures is under consideration. Combinations of liquid or compressed storage with solid storage technologies are

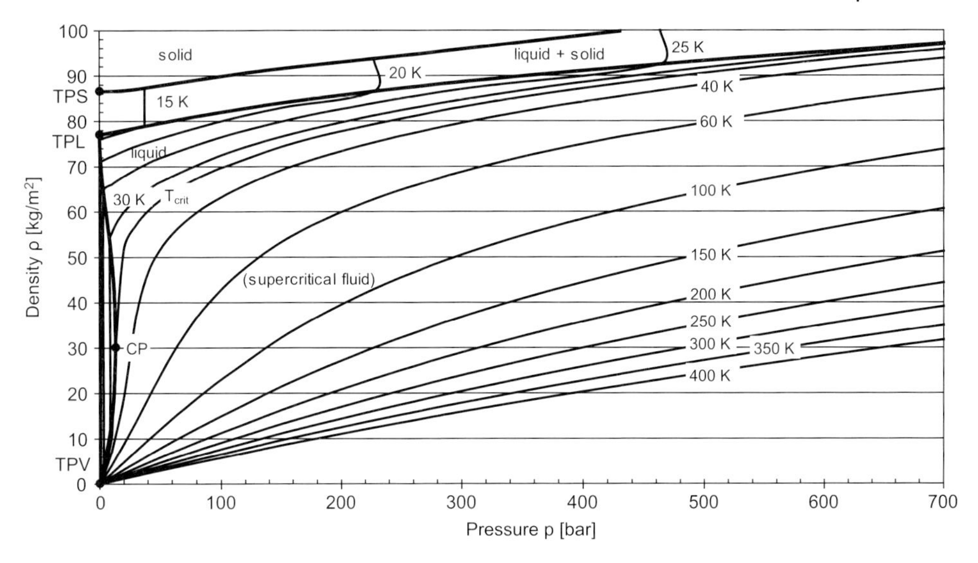

Figure 1.20 Density of hydrogen as a function of pressure and temperature.

also being investigated, for example, to store the boil-off of a liquid tank system in solid state absorbers.

The density of any substance depends on pressure and temperature. For gases at ambient conditions the influence of the pressure prevails, but with temperatures approaching the critical value, the influence of temperature increases. The potential and limits in terms of density can be judged from the state diagrams for hydrogen. Figure 1.20 shows density versus pressure with lines of constant temperature, Figure 1.21 density versus temperature with lines of constant pressure, both especially for low temperatures. The diagrams with critical point CP, triple point for solid TPS, liquid TPL and vapor TPV, and lines of phase transition show the potential for high hydrogen densities at low temperatures and high pressures.

1.5.1
Supercritical Storage

The state of a substance at temperatures and pressures above the critical values is called supercritical fluid. The properties are in between those of the liquid and gaseous state. Hydrogen is a supercritical fluid at temperatures above 33.2 K and pressures above 31.1 bar. As can be seen in Figure 1.21, the density at 350 bar and 35 K lies around 80 kg m^{-3} and thus above the density for the saturated liquid. The hybrid storage technology combining elements of cryogenic liquid and compressed storage is called cryo-compressed storage. It promises high energy densities. As there is no change of phase, evaporation can be avoided, pressure build-up time is assumed to increase, boil-off losses to diminish [7]. Due to heat input and warming, pressure will, nevertheless, rise in the tank and eventually reach a limit, where a boil-off valve has to open. Challenges for tank system design and refueling infrastructure are high.

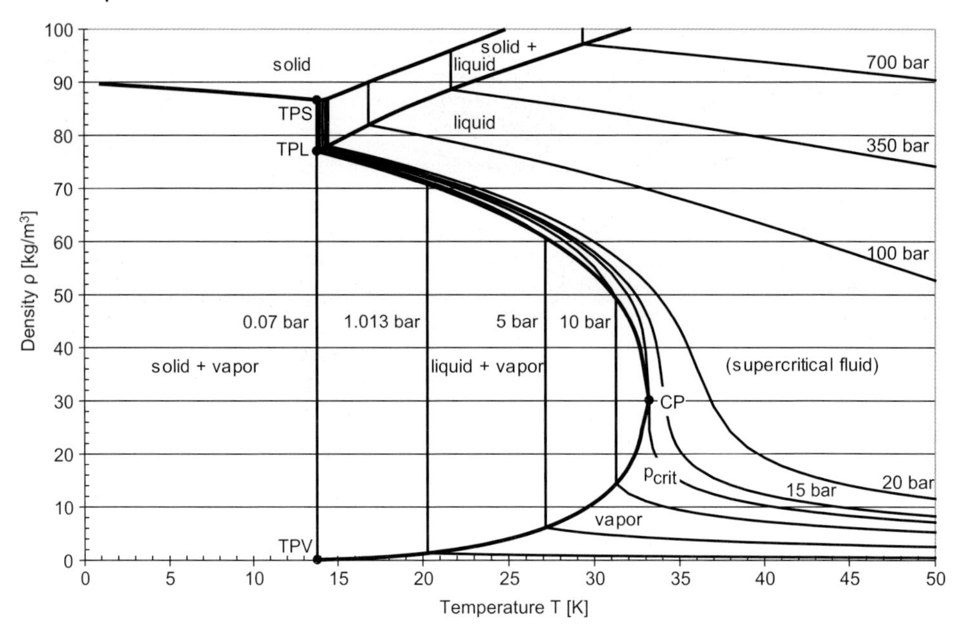

Figure 1.21 Density of hydrogen as a function of temperature and pressure.

1.5.2 Hydrogen Slush

Slush hydrogen is a two-phase solid–liquid cryogenic fluid consisting of solid hydrogen particles in liquid hydrogen at the triple point (0.07 bar, 13.8 K). The density of hydrogen slush at a solid mass fraction of 50% is halfway between the density of the liquid at the triple point and that of solid hydrogen, and thus about 16% higher than the density of the liquid [22]. As the energy for sublimation exceeds the energy for vaporization, a container of hydrogen slush has a much longer pressure build-up time compared with a container of liquid hydrogen. Due to its higher density, slush has been investigated as a rocket fuel, other applications are under consideration. Production of hydrogen slush is complicated, a number of production technologies based on liquid helium and cooling by expansion are known [10], all being on a laboratory scale.

1.6 Comparison of Energy Densities

Figure 1.22 shows the dependence of density and of volumetric energy density on pressure for cryogenic liquid and gaseous compressed hydrogen. It is obvious that the density of liquid hydrogen stored at pressures between 2 and 4 bar is at least 50% higher than the density of gaseous hydrogen stored at 700 bar. At the same time this

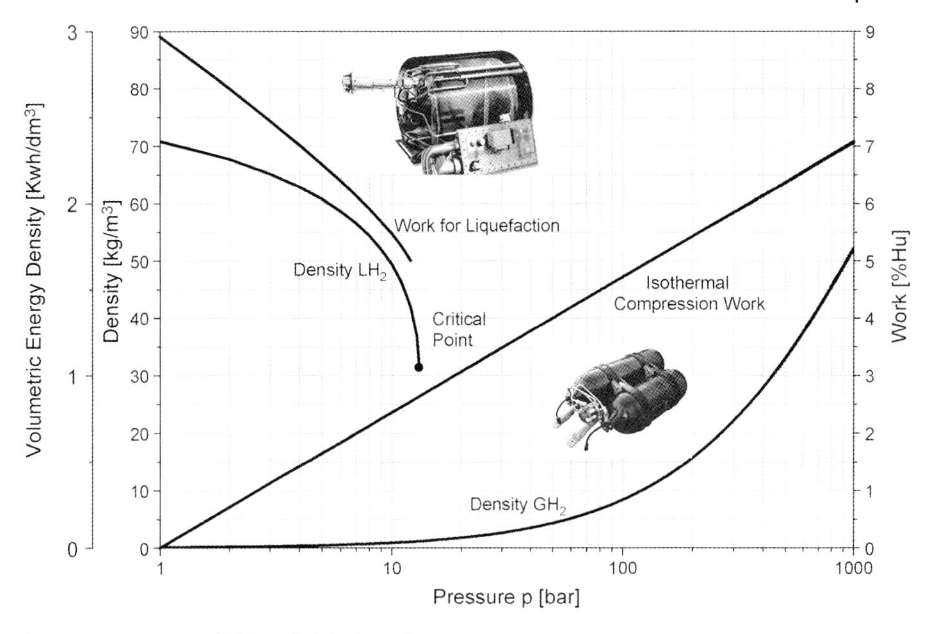

Figure 1.22 Density of LH_2 and GH_2 dependent on pressure.

denotes the physical limits of the storage densities of pure hydrogen. Without counting the volume of the storage system, the maximum volumetric energy density of liquid hydrogen at 2 bar is 2.3 kW h dm^{-3}, of gaseous hydrogen at ambient temperature and 700 bar 1.3 kW h dm^{-3}. The graph also shows the minimum work required for the liquefaction and isothermal compression of hydrogen in percentages of the calorific value H_u of 120 MJ kg^{-1}. The liquefaction and compression work represented here does not take into account the process efficiencies that lie around 50% for compressors and around 30% for liquefaction.

A comparison between the volumetric and gravimetric energy densities of state-of-the-art available energy storage systems is given in Figure 1.23 and Figure 1.24 [7].

Figure 1.23 shows the volumetric energy densities for compressed hydrogen at 350 and 700 bar, for liquid hydrogen, for solid hydrogen storage in metal hydrides, for Li ion batteries, for compressed and liquefied natural gas, and for gasoline. Where available, values are given for the pure substance and for the whole storage system. Liquid hydrocarbons like diesel and gasoline with their light and cheap tank systems allow by far the highest energy densities and thus longest driving ranges in a vehicle.

The corresponding comparison of gravimetric energy densities is given in Figure 1.24. Even though hydrogen has the highest gravimetric energy density of all pure fuels at 33.3 kW h kg^{-1}, because of the heavy tank systems it is still far from reaching the high gravimetric energy density of gasoline storage. Energy densities of available solid hydrogen storage systems or batteries are smaller by an order of magnitude.

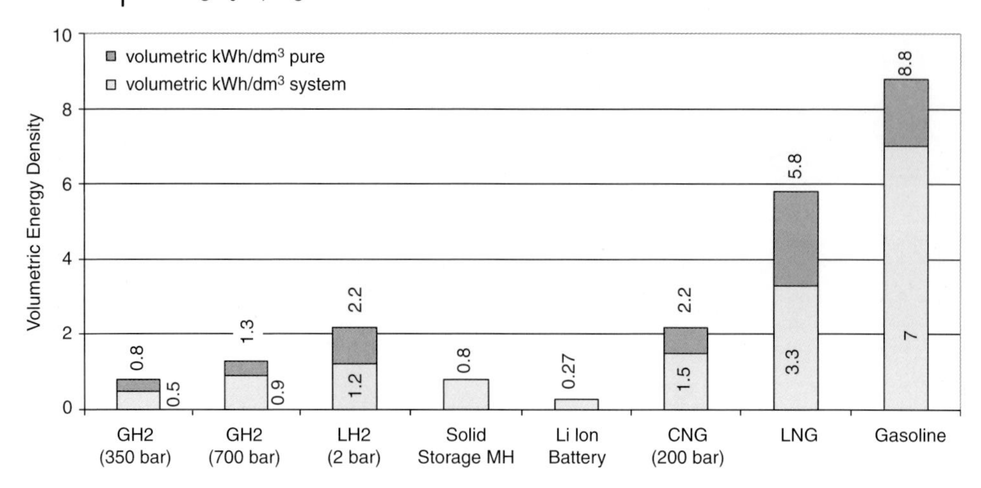

Figure 1.23 Volumetric energy density of storage systems.

If a CO_2-free energy conversion is required, only hydrogen and electrical energy can fulfill this demand. In spite of progress in battery technology, at present electrical energy densities are still quite low. Of course, both hydrogen and electricity would have to be produced in a CO_2-free way.

Another comparison of volumetric versus gravimetric storage density including solid storage systems is given in Figure 1.25 [26]. It is obvious that, with regard to volumetric storage density, storage of hydrogen in compounds has the greater potential. More hydrogen per unit can be stored in compounds than in the pure form. Gasoline and diesel actually are chemical hydrogen compounds with high volumetric and gravimetric energy densities, but hydrogen cannot easily be separated from the carbon.

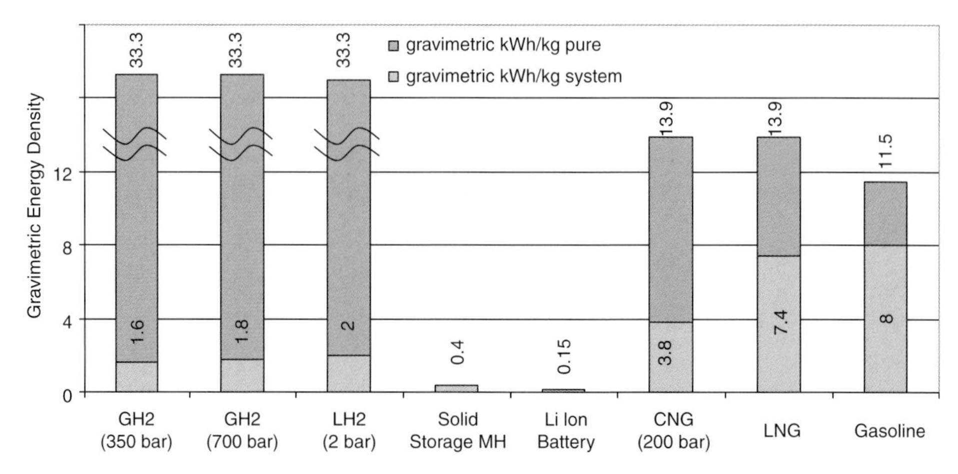

Figure 1.24 Gravimetric energy density of storage systems.

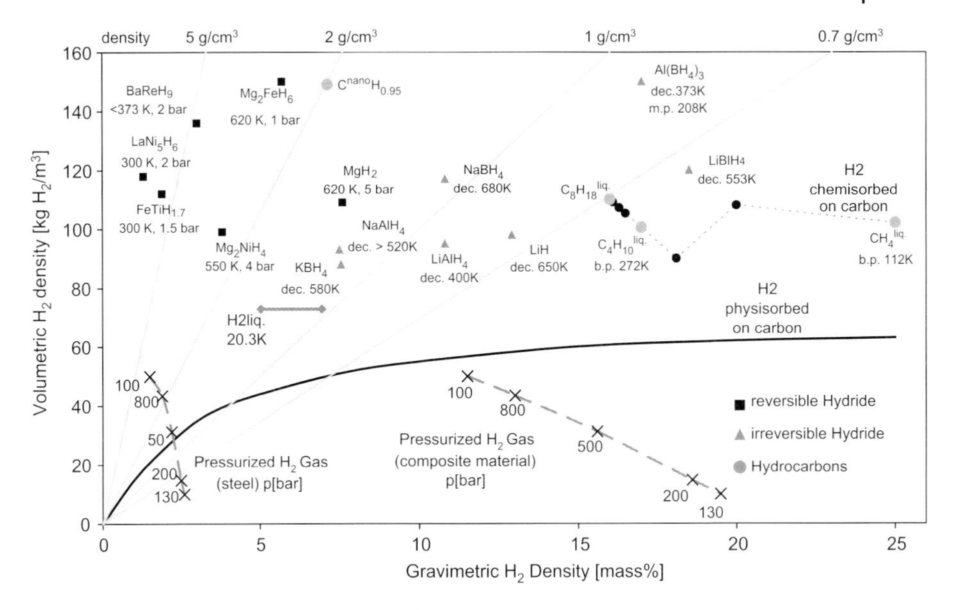

Figure 1.25 Density of hydrogen storage technologies.

Until now most solid storage systems have been on a laboratory scale only. Apart from the weight and volume of the storage systems, topics of research are the conditions of charging and discharging the system (pressure, temperature, heat transfer, time), the potential number of charging cycles (lifetime) and of course costs.

1.7 Conclusion

Hydrogen in a pure form can be stored as a highly compressed gas at up to 700 bar, cryogenically liquefied at −253 °C or in hybrid form.

Gaseous hydrogen storage takes place in a closed system without losses for extended periods. Tank vessels and infrastructure require consideration of questions regarding material choice, component dimensioning and safety, but resemble established technologies applied to compressed natural gas. Type IV containers of composite materials are commercially available for 350 and 700 bar. The hydrogen density is 23.3 $kg\,m^{-3}$ at 350 bar and 39.3 $kg\,m^{-3}$ at 700 bar and 25 °C. System energy densities of 1.8 $kW\,h\,kg^{-1}$ and 0.9 $kW\,h\,dm^{-3}$ at 700 bar can be achieved. The physical limit of the volumetric energy density at 700 bar is 1.3 $kW\,h\,dm^{-3}$ for pure hydrogen. The energy required for compression is up to 15% of the fuel energy content.

Higher storage densities are possible with liquid hydrogen: the density at 2 bar is 67.67 $kg\,m^{-3}$. The physical limit of the volumetric energy density at 2 bar is thus 2.3 $kW\,h\,dm^{-3}$ for pure hydrogen. However its very low boiling point at −253 °C means that the generation of liquid hydrogen is complex and requires 20 to 30% of its energy content. The storage of liquid hydrogen is technically challenging. Containers with high levels of insulation are used, consisting of an inner tank and an outer container with an insulating vacuum between them. Nevertheless, heat transfer cannot be ruled out completely. As a result of inevitable inward heat leakage, hydrogen evaporates in the container leading to increases in pressure and temperature. Liquid hydrogen containers must therefore always be equipped with a suitable pressure relief system and safety valves. Liquid storage takes place in an open system in which released hydrogen has to be dealt with by means of catalytic combustion, dilution or alternative consumption. Evaporation losses on today's tank installations are between 0.3 and 3% per day. A highly sophisticated and expensive production and processing system is necessary in order to minimize losses caused by diffusion, evaporation and impurity. With today's available liquid hydrogen storage systems energy densities of $2\,\mathrm{kW\,h\,kg^{-1}}$ and $1.2\,\mathrm{kW\,h\,dm^{-3}}$ can be achieved.

From the application point of view storage of compressed hydrogen at 700 bar offers an acceptable energy density at affordable costs. Type IV tank systems are quite mature but also do not offer much potential for further improvement. With a higher effort in production and storage, higher energy densities can be achieved with liquid hydrogen storage. This is applied if driving range is decisive or if large quantities are used, as in centralized production and distribution or in space applications as rocket fuel. Hydrogen slush and cryo-compressed storage are issues of research.

In order to proceed with the application of CO_2-free energy carriers in vehicles, both compressed and liquid hydrogen applications are ready for the market from the technical point of view. They can complement electric drive trains that may be convenient for short ranges. Both electricity and hydrogen have to be produced, a forceful expansion of regenerative energy production is mandatory. Due to higher costs compared with fossil fuels, the widespread use of hydrogen requires political support, for example, through CO_2 taxation.

References

1 Dynetek Industries Ltd., http://www.dynetek.com.

2 Eichlseder, H., Klell, M., Schaffer, K., Leitner, D., and Sartory, M. (2009) Potential of Synergies in a Vehicle for Variable Mixtures of CNG and Hydrogen, SAE paper 2009-01-1420.

3 Eichlseder, H. and Klell, M. (2008) *Wasserstoff in der Fahrzeugtechnik, Erzeugung, Seicherung und Anwendung [Hydrogen in Vehicle Technology, Production, Storage and Application]*, Vieweg + Teubner Verlag, Wiesbaden, ISBN 9783834804785.

4 Eichner, Th. (2005) Kryopumpe für Wasserstoff [Cryo-pump for Hydrogen]. Diploma thesis, Technische Universität Graz.

5 Emans, M., Mori, D., and Krainz, G. (2007) Analysis of back-gas behaviour of an automotive liquid hydrogen storage system during refilling at the filling station. *Int. J. Hydrogen Energy*, **32**, 1961–1968.

6 Enke, W., Gruber, M., Hecht, L., and Staar, B. (2007) Der bivalente V12-Motor des BMW Hydrogen 7 [Bivalent V12 Engine of the BMW Hydrogen 7]. *Motortechnische Zeitschrift MTZ* 68, **06**, 446–453.

7 EU FP6 Integrated Project STORHY, http://www.storhy.net.
8 Gstrein, G. and Klell, M. (2004) *Stoffwerte von Wasserstoff [Properties of Hydrogen]*, Institute for Internal Combustion Engines and, Thermodynamics, Graz University of Technology.
9 Gursu, S., Sherif, S.A., Veziroglu, T.N., and Sheffield, J.W. (1993) Analysis and optimization of thermal stratification and self-pressurization effects in liquid hydrogen storage systems – Part 1: Model development. *J. Energy Resources Technology*, **115**, 221–227.
10 Haberbusch, M. and McNelis, N. (1996) Comparison of the Continuous Freeze Slush Hydrogen Production Technique to the Freeze/Thaw Technique. NASA Technical Memorandum 107324.
11 Kindermann, H. (2006) Thermodynamik der Wasserstoffspeicherung. [Thermodynamics of Hydrogen Storage]. Diploma thesis, HyCentA Graz, Montanuniversität Leoben.
12 Klell, M., Brandstätter, S., Jogl, C., and Sartory, M. (2008) Werkstoffe für Wasserstoffanwendungen [Materials for Hydrogen Application]. Report 3a2008 HyCentA Research GmbH, http://www.hycenta.at.
13 Klell, M. and Sartory, M. (2008) Sicherheit und Standards für Wasserstoffanwedungen [Safety and Standards for Hydrogen Applications]. Report 4a2008 HyCentA Research GmbH, http://www.hycenta.at.
14 Klell, M., Zuschrott, M., Kindermann, H., and Rebernik, M. (2006) Thermodynamics of hydrogen storage. 1st International Symposium on Hydrogen Internal Combustion Engines, Report 88, Institute for Internal Combustion Engines and Thermodynamics, Graz University of Technology, Graz.
15 Kurzweil, P. (2003) *Brennstoffzellentechnik [Fuel Cell Technology]*, Vieweg Verlag, Wiesbaden, ISBN 3528039655.
16 Léon, A. (ed.) (2008) Hydrogen technology, in *Mobile and Portable Applications*, Springer-Verlag, Berlin, Heidelberg, ISBN 9783540790273.
17 Lin, C.S., Van Dresar, N.T., and Hasan, M. (2004) A pressure control analysis of cryogenic storage systems. *J. Propul. Power*, **20** (3), 480–485.
18 Linde Gas Gmbh, http://www.linde-gas.at.
19 Mackay, K.M. (1973) The element hydrogen, ortho- and para-hydrogen, atomic hydrogen, in *Comprehensive Inorganic Chemistry*, **1** (ed. A.F., Trotman-Dickinson), Pergamon Press, Oxford, pp. 1–22.
20 MAGNA STEYR Fahrzeugtechnik AG & Co KG, http://www.magnasteyr.com.
21 National Institute of Standards and Technology NIST, http://www.nist.gov.
22 Ohira, K. (2004) Development of density and mass flow rate measurement technologies for slush hydrogen. *Cryogenics*, **44**, 59–68.
23 Pischinger, R., Klell, M., and Sams, Th. (2009) *Thermodynamik der Verbrennungskraftmaschine [Thermodynamics of the Internal Combustion Engine]*, 3rd edn, Springer Verlag, Wien New York, ISBN 978-3-211-99276-0.
24 Quack, H. (2001) Die Schlüsselrolle der Kryotechnik in der Wasserstoff-Energiewirtschaft [Key Role of Cryo-technology in Hydrogen Energy Sector]. Wissenschaftliche Zeitschrift der Technischen Universität Dresden, 50 Volume 5/6, 112–117.
25 Riedel, E. and Janiak, Ch. (2007) *Anorganische Chemie [Inorganic Chemistry]*, Walter de Gruyter, Berlin New York, ISBN 9783110189032.
26 Schlapbach, L. and Züttel, A. (2001) Hydrogen storage-materials for mobile applications. *Nature*, **414**, 23–31.
27 Scurlock, R. (2006) *Low-Loss Storage and Handling of Cryogenic Liquids: The Application of Cryogenic Fluid Dynamics*, Kryos Publications, Southampton, UK, ISBN 9780955216605.
28 Turns, St. (2006) *Thermodynamics, Concepts and Applications*, Cambridge University Press, New York, ISBN 9780521850421.
29 Züttel, A., Borgschulte, A. and Schlapbach, L. (eds) (2008) *Hydrogen as a Future Energy Carrier*, Wiley-VCH Verlag, Weinheim, ISBN 9783527308170.

2
Physisorption in Porous Materials

Barbara Panella and Michael Hirscher

2.1
Introduction

Physisorption or physical adsorption is the mechanism by which hydrogen is stored in the molecular form, that is, without dissociating, on the surface of a solid material. Responsible for the molecular adsorption of H_2 are weak dispersive forces, called van der Waals forces, between the gas molecules and the atoms on the surface of the solid. These intermolecular forces derive from the interaction between temporary dipoles which are formed due to the fluctuations in the charge distribution in molecules and atoms. The combination of attractive van der Waals forces and short range repulsive interactions between a gas molecule and an atom on the surface of the adsorbent results in a potential energy curve which can be well described by the Lennard-Jones Eq. (2.1).

$$V(r_i) = 4\varepsilon\left[\left(\frac{\sigma}{r_i}\right)^{12} - \left(\frac{\sigma}{r_i}\right)^{6}\right] \tag{2.1}$$

where ε is the depth of the energy well, r_i the distance between an H_2 molecule and a generic atom i on the surface and σ the value of r_i for which the potential is zero [1]. The minimum of the potential curve occurs at a distance from the surface which is approximately the sum of the van der Waals radii of the adsorbent atom and the adsorbate molecule. For microporous materials ($r_{pore} < 1$ nm) [2] where the pore dimensions are comparable to the diameter of the adsorbate molecules this interaction potential is also influenced by the pore shape and the pore dimensions.

Adsorption is an exothermic process, that is, heat is released during the adsorption of gas molecules on a surface. Owing to the low polarizability of the H_2 molecule this type of interaction is very weak and leads to low values of the heat of adsorption, typically in the range of 1–10 kJ mol^{-1} [3]. Compared to chemisorption which involves a change in the chemical nature of the H_2 molecule by, for example, dissociation, physisorption has approximately a ten times smaller heat of adsorption. Chemical hydrides and complex hydrides, which store hydrogen chemically, both have the disadvantage of producing an excessive amount of heat during fast

Handbook of Hydrogen Storage. Edited by Michael Hirscher

ISBN: 978-3-527-32273-2

refueling, leading to very high temperatures in the tank. This makes on-board hydrogen refilling with these systems a presently unresolved problem. Therefore, the low heat of adsorption released during physisorption is an advantage for mobile storage technologies which require on-board refueling. However, due to the weak interaction between molecular hydrogen and solids, the amount of hydrogen stored at room temperature is very low, while it can consistently increase at low temperatures, typically 77 K, that is, the temperature of liquid nitrogen. Bhatia and Myers [4] recently analyzed the thermodynamic requirement for an adsorbent to store molecular hydrogen at room temperature at a reasonable charging pressure of 30 bar and to release it reversibly at 1.5 bar. The resulting optimum average enthalpy of adsorption is $\Delta H^0_{opt} = -15$ kJ mol^{-1} which is obtained assuming a Langmuir model for the adsorption enthalpy and an average entropy change of $\Delta S^0 = -66.5$ J mol^{-1} K^{-1}. Less negative enthalpies than ΔH^0_{opt} would lead to a lower optimum temperature of adsorption than room temperature, while more negative values would lead to stronger adsorption at the discharge pressure and the gas release would therefore decrease at 1.5 bar. However, it should be considered that this optimum value is very sensitive to the estimate of the entropy change and that lower ΔS^0 were also obtained leading to more negative ΔH^0_{opt} [5, 6]. Enthalpy values (ΔH) for hydrogen adsorption in porous materials are typically lower than 15 kJ mol^{-1}, as shown in Table 2.1.

Physisorption is a completely reversible process, which means that the hydrogen can be easily adsorbed and released during several cycles without any losses. Furthermore, in general, no activation energy is involved in the molecular adsorption of hydrogen, which leads to very fast kinetics of the adsorption and desorption process. Fast kinetics and reversibility are important requisites for a mobile hydrogen storage system hence physisorption may, from this point of view, be an optimal mechanism for storing H_2.

Table 2.1 Heat of adsorption of hydrogen on different classes of porous materials.

Materials	$\|\Delta H$ (kJ mol^{-1})	References
Carbon materials and organic polymers		
Activated carbon	5–5.8	[4, 7]
Single walled carbon nanotubes	3.6–9	[8, 9]
Hypercrosslinked Polymers	5–7.5	[10, 11]
Coordinatiopn polymers/MOFs		
Metal-organic frameworks MOF-5	3.8–5.2	[12–14]
Metal-organic frameworks IRMOF-8	6.1	[12]
Metal-organic framework MIL-101	9.3–10	[15]
Prussian blue analogues $M_3[(Co(CN)_6]_2$	4.7–9.0	[14, 16]
Zeolites		
HY	5.7–6.1	[17]
NaY	5.9–6.3	[17]
Na-ZSM-5	11	[18]
Li-ZSM-5	6.5	[5]

Since in physisorption the H_2 is adsorbed only on the surface of a solid, not in the bulk, only porous materials with a high specific surface area (SSA) are suitable materials for molecular hydrogen storage. Compared to metal hydrides and complex hydrides, hydrogen bulk diffusion and restructuring of the solid, as well as the formation of chemical bonds do not take place during this storage mechanism. Therefore, porous materials used for physisorption have typically a long lifetime during cycling.

The adsorption capacity of a porous material is best described through its adsorption isotherm. In contrast to metal hydrides and complex hydrides, where pressure–composition isotherms are generally considered, for porous materials the isotherm reports the hydrogen uptake versus the hydrogen pressure. The hydrogen uptake is typically expressed as gravimetric storage capacity, wt%, referred to the mass of the adsorbent (m_s) and of the adsorbed hydrogen, m_{H2}, (2.2).

$$\text{wt\%} = \frac{m_{H_2}}{(m_{H_2} + m_s)}\% \tag{2.2}$$

or as volumetric storage capacity (vol%), related to the volume occupied by the solid (V_s) (2.3).

$$\text{vol\%} = \frac{m_{H_2}}{V_s}\% \tag{2.3}$$

These are storage capacities referred to the material and not to the total storage system, which would include the tank and all mechanical parts which are necessary for a storage device.

Apart from considering the gravimetric or the volumetric uptake, two ways to express the amount of hydrogen stored in porous materials exist: the absolute storage capacity and the excess storage capacity. Experiments measure the excess storage capacity which is defined as the difference between the quantity of gas which is stored at a temperature, T, and a pressure, P, in a volume containing the adsorbent and the quantity which would be stored under the same conditions in the absence of gas–solid interactions [19]. In order to mimic the absence of gas–solid interactions and measure the excess hydrogen adsorption experimentally, the volume of non-adsorbed gas in the system is typically measured with helium gas which is assumed to be a non-adsorbing gas [20] (Figure 2.1).

The absolute adsorption is predicted from theoretical calculations and is the amount of hydrogen which is adsorbed in the porous material, not considering the gas phase. The relation between the excess adsorption (N_{ex}) and the absolute adsorption (N_{ads}) can be esily derived considering a typical adsorption experiment [21]: The amount of gas adsorbed on a sample (N_{ads}) is expressed as the total amount of gas introduced in the sample cell (N_i) minus the free molecules in the gas phase. (2.4):

$$N_{ads} = N_i - V_g \varrho_g \tag{2.4}$$

where V_g is the free volume occupied by the gas including the pore volume of the sample and ϱ_g is the gas density. Since the real free volume is not known, experiments

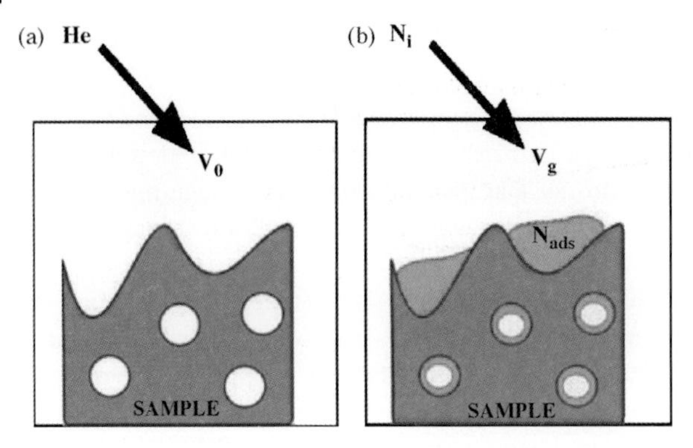

Figure 2.1 Schematic diagram showing the He volume, V_0, of a porous material in the absence of the gas-solid interaction (a), and the gas-volume, V_g, in the presence of the gas–solid interaction (b).

report the excess adsorption, where the free volume is measured by He gas expansion (V_0) (2.5):

$$N_{ex} = N_i - V_0 \varrho_g \tag{2.5}$$

Therefore the relation between experimentally measured excess adsorption and the absolute adsorption can be expressed as in (2.6):

$$N_{ex} = N_{ads} - \varrho_g (V_0 - V_g) = N_{ads} - \varrho_g (V_a) \tag{2.6}$$

where ($V_0 - V_g$), is the difference between the real free volume and the free volume measured with He gas in the absence of an adsorbed phase. This difference corresponds to the volume occupied by the adsorbed phase (V_a). Therefore the absolute amount could, in principle, be determined from experimental excess adsorption values if the volume of the adsorbed phase was known. Sometimes for microporous materials V_a is defined as the pore volume of the solid, since for very narrow pores the density distribution of H_2 everywhere is greater than the gas density [22]. In reality, for most porous materials which possess a distribution of pore sizes the determination of V_a is less simple. Equation (2.6) can be reformulated considering the density of the adsorbed phase, ϱ_a, (2.7) [21, 23]:

$$N_{ex} = N_a \left(1 - \frac{\varrho_g}{\varrho_a}\right) \tag{2.7}$$

Especially under conditions for which the density of the gas phase is comparable to or higher than the density of the adsorbed phase ($\varrho_g / \varrho_a \sim 1$), that is, low temperatures and high pressures, the excess adsorption differs from the absolute adsorption.

The typical absolute hydrogen adsorption isotherm for porous materials at 77 K is a Langmuir or type I isotherm according to the IUPAC classification [2]. This isotherm is characterized by a steep increase in the H_2 uptake at low pressure, while at higher pressures the storage capacity reaches a plateau and a further increase in the hydrogen pressure does not lead to an increase in the storage capacity (Figure 2.2).

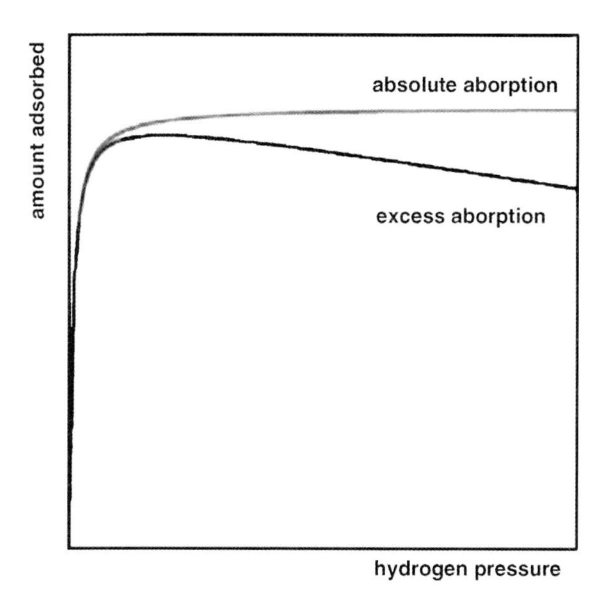

Figure 2.2 Typical shapes of the absolute adsorption isotherm and excess absorption isotherm. The inflection of the excess adsorption isotherm can occur at high pressures.

At the plateau the maximum uptake value of the adsorbent is therefore reached. The type I or Langmuir isotherm describes the formation of a single monolayer on the surface of the porous material. No multilayers of hydrogen are formed at 77 K, because the interaction strength between single layers is too weak at temperatures higher than the critical temperature of H_2. Therefore a type I isotherm is typically obtained for porous materials at adsorption temperatures higher than the critical temperature of the gas.

In contrast to the absolute adsorption, the excess adsorption isotherm, measured experimentally, does not show a plateau at high pressures, but a maximum (2.7) (Figure 2.2). A further increase in the pressure results in a decrease in the excess adsorption, that is, when the gas adsorbed in the pores saturates while the gas density keeps increasing [19]. Therefore the excess adsorption can be zero or even negative at very high pressures and low temperatures. This does not correspond to a real physical decrease in the hydrogen uptake of the material which instead reaches a constant value at high pressure. This is simply an effect of measuring the void volume of the sample using helium gas in the absence of an adsorbed phase. At moderate pressures the absolute and the excess adsorption do not differ greatly and both can be described by a type I isotherm. In the next sections the uptake values determined experimentally are always excess storage capacities, even if omitted.

At room temperature only a small amount of hydrogen is stored in porous materials and the adsorption isotherm typically does not show any saturation described by a plateau. Instead a linear increase in the hydrogen uptake is observed up to relatively high pressure. Therefore, at room temperature the adsorption is better described by a linear Henry type isotherm [1].

2.2 Carbon Materials

Carbon exists in different allotropic modifications, that is, in forms with different chemical structures. The best known are diamond and graphite, both of which are non-porous.

However, a great variety of porous carbon nanostructures exist which mainly consist of benzene-like carbon hexagons with sp^2-hybridized carbon atoms. These structures differ from each other in the way these hexagons are arranged in the material. In principle, these materials can be divided into two main classes: those which possess a long-range order of the carbon hexagons, like carbon nanotubes or nanofibers, and those which have irregular structures, like activated carbon [24]. Most of the carbon nanostructures possess an ordered array, typically of short range, surrounded by disordered parts [25].

Activated carbons are amorphous carbonaceous materials possessing a high degree of porosity and a high specific surface area. They are mainly prepared by a two-step pyrolysis of carbonaceous raw materials with the carbonization process followed by activation at higher temperatures, either by reaction with an oxidizing gas or with inorganic chemicals. The result of this preparation is a material with irregularly arranged aromatic carbon sheets which are cross-linked in a random manner by a network of aliphatic carbon. This gives a continuum of interlayer spaces ranging from distances similar to the spacing in graphite up to sizes which can form micropores [24, 26]. Regular grade activated carbons have specific surface areas of 700–1800 $m^2\,g^{-1}$ but high porosity activated carbons like AX-21 can have SSA up to 3000 $m^2\,g^{-1}$ [27].

A way to produce porous carbons with controlled pore size distribution involves the use of an ordered inorganic porous template (e.g., mesoporous silica or zeolites) where the carbon precursor is deposited and carbonized. After removal of the inorganic template the carbon sample possesses a regular pore structure which is the replica of the inorganic template [28–31]. Depending on the synthesis procedure and on the carbon precursor, some of these porous carbons can possess highly graphitic pore walls [30]. Another class of carbon materials with controlled pore size, so-called carbide-derived carbons, can be produced by high temperature chlorination of metallic carbides [32, 33]. During chlorination the metals are removed as volatile chlorides leaving a highly porous amorphous carbon material.

Carbon nanotubes, which were discovered in 1991 by Iijima [34], are the best example of carbon nanostructures possessing a long-range order in the carbon atom positions. They can be considered as rolled graphene sheets with an inner diameter of around 1 nm and a length of 10–100 μm. Single-walled carbon nanotubes (SWCNTs) consist of one single graphene layer which can be closed at the ends by two fullerene-like hemispheres. Since a graphene sheet can be rolled in different ways, three different types of SWCNTs exist: armchair, zigzag and chiral [35]. The rolling direction of the graphene layer also determines whether single-walled carbon nanotubes are metallic or semiconducting. Because of the van der Walls interactions between single tubular nanostructures SWCNTs typically exist in a bundle of several tubes. Multi-walled carbon nanotubes (MWCNTs) consist of up to 50 concentric tubes with interlayer distances close to that of graphite. Adsorption of hydrogen between the single tubes in MWCNTs would require the stretching of the strong C–C bond within a graphene sheet and seems, therefore, impossible. Graphitic nanofibers are a recently developed allotropic form of carbon [36]. They are formed from the thermal decomposition of hydrocarbons on catalysts. These nanofibers are composed of graphitic platelets stacked together in various orientations to give tubular, platelet and herringbone structures [37]. As for MWCNTs the interlayer distance between the single graphitic platelets is comparable to that in graphite.

Ten years ago, both carbon nanotubes and nanofibers attracted great interest for hydrogen storage at room temperature due to several reports of exceptional hydrogen storage capacities in these carbon nanostructures [38, 39]. Later these high uptake values, which could never be reproduced, were attributed to experimental errors or to the small amounts of pure sample available for performing the uptake measurements [40–42]. Reproducible investigations showed that the hydrogen storage capacity of both purified single-walled carbon nanotubes and graphitic nanofibers does not exceed 0.6 wt% at room temperature [7, 43–46]. Maximum storage values of 2.7 wt% were found in carbon materials only at extremely high hydrogen pressures of 500 bar [46]. However, at low temperatures the storage capacity of carbon materials increases and, therefore, more recent investigations of reversible hydrogen adsorption are mainly focused at 77 K.

For all microporous carbon materials the hydrogen storage capacity is proportional to the specific surface area or the pore volume of the micropores at both 77 K and room temperature [7, 47, 48]. Figure 2.3 shows the hydrogen storage capacity of several carbon nanostructures, such as purified SWCNTs, porous carbon with uniform pore size and activated carbons, versus their BET (Brunauer–Emmet–Teller) specific surface area. Considering the BET model for the specific surface the slope of the curve gives a value of 2×10^{-3} wt% m^{-2} g at 77 K and 0.23×10^{-3} wt% m^{-2} g at RT. This result is in perfect agreement with the empirical "Chahine's rule" obtained for different activated carbons [49]. A similar correlation was obtained considering the theoretical model proposed by Züttel *et al.*, where a monolayer possessing the density of liquid hydrogen is ideally adsorbed with a closed packed two-dimensional geometry on both sides of a perfect graphene sheet [50]. This theoretical model leads to a maximum hydrogen surface density on carbon materials of 2.28 wt% m^{-2} g. Both the BET model for the experimental determination of the SSA and the theoretical model proposed by Züttel are based on simple approximations. Both methods are in good agreement and allow prediction of the storage capacity of carbon materials from a simple measurement of the specific surface area. The hydrogen storage capacity was also shown to correlate very well with the micropore volume of carbon materials [48, 51].

Among all carbon materials, high grade activated carbon and porous carbons are the best adsorbents for hydrogen at 77 K, because they possess the highest specific surface area. Chahine and Bose reported a storage capacity of 5 wt% at 77 K and 20 atm for AX-21, an activated carbon with a high specific surface area of 3000 m^2 g^{-1} [27]. A comparably high storage capacity (4.5 wt%) was obtained for a similar type of activated carbon [7].

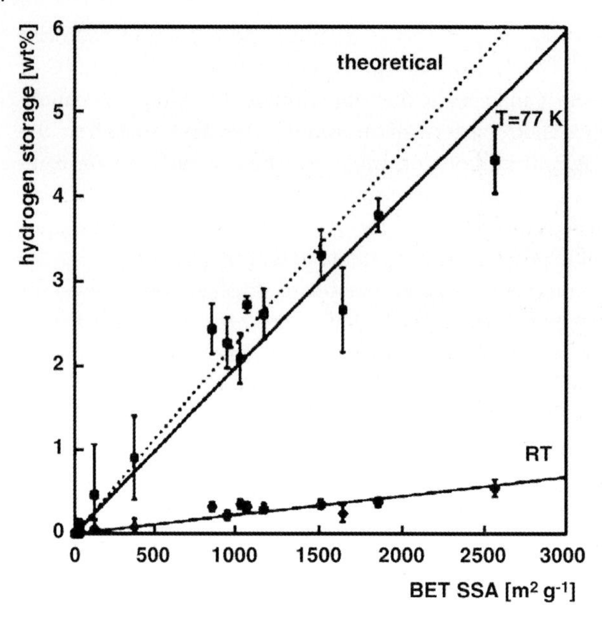

Figure 2.3 Hydrogen storage capacity at RT (diamonds) and 77 K (squares) of different carbon materials versus their BET specific surface area [52]. The dotted line represents the theoretical curve according to [50].

Activated carbon is, however, a disordered structure with a broad distribution of pore sizes which cannot be precisely controlled during the synthesis. In contrast, templated porous carbons mainly possess pores in a very narrow range of dimensions [53]. Recently, Yang and Mokaya reported the hydrogen storage capacity of a zeolite-like carbon material possessing an extremely high specific surface area ($3200\,m^2\,g^{-1}$) and pores in the range of 0.5–0.9 nm [54]. This amorphous carbon material was prepared by chemical vapor deposition (CVD) of acetonitrile as the carbon precursor on β-zeolite used as a template. The use of the ordered zeolite template allows one to control the pore size distribution, particularly in the micropore range. The result is a material capable of storing up to 6.9 wt% of hydrogen at 77 K and 20 bar and possessing a heat of adsorption of 4–8 $kJ\,mol^{-1}$, higher than in activated carbon due to the presence of optimum sized pores. The influence of the pore size on the carbon–H_2 interaction was studied on a series of carbide-derived porous carbons with controlled pore dimensions [9]. The authors showed that the hydrogen uptake at 1 atm and 77 K correlates linearly with the SSA of pores smaller than 1 nm, while no correlation is found for the SSA of pores larger than 1 nm. Since the hydrogen uptake at low pressures is determined by the heat of adsorption [9], this result clearly shows that small pores (<1 nm) lead to an enhanced H_2–carbon interaction and therefore to an increase in the hydrogen storage capacity at low pressures. At hydrogen pressures high enough to reach the saturation H_2-uptake, however, a correlation with the total SSA should be expected (as in Figure 2.3).

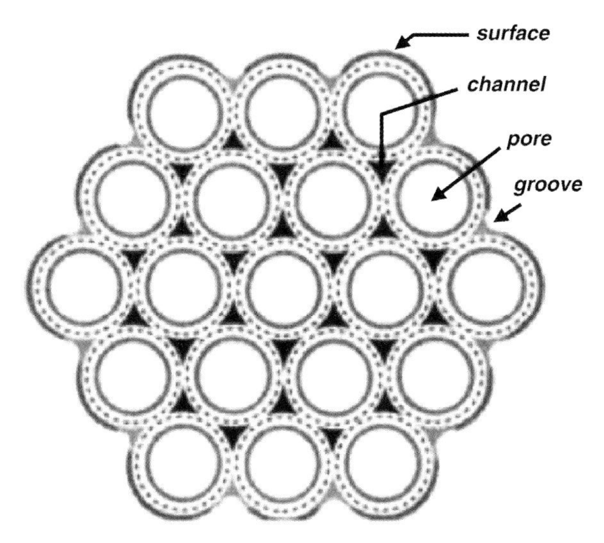

Figure 2.4 Cross-section of a bundle of 19 SWCNTs. The four potential adsorption sites are indicated [55]. Copyright (2002), with permission from Elsevier.

While in the past SWCNTs erroneously attracted great interest for hydrogen storage at room temperature, at 77 K they show a modest storage capacity (~2 wt%) compared to activated carbon and porous carbons. However, due to their long-range ordered structure they are intensively studied in order to understand the interaction between H_2 and different adsorption sites in these carbon materials. A bundle of several single-walled carbon nanotubes possesses different potential adsorption sites for hydrogen [55]: the outer surface, the pores for opened nanotubes, the grooves between two tubes and the interstitial channels, that is, the cavity between three nanotubes. The adsorption energy for hydrogen in these sites increases in the order: $E(\text{channels}) > E(\text{grooves}) > E(\text{pores}) > E(\text{surface})$ (Figure 2.4) [55].

Owing to the small size of the interstitial channels it is probable that hydrogen molecules cannot access these sites [56]. Additionally, the very small pore volume associated with these sites would give a very small contribution to the hydrogen storage, even if these were accessible [57, 58]. At least two different adsorption sites for hydrogen in single-walled carbon nanotubes could be identified by inelastic neutron scattering (INS) [59], by low-temperature hydrogen adsorption [60], by thermal desorption spectroscopy [61], and by Raman spectroscopy [62, 63]. As an example, using INS Georgiev *et al.* determined two energetically different adsorption sites for hydrogen in closed SWCNTs [59]. The stronger adsorption sites were assigned to the grooves or interstitial sites, if these are considered to be accessible, which are already occupied at low coverage. At higher hydrogen loadings the molecules also occupy the outer surface where they are more weakly adsorbed.

2.3 Organic Polymers

Organic polymers attracted considerable interest for hydrogen storage when uptake capacities of 6–8 wt% were reported for polyaniline (PAni) and polypirrole (Ppy) at room temperature [64, 65]. Unfortunately, these results could not be reproduced independently, instead room temperature values of less than 0.5 wt% at pressures up to 94 bar were subsequently observed for similar samples [66, 67].

Organic polymers with high porosity have recently been developed and studied for physisorption of H_2 at low temperatures. Microporous organic polymers are low density, ordered structures obtained by polymerization of organic molecules. Two types of porous polymers are presently being investigated for hydrogen storage:

1) The polymers of intrinsic microporosity (PIMs). These are rigid macromolecules composed of a fused-ring subunit which cannot pack space efficiently and hence form microporous materials (Figure 2.5) [68].
2) Hypercrosslinked polymers (HCPs) which are obtained by polymerization of monomers with crosslinking molecules to give a highly microporous material (Figure 2.6) [11].

The advantages of microporous organic molecules are that they consist of very light atoms (C, H, N, O), and that they can possess high specific surface areas and

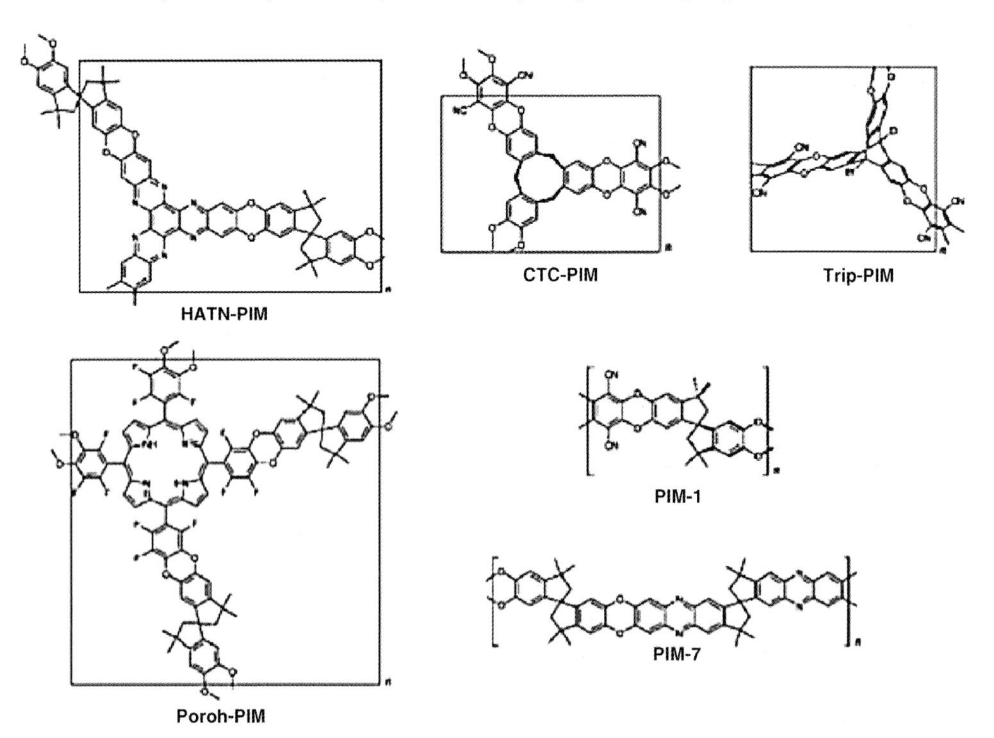

Figure 2.5 Molecular building units of polymers with intrinsic microporosity [69]. Reproduced with permission of the PCCP Owner Societies.

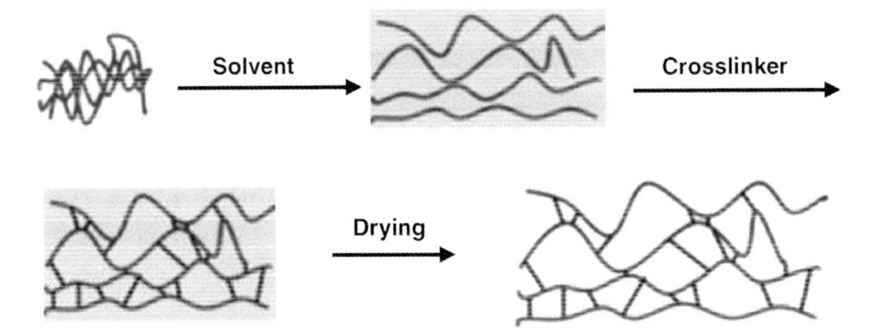

Figure 2.6 Schematic diagram of the hypercrosslinking process [11]. In the solvent the polymers start to swell and are separated from each other. The crosslinking molecules create the connection between the separated polymer chains. When the solvent is removed the microporosity is maintained [11]. Reproduced with permission of the Royal Society of Chemistry.

controlled porosity. A large number of synthetic routes exist which allow introduction of different functionalities to the polymer to enhance the interaction with hydrogen molecules [10]. Additionally, the synthesis of polymers is based on well established technologies which allow large scale production of these materials.

PIMs have been shown to store up to 2.7 wt% of hydrogen at 77 K and 10 bar with the best material being a polymer with triptycene as subunit [69, 70]. Budd *et al.* also showed that the H_2 adsorption behavior depends strongly on the micropore distribution, that is, the higher the concentration of ultramicropores the larger the hydrogen uptake at low pressure and 77 K [69]. For hypercrosslinked polymers even higher specific surface areas and storage capacities could be obtained. Wood *et al.* reported for a HCP prepared by self-condensation of bischloromethyl monomers a BET specific surface area of $1900\,m^2\,g^{-1}$ and an uptake of 3.7 wt% at 77 K and 15 bar [10]. Interestingly, they measured, for a series of these HCPs, some hysteresis in the adsorption isotherms dependent on the average pore width of the polymer. They attributed this effect, which is largest for the material with the smallest pore size, to the kinetic trapping of H_2 in the porous network of the polymer. The small dimension of the pores was indicated to have a positive effect also on the heat of adsorption which is highest again for the sample with the smallest pores ($7.25\,kJ\,mol^{-1}$).

By infrared studies Spoto *et al.* showed that the adsorption of hydrogen on a porous poly(styreneco-divinylbenzene)polymer involves the specific interaction of the H_2 molecule with the phenyl ring with an interaction energy of $\sim 0.4\,kJ\,mol^{-1}$ which is comparable with the interaction energy between H_2 and activated carbon [71]. This interaction may be increased in narrower pores where hydrogen can interact with more than one pore wall, as suggested by Wood *et al.* [10].

The hydrogen storage values of PIMs and HCPs are still smaller than for many porous carbon samples. However, PIMs and HCPs have only recently been investigated for H_2 adsorption and further modifications of these materials can lead to an enhancement of their hydrogen storage capacity at cryogenic temperatures.

2.4
Zeolites

Zeolites are three-dimensional crystalline aluminosilicate structures built of TO_4 tetrahedrons sharing all four corners, where T typically indicates Si^{4+} and Al^{3+} ions. The general formula of zeolites is $M_{x/z}[(AlO_2)_x(SiO_2)_y]\cdot mH_2O$, where M is the non-framework exchangeable cation. The T–O–T bonds are very flexible, so that the tetrahedral units can be linked in a great number of different network topologies. In addition, due to this flexibility the geometry of the framework is able to adapt under diverse conditions of temperature, pressure and chemical surroundings [24]. To maintain the electroneutrality of the structure, for every Si^{4+} substituted with an aluminum ion there is an additional extra-framework metal ion (M) adsorbed in the structure. The additional cations are usually alkali and alkaline earth metal ions. The ion exchange capacity of these materials depends on both the size of the accommodating pores and on the charge of the cation. The presence of strong electrostatic forces inside the channels and pores can produce strong polarizing sites in the free volumes of the zeolite. The electrostatic field is produced by the extra-framework metal ions and it increases with increasing charge and decreasing size. Some zeolites occur in nature, however, most of them are produced synthetically, usually by hydrothermal crystallization of aluminosilicate gels. Zeolites can have a very open microporous structure with different framework types depending on the assembly of the tetrahedral building units. However, since some of the cages in these frameworks are not accessible to gas molecules part of the void volume in the structure does not contribute to hydrogen storage.

Zeolites have been investigated for several years for hydrogen storage applications. Already in 1995 Weitkamp *et al.* showed that the hydrogen storage capacity in different zeolites at room temperature is less than 0.1 wt% [72]. However, at 77 K the hydrogen uptake can be increased, obtaining a maximum storage capacity of 2.19 wt% at 15 bar for CaX zeolite [73]. Using an atomistic model for hydrogen packing in the micropores of zeolites it could be determined that the maximum theoretical H_2 storage capacity in zeolitic frameworks is limited to 2.86 wt% [74]. Considering that this calculated value was obtained under extreme physical conditions that is, high pressures or very low temperatures, real maximum values can be even lower. These storage capacities are too low for any possible hydrogen storage technology. Nevertheless, zeolites are interesting for the adsorption since they have strong electrostatic forces in their channels, derived from the exposed cations. Hence, an understanding concerning the interaction between zeolites and H_2 molecules could allow the development of new materials possessing strong polarizing centers for hydrogen physisorption.

Similarly as for carbon materials, the maximum hydrogen uptake in different zeolites shows a good correlation with the BET specific surface area measured by nitrogen adsorption [73]. Beyond the specific surface area the interaction of hydrogen molecules with zeolites is strongly influenced by the type and the concentration of cations present in the framework channels. This is well reflected by, for example, the increase in the isosteric heat of adsorption in zeolites with increasing aluminum concentration, irrespective of the framework structure [75]. Since a higher Al concentration leads to a higher amount of cations in the framework the stronger interaction with H_2 can be directly correlated to the presence of polarizing cationic centers. The influence of the electrostatic potential of the cations on the interaction with hydrogen molecules in zeolites was studied by inelastic neutron scattering [76]. For three X-zeolites, possessing the same structure but different exchanged cations (Na^+, Ca^{2+}, Zn^{2+}), a direct correlation between the polarizing potential of the positive ion, expressed as the ratio between the charge and the radius of the ion (z/r), and the vibrational frequency of the H_2 molecule was observed. This shows that specific interaction between H_2 and the cations takes place and that the interaction increases with the polarizing potential of the cation. The positive effect on the interaction with hydrogen molecules of using strong polarizing cations was also observed in a Mg^{2+}-exchanged Y zeolite which possesses an extremely high adsorption enthalpy of $-17\ kJ\ mol^{-1}$ due to the high z/r ratio of Mg ions [77].

However these strong binding sites already saturate at low hydrogen concentrations [76], and no influence of the cationic centers on the maximum storage capacity is observed at high pressures. Under these conditions the number of adsorption sites is mainly determined by the surface area.

2.5 Coordination Polymers

Metal organic frameworks (MOFs) are a new class of porous coordination polymers consisting of metal ions, metal oxide clusters or inorganic parts of larger dimensionality like 1D chains or 2D layers, coordinated by organic ligands in a rigid three-dimensional lattice (Figure 2.7) [78, 79]. The reticular design of these materials derives from the precise coordination geometry between metal centers and organic ligands with multiple bonds which ensure the rigid topology of the framework [80]. Since both the metal centers and the organic ligand can be changed, a huge variety of MOFs with different composition and different structures can be produced.

MOFs can possess very high specific surface areas and can, therefore, be applied for catalysis, gas separation and gas storage [80, 81]. In 2003 Yaghi *et al.* reported the first measurement of hydrogen storage in MOFs performed on two Zn-based porous coordination polymers called MOF-5 and IRMOF-8 (isoreticular metal organic framework). MOF-5 is a cubic three-dimensional structure with Zn_4O nodes at each corner of a cube coordinated by six terephthalate ions (Figure 2.8) [82]. IRMOF-8 is supposed to possess a similar cubic structure with naphthalenebicarboxylate as coordinating ligand which leads to larger pores than in MOF-5. Catenation can also occur in IRMOF-8, that is, the growth of one network into the other, which finally leads to smaller pores than the non-catenated structure [81].

The first investigations on these materials showed exceptionally high storage capacities corresponding to 1 wt% at RT and 20 bar and 4.5 wt% at 77 K and 0.8 bar [84]. Later, these high uptake values were attributed to the adsorption of some impurity gases [85]. Nevertheless, maximum storage capacities of 4.5–5.2 wt%

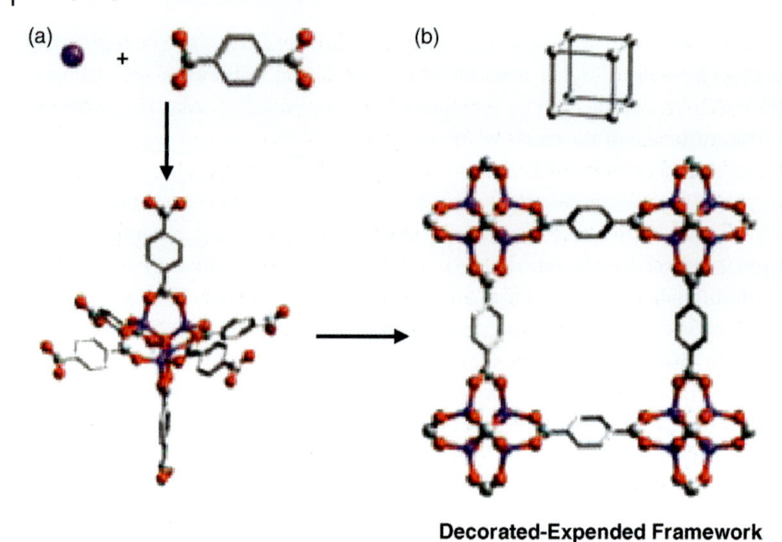

Figure 2.7 (a) Assembly of the central metal ion and the organic ligand which form a cluster subunit. These subunits are connected to each other to give an extended three-dimensional framework. (b) Part of the unit cell of the framework [78]. Reprinted with permission from Accounts of Chemical Research, Copyright (2001) American Chemical Society.

at approximately 50 bar and 77 K were recently reported for MOF-5 independently by three different groups [12, 13, 86]. Contemporaneously, with the group of Yaghi, Férey *et al.* reported on hydrogen uptake in Cr- and Al-based MOFs called MIL-53 (Matériaux de l'Institut Lavoisier n. 53) which possess a storage capacity up to 3.8 wt% at 77 K [87]. These two pioneering studies on MOFs initiated the great interest in them as hydrogen storage materials based on physisorption.

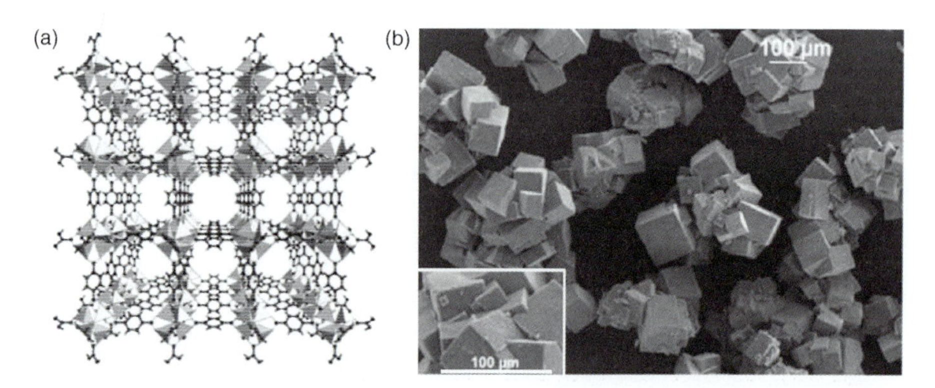

Figure 2.8 (a) Three-dimensional framework of MOF-5 [83]. The gray tetrahedra represent a ZnO_4 cluster with the Zn ion at the center of the tetrahedron and the oxygen atoms deriving the carboxylate ligand and from the central oxygen atom. One vertex consists of a $Zn_4O(CO_2)_6$ octahedral unit. (Picture obtained from the crystallographic data [82] with the program "diamond"). (b) SEM picture of the cubic crystals of MOF-5 [13].

In particular MOF-5 was intensively studied for H_2 adsorption due to its very high specific surface area and simple structure. However, it was observed that the hydrogen uptake of MOF-5 is very sensitive to the procedure used for its synthesis. The measured storage capacities at 77 K ranged from less than 1.6 wt% for a low surface area MOF-5 sample (BET SSA = 572 $m^2\,g^{-1}$) [88, 89] prepared by the so-called Huang's synthesis [90] to 7 wt% when the sample was prepared in an inert atmosphere with complete absence of water and moisture, leading to a material with a BET specific surface area of 3800 $m^2\,g^{-1}$ [91]. These discrepancies were attributed to different factors related to the synthesis procedure, such as: (i) the decomposition of MOF-5 and formation of a non-porous second phase in the presence of moisture [91], (ii) the interpenetration of two MOF-5 networks to give a less porous material, and (iii) the presence of $Zn(OH)_2$ species which partly occupy the pores of MOF-5 [92]. Figure 2.9 shows the three different adsorption isotherms for MOF-5, prepared by the Huang synthesis [88], prepared by a large scale synthesis developed by BASF [13, 81], and prepared in the absence of moisture and water [91].

Metal organic frameworks are the crystalline materials with the lowest density. As an example, MOF-5 possesses a crystal density of only 0.61 $g\,cm^{-3}$ [93]. The volumetric hydrogen storage capacity of MOFs, which is referred to the volume of the material, is therefore affected by these low densities. In spite of this, Müller *et al.* from BASF recently showed the real potential of MOFs for industrial gas storage applications [81]. They showed that a container filled with MOF-5, IRMOF-8 or a porous Cu-based MOF exceeds, at 77 K, the volumetric hydrogen storage capacity (referred to the whole system) of a pressurized empty container filled with hydrogen. The gain in the storage capacity of a container filled with these MOFs increases at

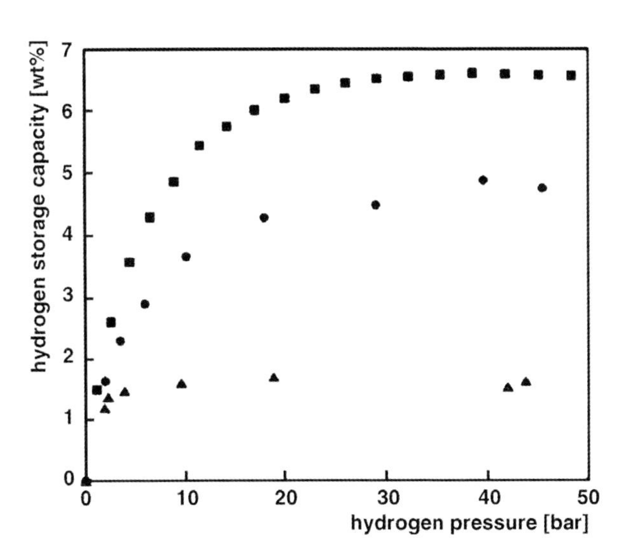

Figure 2.9 Hydrogen adsorption isotherm of MOF-5 with low specific surface area (triangles) [88], of MOF-5 prepared by BASF (circles) [13] and of MOF-5 prepared avoiding exposure to air (square) [91] (data corrected to give the wt% storage capacity as defined in Eq. (2.2)).

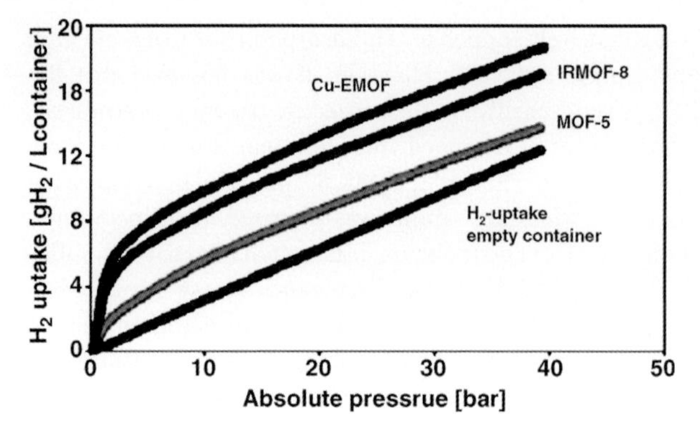

Figure 2.10 Volumetric hydrogen storage capacity at 77 K of a container filled with IRMOF-8, MOF-5, and an electrochemically produced Cu-MOF, compared with a pressurized empty container [81]. Reproduced with permission of the Royal Society of Chemistry.

pressures up to 10 bar. At higher pressures the MOFs become saturated, that is, a further increase in the hydrogen pressure does not lead to an increase in the storage capacity of the material, and the additional increase in the volumetric hydrogen density is due to non-adsorbed H_2 gas in the container (Figure 2.10).

After MOF-5 many other MOFs were studied for hydrogen storage with the best materials showing storage capacities higher than 7 wt% at 77 K. The storage capacity of MOF-177, which consists of $Zn_4O(CO_2)_6$ clusters, like MOF-5, coordinated by the tritopic linker BTB (benezene tribenzoate) [94], was independently confirmed by gravimetric and volumetric measurements and corresponds to 7.5 wt% [86, 95]. This value is higher than for any other porous material which adsorbs hydrogen in the molecular form. Similarly, high uptake values of 6.7 and 7 wt% were reported for two MOFs possessing an NbO-type framework topology and consisting of Cu(II)-dimeric paddle-wheel units coordinated by terphenyl and quaterphenyl connectors [96]. All these MOFs with exceptionally high maximum hydrogen storage capacities also have extremely high specific surface areas, which fact is responsible for the good adsorption properties of these materials [15, 95, 96]. Indeed, it was shown that a direct correlation exists between the maximum hydrogen storage capacity of MOFs at 77 K and their specific surface area (Figure 2.11) [13, 86]. This correlation was found both considering the BET and Langmuir model for the specific surface area applied to nitrogen adsorption isotherms at 77 K. Both models cannot appropriately describe the SSA of microporous materials, since the Langmuir model does not consider multilayer adsorption of N_2 which could occur in larger pores and the classical pressure range of application for the BET model ($0.05 < p/p_0 < 0.3$) is often higher than the saturation pressure of microporous materials [97]. Nevertheless, the correlation found between maximum hydrogen uptake and SSA indicates that a large specific surface area is a necessary prerequisite for reaching high maximum storage capacities. A similar correlation was obtained also from Grand Canonical Monte Carlo simulations of hydrogen adsorption on a series of MOFs possessing the same framework topology but different ligands [98]. The investigations showed that at pressures of 30 bar and at 77 K the absolute hydrogen uptake is linearly dependent on the accessible surface area calculated considering the framework geometries of the MOFs.

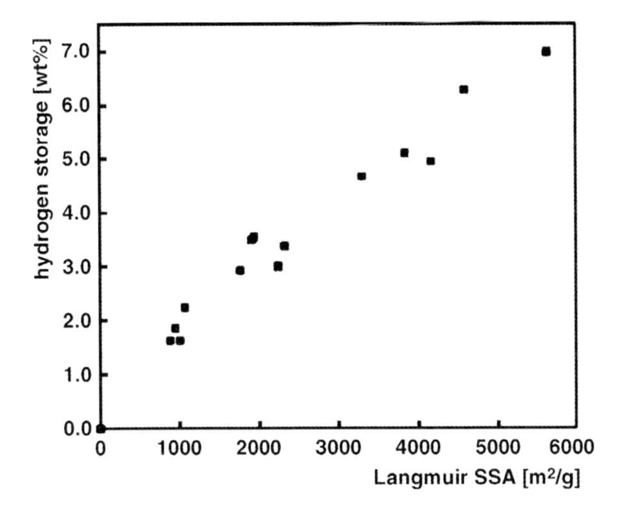

Figure 2.11 Correlation between the hydrogen storage capacity and the apparent Langmuir specific surface area of different MOFs obtained from reference [86] and references [13, 88]. The data of Wong-Foy *et al.* [86] were modified considering the different definition of wt% storage (mH_2/ms%) compared to Eq. (2.2).

However, in this case, higher storage capacities than the experimental excess uptake values were predicted. The same simulations also show that at low hydrogen pressure the storage capacity does not correlate with the specific surface area but depends on the heat of adsorption [98]. A viable strategy to synthesize MOFs which reach a high hydrogen storage capacity already at relatively low pressure consists therefore in creating frameworks not only with a high specific surface area but also with a strong affinity to the hydrogen molecules. Hence, different techniques have been applied to understand the relation between the structure of MOFs and the interaction strength with hydrogen. Using inelastic neutron scattering spectroscopy, which is sensitive to the chemical environment of adsorbed hydrogen, Rowsell *et al.* observed multiple adsorption sites for H_2 in different isoreticular MOFs [99]. These were attributed to primary binding sites composed of the inorganic $Zn_4O(CO_2)_6$ units, which are influenced by the organic ligand, and to weaker adsorption sites in the proximity of the organic ligand. Rietveld structural refinement of powder diffraction data also indicates that the inorganic clusters are the first adsorption sites for hydrogen molecules in MOF-5 [100]. Similarly, molecular simulations predict that, in the series of IRMOFs, hydrogen at low pressure first occupies the corner regions where the metal clusters are positioned, followed by the inner spaces of the cavities, that is, the pores [101]. The finding that metal centers may influence the hydrogen uptake led to the development of new MOFs with coordinatively unsaturated metal centers or so-called open metal sites. These can be obtained by removal of a terminal ligand from the metal site without collapse of the framework [102]. These sites are supposed to strongly polarize the H_2 molecules and, therefore, be responsible for high storage capacities already at low pressures.

Though the metal sites show the highest interaction energy towards hydrogen molecules, the number of adsorption sites close to these centers is limited and, therefore, saturated already at very low hydrogen concentrations [103]. Additionally, theoretical calculation shows that the polarization effect of H_2 by fully coordinated metal sites plays a minor role in the adsorption [104, 105], while the pore dimensions of the cavities in the framework have a strong influence on the hydrogen storage at low pressures [106]. Indeed, in pores possessing a diameter of few H_2 molecular radii, the van der Waals potential of the pore walls can overlap more effectively than in larger pores, thus enhancing the strength of interaction with hydrogen. As a consequence MOFs with similar building units and network topology show, at low pressure, a storage capacity which is higher for the frameworks with smaller pores [96]. Moreover, thermal desorption studies of hydrogen adsorbed at 20 K in different MOFs indicate that the affinity between hydrogen and the MOFs is more influenced by the pore size than by the building units of the MOFs [107]. This result is obtained at hydrogen concentrations which are close to the maximum storage capacities; that is, at technologically relevant uptake values. Furthermore, measurements of the heat of adsorption for hydrogen in different MOFs have been performed over a wide range of surface coverages. Again MOFs with smaller cavities show a higher heat of adsorption for hydrogen [108].

The combined effect of strongly polarizing metal-centers [109–114] and small pores on the hydrogen adsorption properties of MOFs indicates a possible strategy to improve them for H_2 storage. One proposed way to decrease the pore size of MOFs is to form catenated frameworks [102]. Catenation occurs when the pores in the network are large enough to accommodate the metal carboxylate clusters of a second network which is build inside the first one. This can be typically obtained for MOFs with long ligands like, for example, pyrene dicarboxylic acid [115]. Two types of catenation can take place: interweaving and interpenetration. In the case of interpenetration a maximal displacement between the two catenated networks is obtained, while interwoven frameworks are minimally displaced and the inorganic subunits are very close to each other [102]. As pointed out by Rowsell and Yaghi, for hydrogen storage interpenetration rather than interweaving is preferred, since, in the latter case, the pores are reduced only to a small extent while the exposed surface is decreased considerably through the thickening of the walls. Interpenentration of two frameworks would be more efficient in pore size reduction. Grand Canonical Montecarlo (GCMC) simulations of hydrogen adsorption show indeed that interpenentration of two frameworks leads to the formation of new strong adsorption sites which derive from the overlap of the van der Waals potentials between two close linkers (Figure 2.12) [103]. However, other GCMC simulations show that for some MOFs the increase in the isosteric heat of adsorption through catenation does not compensate for the loss of free volume. This leads to lower H_2 uptake values for a catenated than a non-catenated framework [106]. Additionally, interpenetration as a catenation mode has the disadvantage of not being stable when solvent molecules are removed from the framework to make the pores accessible [102]. Therefore, decreasing the pore size by catenation has not yet given considerable advantages for hydrogen storage.

Increasing the affinity to hydrogen by restricting the pores in the framework while at the same time maintaining a large pore volume remains until now one of the main challenges in the synthesis of novel MOFs.

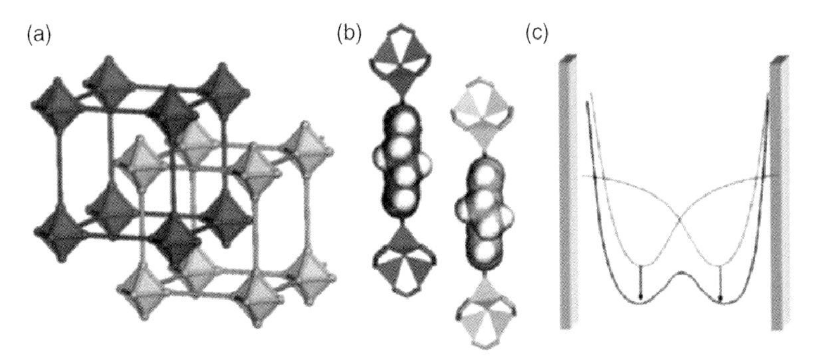

Figure 2.12 (a) Interpenetration of two frameworks as catenation mode; (b) schematic representation of the new adsorption site between two close linkers formed by interpenetration, (c) overlap of the two van der Waals potentials from two solid plates representing the linkers [103]. Reprinted with permission from The Journal of Physical Chemistry B, Copyright (2006) American Chemical Society.

Apart from metal–organic hybrid compounds other inorganic coordination polymers with open frameworks exist. One example is cyano-bridged coordination polymers and among them the best known possessing the simplest structures are Prussian blue analogues. These compound possess a simple cubic framework of the type $M_A[M_B(CN)_6]$ in which octahedral $[M_B(CN)_6]^{n-}$ are coordinated by nitrogen bound M_A^{n+} ions as shown in Figure 2.13 [16, 116]. After removal of the water molecules coordinated to the nitrogen bound cations these cubic frameworks contain coordinatively-unsaturated M_A^{n+} metal centers.

For Prussian blue analogues relatively high storage capacities of up to 1.8 wt% at low pressures (approximately 1 bar) and 77 K and adsorption isotherms with a steep increase in the hydrogen uptake were obtained [14, 117, 118]. However, the maximum storage capacity does not differ strongly from the low-pressure hydrogen uptake and a highest value of only 2.3 wt% was obtained for the best materials, $Cu_2[Fe(CN)_6]$ [16]. As for MOFs, the role of unsaturated coordination positions and cavity size on the interaction with hydrogen molecules has also been intensively investigated for cyano-bridged frameworks. To probe the role of so-called open metal sites, the hydrogen storage capacity of a series of dehydrated Prussian blue analogues with different concentrations of vacancies, that is, increasing concentration of unsaturated metal centers, was measured [16]. The results showed that open metal sites are not necessarily required for H_2 adsorption and that the enthalpy of adsorption is not influenced by the concentration of vacancies. Nevertheless, an increase in the maximum uptake with increasing vacancies in the lattice was observed. This increase can also be well correlated with an increase in the specific surface area of the frameworks rather than to the concentration of vacancies. In a series of dehydrated cyano-bridged compounds of the type $A_2Zn_3[Fe(CN)_6]_2{\cdot}xH_2O$ (A = H, Li, Na, K, Rb), which contain alkali metal cations (A^+) in the cavities, no correlation between the enthalpy of adsorption and the polarizing potential of A^+ was found [117]. Hence, the high adsorption enthalpies of up to 9 kJ mol^{-1} found for these materials may be due to the small size of the pores, which is partially reduced by the cation located inside.

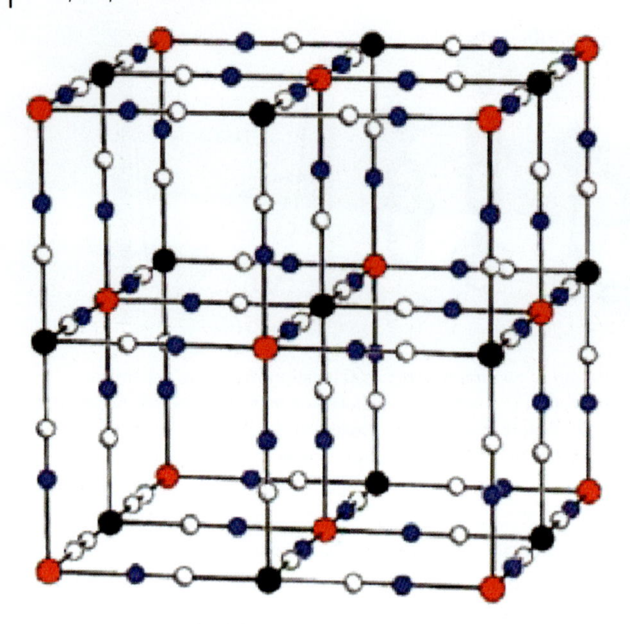

Figure 2.13 Unit cell of the Prussian blue analogue structure $M_A[M_B(CN)_6]$ where the red, black, white and blue spheres represent M_A, M_B, C and N, respectively [16]. Copyright (2007), with permission of Elsevier.

The importance of the pore size in Prussian blue analogues is supported by differential pair distribution function analysis of X-ray and neutron scattering data of hydrogen- and deuterium-loaded $Mn_3[Co(CN)_6]_2$ [119]. This shows that no evidence for adsorption interactions with unsaturated metal sites exists and that the hydrogen molecules are disordered about the center of the pores defined by the cubic framework. In conclusion, experimental results indicate that optimum pore dimensions in Prussian blue analogues are predominantly responsible for the heat of adsorption at low loadings rather than the polarizing effect of open metal sites.

2.6 Conclusions

In this chapter, the physisorption of hydrogen molecules in porous materials as possible hydrogen storage systems has been reviewed. Owing to the weak interaction between H_2 molecules and the adsorbent, high storage capacities are typically reached only at low temperatures (77 K). The advantages of this storage mechanism compared to chemisorption are the fast kinetics, the complete reversibility and the relatively small amount of heat produced during on-board refueling.

Different classes of porous materials possessing different structure and composition have been discussed for hydrogen storage applications. For all materials a direct correlation between the maximum hydrogen uptake at 77 K and the specific surface area is found. For a long time, zeolites were the best crystalline materials and activated carbons the best disordered materials for gas storage by physisorption. Recently, the new classes of crystalline metal organic frameworks and novel templated porous carbons have been developed which possess specific surface areas exceeding those of activated carbons. Presently, excess hydrogen storage capacities of up to 7 wt% at 77 K are reached for the best of these materials.

The adsorption sites for hydrogen in different porous materials have been intensively investigated as a function of the hydrogen concentration. In the range of technologically relevant hydrogen concentrations the pore size determines the interaction between hydrogen and the porous solid, that is, the heat of adsorption. In metal organic frameworks metal centers are additionally preferential adsorption sites for hydrogen at low H_2 concentrations. Therefore, the combination of a high specific surface area and small pores is necessary to develop new porous materials for hydrogen physisoption at low temperatures. This will enable the building of a cryo-adsorption tank for mobile applications, which operates at pressures below 3.5 MPa, allowing a conformable storage in contrast to high-pressure cylinders.

References

1 Rouquérol, J.R.F. and Sing, K. (1999) Adsorption by powders and porous solids: Principles, in *Methodology and Applications*, Academic Press, San Diego, London, Boston, New York, Sydney, Tokyo. Toronto.

2 Sing, K.S.W., Everett, D.H., Haul, R.A.W., Moscou, L., Pierotti, R.A., Rouquérol, J., and Siemieniewska, T. (1985) *Pure Appl. Chem.*, **57**, 603–619.

3 Züttel, A. (2004) *Naturwissenschaften*, **91**, 157–172.

4 Bhatia, S.K. and Myers, A.L. (2006) *Langmuir*, **22**, 1688–1700.

5 Aréan, C.O., Manoilova, O.V., Bonelli, B., Delgado, M.R., Palomino, G.T., and Garrone, E. (2003) *Chem. Phys. Lett.*, **370**, 631–635.

6 Garrone, E., Bonelli, B., and Aréan, C.O. (2008) *Chem. Phys. Lett.*, **456**, 68–70.

7 Panella, B., Hirscher, M., and Roth, S. (2005) *Carbon*, **43**, 2209–2214.

8 Poirier, E., Chahine, R., Bénard, P., Lafi, L., Dorval-Douville, G., and Chandonia, P.A. (2006) *Langmuir*, **22**, 8784–8789.

9 Yushin, G., Dash, R., Jagiello, J., Fischer, J.E., and Gogotsi, Y. (2006) *Adv. Funct. Mater.*, **16**, 2288–2293.

10 Wood, C.D., Tan, B., Trewin, A., Niu, H.J., Bradshaw, D., Rosseinsky, M.J., Khimyak, Y.Z., Campbell, N.L., Kirk, R., Stöckel, E., and Cooper, A.I. (2007) *Chem. Mater.*, **19**, 2034–2048.

11 Germain, J., Fréchet, J.M.J., and Svec, F. (2007) *J. Mater. Chem.*, **17**, 4989–4997.

12 Dailly, A., Vajo, J.J., and Ahn, C.C. (2006) *J. Phys. Chem. B*, **110**, 1099–1101.

13 Panella, B., Hirscher, M., Pütter, H., and Müller, U. (2006) *Adv. Funct. Mater.*, **16**, 520–524.

14 Kaye, S.S. and Long, J.R. (2005) *J. Am. Chem. Soc.*, **127**, 6506–6507.

15 Latroche, M., Surble, S., Serre, C., Mellot-Draznieks, C., Llewellyn, P.L., Lee, J.H., Chang, J.S., Jhung, S.H., and Férey, G. (2006) *Angew. Chem. Int. Ed.*, **45**, 8227–8231.

16 Kaye, S.S. and Long, J.R. (2007) *Catal. Today*, **120**, 311–316.

17 Jhung, S.H., Lee, J.S., Yoon, J.W., Kim, D.P., and Chang, J.S. (2007) *Int. J. Hydrogen Energy*, **32**, 4233–4237.

18 Spoto, G., Gribov, E., Bordiga, S., Lamberti, C., Ricchiardi, G., Scarano, D.,

and Zecchina, A. (2004) *Chem. Commun.*, 2768–2769.
19 Zhou, W., Wu, H., Hartman, M.R., and Yildirim, T. (2007) *J. Phys. Chem. C*, **111**, 16131–16137.
20 Myers, A.L. and Monson, P.A. (2002) *Langmuir*, **18**, 10261–10273.
21 Ozdemir, E., Morsi, B.I., and Schroeder, K. (2003) *Langmuir*, **19**, 9764–9773.
22 Do, D.D. and Do, H.D. (2003) *Carbon*, **41**, 1777–1791.
23 Ozdemir, E. (2004) Ph.D. thesis: Chemistry of the adsorption of carbon dioxide by argonne premium coals and a model to simulate CO_2 sequestration in coal seams. University of Pittsburg, Pittsburgh.
24 Sing, K.S.W., Schüth, F., and Weitkamp, J. (2002) *Handbook of Porous Solids*, vol. **3**, Wiley-VCH, Weinheim.
25 Savage, G. (1993) *Carbon-Carbon Composites*, Chapman & Hall, London.
26 Bansal, M.G.R.C. (2005) Activated carbon adsorption, Taylor & Francis, Boca Raton, London, New York, Singapore.
27 Chahine, R. and Bose, T.K. (1994) *Int. J. Hydrogen Energy*, **19**, 161–164.
28 Ryoo, R., Joo, S.H., and Jun, S. (1999) *J. Phys. Chem. B*, **103**, 7743–7746.
29 Kyotani, T., Ma, Z.X., and Tomita, A. (2003) *Carbon*, **41**, 1451–1459.
30 Kim, T.W., Park, I.S., and Ryoo, R. (2003) *Angew. Chem. Int. Ed.*, **42**, 4375–4379.
31 Lee, J., Kim, J., and Hyeon, T. (2006) *Adv. Mater.*, **18**, 2073–2094.
32 Dash, R.K., Nikitin, A., and Gogotsi, Y. (2004) *Microporous Mesoporous Mater.*, **72**, 203–208.
33 Gogotsi, Y., Dash, R.K., Yushin, G., Yildirim, T., Laudisio, G., and Fischer, J.E. (2005) *J. Am. Chem. Soc.*, **127**, 16006–16007.
34 Iijima, S. (1991) *Nature*, **354**, 56–58.
35 Dresselhaus, M.S., Dresselhaus, G., and Saito, R. (1995) *Carbon*, **33**, 883–891.
36 De Jong, K.P. and Geus, J.W. (2000) *Cat. Rev.-Sci. Eng.*, **42**, 481–510.
37 Rodriguez, N.M., Chambers, A., and Baker, R.T.K. (1995) *Langmuir*, **11**, 3862–3866.
38 Chambers, A., Park, C., Baker, R.T.K., and Rodriguez, N.M. (1998) *J. Phys. Chem. B*, **102**, 4253–4256.
39 Dillon, A.C., Jones, K.M., Bekkedahl, T.A., Kiang, C.H., Bethune, D.S., and Heben, M.J. (1997) *Nature*, **386**, 377–379.
40 Hirscher, M., Becher, M., Haluska, M., Dettlaff-Weglikowska, U., Quintel, A., Duesberg, G.S., Choi, Y.M., Downes, P., Hulman, M., Roth, S., Stepanek, I., and Bernier, P. (2001) *Appl. Phys. A*, **72**, 129–132.
41 Tibbetts, G.G., Meisner, G.P., and Olk, C.H. (2001) *Carbon*, **39**, 2291–2301.
42 Hirscher, M. and Becher, M. (2003) *J. Nanosci. Nanotech.*, **3**, 3–17.
43 Ritschel, M., Uhlemann, M., Gutfleisch, O., Leonhardt, A., Graff, A., Taschner, C., and Fink, J. (2002) *Appl. Phys. Lett.*, **80**, 2985–2987.
44 Schimmel, H.G., Nijkamp, G., Kearley, G.J., Rivera, A., de Jong, K.P., and Mulder, F.M. (2004) *Mater. Sci. Eng. B*, **108**, 124–129.
45 Rzepka, M., Lamp, P., and de la Casa-Lillo, M.A. (1998) *J. Phys. Chem. B*, **102**, 10894–10898.
46 Jordá-Beneyto, M., Suárez-García, F., Lozano-Castelló, D., Cazorla-Amorós, D., and Linares-Solano, A. (2007) *Carbon*, **45**, 293–303.
47 Nijkamp, M.G., Raaymakers, J., van Dillen, A.J., and de Jong, K.P. (2001) *Appl. Phys. A*, **72**, 619–623.
48 Texier-Mandoki, N., Dentzer, J., Piquero, T., Saadallah, S., David, P., and Vix-Guterl, C. (2004) *Carbon*, **42**, 2744–2747.
49 Chahine, T.K.B.R. (1996) 11th World Hydrogen Energy Conference. Hydrogen Energy Progress XI, International Association for Hydrogen Energy, Stuttgart (eds T.N. Veziroglu *et al.*) pp. 1259–1263.
50 Züttel, A., Sudan, P., Mauron, P., and Wenger, P. (2004) *Appl. Phys. A*, **78**, 941–946.
51 Zhao, X.B., Xiao, B., Fletcher, A.J., and Thomas, K.M. (2005) *J. Phys. Chem. B*, **109**, 8880–8888.
52 Hirscher, M. and Panella, B. (2005) *Ann. Chim.-Sci. Mater.*, **30**, 519–529.
53 Pacula, A. and Mokaya, R. (2008) *J. Phys. Chem. C*, **112**, 2764–2769.
54 Yang, Z.X., Xia, Y.D., and Mokaya, R. (2007) *J. Am. Chem. Soc.*, **129**, 1673–1679.
55 Pradhan, B.K., Sumanasekera, G.U., Adu, K.W., Romero, H.E., Williams, K.A.,

and Eklund, P.C. (2002) *Physica B*, **323**, 115–121.

56 Talapatra, S., Zambano, A.Z., Weber, S.E., and Migone, A.D. (2000) *Phys. Rev. Lett.*, **85**, 138–141.

57 Bérnard, P., Chahine, R., Chandonia, P.A., Cossement, D., Dorval-Douville, G., Lafi, L., Lachance, P., Paggiaro, R., and Poirier, E. (2007) *J. Alloys Compd.*, **446**, 380–384.

58 Williams, K.A. and Eklund, P.C. (2000) *Chem. Phys. Lett.*, **320**, 352–358.

59 Georgiev, P.A., Ross, D.K., De Monte, A., Montaretto-Marullo, U., Edwards, R.A.H., Ramirez-Cuesta, A.J., Adams, M.A., and Colognesi, D. (2005) *Carbon*, **43**, 895–906.

60 Wilson, T., Tyburski, A., DePies, M.R., Vilches, O.E., Becquet, D., and Bienfait, M. (2002) *J. Low. Temp. Phys.*, **126**, 403–408.

61 Panella, B., Hirscher, M., and Ludescher, B. (2007) *Microporous Mesoporous Mater.*, **103**, 230–234.

62 Panella, B. and Hirscher, M. (2008) *Phys. Chem. Chem. Phys.*, **10**, 2910–2917.

63 Williams, K.A., Pradhan, B.K., Eklund, P.C., Kostov, M.K., and Cole, M.W. (2002) *Phys. Rev. Lett.*, **88**, 165502-1–165502-4.

64 Cho, S.J. (2002) *Fuel Chem. Div. Prepr.*, **2002**, 790.

65 Cho, S.J., Choo, K., Kim, D.P., and Kim, J.W. (2007) *Catal. Today*, **120**, 336–340.

66 Panella, B., Kossykh, L., Dettlaff-Weglikowska, U., Hirscher, M., Zerbi, G., and Roth, S. (2005) *Synth. Met.*, **151**, 208–210.

67 Jurczyk, M.U., Kumar, A., Srinivasan, S., and Stefanakos, E. (2007) *Int. J. Hydrogen Energy*, **32**, 1010–1015.

68 McKeown, N.B., Budd, P.M., and Book, D. (2007) *Macromol. Rapid. Commun.*, **28**, 995–1002.

69 Budd, P.M., Butler, A., Selbie, J., Mahmood, K., McKeown, N.B., Ghanem, B., Msayib, K., Book, D., and Walton, A. (2007) *Phys. Chem. Chem. Phys.*, **9**, 1802–1808.

70 Ghanem, B.S., Msayib, K.J., McKeown, N.B., Harris, K.D.M., Pan, Z., Budd, P.M., Butler, A., Selbie, J., Book, D., and Walton, A. (2007) *Chem. Commun.*, 67–69.

71 Spoto, G., Vitillo, J.G., Cocina, D., Damin, A., Bonino, F., and Zecchina, A. (2007) *Phys. Chem. Chem. Phys.*, **9**, 4992–4999.

72 Weitkamp, J., Fritz, M., and Ernst, S. (1995) *Int. J. Hydrogen Energy*, **20**, 967–970.

73 Harris, I.R., Langmi, H.W., Book, D., Walton, A., Johnson, S.R., Al-Mamouri, M.M., Speight, J.D., Edwards, P.P., and Anderson, P.A. (2005) *J. Alloys Compd.*, **404–406**, 637–642.

74 Vitillo, J.G., Ricchiardi, G., Spoto, G., and Zecchina, A. (2005) *Phys. Chem. Chem. Phys.*, **7**, 3948–3954.

75 Jhung, S.H., Yoon, J.W., Lee, S., and Chang, J.S. (2007) *Chem. Eur. J.*, **13**, 6502–6507.

76 Ramirez-Cuesta, A.J., Mitchell, P.C.H., Ross, D.K., Georgiev, P.A., Anderson, P.A., Langmi, H.W., and Book, D. (2007) *J. Mater. Chem.*, **17**, 2533–2539.

77 Palomino, G.T., Carayol, M.R.L., and Aréan, C.O. (2006) *J. Mater. Chem.*, **16**, 2884–2885.

78 Eddaoudi, M., Moler, D.B., Li, H.L., Chen, B.L., Reineke, T.M., O'Keeffe, M., and Yaghi, O.M. (2001) *Acc. Chem. Res.*, **34**, 319–330.

79 Férey, G. (2007) *Stud. Surf. Sci. Catal.*, **170**, 1051–1058.

80 Férey, G. (2008) *Chem. Soc. Rev.*, **37**, 191–214.

81 Mueller, U., Schubert, M., Teich, F., Puetter, H., Schierle-Arndt, K., and Pastré, J. (2006) *J. Mater. Chem.*, **16**, 626–636.

82 Li, H., Eddaoudi, M., O'Keeffe, M., and Yaghi, O.M. (1999) *Nature*, **402**, 276–279.

83 Panella, B. and Hirscher, M. (2009) *Encyclopedia of Electrochemical Power Sources* (eds C.K.D.J. Garche, P.T. Moseley, Z. Ogumi, D.A.J. Rand and B. Scrosati), Elsevier, 493–496.

84 Rosi, N.L., Eckert, J., Eddaoudi, M., Vodak, D.T., Kim, J., O'Keeffe, M., and Yaghi, O.M. (2003) *Science*, **300**, 1127–1129.

85 Rowsell, J.L.C., Millward, A.R., Park, K.S., and Yaghi, O.M. (2004) *J. Am. Chem. Soc.*, **126**, 5666–5667.

86 Wong-Foy, A.G., Matzger, A.J., and Yaghi, O.M. (2006) *J. Am. Chem. Soc.*, **128**, 3494–3495.

87 Férey, G., Latroche, M., Serre, C., Millange, F., Loiseau, T., and Percheron-Guégan, A. (2003) *Chem. Commun.*, 2976–2977.

88 Panella, B. and Hirscher, M. (2005) *Adv. Mater.*, **17**, 538–541.

89 Bordiga, S., Vitillo, J.G., Ricchiardi, G., Regli, L., Cocina, D., Zecchina, A., Arstad, B., Bjørgen, M., Hafizovic, J., and Lillerud, K.P. (2005) *J. Phys. Chem. B*, **109**, 18237–18242.

90 Huang, L.M., Wang, H.T., Chen, J.X., Wang, Z.B., Sun, J.Y., Zhao, D.Y., and Yan, Y.S. (2003) *Microporous Mesoporous Mater.*, **58**, 105–114.

91 Kaye, S.S., Dailly, A., Yaghi, O.M., and Long, J.R. (2007) *J. Am. Chem. Soc.*, **129**, 14176–14177.

92 Hafizovic, J., Bjørgen, M., Olsbye, U., Dietzel, P.D.C., Bordiga, S., Prestipino, C., Lamberti, C., and Lillerud, K.P. (2007) *J. Am. Chem. Soc.*, **129**, 3612–3620.

93 Eddaoudi, M., Kim, J., Rosi, N., Vodak, D., Wachter, J., O'Keeffe, M., and Yaghi, O.M. (2002) *Science*, **295**, 469–472.

94 Chae, H.K., Siberio-Pérez, D.Y., Kim, J., Go, Y., Eddaoudi, M., Matzger, A.J., O'Keeffe, M., and Yaghi, O.M. (2004) *Nature*, **427**, 523–527.

95 Furukawa, H., Miller, M.A., and Yaghi, O.M. (2007) *J. Mater. Chem.*, **17**, 3197–3204.

96 Lin, X., Jia, J.H., Zhao, X.B., Thomas, K.M., Blake, A.J., Walker, G.S., Champness, N.R., Hubberstey, P., and Schröder, M. (2006) *Angew. Chem. Int. Ed.*, **45**, 7358–7364.

97 Düren, T., Millange, F., Férey, G., Walton, K.S., and Snurr, R.Q. (2007) *J. Phys. Chem. C*, **111**, 15350–15356.

98 Frost, H., Düren, T., and Snurr, R.Q. (2006) *J. Phys. Chem. B*, **110**, 9565–9570.

99 Rowsell, J.L.C., Eckert, J., and Yaghi, O.M. (2005) *J. Am. Chem. Soc.*, **127**, 14904–14910.

100 Yildirim, T. and Hartman, M.R. (2005) *Phys. Rev. Lett.*, **95**, 215504-1–215504-4.

101 Yang, Q.Y. and Zhong, C.L. (2005) *J. Phys. Chem. B*, **109**, 11862–11864.

102 Rowsell, J.L.C. and Yaghi, O.M. (2005) *Angew. Chem. Int. Ed.*, **44**, 4670–4679.

103 Jung, D.H., Kim, D., Lee, T.B., Choi, S.B., Yoon, J.H., Kim, J., Choi, K., and Choi, S.H. (2006) *J. Phys. Chem. B*, **110**, 22987–22990.

104 Kuc, A., Heine, T., Seifert, G., and Duarte, H.A. (2008) *Theor. Chem. Acc.*, **120**, 543–550.

105 Klontzas, E., Mavrandonakis, A., Froudakis, G.E., Carissan, Y., and Klopper, W. (2007) *J. Phys. Chem. C*, **111**, 13635–13640.

106 Frost, H. and Snurr, R.Q. (2007) *J. Phys. Chem. C*, **111**, 18794–18803.

107 Panella, B., Hönes, K., Müller, U., Trukhan, N., Schubert, M., Pütter, H., and Hirscher, M. (2008) *Angew. Chem. Int. Ed.*, **47**, 2138–2142.

108 Schmitz, B., Müller, U., Trukhan, N., Schubert, M., Férey, G., and Hirscher, M. (2008) *Chem. Phys. Chem.*, **9**, 2181–2184.

109 Liu, Y., Kabbour, H., Brown, C.M., Neumann, D.A., and Ahn, C.C. (2008) *Langmuir*, **24**, 4772–4777.

110 Moon, H.R., Kobayashi, N., and Suh, M.P. (2006) *Inorg. Chem.*, **45**, 8672–8676.

111 Dincă, M. and Long, J.R. (2005) *J. Am. Chem. Soc.*, **127**, 9376–9377.

112 Vitillo, J.G., Regli, L., Chavan, S., Ricchiardi, G., Spoto, G., Dietzel, P.D.C., Bordiga, S., and Zecchina, A. (2008) *J. Am. Chem. Soc.*, **130**, 8386–8396.

113 Forster, P.M., Eckert, J., Heiken, B.D., Parise, J.B., Yoon, J.W., Jhung, S.H., Chang, J.S., and Cheetham, A.K. (2006) *J. Am. Chem. Soc.*, **128**, 16846–16850.

114 Dincă, M. and Long, J.R. (2008) *Angew. Chem. Int. Ed.*, **47**, 6766–6779.

115 Rowsell, J.L.C. and Yaghi, O.M. (2006) *J. Am. Chem. Soc.*, **128**, 1304–1315.

116 Maspoch, D., Ruiz-Molina, D., and Veciana, J. (2007) *Chem. Soc. Rev.*, **36**, 770–818.

117 Kaye, S.S. and Long, J.R. (2007) *Chem. Commun.*, 4486–4488.

118 Chapman, K.W., Southon, P.D., Weeks, C.L., and Kepert, C.J. (2005) *Chem. Commun.*, 3322–3324.

119 Chapman, K.W., Chupas, P.J., Maxey, E.R., and Richardson, J.W. (2006) *Chem. Commun.*, 4013–4015.

3
Clathrate Hydrates

Alireza Shariati, Sona Raeissi, and Cor J. Peters

3.1
Introduction

Hydrogen fuel, if provided via a green method, is renewable and environmentally friendly. However, the lack of practical storage methods has restricted its use to such an extent that hydrogen storage is currently a crucial obstacle in the development of a hydrogen economy. The challenge, particularly for mobile applications, is that the hydrogen storing system needs to be lightweight and compact. The two most popular methods currently being used are the storage of hydrogen as compressed gas under high pressures, or as a cryogenic liquid. To approach the goals required to develop a hydrogen economy, researchers are investigating a range of different techniques.

A practical hydrogen storage method must satisfy a number of requirements [1]:

1) high hydrogen content per unit mass
2) high hydrogen content per unit volume
3) moderate synthesis pressure (preferably less than 400 MPa, the pressure that can be reached by a simple compressor)
4) near ambient pressure and moderate temperature for storage
5) fast and efficient hydrogen release
6) environmentally friendly byproducts, if any.

The most common ways of storing hydrogen fuel as liquid hydrogen and compressed hydrogen gas have the drawbacks that the fuel needs to be stored at extremely low temperatures (20 K for liquid hydrogen) or at high pressures (35 MPa for compressed hydrogen) [2].

To overcome such issues, much attention is currently being given to storing hydrogen in solid-state materials. Recently emerging materials include doped carbon-based nanostructures [3], metal organic frameworks [4], metallic hydrides [5], complex hydrides and destabilized hydrides [6, 7]. However, no material so

Handbook of Hydrogen Storage. Edited by Michael Hirscher

ISBN: 978-3-527-32273-2

far has been successful in combining high volumetric and gravimetric density and the ability to reversibly absorb and desorb hydrogen with kinetics and thermodynamics that satisfy the requirements of the Department of Energy (DOE).

This study focuses on the potential use of clathrates as a novel hydrogen storage technique. A *clathrate,* also called a *clathrate compound* or a *cage compound,* is a material consisting of a lattice of one type of molecule trapping and containing a second type of molecule. Powell [8] was the first to call such compounds clathrates. In a paper published in the Journal of the Chemical Society.

Quinol (hydroquinone) crystal was the first composite to be called a clathrate. Nowadays, this term has been adopted for many complexes which consist of a host molecule (forming the basic frame) and a guest molecule (set in the host molecule by interaction). The clathrate that is of interest to this study is the clathrate hydrate, also referred to as gas hydrate. Clathrate hydrates were discovered in 1810 by Sir Humphrey Davy. In his lecture to the Royal Society in 1810, he said that he had found, by several experiments, that the solution of chlorine gas in water freezes more readily than pure water [9].

Clathrate hydrates are water-based crystalline solids, physically resembling ice, in which small nonpolar molecules (typically gases) are trapped inside cages of hydrogen-bonded water molecules. Without the support of the trapped molecules, the lattice structure of the clathrate hydrates would collapse into the conventional ice crystal structure or liquid water. Most low molecular weight gases, as well as some higher hydrocarbons and organic components will form clathrate hydrates at suitable temperatures and pressures. Clathrate hydrates are not chemical compounds as the sequential molecules are never bonded to the lattice. The formation and decomposition of the clathrate hydrates are ordinary phase transitions, not chemical reactions.

Clathrate hydrates have been found to occur naturally in large quantities. Around $120 \times 10^{15}\,m^3$ (at STP) of methane is estimated to be trapped in deposits of the deep ocean floor [10]. Clathrate hydrates are also suspected to occur in large quantities on some outer planets, moons, and trans-Neptunian objects [11]. In the petroleum industry, hydrocarbon clathrate hydrates are a cause of problems because they can form inside gas pipelines, often resulting in plugging. Deep sea deposition of carbon dioxide clathrate hydrate has been proposed as a method to remove this greenhouse gas from the atmosphere and control climate change [12].

3.2
Clathrate Hydrate Structures

Gas hydrates usually form two crystallographic structures – structure Type I and structure Type II. Rarely, a third structure may be observed (Type H). Figure 3.1 shows these three structures of clathrate hydrates and Table 3.1 presents their structural properties.

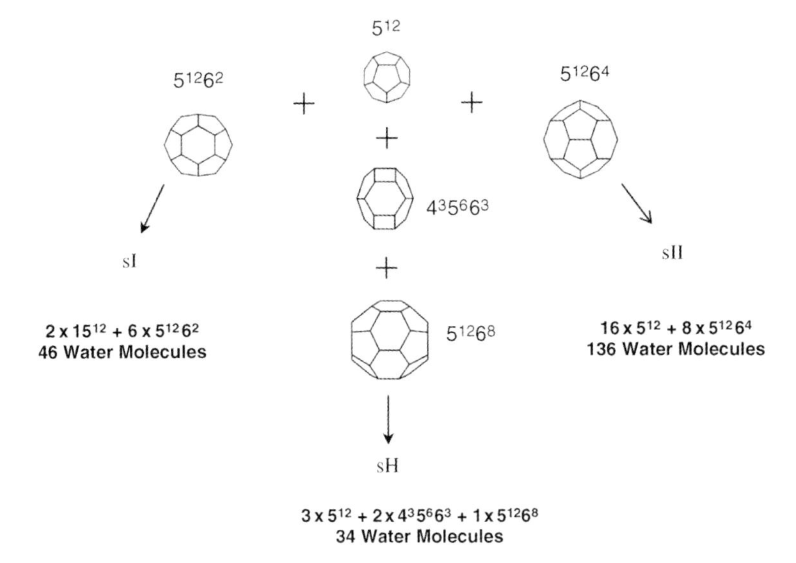

Figure 3.1 Cages building the different gas hydrate structures.

The unit cell of Type I consists of 46 water molecules, forming two types of cages – small and large. There are two small cages and six large ones in every unit cell. The small cage has the shape of a pentagonal dodecahedron (5^{12}) and the large one is a tetradecahedron, which has 12 pentagonal and 2 hexagonal faces ($5^{12}6^2$). Typical guest molecules forming Type I hydrates are CO_2 in carbon dioxide clathrates and CH_4 in methane clathrates.

The unit cell of Type II structures consists of 136 water molecules, also forming two types of cages – small and large. Every unit cell is made of sixteen small cages and eight large ones. The small cage has again the shape of a pentagonal dodecahedron (5^{12}) but the large one is a hexadecahedron ($5^{12}6^4$). Common Type II hydrates include those formed by O_2 and N_2 as guest gases. Sloan and Koh [9] provide a comprehensive list of guest molecules forming structure I (sI) and structure II (sII) clathrate hydrates.

Table 3.1 Structural properties of different hydrate structures.

Hydrate structure	Structure I		Structure II		Structure H		
Cavity size	**Small**	**Large**	**Small**	**Large**	**Small**	**Medium**	**Large**
Cavity shape	5^{12}	$5^{12}6^2$	5^{12}	$5^{12}6^4$	5^{12}	$4^35^66^3$	$5^{12}6^8$
Number of cavities per unit cell	2	6	16	8	3	2	1
Number of H_2O molecules per unit cell	46		136		34		

The unit cell of Type H contains 34 water molecules, forming three types of cages –small, medium, and large. In this case, the unit cell consists of three small cages of type 5^{12}, two medium ones of type $4^35^66^3$ and one large one of type $5^{12}6^8$. The formation of Type H requires the cooperation of two different guest molecules (large and small) to be stable. It is the large cavity that allows structure H hydrates to fit in large molecules (e.g., butane, organic compounds), given the presence of other smaller gases to fill and support the remaining cavities. Structure H (sH) hydrates were suggested to exist in the Gulf of Mexico because of the supply of thermogenically-produced heavy hydrocarbons [9].

Among tens of compounds which are known to form clathrate hydrates with water molecules, the majority form either sI or sII, with a few exceptions such as dimethyl ether (DME) forming sH [13]. Since these exceptions are not related to hydrogen storage, they are not described here in detail. The reader is referred to the review of Loveday and Nelmes [14] for further information. Sloan and Koh [9] presented an extensive list of hydrate-forming compounds.

3.3 Hydrogen Clathrate Hydrate

One of the paths currently being investigated by researchers for hydrogen storage is the use of clathrate hydrates, that is, storing hydrogen in the solid state in the form of guest molecules within hydrates. Hydrogen molecules were historically considered to be too small to stabilize the structures of clathrate hydrates [15]. However, in 1999, Dyadin *et al.* [16] discovered for the first time that hydrogen can form clathrate hydrates at high pressures (up to 1.5 GPa). This was followed by structural studies performed by Mao *et al.* [17] using high-pressure Raman, infrared, X-ray, and neutron studies, where hydrogen was shown to have multiple occupancy within the cavities of a structure II hydrate at high pressures (300 MPa at 350 K). Mao *et al.* [17], and Mao and Mao [2] suggested that pure hydrogen hydrate is stabilized with double and quadruple occupancy of hydrogen molecules in the small and large cages of sII hydrate, respectively. According to Mao and Mao [2], this corresponds to up to 5.3 mass% hydrogen storage capacity within the hydrate structure. However, more recently, Lokshin *et al.* [18] reported, from neutron diffraction studies, that D_2 molecules only singly occupy the small cages of D_2 hydrate. The large cavity occupancy was also found to vary between two and four molecules, depending on pressure and temperature conditions.

Although the above-mentioned studies indicate the possibility of hydrogen storage in hydrates, the high pressures required for stable structures seem to limit such applications from an economic point of view. In an attempt to lower the pressures necessary for hydrogen hydrate formation, Florusse *et al.* [19] tried to stabilize the hydrogen clathrate hydrate structure by using a second guest molecule or *promoter.* They demonstrated that hydrogen can be stabilized in the clathrate framework at pressures over two orders of magnitude lower than for pure hydrogen hydrate by using tetrahydrofuran (THF) as the second guest molecule. Their selection of THF as the promoter was based on the facts that: (i) it is completely water soluble and (ii) it is capable of independently forming structure sII hydrates without the presence of other types of guest molecules. The formation of a binary H_2/THF hydrate will stabilize H_2 within the sII framework, with THF occupying the large cavities at pressures two orders of magnitude lower than that necessary for pure hydrogen hydrate. Reducing the formation pressures of hydrogen hydrate is a critical initial step towards the realization of a practical hydrogen storage method. The existence of hydrogen in the solid phase was confirmed by Florusse *et al.* [19] using X-ray powder diffraction (XRPD). However, the presence of THF dramatically decreases the vacant sites of the clathrate structure for hydrogen storage. Florusse *et al.* [19] predicted that with partial occupancy of the large cavities by THF, the storage capacity of hydrogen in sII hydrates can go up to 4 mass%, by assuming double and quadruple hydrogen occupancies for small and large cavities, respectively.

In 2005, Lee *et al.* [20] used Raman measurements to show that stoichiometric THF/H_2 hydrate (formed from 5.6 mol% THF solution) contained two hydrogen molecules per small cavity. Lee *et al.* [20] also introduced the concept of "tuning," where the THF concentration was decreased to create large vacant cavities in which hydrogen could reside. Following this approach, Lee *et al.* [20] claimed to have formed binary hydrates from a 0.15 mol% THF solution at only 12.0 MPa and 270 K, in which the hydrogen content was reported to be about 4.0 wt%. While the results of Lee *et al.* [20] offer significant promise for developing clathrate hydrates as hydrogen media, they have yet to be independently duplicated.

In fact, more recent detailed studies of H_2–THF clathrate hydrate directly contradict the findings of Lee *et al.* [20]. Strobel *et al.* [21] similarly utilized volumetric measurements in conjunction with Raman and NMR data for the binary system of H_2–THF. They concluded that small cavities can only accommodate single H_2 molecules, and that, irrespective of initial aqueous THF concentration and/or formation conditions, large cavities are always fully occupied by THF. They reported, therefore, that a maximum hydrogen content of around 1 mass% was possible for pressures less than 60 MPa, a value considerably lower than that claimed by Lee *et al.* [20]. The result of such contradictory findings is that the true phase behavior of binary H_2–THF hydrates remains to be identified through future studies on the topic.

In order to gain a comprehensive insight into the phase behavior of the system H_2–THF–H_2O, Rovetto *et al.* [22] experimentally determined the effect of the THF concentration on the equilibrium conditions of the hydrate phase. They explained how the economy/efficiency of hydrogen storage using gas hydrate deals with a compromise between the amount of hydrogen stored or the gravimetric density of the material, and the desired reduction in pressure. Their measurements showed that a maximum in equilibrium temperature occurs when the molar concentration of THF is around 5.6 mol% for all measured pressures (in the range of 2.5–14.5 MPa). This THF concentration corresponds to full occupation of the large cavities of the sII structure of the gas hydrate by the promoter, THF. On the other hand, the hydrate structure has the lowest storage capacity for hydrogen at this concentration of the promoter due to the unavailability of the large cavities. Assuming double occupancy of the small cages of the sII structure by hydrogen, Rovetto *et al.* [22] reported that the total hydrogen storage capacity is decreased to around 2.4 mass%. By decreasing the promoter concentration, the hydrogen storage capacity inside the hydrate structure increases due to the availability of some empty large cages of the sII structure for hydrogen occupancy, while the equilibrium temperature decreases at constant pressure. They predicted that if only 50% of the large cavities of the sII structure are occupied by THF, the hydrogen content can be as high as 3.8 mass% by assuming quadruple occupancy of hydrogen in the large cavities. Figure 3.2 shows the effect of promoter concentration on the equilibrium conditions of the hydrate phase.

In another study carried out in 2007, Anderson *et al.* [23] experimentally determined the phase behavior of the system H_2–THF–H_2O as a function of aqueous THF concentration within a temperature range of 260–290 K and at pressures up to 45 MPa. This is a wider range of temperatures and pressures than in Rovetto *et al.*'s study on the same system. Similar to the findings of Rovetto *et al.* [22], they found that H_2–THF clathrates showed maximum thermal stability at the stoichiometric value of 5.56 mol% initial THF concentration.

In order to investigate Lee *et al.*'s [20] claims about the hydrogen content of clathrates formed from THF concentrations below the eutectic composition (approximately 1.0 mol% at 1 atm), Anderson *et al.* [23] experimentally prepared

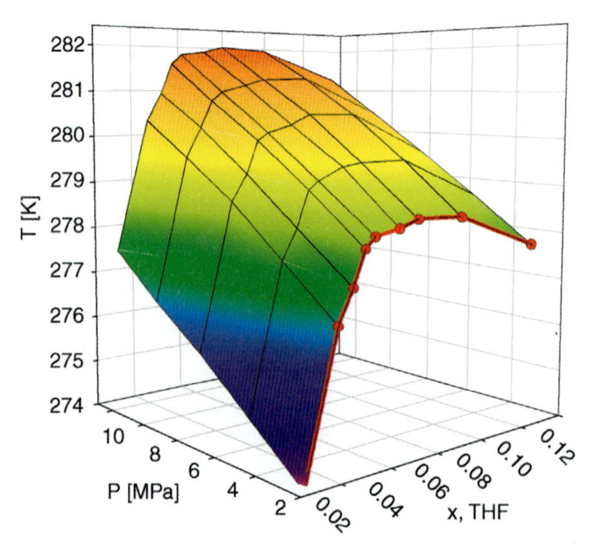

Figure 3.2 Effect of the promoter concentration on the equilibrium conditions of the hydrate phase temperature at any value of the pressure [22].

hydrogen hydrate from a 0.2 mol% THF solution and stabilized it in equilibrium with ice at 270 K and 30 MPa. Based on these measurements, they concluded that, at concentrations lower than the eutectic point, ice is the more stable phase at lower pressures. Therefore, their results did not support the conclusions of Lee *et al.* [20]. Instead, Anderson *et al.*'s data strongly suggested the formation of clathrates with a stoichiometric THF-to-water ratio of 1 : 17. This means that the large cavities were filled by THF, irrespective of the initial aqueous THF concentration. Anderson *et al.* [23] found no evidence of H_2 entering and stabilizing the large sII cavity under the conditions tested, that is, at relatively low pressures. Instead, there are indications that when the aqueous THF concentration is reduced, the THF-to-water stoichiometry is retained at its original stoichiometric value with less clathrate hydrate formed, that is, all remaining free water is then present as ice at equilibrium below the eutectic point.

Therefore, with all large cavities filled by THF, only small dodecahedral cages should be available for the accommodation of hydrogen. Lee *et al.* [20] suggested that these cavities could host two H_2 molecules at the considered pressures. Anderson *et al.* [23] measured the composition of clathrates directly by a mercury isolation technique. Compositional analyses were carried out on two different hydrates formed: (i) from a stoichiometric (5.56 mol%) THF solution in the hydrate + gas region, and (ii) from a 0.2 mol% solution in the hydrate + gas + ice region. For the 5.56 mol% solution at 283 K and 30 MPa, clathrates were found to be of the composition $1.8H_2{\cdot}THF{\cdot}17H_2O$. This is consistent with a 90% occupancy of the small sII dodecahedral cavities by single H_2 molecules and is equivalent to 0.95 mass % of the single occupancy by hydrogen of the small cavities.

With the initial intention of making compositional analyses of the clathrate formed from the 0.2 mol% THF solution, Anderson *et al.* [23] had no way of separating the ice phase from the hydrate during their mercury isolation technique. Thus, only the number of moles of hydrogen released from all the solid (ice and hydrate phases) could be measured. Because of this, Anderson *et al.* [23] made an assumption regarding the composition of the clathrates in terms of THF-to-water ratio. They reasoned that there are indications that clathrate hydrates are consistently stoichiometric in this respect, having a THF-to-water ratio of 1 : 17. Using this ratio, Anderson *et al.* [23] determined a hydrate formula of $1.6H_2{\cdot}THF{\cdot}17H_2O$ from the compositional data. This is equal to 80% occupancy of small cages by single H_2 molecules and gives a clathrate of 0.83% H_2 by mass.

Anderson *et al.*'s findings regarding single H_2 occupancy of the small sII cages are further supported by recent high-pressure neutron diffraction studies of single guest (D_2) and binary H_2–THF sII clathrates [18, 24]. Based on the above-mentioned studies and analyses, it is strongly suggested that Lee *et al.*'s "tuning" concept (i.e., that reducing THF concentration increases H_2–THF clathrate hydrogen content) is not viable. Indeed, the data generated by Anderson *et al.* [23] support the conclusions of Strobel *et al.* [21] who suggested that the amount of hydrogen stored in the stoichiometric hydrate increases with pressure and exhibits asymptotic (Langmuir) behavior to a maximum of approximately 1.0 wt% H_2. This hydrogen concentration corresponds to one hydrogen molecule occupying each of the small 5^{12} cavities and one THF molecule in each large $5^{12}6^4$ cavity in each hydrate unit cell. So, the final conclusion from the studies of Strobel *et al.* [21] and Anderson *et al.* [23] is that small cavities can accommodate one H_2 molecule only, with large cavities being occupied by THF, irrespective of the initial aqueous THF concentration.

In order to increase the hydrogen storage capacity of the clathrate hydrates, a number of studies have recently been performed. Rovetto *et al.* [22] considered promoters other than THF in order to ease the equilibrium conditions of the hydrogen hydrate in terms of equilibrium temperature and pressure. For this purpose, they selected for experimental investigation, different cyclic organic compounds which had similarities in their chemical structure. The selected compounds were 1,3-dioxolane (Diox), 2,5-dihydrofuran (DHF), tetrahydropyran (THP), furan (F), and tetrahydrofuran (THF). Their results showed that THF has the highest promoting effect among the studied promoters. Figure 3.3 shows the equilibrium conditions of their studied systems. From these results, it is clear that the promoting effect decreases in the order THF > F > THP > DHF > Diox.

Additionally, molecules like propane have also been investigated [25]. The experimental results indicate that the hydrate formation pressures of hydrogen methane and hydrogen + propane gas mixtures were higher than those for pure methane and pure propane.

One particularly interesting binary sII hydrate was formed with hydrogen using cyclohexanone (CHONE) as the second (large) guest [26]. Unlike THF, propane, or other various furan-type sII-forming molecules, CHONE is incapable of stabilizing

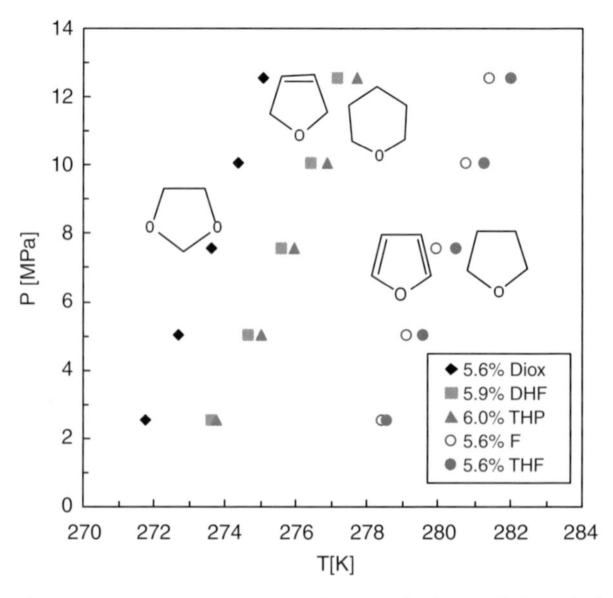

Figure 3.3 Pressure–temperature diagram of hydrogen clathrate hydrate formation with different promoters [22].

a sII hydrate on its own. Due to the large size of CHONE, a second "help-gas" molecule is required for stability [27]. This second molecule is thought to add stabilization through interactions with the small cavity to help decrease unfavorable distortions to the large cavity caused by the presence of the large molecule.

Another method for enhancing hydrogen storage which was studied theoretically by Daschbach *et al.* [28], was the enclathration of hydrogen in a nonaqueous solid crystal to produce a clathrate material or a so-called organic clathrate. They considered hydroquinone (HQ) for hydrogen storage. Using molecular dynamic simulations, Daschbach *et al.* [28] suggested that it should be possible to load hydrogen into the hydroquinone clathrate structure. Strobel *et al.* [29] proposed to increase hydrogen storage using the concept of chemical–clathrate hybrid technology to combine molecular hydrogen contained within the clathrate cages with hydrogen that is chemically bound to the host lattice. They studied this concept also using HQ clathrate, and successfully synthesized the HQ + H_2 clathrate by recrystallization from a saturated ethanol solution in the presence of hydrogen. The clathrate phase of HQ has cavities composed of six HQ molecules terminated at the top and bottom by rings of six hydroxy groups. These cavities have a radius of about 4.5 Å [29]. Using Raman spectroscopy, the presence of H_2 within the HQ clathrate lattice was confirmed. Their preliminary tests demonstrated the enhanced stability of the HQ + H_2 clathrate when compared with the H_2O + H_2 clathrate hydrate. They realized that after 10 min of exposure to ambient conditions, the HQ + H_2 clathrate maintained significant H_2 Raman peak intensity. In the case of H_2O + H_2 clathrate hydrate, exposure to ambient conditions resulted in the immediate dissociation of the clathrate and release of all the contained hydrogen.

The large cages of sII clathrate hydrates have a radius of about 4.7 Å and were shown experimentally to contain up to four hydrogen molecules. Knowing that HQ clathrate cavities have a radius of 4.5 Å, the possibility of enclathration of four hydrogen molecules still seems very reasonable [30]. With four H_2 molecules per cavity, this novel clathrate would contain about 2.4 wt% hydrogen, counting only molecules contained within the cavities (excluding atomic hydrogen in the host framework). Strobel *et al.* [30] proposed, as the next step in the chemical–clathrate hybrid process, to chemically dehydrogenate the host material, HQ. For this purpose, they separated the two hydrogen atoms from the two hydroxy groups of HQ. It was proposed to do this by an oxidation reaction to form benzoquinone. As they explained, if each of the clathrate cavities in the HQ clathrate structure were to contain four hydrogen molecules, and the redox reaction were to react to completion, the hydrogen storage capacity of this material would be about 4.2 wt%. Compared with the maximum storage of 3.8 wt% for clathrate hydrates (pure sII), this storage scheme may possibly be promising, and many other chemical components may prove to be more favorable than HQ. They did not mention the operational conditions of HQ clathrat–chemical hybrid hydrogen storage. Further investigation is necessary to evaluate the economic and practical possibilities of such processes.

With the goal of overcoming the problem of the occupation of the vacant cages of structure II hydrates by promoter molecules, Duarte *et al.* [31] proposed to increase the storage capacity of the clathrate hydrates of hydrogen by synthesizing structure H clathrate hydrate using different promoters. As shown earlier in Figure 3.1, structure H of clathrate hydrates has three small 5^{12} cages, two medium-sized $4^3 5^6 6^3$ cages, and only one large $5^{12}6^8$ cage per unit cell. Each unit cell of structure H requires 34 water molecules. To achieve their objective, Duarte *et al.* [31] examined different potential promoters, including 1,1-dimethylcyclohexane (DMCH), methyl *tert*-butyl ether (MTBE) and methylcyclohexane (MCH). They considered these promoters as H formers due to their molecular size. Their diameters (8.4 Å for DMCH, 7.8 Å for MTBE, and 8.59 Å for MCH) are large enough to stabilize the large cages of structure H. Spectroscopic evidence for this conclusion is provided by Strobel *et al.* [29, 30]. Duarte *et al.* [31] found that DMCH, with a diameter of 8.4 Å, was able to stabilize the clathrate hydrate of hydrogen at 50.0 MPa and 274.7 K. If the temperature rises to 279.5 K, the required pressure increases to 95.0 MPa. Other promoters like MTBE and MCH could also stabilize hydrogen clathrate hydrate in structure H. However, the minimum pressures at which these two clathrate hydrates were found to be stable were as high as 70.1 MPa at 269.2 K and 83.1 MPa at 274.0 K, respectively. Figure 3.4 shows the *p*–*T* diagram of the systems $H_2O + H_2 +$ promoter. The equilibrium data for the pure H_2 hydrate [16] and the THF + H_2 hydrate [19] are also included in this figure.

Duarte *et al.* [31] considered single hydrogen occupancy for the small 5^{12} cages of structure H. In addition, complete occupancy of the large $5^{12}6^8$ cages by the promoter

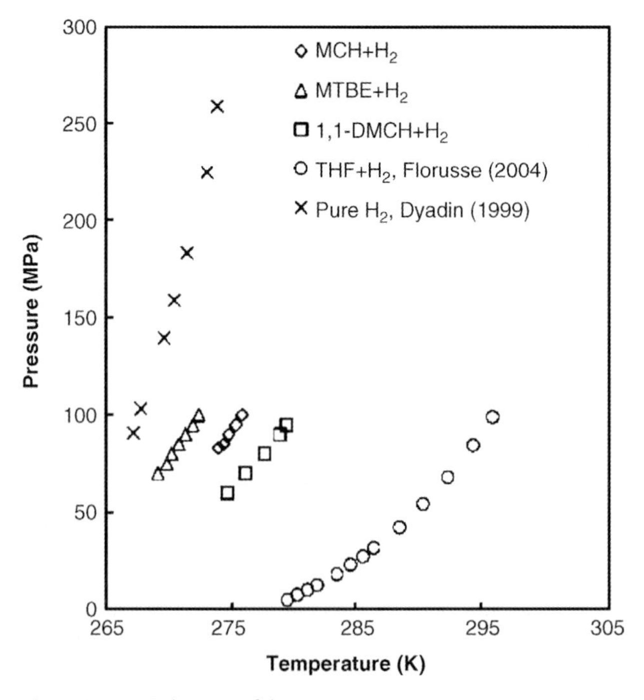

Figure 3.4 *p*–*T* diagram of the system $H_2O + H_2 +$ promoter [31].

was assumed and as the medium-sized $4^35^66^3$ cage of H is slightly larger than the 5^{12} cage, single hydrogen occupancy was also assumed for this cage. This scenario leads to an estimate of about 1.4 wt% of H_2 storage in structure H, which is almost a 40% higher storage capacity than that in structure sII. Therefore, it could be concluded that the $H_2/(H_2O$ + promoter) molar ratio in structure H can increase up to 1 : 7, which is an improvement in the gravimetric density of clathrate hydrates in comparison with the 1 : 9 ratio of structure sII in the H_2 + THF system.

3.4
Kinetic Aspects of Hydrogen Clathrate Hydrate

As discussed in previous sections of this chapter, the molecular occupancy and the capacity for storing hydrogen in clathrate hydrates are two important issues that are still the subject of discussion among several research groups. The storage capacity of hydrates must be maximized to the requirements of new energy applications. Particularly in the transport sector, the goals set by some energy agencies by the end of 2010 require storage capacities up to 6 wt% of hydrogen, together with environmentally clean and safe methods for producing and storing hydrogen. In addition to this, another extremely important issue for large-scale applications for hydrogen storage and transportation concerns the kinetics of formation and decomposition of gas hydrates, that is, the rates at which hydrogen can be practically stored and released. In the area of hydrogen hydrates, the challenge goes beyond the time-independent equilibrium thermodynamics, since time-dependent kinetic measurements and issues are no less challenging. The essential questions for better knowledge of hydrogen clathrate hydrates are: (i) what is the time necessary for hydrogen hydrates to form? and (ii) how fast will they dissociate?

Published information on the kinetics of hydrogen hydrate formation is very scarce. Lee *et al.* [20] made a brief reference to this subject, demonstrating that the presence of a dispersed phase should improve the kinetics. In 2007, Duarte *et al.* [32] extensively studied the kinetics of hydrogen hydrate formation of the ternary system H_2 + THF + H_2O. Based on their experiments, they provided the pressure and temperature path during formation and decomposition of hydrogen hydrates. Figure 3.5 shows the nucleation, growth, and dissociation of hydrogen hydrate formation studied by Duarte *et al.* [32]. As they pointed out, without metastability, hydrate formation should begin at Point A in Figure 3.5. However, the system pressure continues to decrease linearly with temperature for a number of hours. At the start of the experiment (Point A), the temperature was stable and the system was pressurized with hydrogen. Cooling and stirring started at this moment. From A to B, as the temperature was lowered, the pressure also decreased linearly with temperature due to contraction of the gas upon cooling. At point B, the hydrate started to be formed at approximately constant temperature, and a sharp decrease in pressure to Point C was observed, due to the entrapment of the gas in the solid phase. The dissociation of the hydrate began when the system was heated to the initial temperature. The pressure started to increase again due to the increase in

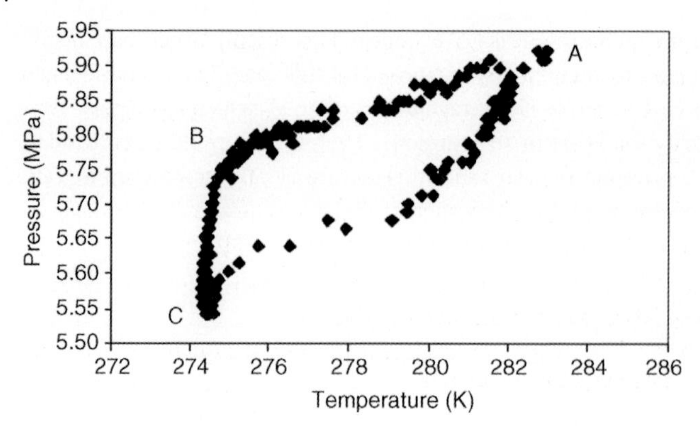

Figure 3.5 The pressure and temperature trace during formation and decomposition of hydrogen hydrates [32].

temperature until a certain temperature where the hydrate dissociate and the gas entrapped in the solid phase was completely released.

The closed loop observed in Figure 3.5 demonstrates the reversibility of the process, which is a great advantage for practical application of these novel materials for on-board hydrogen storage.

In 2008, in addition to some phase behavior experiments, Talyzin [33] carried out some experiments on the kinetics of hydrogen hydrate formation and dissociation for the system H_2 + THF + H_2O. It was concluded that due to the slow kinetics and small hydrogen capacities of THF + H_2O hydrogen clathrates, this material is not suitable for application as a hydrogen-storing material. However, it was mentioned that clathrate hydrates, nevertheless, remain rather promising materials for hydrogen adsorption since the THF can possibly be replaced by more efficient solvents. Moreover, from preliminary research it became apparent that by applying a spray technique, hydrogen hydrates can be formed instantaneously.

3.5
Modeling of Hydrogen Clathrate Hydrates

The storage of hydrogen in hydrates involves mechanisms similar, but not identical to those of physical adsorption mechanisms. These are governed mostly by intermolecular van der Waals forces and stronger forces, such as hydrogen bonding, which hold together the host framework. In order to select hydrogen-storage materials tailored to the needs of a hydrogen-based economy, knowledge of all such forces is crucial since they determine the extent of hydrogen retention in the clathrate structure. To establish the hierarchy of molecular forces, it is necessary to carry out detailed experimental studies of stable molecular structures by X-ray and neutron scattering and determine their characteristic vibrational spectra by neutron scattering, Raman, and IR spectroscopy. Additionally, theoretical calculations give further insight and guidelines for development of optimum hydrogen storage materials. Given the structural complexity of the clathrate hydrates, theoretical work should rely on a combination of different computational techniques that can balance the system size and the accuracy needed [1].

The theoretical developments for hydrogen storage in molecular clathrates have been discussed extensively by Struzhkin *et al.* [1]. They explained that the complete theoretical description of hydrogen clathrate hydrates is usually based on a combination of different techniques including density functional theory (DFT) [34], molecular dynamicss (MD) [35], and similar approaches.

A statistical mechanical model in conjunction with *ab initio* quantum chemical calculations of the thermodynamic stability of H_2 guests in small and large sII clathrates was recently reported by Patchkovskii and Tse [36]. Their model uses DFT and Möller-Plesset (MP2) calculations to estimate the encapsulation energies of hydrogen molecules in small and large isolated rigid water cages in which the positions of the hydrogen guests have been optimized. The positions of the oxygen atoms for the water cages in their model were determined from the X-ray structure. Water hydrogen atoms were placed at 1.0 Å distance from the oxygen atoms. They calculated encapsulation energies for up to three hydrogen guests in the small cages and up to five hydrogen guests in the large cages. The model predicts that at a pressure of 2 kbar and a temperature of 250 K, the most stable configuration encapsulates two guests in the small cages and four in the large cages.

Alavi *et al.* [37] used a molecular dynamics approach to study the stability of sII hydrogen clathrates with different H_2 guest occupancies. Their simulations were done at pressures of 2.5 kbar and 1.013 bar, and for temperatures ranging from 100 to 250 K. The modeling results showed that at 100 K and 2.5 kbar the most stable configurations have single occupancy in the small cages and quadruple occupancy in the large cages. Further, the optimum occupancy for the large cages decreased as the temperature was increased. They concluded that double occupancy in the small cages increases the energy of the structures and causes tetragonal distortion in the unit cell. In a later study, the same authors studied theoretically the effect of multiple cage occupation and promotion with THF [38] in type sII clathrates, as well as type H clathrates promoted with methyl *tert*-butyl ether (MTBE) [39]. These investigations were focused primarily on the effect that multiple H_2 occupation would have on the calculated total potential energy, unit cell volume, and thermal expansion coefficients of the clathrates. Their results showed that substitution of the THF or MTBE molecules in the large cages with 4 hydrogen molecules increases the configurational energy of the unit cell. Common model potentials were used for the interactions with the promoter molecules. The SPC/E model [40] was used for the water–water interactions. This model, however, has known deficiencies in describing liquid–water and ice–water interactions [41–43]. There are no reports in the literature validating the use of these potentials for describing pure or promoted hydrogen clathrate hydrates, nor any investigations on the sensitivity of the results to the nature of the potentials.

In 2007, Frankcombe and Kroes [44] tested a range of model interaction potentials in order to identify the potentials that do succeed at reproducing experimentally measured properties of hydrogen-containing clathrates. Although they found that no water–water interaction models can reproduce the behavior expected at the studied temperature and pressure conditions, they found that the TIP5P potential [45] gave the closest match to experimentally observed clathrate stabilities.

Katsumasa *et al.* [46] examined the cage occupancy of hydrogen clathrate hydrates using grand canonical Monte Carlo (GCMC) simulations for wide ranges of temperature and hydrogen pressures. Their simulations were carried out with a fixed number of water molecules and a fixed chemical potential for the guest hydrogen molecules. They extended the method to a clathrate hydrate under a constant guest component pressure. They showed that a hybrid Monte Carlo approach is suitable for estimating the occupancy of guests in clathrate hydrates. It was also found that the smaller cages in structure sII are capable of accommodating only a single guest molecule, even at pressures as high as 500 MPa, which agrees with recent experimental observations. The larger cage was found to encapsulate at most four hydrogen molecules. This leads to a highest occupancy ratio of 0.35. The obtained result is reasonable considering the size of a hydrogen molecule.

In a sequence of three papers, Papadimitriou *et al.* [47–49] performed Monte Carlo molecular simulations in the Grand Canonical Ensemble to study the multiple occupancy of argon and hydrogen in the various cavities of structures II and H.

It was found that in sII the small cavity always contained only one H_2 molecule, whereas the large cavity of this structure could accommodate up to four hydrogen molecules. In the presence of THF, all large cavities turned out to be occupied by the promoter molecule, regardless of its concentration in the aqueous phase.

Similar molecular simulations for sH hydrate showed that both the small and medium sized cavities could accommodate only one hydrogen molecule, whereas in the large cavity of sH, as many as eight hydrogen molecules could be hosted.

From the above discussions, it can be concluded that the accurate determination, born experimentally and theoretically, of H_2 occupancy in the small cages of hydrate structures under different conditions of pressure and temperature is crucial to establish whether H_2 hydrate will be able to compete as a suitable material for hydrogen storage.

3.6 Future of Hydrogen Storage

To converge to a practical and cost-efficient solution regarding hydrogen fuel storage, much research remains to be done. This is even more important in the field of hydrogen hydrates, which seriously lacks the basic knowledge necessary. As explained above, scientists do not even agree yet on the number of guest molecules occupying the cages of hydrates. This is why Schlapbach and Zuttel [50] suggest that the key to the fabrication of proper hydrogen storage materials lies in further characterization of the fundamental nature and strength of hydrogen bonding interactions with a variety of host materials. Fortunately, with the discovery of hydrogen clathrate hydrates in 1999 and the spark of the idea of hydrogen storage in hydrates, the current decade is witnessing some activity by various groups. It is expected that research will intensify in this area in the future.

There is much scope to investigate different potential promoters, which not only increase hydrate stability and lower storage pressures, but may also be able to alter the hydrate structure to other forms that have higher storage capacities. Fundamental research must be coupled with detailed economic studies to ensure optimum combinations of storage temperatures and pressures, among the wide range of possible combinations. The possibility of storing natural gas hydrates under self-preserving conditions at temperatures up to 75 °C higher than the equilibrium temperatures has been well-documented. Investigation of the existence or lack of such a phenomenon for hydrogen hydrates may lead to significant design improvements and cost reductions. Calorimetric studies are also necessary for economic evaluations of hydrogen hydrate formation and dissociation energy demands. A proper selection of the form of storage of hydrates, that is, bulk, solid, slurry, powder, pellets, and so on, will ensure the optimum selection between maximum gas density, hydrate stability, and a fast-enough rate of release of gas. The form of hydrates will also affect the type of storage; whether a fixed refillable tank within the vehicle, or removable tanks that are refilled in the factory. The overall cost of hydrate storage is heavily dependent on production costs. In this respect, an increased rate of production can significantly decrease the expenses. This can be achieved by different techniques: gas–water–hydrate mixture stirring at hydrate formation conditions, hydrate formation from fine ice powder, gas bubbling through a water column, water spraying in a gas atmosphere, using different additives in water to accelerate hydrate nucleation, using small amounts of surfactants (in the ppm range) to increase liquid–vapor contact, or a combination of the above [51].

References

1 Struzhkin, V.V., Militzer, B., Mao, W.L., Mao, H.K., and Hemley, R.J. (2007) Hydrogen storage in molecular clathrates. *Chem. Rev.*, **107**, 4133.

2 Mao, W.L. and Mao, H.K. (2004) Hydrogen storage in molecular compounds. *PNAS*, **101**, 708.

3 Deng, W.Q., Xu, X., and Goddard, W.A. (2004) New Alkali Doped Pillared Carbon Materials Designed to Achieve Practical Reversible Hydrogen Storage for Transportation. *Phys. Rev. Lett.*, **92**, 166103.

4 Yildirim, T. and Hartman, M.R. (2005) Direct observation of hydrogen adsorption sites and nanocage formation in metal-organic frameworks. *Phys. Rev. Lett.*, **95**, 215504.

5 Gagliardi, L. and Pyykko, P. (2004) How many hydrogen atoms can be bound to a metal? Predicted MH12 species. *J. Am. Chem. Soc.*, **126**, 15014.

6 Fichtner, M. (2005) Nanotechnological aspects in materials for hydrogen storage. *Adv. Eng. Mater.*, **7**, 443.

7 Bogdanovic, B., Felderhoff, M., Pommerin, A., Schuth, T., and Spielkamp, N. (2006) Advanced hydrogen-storage materials based on Sc-, Ce-, and Pr-doped $NaAlH_4$. *Adv. Mater.*, **18**, 1198.

8 Powell, H.M. (1948) The structure of molecular compounds. 4. Clathrate compounds. *J. Chem. Soc.*, 61–73.

9 Sloan, E.D. and Koh, C.A. (2008) *Clathrate Hydrates of Natural Gases*, 3rd edn, Taylor & Francis Group, Boca Raton.

10 Klauda, J.B. and Sandler, S.I. (2005) Global distribution of methane hydrate in ocean sediments. *Energ. Fuel*, **19**, 459.

11 Loveday, J.S., Nelmes, R.J., Guthrie, M., Belmonte, S.A., Allan, D.R., Klug, D.D., Tse, J.S., and Handa, Y.P. (2001) Stable methane hydrate above 2 GPa and the source of Titan's atmospheric methane. *Nature*, **410**, 661.

12 Gabitto, J., Riestenberg, D., Lee, S., Liang, L.Y., and Tsouris, C. (2004) Ocean disposal of CO_2: Conditions for producing sinking CO_2 hydrate. *J. Disp. Sci. Technol.*, **25**, 703.

13 Udachin, K.A., Ratcliffe, C.I., and Ripmeester, J.A. (2001) A dense and efficient clathrate hydrate structure with unusual cages. *Angew. Chem. Int. Ed.*, **40**, 1303.

14 Loveday, J.S. and Nelmes, R.J. (2008) High-pressure gas hydrates. *Phys. Chem. Chem. Phys.*, **10**, 937.

15 Holder, G.D., Stephenson, J.L., John, V.T., Kamath, V.A., and Malekar, S. (1983) Formation of clathrate hydrates in hydrogen-rich gases. *Ind. Eng. Chem. Process Des. Dev.*, **22**, 170.

16 Dyadin, Y.A., Larionov, G., Aladko, E.Y., Manakov, A.Y., Zhurko, F.V., Mikina, T.V., Komarov, V.Y., and Grachev, E.V. (1999) Clathrate formation in water-noble gas (hydrogen) systems at high pressures. *J. Struct. Chem.*, **40**, 70.

17 Mao, W.L., Mao, H.K., Goncharov, A.F., Struzhkin, V.V., Guo, Q., Hu, J., Shu, J., Hemley, R.J., Somayazulu, M., and Zhao, Y. (2002) Hydrogen clusters in clathrate hydrate. *Science*, **297**, 2247.

18 Lokshin, K.A., Zhao, Y., He, D., Mao, W.L., Mao, H.K., Hemley, R.J., Lobanov, M.V., and Greenblatt, M. (2004) Structure and dynamics of hydrogen molecules in the novel clathrate hydrate by high pressure neutron diffraction. *Phys. Rev. Lett.*, **93**, 125503.

19 Florusse, L.J., Peters, C.J., Schoonman, J., Hester, K.C., Koh, C.A., Dec, S.F., Marsh, K.N., and Sloan, E.D. (2004) Stable low-pressure hydrogen clusters stored in a binary clathrate hydrate. *Science*, **306**, 469.

20 Lee, H., Lee, J.W., Kim, D.Y., Park, J., Seo, Y.T., Zeng, H., Moudrakovski, I.L., Ratcliffe, C.I., and Ripmeester, J.A. (2005) Tuning clathrate hydrates for hydrogen storage. *Nature*, **434**, 743.

21 Strobel, T.A., Taylor, C.J., Hester, K.C., Dec, S.F., Koh, C.A., Miller, K.T., and Sloan, E.D. (2006) Molecular hydrogen storage in binary THF-H_2 clathrate hydrates. *J. Phys. Chem. B*, **110**, 17121.

22 Rovetto, L.J., Shariati, A., Schoonman, J., and Peters, C.J. (2006) Storage of hydrogen in low-pressure clathrate hydrates. Proceedings of the 22nd European Symposium on Applied Thermodynamics (ESAT 2006), June 28–July 1, Elsinore, Denmark.

23 Anderson, R., Chapoy, A., and Tohidi, B. (2007) Phase relations and binary clathrate hydrate formation in the system H_2-THF-H_2O. *Langmuir*, **23**, 3440.

24 Hester, K.C., Strobel, T.A., Sloan, E.D., Koh, C.A., Hug, A., and Schultz, A.J. (2006) Molecular hydrogen occupancy in binary THF-H2 clathrate hydrates by high resolution neutron diffraction. *J. Phys. Chem. B*, **110**, 14024.

25 Zhang, S.X., Chen, G.J., Ma, C.F., Yang, L.Y., and Guo, T.M. (2000) Hydrate formation of hydrogen + hydrocarbon gas mixtures. *J. Chem. Eng. Data*, **45**, 908.

26 Strobel, T.A., Hester, K.C., Sloan, E.D., and Koh, C.A. (2007) A hydrogen clathrate with cyclohexanone: Structure and stability. *J. Am. Chem. Soc.*, **129**, 9544.

27 Ripmeester, J.A., Tse, J.S., Ratcliffe, C.I., and Powell, B.M. (1987) A new clathrate hydrate structure. *Nature*, **325**, 135.

28 Daschbach, J.L., Chang, T.M., Corrales, L.R., Dang, L.X., and McGrail, P. (2006) Molecular Mechanisms of hydrogen-loaded β-hydroquinone clathrate. *J. Phys. Chem. B*, **110**, 17291.

29 Strobel, T.A., Kim, Y., Koh, C.A., and Sloan, E.D. (2008) Clathrates of hydrogen with application towards hydrogen storage. Proceedings of the 6th International Conference on Gas Hydrates (ICGH 2008), July 6–10, Vancouver, Canada.

30 Strobel, T.A., Koh, C.A., and Sloan, E.D. (2008) Water cavities of sH clathrate hydrate stabilized by molecular hydrogen. *J. Phys. Chem. B*, **112**, 1885.

31 Duarte, A.R.C., Shariati, A., Rovetto, L.J., and Peters, C.J. (2008) Water cavities of sH

clathrate hydrate stabilized by molecular hydrogen: Phase equilibrium measurements. *J. Phys. Chem. B*, **112**, 1888.

32 Duarte, A.R.C., Zevenbergen, J., Shariati, A., and Peters, C.J. (2007) Kinetics of formation and dissociation of sH hydrogen clathrate hydrates. Proceedings of the 5th International Symposium on High Pressure Process Technology and Chemical Engineering, June 24–27, Segovia, Spain.

33 Talyzin, A. (2008) Feasibility of H_2-THF-H_2O clathrate hydrates for hydrogen storage applications. *Int. J. Hydrogen Energy*, **33**, 111.

34 Parr, R.G. and Yang, W. (1989) *Density Functional Theory of Atoms and Molecules*, Oxford University Press, Oxford.

35 Haile, J.M. (1997) *Molecular Dynamics Simulation: Elementary Methods*, Wiley, New York.

36 Patchkovskii, S. and Tse, J.S. (2003) Thermodynamic stability of hydrogen clathrates. *PNAS*, **100**, 14645.

37 Alavi, S., Ripmeester, J.A., and Klug, D.D. (2005) Molecular-dynamics study of structure II hydrogen clathrates. *J. Chem. Phys.*, **123**, 024507.

38 Alavi, S., Ripmeester, J.A., and Klug, D.D. (2006) Molecular-dynamics simulations of binary structure II hydrogen and tetrahydrofuran clathrates. *J. Chem. Phys.*, **124**, 014704.

39 Alavi, S., Ripmeester, J.A., and Klug, D.D. (2006) Molecular-dynamics simulations of binary structure H hydrogen and methyl-tert-butylether clathrate hydrates. *J. Chem. Phys.*, **124**, 204707.

40 Berendsen, H.J.C., Grigera, J.R., and Straatsma, T.P. (1987) The missing term in effective pair potentials. *J. Phys. Chem.*, **91**, 6269.

41 Motakabbir, K.A. and Berkowitz, M. (1990) Isothermal compressibility of SPC/E water. *J. Phys. Chem.*, **94**, 8359.

42 Baez, L.A. and Clancy, P. (1994) Existence of a density maximum in extended simple point-charge water. *J. Chem. Phys.*, **101**, 9837.

43 Chialvo, A.A. and Cummings, P.T. (1996) Microstructure of ambient and super critical water. Direct comparison between simulation and neutron scattering experiments. *J. Phys. Chem.*, **100**, 1309.

44 Frankcombe, T.J. and Kroes, G.J. (2007) Molecular dynamics simulations of type-sII hydrogen clathrate hydrate close to equilibrium conditions. *J. Phys. Chem. C*, **111**, 13044.

45 Mahoney, M.W. and Jorgensen, W.L. (2000) A five-site model for liquid water and the reproduction of the density anomaly by rigid, nonpolarizable potential functions. *J. Chem. Phys.*, **112**, 8910.

46 Katsumasa, K., Koga, K., and Tanaka, H. (2007) On the thermodynamic stability of hydrogen clathrate hydrates. *J. Chem. Phys.*, **127**, 044509.

47 Papadimitriou, N.I., Tsimpanogiannis, I.N., Papaioannou, A.Th., and Stubos, A.K. (2008) Monte Carlo study of sII and sH argon hydrates with multiple accupancy of cages. *Mol. Simul.*, **34**, 1311.

48 Papadimitriou, N.I., Tsimpanogiannis, I.N., Papaioannou, A.Th., and Stubos, A.K. (2008) Evaluation of the hydrogen-storage capacity of pure H_2 and binay H_2-THF hydrates with Monte Carlo simulations. *J. Chem. Phys. C*, **112**, 10294.

49 Papadimitriou, N.I., Tsimpanogiannis, I.N., Peters, C.J., Papaioannou, A.Th., and Stubos, A.K. (2008) Hydrogen storage in sH hydrates: a Monte Carlo study. *J. Chem. Phys. B*, **112**, 14206.

50 Schlapbach, L. and Zuttel, A. (2001) Hydrogen-storage materials for mobile applications. *Nature*, **414**, 353.

51 Yakushev, V. (2002) Production of dense (low-porous) natural gas hydrate samples, 4th International Conference On Gas Hydrates, May 19–23, Yokohama, Japan.

4
Metal Hydrides

Jacques Huot

4.1
Introduction

Metal hydrides are promising candidates for many stationary and mobile hydrogen storage applications. Presently, the most common use of metal hydrides is as anode material in commercial nickel-metal hydrides (Ni-MH) rechargeable batteries which have replaced conventional nickel-cadmium batteries in many applications. A wide variety of Ni-MH batteries are now on the market with cell sizes ranging from 30 mAh to 250 Ah [1, 2]. Some other applications are: hydrogen compression [3], aircraft fire-detectors [4], isotope separation [5], and hydrogen getters for microelectronic packages. Hydride formation is also used in the HDDR process (hydrogenation–disproportionation–desorption–recombination) for the synthesis of magnetic materials such as $Nd_2Fe_{14}B$ [6]. A striking application is as a component of cryocoolers for the ESA Planck mission [7]. Another exciting new application is the use of metal hydrides for switchable mirrors [8]. As we can see, metal hydrides have a wide range of applications. However, most research and development is targeted towards two applications: batteries and hydrogen storage. The former is now widely exploited but the latter is still facing many technical problems. The main advantages of storing hydrogen in a metal hydride are the high hydrogen volumetric densities (sometimes higher than in liquid hydrogen) and the possibility to absorb and desorb hydrogen with a small change in hydrogen pressure [9].

The physics of hydrogen in elemental metals was reviewed at the end of the 1970s in the books *Hydrogen in Metals*, volumes I and II [10, 11]. The 1970s also saw the emergence of the important field of intermetallic compounds as hydride materials. An in-depth review of this field can be found in *Hydrogen in Intermetallic Compounds* volumes I, and II [12, 13]. A review of the properties and applications of hydrogen in metals can be found in the third volume of *Hydrogen in Metals* [14]. The basic properties of the metal–hydrogen system are discussed from a fundamental point of view by Fukai [15]. Although initial studies on nanocrystalline and amorphous metal hydrides were initiated around the 1980s [16], the real emergence of nanocrystalline metal hydrides as a new class of

Handbook of Hydrogen Storage. Edited by Michael Hirscher

ISBN: 978-3-527-32273-2

hydrides came only in the early 1990s [17]. There are a number of recent review papers targeted on metal hydrides for hydrogen storage [18–31].

Even though the first metal hydride was discovered more than 100 years ago, intensive research on this class of materials is relatively new. Investigation of the hydrogen–metal interactions is very interesting from a fundamental point of view as well as for practical applications. In this chapter, we focus on the materials and properties that are especially important for hydrogen storage applications.

4.2
Elemental Hydrides

Metal hydrides can be defined as a concentrated single-phase compound between a host metal and hydrogen [32]. The first metal hydride was discovered by Graham who observed that palladium absorbed a large amount of hydrogen [33]. Only a few simple hydrides were known before the twentieth century and hydride chemistry did not become active until the time of World War II [34]. Simple binary metal hydrides can be grouped into three basic types according to the nature of the metal–hydrogen bond [35].

4.2.1
Ionic or Saline Hydrides

This group include the binary hydrides of all alkali metals and of alkaline earth metals from calcium through barium [36]. In these compounds, hydrogen exists as a negatively charged ion (H^-) and can be considered as a member of the halogen series. Therefore, many physical properties such as hardness, brittleness, optical properties and crystal structures are similar to the corresponding halides. The alkali metal hydrides have a sodium chloride structure, while the alkaline earth hydrides have a barium chloride structure [37]. Typical binary ionic hydrides are sodium hydride NaH and calcium hydride CaH_2. Saline hydrides have high conductivities just below or at the melting point. Complex ionic hydrides such as lithium aluminum hydride $LiAlH_4$ and sodium borohydride $NaBH_4$ are used commercially as reducing agents. In general, the binary ionic hydrides are too stable for hydrogen storage application with the exception of magnesium [37]. However, magnesium hydride is not a true ionic hydride. In magnesium hydride the interaction between hydrogen and magnesium is partly ionic and partly covalent. Thus, magnesium hydride should be considered to be a transition hydride between ionic and covalent hydrides.

4.2.2
Covalent Hydrides

Covalent metal hydrides are compounds of hydrogen and non-metals. Here, atoms of similar electronegativities share electron pairs. In general, covalent hydrides have low melting and boiling points. Because of the weak van der Waals forces between molecules, most covalent hydrides are liquid or gaseous at room temperature and those that are solid are thermally unstable [38]. Examples of covalent hydrides are water (H_2O), hydrogen sulfide (H_2S), silane (SiH_4), aluminum borohydride $Al(BH_4)_3$, methane (CH_4) and other hydrocarbons. Covalent hydrides cannot be formed by direct reaction of hydrogen gas and the element, complex chemical reactions must be used to synthesize them [35]. Because of this difficulty in synthesis, covalent hydrides are not good candidates for hydrogen storage applications.

4.2.3 Metallic Hydrides

Most of the hydrides that could be used for hydrogen storage are metallic in nature [37]. Metallic hydrides are formed by transition metals including the rare earth and actinide series. In these hydrides, hydrogen acts as a metal and forms a metallic bond. They have high thermal and electrical conductivities. However, unlike metals, they are quite brittle [38]. The prevailing explanation is that in the metallic hydrides the 1s electron on the hydrogen atom participates in the conduction band of the metal to generate new M–H bonding states [20, 39]. Metallic hydrides have a wide variety of stoichiometric and non-stoichiometric compounds and are formed by direct reaction of hydrogen with the metal or by electrochemical reaction. Examples of metallic hydrides are TiH_2 and ThH_2.

The above division should not be taken too literally. In fact, most of the metal hydrides have a mixture of different types of bonding. For example, in LiH the bonds are mainly ionic but still have a significant covalent component [35].

4.3 Thermodynamics of Metal Hydrides

4.3.1 Introduction

As most of the hydrides discussed here are formed by direct reaction with gaseous hydrogen, it is important first to discuss the thermodynamics of hydride formation. In this section, the thermodynamic aspect of hydride formation will be briefly sketched. In-depth treatment of the thermodynamics of metal–hydrogen systems can be found in the literature [32, 40–44].

When exposed to hydrogen, a hydride-forming metal (M) will form a metal hydride following the reaction:

$$M + \frac{x}{2} H_2 \Leftrightarrow MH_x + Q \tag{4.1}$$

where Q is the heat of reaction of hydride formation. The thermodynamic aspect of this reaction is conveniently described by pressure–composition isotherms (PCI). A schematic PCI curve is shown in Figure 4.1.

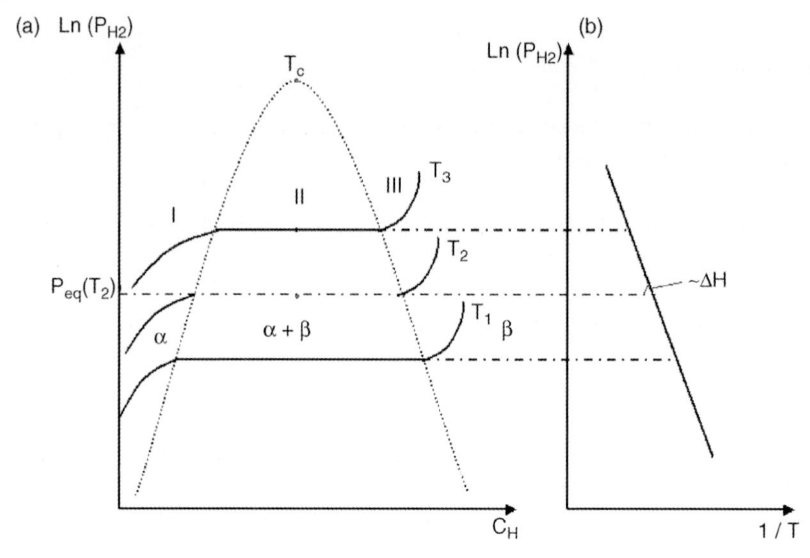

Figure 4.1 (a) Schematic of a pressure–composition isotherm. α is the solid solution of hydrogen and β is the hydride phase; (b) van't Hoff plot giving the enthalpy of hydride formation Δ*H*. (Adapted from [32]).

At low concentration ($x \ll 1$) hydrogen first dissolves in the metal lattice and forms a solid solution phase (α phase). Hydrogen is then randomly distributed in the metal host lattice and the concentration varies slowly with temperature. The α phase has the same crystal structure as the bare metal. The condition for thermodynamic equilibrium is given by:

$$\frac{1}{2}\mu_{H_2}(p, T) = \mu_H(p, T, c_H) \tag{4.2}$$

where μ_{H_2} is the chemical potential of molecular hydrogen, μ_H is the chemical potential of atomic hydrogen in solution in the metal and c_H is the hydrogen concentration ($c_H \equiv H/M$) where H and M are, respectively, the number of hydrogen and metals atoms in the unit cell.

As the hydrogen pressure increases, the concentration also increases until the attractive H–H interaction becomes important [32]. At this point, nucleation of a higher concentration phase (β phase) occurs. The system now has three phases (α, β and hydrogen gas) and two components (metal and hydrogen). From the Gibbs phase rule, the degree of freedom (f) is:

$$f = C - P + 2 \tag{4.3}$$

Where C is the number of components and P is the number of phases. Therefore, in the two-phase region the concentration increases while the hydrogen pressure is constant. Once the pure β phase is reached (complete disappearances of the α phase) the system has two degrees of freedom. Hydrogen enters into solid solution in the β phase and the hydrogen pressure again rises with concentration.

4.3.2
Low Concentration

At low pressures ($P < 100$ bar), hydrogen can be considered as an ideal gas and its chemical potential is given by [45]:

$$\frac{1}{2}\mu_{H_2} = \frac{1}{2}H^0_{H_2} - \frac{1}{2}TS^0_{H_2} + RT\ln p^{1/2}_{H_2} \tag{4.4}$$

Where $H^0_{H_2}$ and $S^0_{H_2}$ are, respectively, the standard state enthalpy and entropy.

On the other hand, the chemical potential of a dissolved H atom is [15]:

$$\mu_S = H_S - TS^{id}_S + RT\ln\left(\frac{c}{b-c}\right) \tag{4.5}$$

Where H_S is the enthalpy, S^{id}_S is the non-configurational part (vibration) of the entropy of hydrogen in solid solution. The third term is the configurational part of the entropy where b is the number of interstitial sites per metal atoms and c is the number of sites occupied by hydrogen atoms. According to Eq. (4.2), the equilibrium condition is then expressed as:

$$\Delta H_S - T\Delta S_S = RT\ln p^{1/2} + RT\ln\left(\frac{b-c}{c}\right) \tag{4.6}$$

Where

$$\Delta H_S = H_S - \frac{1}{2}H^0_{H_2} \tag{4.7}$$

$$\Delta S_S = S^{id}_S - \frac{1}{2}S^0_{H_2} \tag{4.8}$$

For very dilute solutions $c \ll b$ (region I of Figure 4.1), the equilibrium is described by Seivert's law which is simply Henry's law for a dissociating solute [46].

$$p^{1/2}_{H_2} = K_S \tag{4.9}$$

Where K_S is a constant given by:

$$K_S = \exp\left\{\frac{1}{RT}[\Delta H_S - T\Delta S_S - RT\ln b]\right\} \tag{4.10}$$

Seivert's law is valid because gaseous hydrogen can be considered an ideal gas and H_2 molecules are dissociated into atoms before becoming dissolved in metals [15]. Figure 4.2 presents the heat of dissolution for alkali, alkaline earth, and transition metals.

We see that the heat of solution has the same trend across the periodic table for the three rows of elements. This is an indication that the heat of solution is determined by the coarse electronic structure of the host metal [15]. In the case of the entropy term,

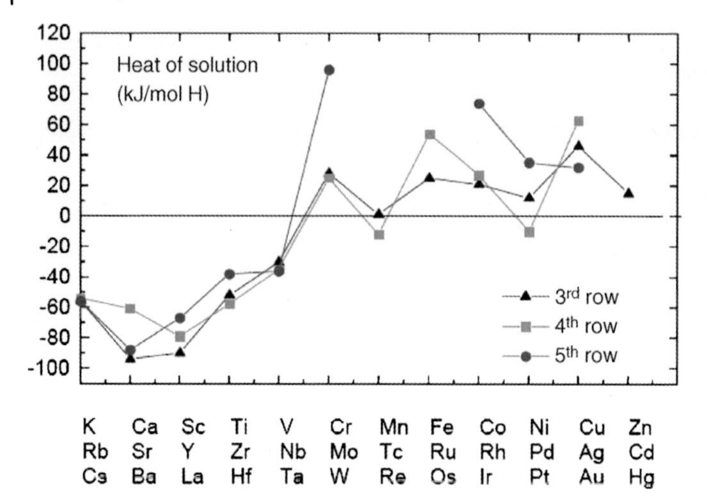

Figure 4.2 Heat of solution of hydrogen in alkali, alkaline earth, and transition metals. (From ref. [42, 47]).

the major contribution is from configurational entropy loss of gaseous hydrogen and is thus roughly constant for all metals ($\Delta S_s \approx -65$ kJ (mol H)$^{-1}$).

4.3.3
High Concentration

In general, the heat of solution is a function of the hydrogen concentration. In the preceding section we derived an expression for infinite dilution. We will now consider the case of high concentration (region II of Figure 4.1). We have seen that the phase rule requires the pressure to be invariant at any temperature below the critical temperature. The equilibrium pressure P_{eq} at the $\alpha \rightarrow \beta$ phase transformation is given by the van't Hoff law.

$$\ln P_{eq} = \frac{\Delta H}{RT} - \frac{\Delta S}{R} \tag{4.11}$$

where ΔH and ΔS are, respectively, the enthalpy and entropy of the $\alpha \rightarrow \beta$ transition. In general, the entropy change is mainly given by the loss of standard entropy of the hydrogen gas as it enters the metal lattice. This means that the entropy term does not depend significantly on the nature of the metal and that the ΔS term can be considered constant [43]. In the calculation of the enthalpy, the H–H interactions should now be taken into account. In the following, we assume that the enthalpy of formation (ΔH) of the β phase varies linearly with hydrogen concentration. The dependence of ΔH on x could be expressed as [15]

$$\frac{\partial \Delta H}{\partial x} = \left(\frac{\partial \Delta H}{\partial V}\right)_x \frac{\partial V}{\partial x} + \left(\frac{\partial \Delta H}{\partial x}\right)_V \tag{4.12}$$

The first term on the right-hand side is the elastic contribution and is expressed as [15]:

$$\left(\frac{\partial \Delta H}{\partial V}\right)_x \frac{\partial V}{\partial x} \cong -\gamma K_0 \frac{v_H^2}{v_0} \equiv -u_{els} \qquad (4.13)$$

where v_0 is the atomic volume, v_H^2 is the volume increase per hydrogen atom (usually between 2 and 3 Å^3 [48]), K_0 is the bulk modulus and γ is given by:

$$\gamma = \frac{2(1-2\sigma)}{3(1-\sigma)} \qquad (4.14)$$

where σ is the Poisson's ratio. Usually, K_0 is nearly constant with hydrogen concentration. Therefore, for a given metal, the elastic contribution could be considered as constant. The contribution of the elastic interaction on the heat of solution was first recognized by Alefeld and can be visualized in the following way [49, 50]. When a hydrogen atom is inserted into an elastic medium, it will produce local distortion of the lattice. The force exerted is generally repulsive, therefore the lattice will expand, decreasing the potential energy of the hydrogen atom and increasing the lattice energy of the lattice. This long-range displacement field falls off with $1/r^2$ and its symmetry depends on the crystal symmetry and also on the local symmetry of the H-atom site. When a second hydrogen atom is inserted, it enters into a pre-expanded lattice. The potential energy of this second hydrogen will therefore be lower than if the first hydrogen atom was absent. The potential energy of the crystal decreases as the two hydrogens get closer, until its short-range repulsion becomes important. The heat of solution is thus lowered in proportion to the hydrogen concentration. A calculation taking into account the discrete nature of the lattice has been made for Nb and Ta [51].

The second term of (4.12) which includes all volume-independent effects could be called the "electronic" contribution ($u_{e\,ln}$). This contribution will be discussed in more detail later. A simple interpretation is that the main part of $u_{e\,ln}$ comes from the extra electrons brought into the lattice by the hydrogen atoms. However, a close inspection of the heat of solution for a number of elements shows that the electronic contribution varies systematically with the group number and that it can be negative. This is in contradiction with the band-filling model [15]. Furthermore, at high hydrogen concentration the many-body interaction should be taken into consideration [52]. In a first principle investigation on alkali, alkaline earth, and transition metal hydrides, Smithson *et al.* [53] have shown that the electronic term is indeed negative and provides the stabilization for the hydride. They confirm that the electronic term becomes more positive across the transition-metal series. Their calculation also shows that the elastic term has approximately the same magnitude as the electronic one.

Before closing this section on thermodynamics, it should be mentioned that two schemes are used for modeling PCI curves [54]. In one of them, mathematical expressions based on the interstitial site occupation of hydrogen are used to calculate pressure as a function of concentration [55, 56]. The other is based on the phase transition between the α and β phases [54, 57, 58]. The advantage of the last scheme is that it can explain hysteresis in the PCI curve.

4.4 Intermetallic Compounds

4.4.1 Thermodynamics

Most of the natural elements can form hydrides. However, as shown in Figure 4.3, their dissociation pressures and temperatures are not suitable for most practical applications (0–100 °C and from 1 to 10 bar of hydrogen pressure). Only vanadium satisfies these criteria but its reversible hydrogen capacity is still too low. Therefore, alloys have to be developed with the goal being to meet the specific criteria of a given practical application. However, the alloy–hydrogen system is more complex than the metal–hydrogen system, mainly because the interstitial sites now have different chemical and geometrical configurations.

The first intermetallic compound that could reversibly absorb hydrogen (ZrNi) was reported by Libowitz *et al.* [60]. A magnesium-based hydride with relatively high capacity (Mg_2Ni) was reported by Reilly and Wiswall in 1968 [61]. Later, room-temperature hydrides such as the ternary hydrides $TiFeH_2$ and $LaNi_5H_6$ were

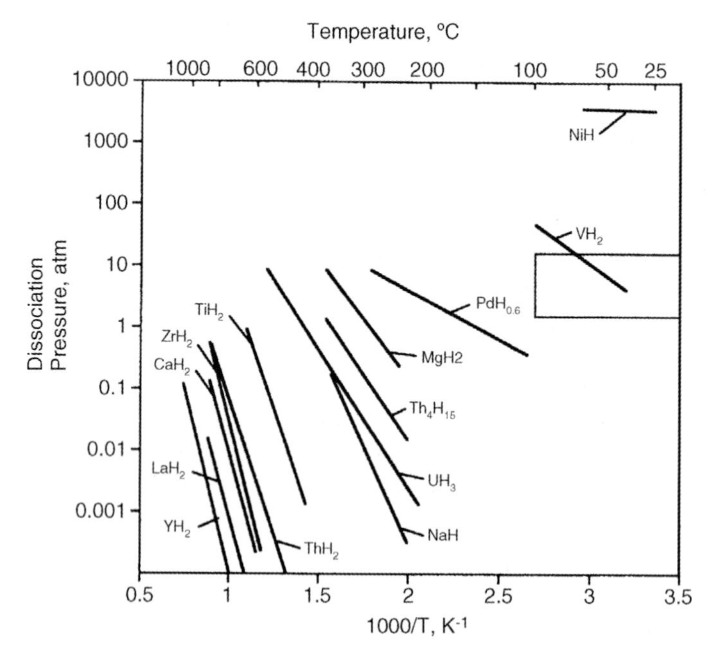

Figure 4.3 Dissociation pressure as a function of temperature for selected elemental hydrides. Limits of the inset box are 0–100 °C and 1–10 atm. From ref. [59].

discovered [62, 63]. Since then, a huge number of intermetallic hydrides have been investigated. Compared to pure metal hydrides, intermetallic hydrides are considered to have a wider range of hydride stability [43].

In most cases, intermetallic compounds are built by alloying a metal which easily forms stable hydrides (A) and another element which *does not* form stable hydrides (B). The intermetallics thus formed could then be grouped according to their stoichiometry such as: AB_5 ($LaNi_5$, $CaNi_5$), AB_2 ($ZrMn_2$, ZrV_2), AB (TiFe) and A_2B (Mg_2Ni).

In some cases, empirical rules can also relate thermodynamic properties to crystal structures. One of the best-known cases is in the AB_5 systems where the equilibrium pressure is linearly correlated to the cell volume. As the cell volume increases, the equilibrium plateau pressure decreases, following a $\ln P_H$ law [64]. However, some exceptions exist to this rule such as in $LaPt_5$ where electronic effects make the smaller unit cell more stable [65]. Nevertheless, generally, for intermetallic compounds the stability of the hydride increases with the size of the interstices [66]. A limitation of this empirical rule is that comparison between different types of intermetallics is impossible. For example, the stabilities of AB_2 alloys cannot be compared with those of AB_5 alloys [43].

In order to meet the requirements of a practical application, a metal hydride must first satisfy the thermodynamic requirements: operation temperature and hydrogen pressure. As the entropy term of Eq. (1.11) is effectively the same for all compounds, this means that the heat of formation (ΔH) is the principal parameter of a given alloy for hydrogen storage applications. Unfortunately, first principles calculations of ΔH for ternary alloys are still lacking [47]. However, semi-empirical models can be applied for some systems and give useful physical insight on the hydride formation. We will briefly discuss here the two principal semi-empirical models: Miedema's model and the semi-empirical band structure model.

4.4.1.1 **Miedema's Model**

Miedema and coworkers have developed a model which gives reasonably good agreement between calculated and experimental heat of formation [67–70]. Consider a binary alloy AB_n where element A easily forms a hydride and element B does not form a hydride. Element A forms a hydride following the reaction

$$A + (x/2)H_2 \leftrightarrow AH_x \tag{4.15}$$

At a given temperature T, this reaction has a hydrogen equilibrium pressure P'. Consider now a solid solution of this metal with a non-hydride forming metal (B). The relative partial molar free energy of A in the alloy is ΔG_A. If the resulting alloy AB_n reacts with hydrogen as

$$AB_n + (x/2)H_2 \leftrightarrow AH_x + nB \tag{4.16}$$

Then, the hydrogen pressure equilibrium of reaction (4.16) is given by

$$P = P' \exp\left(\frac{-2\Delta G_A}{xRT}\right) \tag{4.17}$$

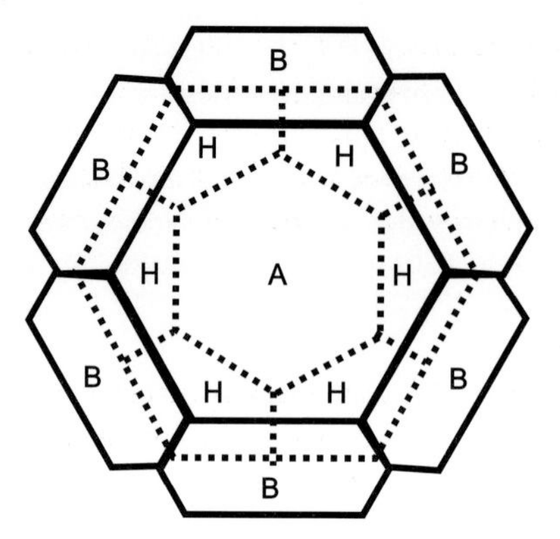

Figure 4.4 Schematic of unit cell of AB_n alloy with and without hydrogen. Atomic cells of hydrogen are indicated by broken lines. Adapted from refs [37, 70].

Since ΔG_A is always negative, P is always higher than P'. Therefore, the formation of a solid solution with a non-forming hydride has the effect of destabilizing the AH_x hydride. The above relation holds when the alloy decomposes upon hydrogenation. However, most of the time, the AB alloy will not decompose upon hydrogenation but instead react as a ternary hydride.

$$AB_n + xH_2 \leftrightarrow AB_nH_{2x} \tag{4.18}$$

In general, the plateau pressure of reaction (4.18) will also be much higher than in reaction (4.15). For example, the dissociation pressure of $ZrNiH_2$ is 10^{10} higher than for ZrH_2 [46]. In Miedema's model, it is assumed that, in a compound, the atomic cells of metals A and B are similar to the atomic cells of the pure A and B metals. A schematic representation of the unit cell is given in Figure 4.4.

Upon hydrogenation the hydrogen atoms will bond with an A atom but they will also be in contact with B atoms. The atomic contact between A and B that was responsible for the heat of formation of the binary compound is lost. The contact surface is approximately the same for A–H and B_n–H thus implying that the ternary hydride AB_nH_{2m} is energetically equivalent to a mechanical mixture of AH_m and B_nH_m [37]. More specifically, this could be explained by two terms: one is due to the mismatch of the electronic density of metals A and B at the boundary of their respective Wigner–Seitz cells, the other term is associated with the difference in chemical potential of the electrons in metals A and B. From these considerations, a semi-empirical relation for the heat of formation of a ternary hydride can be written as [70]:

$$\Delta H(AB_nH_{2x}) = \Delta H(AH_x) + \Delta H(B_nH_x) - \Delta H(AB_n) \tag{4.19}$$

Relation (4.19) is usually called Miedema's *rule of reversed stability* and states that the heat of formation of a ternary hydride is the difference between the sum of the heat of formation of the elemental hydride and the alloy enthalpy of formation. Because atom A is hydride forming, the first term of the right-hand side is negative and has a large absolute value while the second term is small (or even positive) and may be neglected. Because of the minus sign associated with the alloy enthalpy of formation, when the alloy is *more stable* (enthalpy more negative) the right-hand term becomes more positive and the ternary hydride becomes *more unstable*. For example, $LaCo_5$, being less stable than $LaNi_5$, the hydride of $LaCo_5$ is more stable than the corresponding $LaNi_5$ hydride. Thus, Miedema's rule is *less stable alloys form more stable hydrides*.

4.4.1.2 **Semi-Empirical Band Structure Model**

In this model, proposed by Griessen and Driessen [71], each metal is characterized by the difference between the Fermi level (E_F) and the energy of the lowest conduction band of the host metal (E_s). For a binary hydride, the heat of formation is then given by:

$$\Delta H = \frac{n_s}{2}[\alpha(E_F - E_s) - \beta] \tag{4.20}$$

Where, from fitting experimental values, the parameters α and β are determined to be equal to 29.62 kJ eV^{-1} $(molH)^{-1}$ and -135 kJ $(molH)^{-1}$, respectively [71, 72]. This model could also be applied to ternary hydrides with an agreement with experimental values that is generally better than Miedema's model. Figure 4.5 shows a comparison between experimental and calculated heats of formation for various ternary

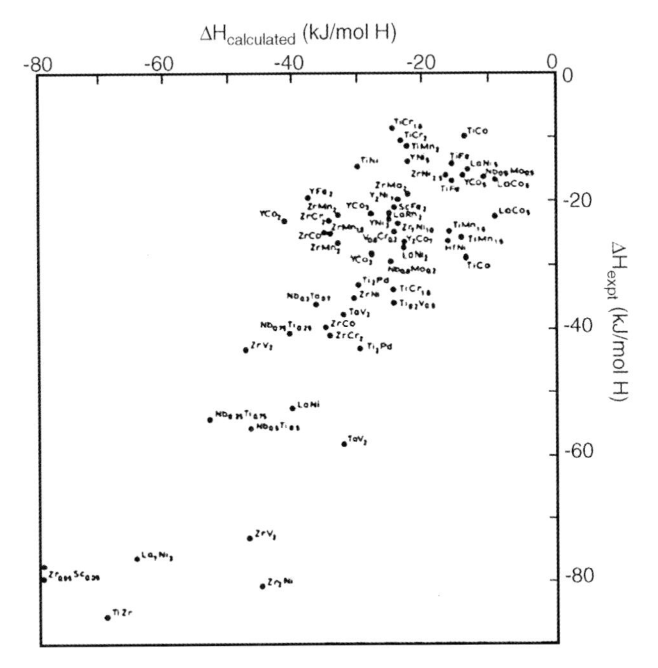

Figure 4.5 Measured heat of formation of ternary hydrides compared to values calculated with the semi-empirical band model From ref. [71].

hydrides [71]. The model gives a fairly good approximation of the experimental heat of formation.

Before closing this section it should be pointed out that the formation of most hydrides of intermetallic compounds is metastable with respect to disproportionation [35]. Again, taking $LaNi_5$ as an example, the disproportionation reaction:

$$LaNi_5 + H_2 \rightarrow LaH_2 + 5Ni \tag{4.21}$$

has a lower free energy than the hydride formation reaction:

$$LaNi_5 + 3H_2 \leftrightarrow LaNi_5H_6 \tag{4.22}$$

Nevertheless, because the disproportionation reaction requires rearrangement of metal atoms while in the hydride reaction the motion of atoms is minimal, the ternary hydride formation is kinetically favored at low temperature [35]. However, at high temperature (around 573 K [73]) the disproportionation reaction is more likely to occur.

4.4.2 Crystal Structure

Ternary hydrides have a diversity of crystal structures with various symmetries. Usually, upon hydrogenation binary alloys will experience lattice expansion and distortion but the crystal structure will stay the same. On the other hand, for some hydrides, such as TiFe, Mg_2Ni, $CaNi_5$ and $LaNi_5$, the crystal structure changes with hydrogen content. From geometrical considerations and calculations for various hydrogen concentrations in AB_2 hydrides of Friauf–Laves phases type, Westlake found the empirical rule that the minimum hole size is 0.4 Å while the minimum bond distance between hydrogen atoms is 2.1 Å [74, 75]. There are only a few exceptions to this rule and the reported discrepancies may be due to the difficulties of structure refinement using neutron diffraction. Gross *et al.* have shown, for $LaNi_5$ and $MgNi_3B_2$, that this criterion is a valuable tool in the search for hydride-forming intermetallic compounds [76].

On hydrogenation, hydrogen atoms will occupy specific interstitial sites. The interstitial sites in three major crystal structures of hydrides are shown in Figure 4.6. Only octahedral (O) and tetrahedral (T) sites are shown because they are the only ones occupied by hydrogen atoms. However, some distinction should be made between the different structures. In the fcc lattice, the T and O sites are, respectively, enclosed in regular tetrahedral and octahedral formed by metal atoms. At low or medium H concentration, the preferred interstitial sites are octahedral. In the hcp lattice, the tetrahedral or octahedral sites become distorted as the ratio of lattice parameters c/a deviates from the ideal value of 1.633. Tetrahedral sites are favored at low H concentration in hcp metals. In fcc and hcp lattices, for each metal atom, there is one octahedral site and 2 tetrahedral sites available for hydrogen after considering the Westlake criterion.

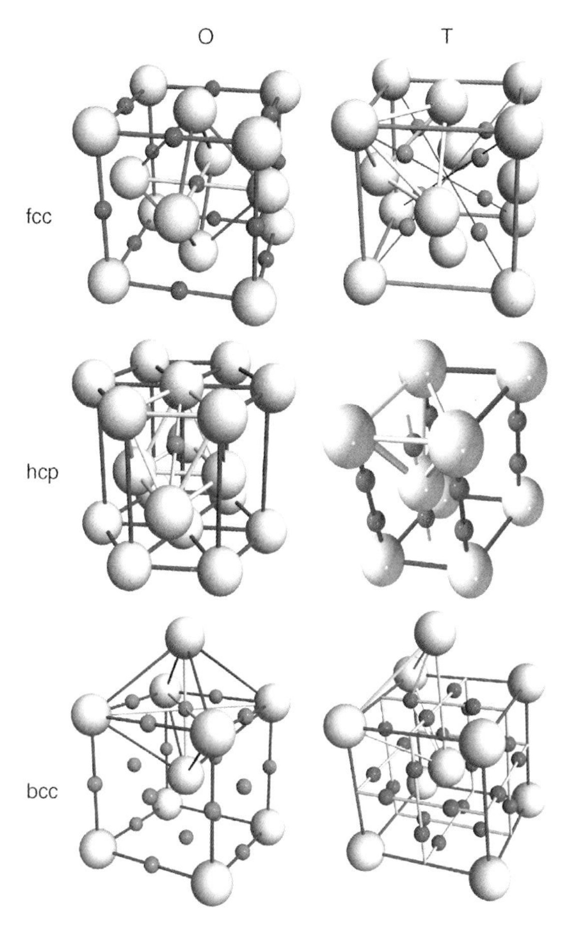

Figure 4.6 Interstitial octahedral (O) and tetrahedral (T) sites in face centered cubic (fcc), hexagonal close packed (hcp), and body-centered cubic (bcc) lattices.

In fcc lattices, the tetrahedral and octahedral sites are surrounded by regular polyhedra. For hcp lattices, the polyhedra are distorted if the ratio of lattice parameters c/a deviates from the ideal value of 1.633. In a bcc lattice, the polyhedra are greatly distorted. For the O sites, two metal atoms are much closer to the interstitial site than the other four metal atoms. Therefore, the O sites are subdivided into O_x, O_y, and O_z sites according to the direction of the fourfold symmetry axis. In the same way, the T sites of a bcc lattice are divided according to their symmetry axis and are denoted T_x, T_y, and T_z. The hydrogen atom has three octahedral and six tetrahedral sites available per metal atom. Hydrogen will preferentially occupy the tetrahedral sites in bcc metals [15]. Upon hydrogenation, the sites will be filled from the lowest to the highest energies. However, some sites will remain empty because of mutual repulsion between H atoms and reorganization of interstitial sublattices due to H occupancy [15].

4.4.3 Electronic Structure

The consequence of hydrogenation on the electronic structure can be broadly classified into four types of effects [32].

1) Modification of the symmetry of electronic states and reduction of the width of the bands due to expansion of the crystal lattice.
2) Appearance of a metal–hydrogen bonding band below the metal d-band. Electrons are transferred from the s–d band to this new band and some metal states could be pulled down below the Fermi level.
3) In hydrides that have more than one hydrogen atom per unit cell, the H–H interaction produces new attributes in the lower portion of the density of states.
4) General upward shift of the Fermi level due to the inequality between the additional electrons brought by hydrogen and the number of new electron states.

Density functional theory is used for band structure calculations of hydrogen storage materials. This method has been applied to a variety of hydrides such as AB_5 [77–79], AB [77], transition metals [53, 80], Laves phases [81], and complex hydrides [82]. Theoretical investigation is not only useful for the prediction of the heat of formation but it could also assess the elastic and mechanical properties of these materials, properties which are usually difficult to measure in the case of hydrides [78].

Recently, Grochala and Peters have shown that the thermal decomposition temperature of many binary hydrides could be correlated with the standard redox potential for the metal cation/metal redox couple and with the standard enthalpy of decomposition [18, 83]. They also showed that, for multinary hydrides, the decomposition temperature could be tuned by a careful choice of the stoichiometric ratio and of the Lewis acid/base character of the constituent elements.

4.5 Practical Considerations

As the object of this book is hydrogen storage, we will consider here some practical aspects of metal hydrides that are important for their large scale utilization. Obviously, the first consideration should be the synthesis of the alloy/hydride. If an alloy is synthesized, then the next problem to face is activation (i.e., first hydrogenation) which should be done in conditions as close as possible to the true thermodynamic equilibrium. For most uses, the absorption/desorption of hydrogen will have to be done in a small temperature/pressure interval, this is one of the main advantages of metal hydrides over high-pressure tanks. However, this condition means that the system should present a flat plateau and a small hysteresis. For the majority of applications, the most important parameter is the amount of hydrogen that can be reversibly stored and extracted from the metal hydride. Evidently, each specific application has a particular requirement in terms of hydrogenation/

dehydrogenation kinetics and life cycle. Finally, the resistance of the alloy to pulverization should be assessed. In the following sections each of these aspects will be discussed.

4.5.1 Synthesis

As more sophisticated metal hydrides are developed (nanocrystalline, multicomponent systems, composites and nanocomposites, graphite/metals or similar hybrid systems, clusters, etc.), it is important to be aware that, for practical applications, a large volume of material should be processed in a fast, inexpensive and reliable way, for example casting. Techniques such as cold vapor deposition may be impossible to scale up but this does not mean they should be discarded as a means of studying new metal hydrides. On the contrary, laboratory techniques allow much better control of the end product and permit the elaboration of new compounds. Once an attractive compound is found then another challenge will have to be faced: scaling up the synthesis. In this respect, it is important for the community of metal hydrides researchers to also study large-scale production techniques in order to make the transition from laboratory to industrial scale easier.

4.5.2 Activation

A very important problem for practical applications of metal hydrides is the fact that the surfaces of metals are usually covered with oxide of various thicknesses, depending on the formation process of each particular metal. This will most probably be the case for all industrial means of production of metal hydrides. This oxide layer acts as a hydrogen barrier and must be broken in order for the gaseous hydrogen to access the bare metal. Therefore, the first hydrogenation is usually performed (for conventional alloys) at high temperature and hydrogen pressure in order to "force" the hydrogen through the oxide layer. Upon hydrogenation, the lattice volume increases significantly while on dehydrogenation the lattice reverts to its original size. This expansion–contraction breaks the metal particles, exposing fresh metal surface and reducing particle size [43]. Significant anisotropic lattice strain and dislocations are also introduced by the activation process [84–87]. In the activation of $LaNi_5$, isotropic and anisotropic lattice strains are introduced only in the hydride phase while the solid solution phase strain is as small as in the unhydrided phase [87]. Moreover, the majority of dislocations are a-type edge dislocations (with Burgers vectors of the 1/3<12-10> types) on (10-10) [88]. The density of these dislocations is quite high and was evaluated to be of the order of $10^{12}\,cm^{-2}$ [88]. The hydride phase is considered to nucleate at lattice defects where the elastic strain is largely relieved [89].

For Mg_2Ni, Chen *et al.* [90] reported that activation is easier after mechanical grinding due to the increase in specific surface area and the creation of new defects. However, they found that mechanical grinding has no positive effect on the activation of $La_2Mg_{18}Ni$. For coarse-grained materials, there have been a number of methods designed to facilitate the activation process. For example, in fluorination treatment the surface of hydrogen storage alloys is modified by reaction in a fluorine-containing solution [91–93]. By removing the surface oxides, increasing the specific surface area and creating more catalytic sites on the surface, fluorination treatment can improve the activation and hydrogen sorption kinetics.

For practical applications, activation is an important factor to take into account while designing the tank system. If activation has to be performed in the tank then it has to be designed for a range of temperatures and pressures greater than the working conditions. On the other hand, filling the tank with activated metal hydrides is difficult since activated alloys are usually more sensitive to air contamination than unactivated metal hydrides.

4.5.3
Hysteresis

Hysteresis is a complex phenomenon in which the absorption plateau is at a higher pressure than the desorption plateau, thus forming a "hysteresis loop", as shown in Figure 4.7. There are, therefore, two sets of "thermodynamic" parameters which correspond to each plateau [41].

One explanation of hysteresis is that the accommodation of elastic and plastic energies is not equal in the hydrogenation and dehydrogenation processes [94–98]. Another model explains the hysteresis in terms of coherent strain [57, 58]. In this model, the metal–hydrogen system is treated as a partially open two-phase structure in equilibrium with a large source of interstitials (large open system), and a coherency strain between the two phases generates a macroscopic barrier. This macroscopic barrier cannot be overcome by thermal fluctuation, thus a metastable phase is locked until the chemical potential reaches a sufficient value to overcome this obstacle. The macroscopic barrier depends on the nature of the alloy and is independent of the way the transformation proceeds [58]. However, a model based on strain coherency implies that lattice expansion between dehydrided and hydrided phases is not too large, a fact not common for typical metal hydrides where expansions of the order of 20% and over are common. Recently, Rabkin and Skripnyuk proposed a pseudo-two-phase equilibrium concept where coexistence between a dehydrided matrix and hydrided particles is possible [99]. This model predicts a decrease in hysteresis with decrease in powder particle size.

Quantitatively, hysteresis is represented by the free energy difference:

$$\Delta G_{H_2}(\text{hyst}) = RT\ln(P_A/P_D) \qquad (4.23)$$

where P_A and P_D are the absorption and desorption pressures, respectively. There are some views that P_D represents the "true" equilibrium plateau pressure, but Flanagan *et al.* provided evidence that the equilibrium pressure lies somewhere between P_A and P_D [41]. Sandrock *et al.* pointed out that hysteresis is not a unique material property and depends on the sample's history and on the test procedure used [100]. For practical applications, hysteresis is an important feature, because it

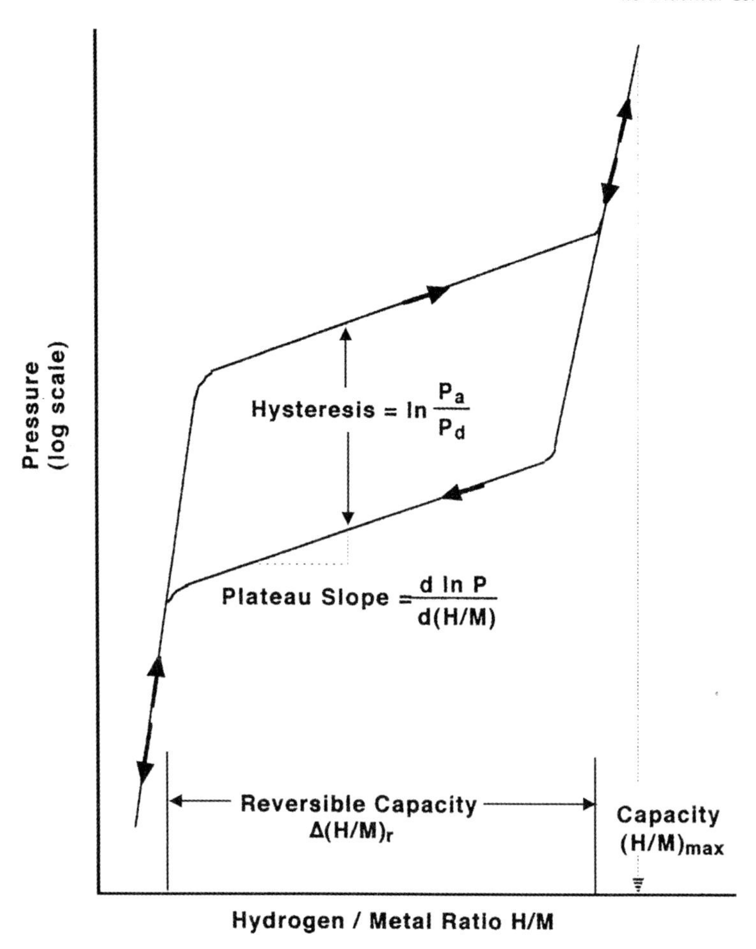

Figure 4.7 Schematic isothermal pressure composition hysteresis loop. From ref. [59].

has an important impact on the service pressure of the storage tank. In fact, it could translate to a loss in the efficiency of the material due to irreversible deformation during absorption and desorption of hydrogen. In most applications, hysteresis should be as small as possible. This can be achieved by element substitution and heat treatment.

4.5.4
Plateau Slope

Another important characteristic of a "real" PCI curve is the plateau slope which is usually represented by the relation [100]:

$$\text{slope} = \frac{\mathrm{d}(\ln P)}{\mathrm{d}(H/M)} \tag{4.24}$$

The plateau slope is not necessarily constant over the whole length of the plateau. The major cause of plateau slope can be attributed to compositional inhomogeneities [56, 101, 102]. For arc-melted alloys, heat treatment could reduce the plateau slope. Usually, the pressure in the middle of the plateau is used to characterize the alloy. As with hysteresis, sloping plateaus have an important technological impact on some applications and reduction of the plateau slope is advisable.

4.5.5 Reversible Capacity

For practical applications the important factor is the reversible capacity $\Delta(H/M)_r$ which is defined as the plateau width. Reversible capacity can be considerably less than the maximum capacity $(H/M)_{max}$ [59]. A classic example is vanadium hydride. At 353 K vanadium forms the so-called monohydride VH_x ($x \approx 1$) at hydrogen pressure of the order of 10^{-6} to 10^{-7} MPa. At much higher pressure (about 1 MPa), the dihydride VH_x ($x \approx 2$) is formed. For practical systems, dehydrogenation of the monohydride is impossible and we could say that the vanadium reversible capacity is between $VH_{\approx 1}$ and $VH_{\approx 2}$. Hydrogen capacity can be reported either in terms of the atomic ratio (H/M) or in weight percent (wt.%). The atomic ratio is the number of hydrogen atoms stored with respect to the number of metallic atoms. Weight percent is the weight of hydrogen stored over the weight of the hydride phase. Another way to report hydrogen capacity that is not often quoted, but has an important technological impact for mobile and portable applications, is the volumetric capacity expressed by the number of hydrogen atoms per unit volume. Usually, the reversible capacity (ΔN_H) and crystal volume are used to calculate this number. This definition does not include the void volume in the crystal (always present to some extent) and, thus, should be seen as an upper value.

4.5.6 Hydrogenation Kinetics

Hydrogenation is a complex process in which a molecule in the gas phase is split into individual atoms which are then bonded in the crystal lattice of the metal hydride. The overall kinetics is limited by the slowest step, the so-called rate-limiting step. For laboratory work and development of new alloys, it is advisable that heat and mass transport do not constitute the rate-limiting step [103]. This means that great care should be given to the design of the sample holder in order to ensure maximum heat conduction and minimal hydrogen flow resistance. On the other hand, for practical applications, heat and mass transport will be of the utmost importance. As an example, let us consider the hydrogenation of magnesium. Currently, the fastest kinetics are achieved by ball-milled MgH_2 composites doped with 1 mol% Nb_2O_5 which, at 150 and 250 °C, can absorb more than 5.0 wt.% within 30 s [104]. This means that a power of 2 kW per mole of hydrogen has to be removed from the alloy in order to keep it at constant temperature. This is a large amount of heat that has to be transferred out of the storage tank. Even for absorption times of 300 s, which is roughly the charging time asked for by the industry, and for a tank storing 5 kg of hydrogen, this means 500 kW of heat power! This simple calculation indicates that for practical applications the hydrogenation kinetics will be controlled by heat transfer, at least for the hydrides with large heats of formation.

For metal hydride characterization, the kinetics curve for a given reaction is the transformed fraction versus time. The dependence of this curve on pressure and temperature should be investigated in order to deduce the rate-limiting steps of the reaction. In the case of hydrogen–metal reaction, the situation is particularly complex because of the heat of reaction, relatively fast reaction rates, poor thermal conductivity of the hydride phase, and embrittlement of the products [103]. Sample size, particle size, surface properties, and the purity of the solid and gas phases are also factors that could have an important impact on the kinetics of a given sample. Moreover, Mintz *et al.* [105] argued that the determination of the intrinsic rate-limiting step could only be performed on massive samples and not on powders. All these facts make the comparison of different experiments a difficult task that should be done with great care.

A detailed description of the ways to analyze kinetic curves and to extract the rate-limiting steps is outside the scope of this section. For in-depth treatment, the interested reader can consult various references [106–117] and the classic book of Christian [118].

4.5.7 Cycle Life

For practical applications it is essential that the sorption properties (capacity, kinetics, reversibility, plateau pressure, etc.) of the hydrogen storage system stay relatively constant during the whole life of the device. Different applications will need a different number of cycles but usually, for most applications, the number of cycles should be between a few hundreds to many thousands. Cycling has been studied mainly for magnesium-based and AB_5 compounds [73, 119–126]. Cycling life depends on many parameters. An important one is, as seen in Section 4.4.1, that hydrides of intermetallic compounds are metastable with respect to disproportionation, thus leading to short cycle life. Other reasons for the reduction in cycle life are: impurities in the hydrogen gas leading to poisoning [127–131], agglomeration of particles [132], structural relaxation and crystalline growth [133], phase change and/or formation of non-hydride phases [134, 135]. Nevertheless, magnesium and magnesium-based compounds could sustain up to a few thousand cycles without drastic changes in hydrogen sorption properties [133, 136–138]. A special way to improve cycling stability is by element substitution [139–145].

4.5.8 Decrepitation

Decrepitation is the self-pulverization of alloy particles into smaller-size powder because of volume change upon hydrogenation and the brittle nature of hydriding alloys [59]. This could change the packing of a material which in turn will change the heat transfer and gas flow inside the storage tank. This effect is amplified by the fact that the intrinsic heat conductivity is usually different for the hydrided and dehydrided state. Decrepitation could be a serious problem because the small particles will fall to the lower part of the tank and, upon hydrogenation, could expand to such an extent as to lead to tank rupture.

4.6 Metal Hydrides Systems

In this section, some metal hydrides systems will be reviewed with a special emphasis on nanocrystalline materials and gas-phase reaction. A similar review for polycrystalline materials has been made by Sandrock [146].

4.6.1 AB_5

Literature on AB_5 alloys is abundant, especially because this class of alloy has interesting electrochemical properties. A wide range of AB_5 compounds can be synthesized because it is relatively easy to substitute elements on the A and B sites. Element A is usually one of the lanthanides, calcium, yttrium or zirconium but for industrial applications mischmetal is mostly used. Mischmetal is a generic name used for an alloy of rare earth elements in various naturally-occurring proportions. The composition varies according to the source but it typically includes approximately 50% cerium and 45% lanthanum, with small amounts of neodymium and praseodymium. The B site is mainly nickel but substitution with other transition elements such as Sn, Si, Ti or Al is common. By substitution on the A and B sites hydrogen storage properties such as plateau pressure and slope, hysteresis, resistance to cycling and contamination can be controlled. Unfortunately, in many instances improving one property may lead to deterioration of another. Thus, optimization by multiple substitution is an active field of research of AB_5 alloys.

As shown in Figure 4.8, ball milling drastically reduces the hydrogen storage capacity of $LaNi_5$ but it can be recovered by annealing [147]. On the other hand, mechanical alloying seems to be a suitable procedure for the synthesis of nanocrystalline AB_5 alloys [148–150]. When milling is performed under an inert atmosphere, hydrogen can be readily absorbed if the milled powder is not exposed to the air [151]. Compared to the unmilled alloy, ball-milled $LaNi_5$ is easier to activate and has faster first hydrogenation kinetics [152]. Moreover, because of the large difference in melting point between La and Ni, mechanical alloying could be a good method for the synthesis of $LaNi_5$ compounds [153].

The plateau pressure increases for nanocrystalline materials with a reduction in hydrogen capacity and a sloping plateau [152, 154, 155]. In reference [154] a high energy shaker mill (SPEX 8000) was used, while for references [152, 155] milling was performed on a Fritsch P7 planetary mill. However, because of batch sizes, grinding

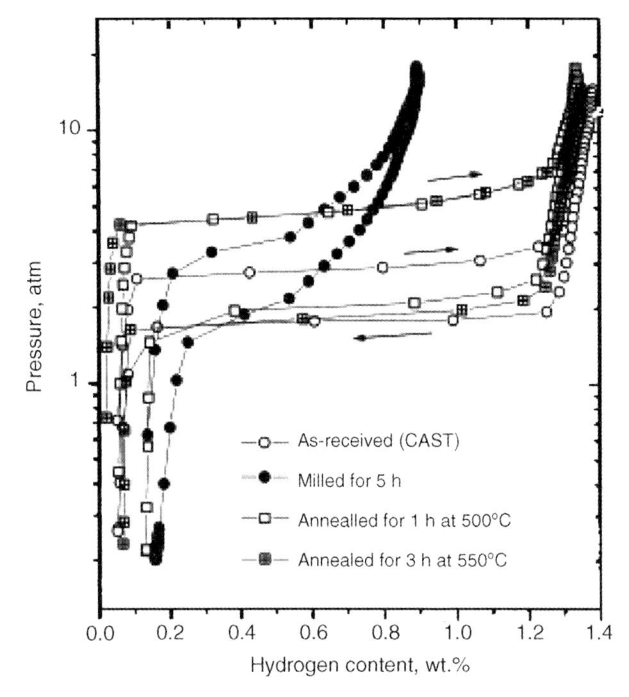

Figure 4.8 PCI curves of $LaNi_5$ after different treatments. From ref. [147].

balls and milling speed, the milling in [155] was more energetic than in [152]. This explains why in [155] the loss of capacity is more important and the plateau slope is greater than in [152]. Fujii *et al.* concluded that no improvement in hydrogen sorption properties is expected from nanocrystalline $LaNi_5$ because of the formation of a too stable hydride in the grain boundaries [155]. A systematic investigation of the effect of milling on a multi-substituted alloy has been performed by Ares *et al.* [156].

4.6.2 TiFe

TiFe was recognized as a metal hydride by Reilly and Wiswall [62]. Although it is a good candidate for hydrogen storage applications despite its rather low total hydrogen capacity (1.9 wt.%), one of the major drawbacks is the need for activation under drastic conditions. Usually, activation is done by heating the alloy at 623–673 K under vacuum or hydrogen, and hydrogenation starts when the alloy is cooled to room temperature [157]. This alloy seems to be the one used in the hydrogen tanks of fuel cell powered submarines [158].

Early work on ball-milled TiFe has shown that nanocrystalline alloys are easier to activate than conventional alloys [159, 160]. The final state after ball milling depends on the oxygen content. High O_2 content (>2.9 atom%) leads to formation of an amorphous TiFe phase [159]. If the oxygen content is larger than 5 atom%, a Ti-rich amorphous phase is formed with imbedded unreacted Fe residual particles [161]. A similar result was reported by Sun *et al.* [162]. Study of nanocrystalline TiFe prepared by ball-milling has shown that this material tends to be oxidized easily, thus forming TiO_2, Fe_2TiO_5, and iron clusters on the nanoparticle surface [163]. Milling intensity is also an important factor for the achievement of a specific structure: at low milling energy the intermetallic compound $Fe_{50}Ti_{50}$ is synthesized while at higher milling intensities a partly amorphous material is formed [164]. The activation step has been shown to be unnecessary when TiFe is milled under argon with addition of a small amount of nickel [165]. Beside being easier to activate, nanocrystalline TiFe has much higher hydrogen solubility at low pressure than conventional TiFe [162, 166]. The simpler activation process was related to the presence on the surface of metallic iron which acts as a catalyst for the dissociation of the hydrogen molecule [163]. Ball-milled nanocrystalline TiFe is formed by two components: crystalline nano-grains and highly disordered (amorphous) grain boundaries [167]. Addition of a small amount of palladium ($<$1 wt.%) eliminates the need for activation and permitted the study of the relaxation effect in nanocrystalline materials [168]. In nanocrystalline TiFe with palladium catalyst, it was found that structural relaxation leads to an increase in the solubility gap and a change in the slope of the plateau in the PCI curve [169]. Tessier *et al.* have shown that the absorption plateau of nanocrystalline $Fe_{50}Ti_{50}$ is narrower and at a lower pressure than the coarse-grained TiFe [170]. The γ phase was also absent in the nanocrystalline $Fe_{50}Ti_{50}$ hydride. They investigated two possible explanations for these effects: elastic stress from the amorphous phase produced during milling and chemical disorder. The spherical shell model of the elastic stress was in closer agreement with the experimental results [170]. A recent investigation by Abe and Kujii showed that TiFe could be synthesized by milling Ti and Fe, followed by annealing at a temperature higher than 773 K [171].

4.6.3
AB_2 Laves Phases

Laves phases consist of three structure types, which are named after the representatives cubic $MgCu_2$ (C15), hexagonal $MgZn_2$ (C14) and hexagonal $MgNi_2$ (C36). Laves phases form the largest group of intermetallics and thus have a wide range of properties. Their stability depends on various factors such as: geometry, packing density, valence electron concentration, or the difference in electronegativity. However, these concepts are of limited value for the prediction of new alloys [172].

Laves phases have been recognized to be attractive hydrogen storage materials, particularly the Zr-based alloys [173–176]. They have relatively good hydrogen storage capacity and kinetics, long cycling life and low cost but they are usually too stable at room temperature and are sensitive to gas impurities [19, 177, 178]. In order to achieve optimum hydrogen storage properties, multicomponent systems $Zr_{1-x}T_x(Mn,Cr)_{2-y}M_y$ where T = Ti, Y, Hf, Sc, Nb and M = V, Mo, Mn, Cr, Fe, Co, Ni, Cu, Al, Si, Ge are usually used. It has been shown in the $Zr_{1-x}Ti_xCr_{2-y}M_y$ system that single phase alloys absorb more hydrogen but have slower kinetics than multiphase alloys [179]. In an attempt to synthesize an alloy containing magnesium and zirconium, Cracco and Percheron-Guégan ball-milled a mixture of the pre-alloy $Zr_{0.9}(Mn_{0.6}Ni_{1.15}V_{0.2}Cr_{0.1})$ with Mg_2Ni [180]. They found that magnesium does not enter the pre-alloy cell but instead forms a nanoscale mixture with the pre-alloy and improves the hydrogen capacity as well as the hydrogenation kinetics [180].

4.6.4
BCC Solid Solution

It was reported that Ti-based Laves phase alloys of AB_2 type are multiphase alloys consisting of BCC and Laves phases, where both phases contribute to hydrogenation [181–183]. In fact, the BCC and Laves phase showed the same equilibrium pressure. For this reason, this new class of alloys has been called *Laves phase-related BCC solid solution* [184, 185]. Early works on $Zr_{0.5}Ti_{0.5}VMn$ have shown that this alloy is formed of three phases: a matrix C14 (composition $Zr_{0.6}Ti_{0.4}V_{0.9}Mn_{1.1}$), BCC colonies and small α-ZrO_2 particles [181, 182]. In subsequent work on the Ti–V–Mn system, microstructure analysis by TEM revealed that the BCC phase consisted of two nanoscale phases with a fine lamellar structure of 10 nm thickness [186, 187]. Satellites in the electron diffraction pattern led to the conclusion that these lamella structures were formed by spinodal decomposition. Hydrogen capacity higher than 2 wt.% was found for the system Ti–Cr–V. Furthermore, this system has a smaller hysteresis than Ti–V–Mn and needs only one activation cycle [188]. Usually, BCC solid solution alloys form two types of hydrides: the first (approximately a monohydride) is very stable and usually cannot be desorbed under practical conditions; the second (dihydride) is mainly responsible for the reversible capacity. Therefore, the challenge is to destabilize the first hydride or increase the intrinsic reversible capacity of the dihydride. An example is given in Figure 4.9 for the Ti–xCr–20V system ($x = 52$–62) [189]. It can be seen that a slight variation of x has a large impact on the PCI curve. Tamura *et al.* found that for Ti–Cr–V systems the monohydride has an fcc or an hcp structure while the dihydride takes an fcc structure with different lattice parameters to those of the monohydride fcc [189, 190].

Recently, many groups have reported BCC alloys of various compositions which could store hydrogen with a maximum capacity close to 4 wt.% and with reversible capacities of more than 2 wt.% [189, 191–193]. It has been shown that the lattice parameter plays an important role in the sorption properties. For example, in Ti–V–Cr–Fe alloy the maximum reversible capacity is reached for a lattice parameter of 3.036 Å [192]. However, the electron to atom ratio e/a (where e is the number of valence electrons and a is the number of atoms) also plays an important role. The critical value seems to be $e/a = 5.25$, for values higher than this the hydrogen capacity decreases rapidly [194]. Because they contain a high proportion of vanadium, these alloys are relatively costly. In order to reduce the price, the effect of Fe substitution in Ti–Cr–Mn–V alloy has been investigated by Yu *et al.* [195]. A more drastic step has been taken by Taizhong *et al.* who replaced vanadium by ferrovanadium (FeV) [196–198]. They reported that the $TiCr_{1.2}(FeV)_{0.6}$ composition, which has BCC solid solution as the matrix and a C14 Laves phase as the second phase, could reach nearly 3.2 wt.% of hydrogen storage capacity and 2.0 wt.% of reversible hydrogen storage capacity.

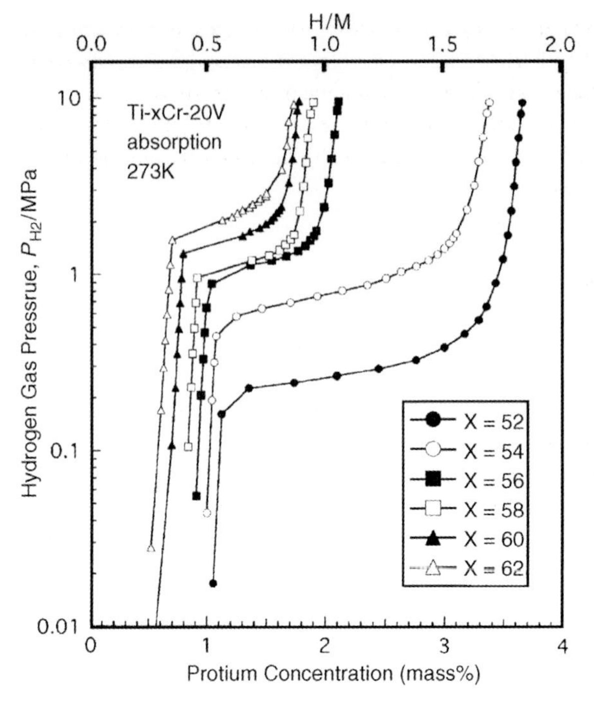

Figure 4.9 PCI absorption curves at 273 K of Ti–xCr–20V ($x = 52$–62) alloys. From ref. [189].

4.7
Nanocrystalline Mg and Mg-Based Alloys

Because of its low cost and large hydrogen storage capacity magnesium is an attractive medium for storing hydrogen. Even though magnesium hydride was observed in the late nineteenth century, the thermodynamic parameters of pure magnesium were reported only in 1955 [199]. Magnesium is hard to activate, its temperature of operation is high (of the order of 573–673 K) and the hydrogen absorption/desorption kinetics are slow. Moreover, due to the low melting point and high vapor pressure of magnesium, preparation of Mg-based alloys by conventional metallurgical procedures is difficult [200]. The literature on Mg and Mg-based alloys is quite important and an exhaustive study of all aspects related to hydrogen storage is impossible. More details can be found in recent reviews [19, 30, 31, 201, 202]. Basically, improvement of hydrogen sorption properties could be achieved by adding a catalyst to make the reaction fast enough for practical applications and/or by changing the thermodynamics of the reaction in order to have a temperature of operation in the desired range. Evidently, these improvements should not be associated with a drastic loss of hydrogen storage capacity. To meet these conditions, the scientific and technical challenges are quite important. However, spectacular improvements in the catalytic aspect have been achieved, particularly with the widespread use of mechanical milling and alloying, which bring nanocrystallinity and higher specific surface area (reduced powder size). Unfortunately, the same type of improvement has not been seen on the thermodynamic side. In this section, we will briefly expose the present state of research on catalytic and thermodynamic aspects of magnesium-based systems. We will close this section with a discussion on new synthesizing techniques.

4.7.1
Hydrogen Sorption Kinetics

First reports on the effect of ball milling magnesium and magnesium hydride showed enhanced hydrogenation kinetics [203–205]. Figure 4.10 illustrates the drastic effect of milling on the hydrogenation kinetics of magnesium hydride. This improvement was explained by an abundance of defects that act as nucleation sites for the hydride phase and by grain boundaries that facilitate hydrogen diffusion in the matrix. Increases in specific surface area also play a significant role [205, 206]. In a series of papers by different research groups, it was shown that particle size is probably the dominant factor of the enhanced kinetics of ball-milled magnesium hydride [207–213].

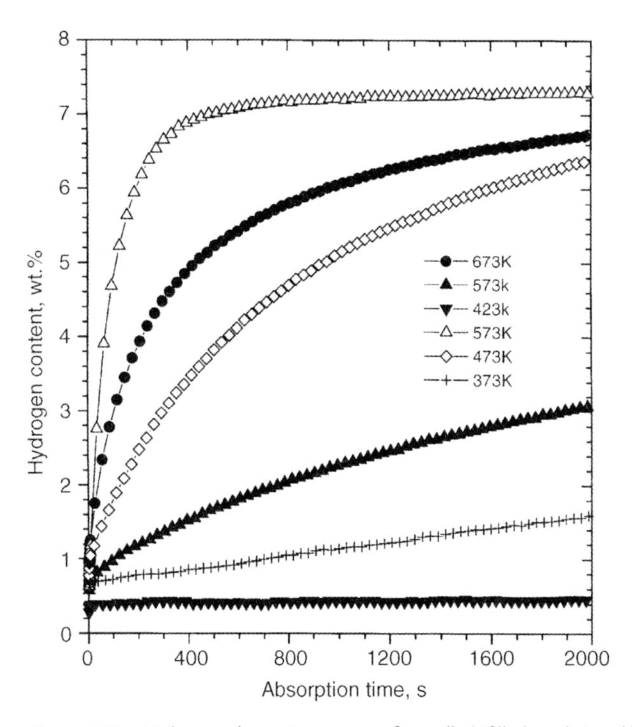

Figure 4.10 Hydrogen absorption curves of unmilled (filled marks) and ball-milled (hollow marks) MgH_2 under a hydrogen pressure of 1.0 MPa. From ref. [205].

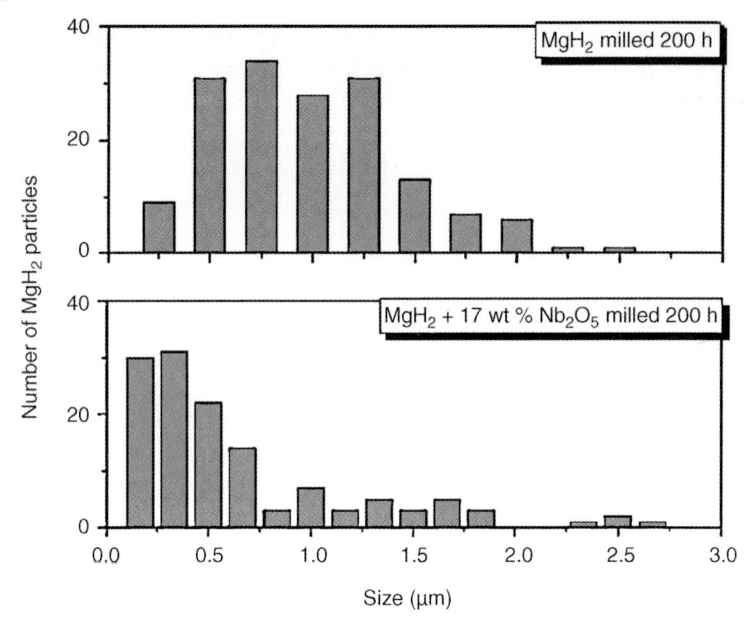

Figure 4.11 Distribution of MgH_2 particle size after 200 h of ball milling with and without Nb_2O_5. Milling performed on a Fritsch P5 planetary mill. From ref. [212].

In order to further increase the hydrogen sorption kinetics, a wide range of catalysts has been tested, such as palladium and other transition metals [204, 214–220], or metal hydrides such as $LaNi_5$ and TiFe [221–223]. In fact, it has been observed that milling magnesium with metal oxides gives as good and even faster hydrogenation kinetics than their metallic counterpart [224, 225]. The literature on this subject is abundant but, recently, it has been recognized that milling magnesium with a metal oxide gives a smaller average particle size than when the milling is performed with a metal [212, 213]. This is clearly seen in Figure 4.11 from the paper of Aguey-Zinsou *et al.* [212] which compares the particle size distribution of magnesium hydride milled for 200 h with and without Nb_2O_5. Moreover, the authors showed that if milling of pure MgH_2 is performed for 700 h the particle size distribution is similar to that of the hydride milled for 200 h with Nb_2O_5 additive. This gave both samples identical hydrogen sorption kinetics. The reduction in particle size is attributed to the role of Nb_2O_5 as a "process control agent" which act as a lubricant, dispersing and cracking agent during milling [212].

However, a true catalytic effect is most probably present in transition-metal doped magnesium. A proof of this is the fact that nanostructured catalyst gives enhanced sorption properties compared to its micro-sized counterparts [226–229]. Hanada *et al.* also showed that after milling the catalyst is homogeneously distributed on a nanometer scale [230]. A possible interpretation of the catalytic effect may be the appearance of ternary magnesium-niobium oxides, which was evidenced by TEM [229, 231] and neutron diffraction [232]. Despite the abundant literature on this subject, more research is needed in order to elucidate the relative effects of particle size, presence of defects, grain boundaries, ternary magnesium-metal oxides, impurities, cycling, and so on. Particular attention should be paid to the thermal history of the compound as exposure to high temperature may change the nanocrystallinity and chemical composition [210].

4.7.2 Reduction of the Heat of Formation

The high temperature of operation of pure magnesium is the main problem for hydrogen storage applications. Therefore, the natural solution is to reduce this temperature by reducing the heat of formation. From Miedema's law of reversed stability (Section 4.4.1.1) a ternary hydride will be less stable (lower temperature of operation) if the binary alloy is stable. The classical example is Mg_2Ni, where the heat of formation of the ternary hydride is reduced to $-64.5\,kJ\,mol^{-1}$ compared to $-74.5\,kJ\,mol^{-1}$ for pure magnesium. However, this changes only slightly the temperature of operation but gives a drastic reduction in hydrogen capacity (3.6 wt.% for Mg_2Ni compared to 7.66 wt.% for Mg). There have been some efforts towards the synthesis of magnesium-based intermetallic alloys but no compound has been found that shows high reversible capacity at an acceptable temperature of operation. As mentioned by Dornheim *et al.* [201], the chances of finding a new Mg-based intermetallic phase that forms a multicomponent hydride are small [201].

As magnesium is immiscible with many late-transition metals, from Miedema's principle the fact that the intermetallic is unstable means that the ternary hydride will be stable. Here, hydrogen effectively acts as a binding element. An example is the formation of Mg_2FeH_6 [233]. This class of metal hydrides could be synthesized by high hydrogen pressure [234], thin films [235] or ball milling [236–239]. In the case of thin films, metastable phases could be stabilized by the high stresses due to the clamping of the material to the substrate [201, 240, 241].

A very interesting new field of study is destabilized systems [242, 243]. As this is the subject of a separate chapter of this book we will only briefly mention here two magnesium-based systems. In the Mg–Al system it has been shown that, under hydrogen, the alloy $Mg_{17}Al_{12}$ undergoes reversible disproportionation reactions with formation of MgH_2 at each step [244–246]. Recently, the same behavior was found in the Mg–Pd system where it was found that Mg_6Pd absorbs hydrogen reversibly in a three-step disproportionation reactions with formation of MgH_2 at each step [247–249].

Up to now, research on ternary metal hydrides based on magnesium and alkaline or alkaline earth metals has not produced alloys with practical hydrogen storage characteristics [250]. However, study of quaternary alloys is a new field that is worth investigating and could give practical hydrogen storage systems [250].

Therefore, we see that despite the fact that destabilization of magnesium by forming intermetallic alloys may be an overwhelming challenge, other avenues are possible and have already shown interesting results. Again more research is needed in this direction.

4.7.3
Severe Plastic Deformation Techniques

In severe plastic deformation (SDP) processes the metallic material is plastically deformed up to very high strain values. This leads to a subdivision of the initially coarse grained microstructure into a granular-type nanostructure containing mainly high angle grain boundaries [251, 252]. Typical SDP techniques are: equal channel angular pressing (ECAP) where a billet is pressed through a die with two channels of equal cross-section intersecting at a certain angle; high-pressure torsion (HPT) where a disk-shaped sample is deformed by pure shear between two anvils that are rotating with respect to each other; multiple forging or cyclic channel die compression (CCDC) which is a variant of ECAP where a change in the axis of the applied strain load is made on successive pressings. The common feature of all these SPD techniques is that the cross-section of the material remains constant during or after SPD processing. Many other SDP techniques are available [252].

A closely related technique to the SPD techniques, and which is sometimes included with them, is cold rolling. Here, the metal is deformed by passing it through rollers at a temperature below its recrystallization temperature. Here, contrary to the other SPD techniques, the cross-section of the material changes after rolling and the grain boundaries have a low angle misorientation [251]. However, for our purpose, all these techniques will be treated as SPD.

The advantages of SPD techniques over high energy milling are a lower concentration of impurities and a lower production cost for large quantities [253].

Application of SPD techniques to metal hydrides is a new field of research. Skripnyuk and Rabin were the first to use ECAP to improve the hydrogen storage properties of Mg-based alloys [254, 255]. The first study on ECAP-processed magnesium alloy ZK90 showed an improvement in sorption kinetics without loss of hydrogen capacity or change in thermodynamic parameters [254]. Figure 4.12 shows comparable results for the Mg–Ni eutectic alloy [255].

Kusadome *et al.* studied the effect of HPT on the hydrogen storage capability of $MgNi_2$, an alloy that is well known for not absorbing hydrogen [256]. They found that a slight amount of hydrogen (0.1 wt.%) could be absorbed in the grain boundaries created by HTP.

In the case of cold rolling only a few studies have also been reported on its use with metal hydrides. In an earlier study of the effects of cold rolling on the hydrogen sorption properties of Ti-22Al-27Nb alloy, Zhang *et al.* found that hydrogen absorption and desorption properties are improved by small deformations [257, 258]. However this improvement is lost upon hydrogen cycling. For a magnesium-based system, the first report was made by Ueda *et al.* who studied the effect of cold rolling followed by heat treatment on a Mg–Ni laminated composite [259]. Interdiffusion by heat treatment resulted in the formation of the intermetallic compound Mg_2Ni and

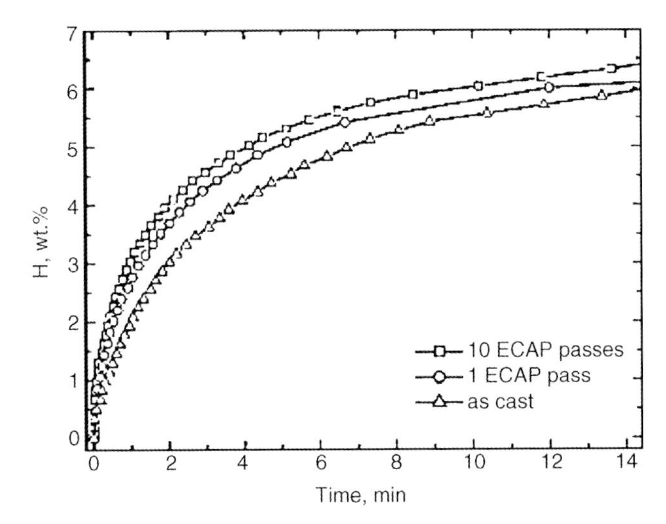

Figure 4.12 Hydrogen desorption kinetics at 573 K of as-cast and ECAP-processed $Mg_{89}Ni_{11}$ eutectic alloy. From ref. [255].

exposure to hydrogen led to complete hydrogenation to Mg_2NiH_4. The same group showed that superlaminated composites Mg/Ti/Ni could absorb hydrogen [260], and that $Mg_{17}Al_{12}$ could be synthesized by a combination of cold rolling and heat treatment [261]. Dufour and Huot have shown for the Mg–Pd system that, compared to ball-milled samples, composites made by cold rolling have an enhanced resistance to air contamination and a much faster activation process [249]. It seems that SPD techniques and cold rolling could be applied for the synthesis and modification of metal hydrides. Preliminary results are encouraging and show that these methods could produce enhanced hydrogen storing materials at low cost.

4.8 Conclusion

In this chapter, the fundamental characteristics of metal hydrides as hydrogen storage materials have been reviewed. Their potential for practical applications is well known but more research need to be done in order to find a system that will meet the industry's criteria. This research could be broadly divided in the following way:

4.8.1 Alloys Development

This comprises both new catalysts and new alloys. Here, new fundamental knowledge is needed in order to fully understand the catalyst mechanism and, more importantly, hydride destabilization. Further discussions on some new directions are the subject of other chapters of this book (complex hydrides, amides, destabilized systems, aluminum hydride, nanoparticles and theoretical modeling).

4.8.2 Synthesis

For practical applications, the synthesis process is a crucial step since large quantities will have to be produced reliably and at low cost. Therefore, more research is needed into new methods of synthesis, particularly for nanomaterials. In this respect, severe plastic deformation techniques could be an option but more thought has to be given to scaling up ball milling.

4.8.3 System Engineering

The last step to bring a hydrogen storing system to the consumer is probably the most important. Here, safety is the first concern but utilization conditions, cycle life, cost, and so on, have to be considered. A crucial step is the design of the tank itself where problems of mass and heat transfer will have to be resolved. We should not forget that all benefits from an optimum alloy and catalyst could be easily destroyed by a poorly designed reservoir. Here, computer simulation will play an important role in the development of a safe and optimal hydrogen tank.

The above classification should not be taken too literally. For example, it is well known that new synthesis processes could also mean new alloys as mechanical alloying showed us. In the same way, tank design will need a better evaluation of the critical parameters such as heat capacities and heat conductivity of nanostructured metal hydrides.

Metal hydrides development touches many fields of research: solid-state physics, surface chemistry, thermodynamics, diffusion, metallurgy, gas–solid reactions, and so on. Therefore, it requires a multidisciplinary approach on a team basis as well as for the individual researcher. Fundamental and applied knowledge still needs to be acquired for the full understanding of this class of materials. The intellectual challenge is exciting and, moreover, the developments will serve to make a better world for future generations.

References

1. Fetcenko, M.A., Ovshinsky, S.R., Reichman, B. *et al.* (2007) *J. Power Sources*, **165**, 544.
2. Bäuerlein, P., Antonius, C., Löffler, J. *et al.* (2008) *J. Power Sources*, **176** (2), 547.
3. Wang, X.H., Bei, Y.Y., Song, X.C. *et al.* (2007) *Int. J. Hydrogen Energy*, **32** (16), 4011.
4. Warren, D.E., Faughnan, K.A., Fellows, R.A. *et al.* (1984) *J. Less-Common Met.*, **104**, 375.
5. Sandrock, G. and Bowman, R.C. (2003) *J. Alloy. Compd.*, **356–357**, 794.
6. Fruchart, D., Bacmann, M., de Rango, P. *et al.* (1997) *J. Alloy. Compd.*, **253–254**, 121.
7. Pearson, D., Bowman, R., Prina, M. *et al.* (2007) *J. Alloy. Compd.*, **446–447**, 718.
8. Huiberts, J.N., Griessen, R., Rector, J.H. *et al.* (1996) *Nature*, **380** (21 March), 231.
9. Reilly, J.J. (1979) *Z. Phys. Chem. N. F.*, **117**, 155.

10 Alefeld, G. and Völkl, J. (1978) *Top. Appl. Phys.*, **28**, 423.
11 Alefeld, G. and Völkl, J. (1978) *Top. Appl. Phys.*, **29**, 385.
12 Schlapbach, L. (1988) *Top. Appl. Phys.*, **63**, 350.
13 Schlapbach, L. (1992) *Top. Appl. Phys.*, **67**, 328.
14 Wipf, H. (1997) *Top. Appl. Phys.*, **73**, 348.
15 Fukai, Y. (2005) *The Metal-Hydrogen System, Springer Series in Materials Science,* 2nd edn, vol. 21, Springer-Verlag, Berlin, p. 497.
16 Kirchheim, R., Sommer, F., and Schluckebier, G. (1982) *Acta. Metall. Mater.*, **30**, 1059.
17 Zaluski, L., Zaluska, A., and Strom-Olsen, J.O. (1997) *J. Alloy. Compd.*, **253–254**, 70.
18 Grochala, W. and Edwards, P.P. (2004) *Chem. Rev.*, **104** (3), 1283.
19 Sakintuna, Billur, Lamari-Darkrim, Farida, and Hirscher, Michael. (2007) *Int. J. Hydrogen Energy*, **32** (9), 1121.
20 Maeland, A.J. (2001) Hydrides for hydrogen storage, in *Recent Advances in Hydride Chemistry* (eds M. Peruzzini and R. Poli), Elsevier, p. 531.
21 Kohlmann, H. (2002) Metal hydrides, in *Encyclopedia of Physical Science and Technology*, vol. 9 (ed. Robert A. Meyers), Academic Press, San Diego, p. 441.
22 Chandra, D., Reilly, J.J., and Chellappa, R. (2006) *JOM*, **58** (2), 26.
23 Bérubé, V., Radtke, G., Dresselhaus, M. *et al.* (2007) *Int. J. Energ. Res.*, **31** (6–7), 637.
24 Schüth, F., Bogdanovic, B., and Felderhoff, M. (2004) *Chem. Commun.*, **2004**, 2249.
25 van den Berg, A.W.C. and Arean, C.O. (2008) *Chem. Commun.*, **2008**, 668.
26 Züttel, A. (2003) *Mater. Today*, **6** (9), 24.
27 Züttel, A. (2004) *Naturwissenschaften*, **91**, 157.
28 Fichtner, M. (2005) *Adv. Eng. Mater.*, **7** (6), 443.
29 Pundt, A. and Kirchheim, R. (2006) *Annu. Rev. Mater. Res.*, **36**, 555.
30 Léon, A. (2009) *Green Energy and Technology*, Springer.
31 Varin, R.A., Czujko, T., and Wronski, Z.S. (2009) *Nanomaterials for Solid State Hydrogen Storage*, Springer.
32 Schlapbach, L., Anderson, I., and Burger, J.P. (1994) Hydrogen in metals, in *Electronic and Magnetic Properties of Metals and Ceramics Part II*, vol. 3B (ed. K.H. Jürgen Buschow), VCH, Weinheim, p. 271.
33 Graham, T. (1866) *Philos. Trans. R. Soc. (London)*, **156**, 399.
34 Sullivan, E.A. (1995) Hydrides, in *Encyclopedia of Chemical Technology* (ed. M. Howe-Grant), John Wiley & Sons, New York.
35 Libowitz, G.G. (1991) presented at the Hydrogen Storage Materials, Batteries, and Electrochemistry, Phoenix, Arizona, Electrochemical Society Proceedings (eds D.A. Corrigan and S. Srinivasan), **92-5**, 3.
36 Libowitz, G.G. (1965) *The Solid-State Chemistry of Binary Metal Hydrides*, W.A. Benjamin, New-York.
37 Ivey, D.G. and Northwood, D.O. (1983) *J. Mater Sci.*, **18** (2), 321.
38 Maeland, A.J. (1977) presented at the Hydrides for Energy Storage, Geilo, Norway, Pergamon (eds A.F. Andresen and A.J. Maeland).
39 Switendick, A.C. (1979) *Z. Phys. Chem. N.F.*, **117**, 89.
40 Flanagan, T.B. (1977) presented at the Hydrides for Energy Storage, Geilo, Norway, Pergamon (eds A.F. Andresen and A.J. Maeland).
41 Flanagan, T.B. and Oates, W.A. (1988) Thermodynamics of Intermetallic Compound-Hydrogen Systems, in *Hydrogen in Intermetallic Compounds I* (ed. L. Schlapbach), Springer Verlag, Berlin, p. 49.
42 Griessen, R. and Riesterer, T. (1988) Heat of formation models, in *Hydrogen in Intermetallic Compounds I* (ed. L. Schlapbach), Springer-Verlag, Berlin, pp. 219.
43 Yamaguchi, M. and Akiba, E. (1994) Ternary hydrides, in *Electronic and Magnetic Properties of Metals and Ceramics Part II*, vol. 3 (ed. K.H. Jürgen Buschow), VCH, Weinheim, p. 333.
44 Fukai, Y. and Okuma, N. (1993) *Jpn. J. Appl. Phys.*, **32**, L1256.
45 Swalin, R.A. (1962) *Thermodynamics of Solids*, John Wiley & Sons, New York.

46 Wiswall, R. (1978) Hydrogen storage in metals, in *Hydrogen in Metals II*, vol. 29 (eds G. Alefeld and J. Völkl), Springer-Verlag, Berlin, pp. 201.

47 Griessen, R. (2007) Science and Technology of Hydrogen in Metals, http://www.nat.vu.nl/CondMat/griessen/STofHinM/STHM.htm.

48 Fukai, Y. (1991) *J. Less-Common Met.*, **172–174**, 8.

49 Alefeld, G. (1972) *Ber. Bunsenges. Phys. Chem.*, **76** (8), 746.

50 Tessier, P. (1995) *Hydrogen Storage in Metastable Fe-Ti*, in Dept. Phys. Ph.D. thesis, McGill, Montreal, Canada, p. 143.

51 Wagner, H. and Horner, H. (1974) *Adv. Phys.*, **23**, 587.

52 Oates, W.A. and Stoneham, A.M. (1983) *J. Phys. F: Met. Phys.*, **13**, 2427.

53 Smithson, H., Marianetti, C.A., Morgan, D. *et al.* (2002) *Phys. Rev. B*, **66**, 144107.

54 Lexcellent, Ch. and Gondor, G. (2007) *Intermetallics*, **15** (7), 934.

55 Feng, F., Geng, M., and Northwood, D.O. (2002) *Comput. Mater. Sci.*, **23**, 291.

56 Lototsky, M.V., Yartys, V.A., Marinin, V.S. *et al.* (2003) *J. Alloy. Compd.*, **356–357**, 27.

57 Schwarz, R.B. and Khachaturyan, A.G. (1995) *Phys. Rev. Lett.*, **74** (13), 2523.

58 Schwarz, R.B. and Khachaturyan, A.G. (2006) *Acta Mater.*, **54** (2), 313.

59 Sandrock, G. (1999) *J. Alloy. Compd.*, **293–295**, 877.

60 Libowitz, G.G., Hayes, H.F., and Gibb, T.R.P. (1958) *J. Phys. Chem.*, **62**, 76.

61 Reilly, J.J. and Wiswall, R.H. (1968) *Inorg. Chem.*, **7** (11), 2254.

62 Reilly, J.J. and Wiswall, R.H. (1974) *Inorg. Chem.*, **13** (1), 218.

63 van Vucht, J.H.N., Kuijpers, F.A., and Bruning, H.C.A.M. (1970) *Philips Res. Rep.*, **25**, 133.

64 Mendelsohn, M.H., Gruen, D.M., and Dwight, A.E. (1977) *Nature*, **269**, 45.

65 Chung, Y., Takeshita, T., McMasters, O.D. *et al.* (1980) *J. Less-Common Met.*, **74**, 217.

66 Lundin, C.E., Lynch, F.E., and Magee, C.B. (1977) *J. Less-Common Met.*, **56**, 19.

67 Miedema, A.R. (1973) *J. Less-Common Met.*, **32**, 117.

68 Miedema, A.R., Buschow, K.H.J., and Van Mal, H.H. (1976) *J. Less-Common Met.*, **49**, 463.

69 Bouten, P.C.P. and Miedema, A.R. (1980) *J. Less-Common Met.*, **71**, 147.

70 van Mal, H.H., Buschow, K.H.J., and Miedema, A.R. (1974) *J. Less-Common Met.*, **35**, 65.

71 Griessen, R. and Driessen, A. (1984) *Phys. Rev. B*, **30** (8), 4372.

72 Griessen, R. (1988) *Phys. Rev. B*, **38** (6), 3690.

73 Sandrock, G.D., Goodell, P.D., Huston, E.L. *et al.* (1989) *Z. Phys, Chem. N. F.*, **164**, 1285.

74 Westlake, D.G. (1983) *J. Less-Common Met.*, **91** (1), 1.

75 Westlake, D.G. (1983) *J. Less-Common Met.*, **90**, 251.

76 Gross, K.J., Zuttel, A., and Schlapbach, L. (1998) *J. Alloy. Compd.*, **274**, 239.

77 Gupta, M. (1999) *J. Alloy. Compd.*, **293–295**, 190.

78 Hector, L.G., Herbst, J.F., and Capehart, T.W. (2003) *J. Alloy. Compd.*, **353**, 74.

79 Gupta, M. (2000) *Int. J. Quantum. Chem.*, **77**, 982.

80 Miwa, K. and Fukumoto, A. (2002) *Phys. Rev. B*, **65** (15), 155114.

81 Nagasako, N., Fukumoto, A., and Miwa, K. (2002) *Phys. Rev. B*, **66** (15), 155106.

82 Hector, L.G. and Herbst, J.F. (2008) *J. Phys.: Condens. Matter*, **20**, 064229.

83 Grochala, W. and Edwards, P.P. (2005) *J. Alloy. Compd.*, **404–406**, 31.

84 Percheron-Guegan, A., Lartigue, C., Achard, J.C. *et al.* (1980) *J. Less-Common Met.*, **74**, 1.

85 Nomura, K., Uruno, H., Ono, S. *et al.* (1985) *J. Less-Common Met.*, **107**, 221.

86 Nakamura, Y., Oguro, K., Uehara, I. *et al.* (2000) *J. Alloy. Compd.*, **298**, 138.

87 Nakamura, Y. and Akiba, E. (2000) *J. Alloy. Compd.*, **308**, 309.

88 Yamamoto, T., Inui, H., and Yamaguchi, M. (2001) *Intermetallics*, **9** (10–11), 987.

89 Tanaka, K., Okazaki, S., Ichitsubo, T. *et al.* (2000) *Intermetallics*, **8**, 613.

90 Chen, C.P., Liu, B.H., Li, Z.P. *et al.* (1993) *Z. Phys. Chem. N. F.*, **181**, 259.

91 Liu, F.-J., Sandrock, G., and Suda, S. (1992) *J. Alloy. Compd.*, **190**, 57.

92 Liu, F.-J., Suda, S., and Sandrock, G. (1996) *J. Alloy. Compd.*, **232**, 232.

93 Li, Z.P. (2001) Hydrogen-metal systems: fluorinated metal hydrides, in *Encyclopedia of Materials: Science and*

Technology (eds K.H. Jürgen Buschow, R.W. Cahn, M.C. Flemings*et al.*), Elsevier, Amsterdam.
94 Balasubramaniam, R. (1997) *J. Alloy. Compd.*, **253–254**, 203.
95 Flanagan, T.B. (2001) The thermodynamics of hydrogen solution in 'perfect' and defective metals alloys, in *Progress in Hydrogen Treatment of Materials* (ed. V.A. Goltsov), Donetsk State Technical University, Donetsk, pp. 37.
96 Flanagan, T.B. and Clewley, J.D. (1982) *J. Less-Common Met.*, **83**, 127.
97 Flanagan, T.B., Park, C.N., and Everett, D.H. (1987) *J. Chem. Educ.*, **64** (11), 944.
98 Flanagan, T.B., Park, C.N., and Oates, W.A. (1995) *Prog. Solid State Chem.*, **23**, 291.
99 Rabkin, E. and Skripnyuk, V.M. (2003) *Scr. Mater.*, **49**, 477.
100 Sandrock, G., Suda, S., and Schlapbach, L. (1992) *Applications. Hydrogen in Intermetallic Compounds II* (ed. L. Schlapbach), Springer-Verlag, Berlin, p. 197.
101 Park, C.N., Luo, S., and Flanagan, T.B. (2004) *J. Alloy. Compd.*, **384**, 203.
102 Luo, S., Park, C.N., and Flanagan, T.B. (2004) *J. Alloy. Compd.*, **384**, 208.
103 Gerard, N. and Ono, S. (1992) Hydride formation and decomposition kinetics, in *Hydrogen in Intermetallic Compounds II*, vol. 67 (ed. L. Schlapbach), Springer-Verlag, Berlin, Chapter 4.
104 Hanada, N., Ichikawa, T., Hino, S. *et al.* (2006) *J. Alloy. Compd.*, **420**, 46.
105 Mintz, M.H. and Bloch, J. (1985) *Prog. Solid State Chem.*, **16**, 163.
106 Goodell, P.D., Sandrock, G.D., and Huston, E.L. (1980) *J. Less-Common Met.*, **73**, 135.
107 Goodell, P.D. and Rudman, P.S. (1983) *J. Less-Common Met.*, **89**, 117.
108 Rudman, P.S. (1983) *J. Less-Common Met.*, **89**, 93.
109 Jung, W.B., Nahm, K.S., and Lee, W.Y. (1990) *Int. J. Hydrogen Energy*, **15** (9), 641.
110 Martin, M., Gommel, C., Borkhart, C. *et al.* (1996) *J. Alloy. Compd.*, **238**, 193.
111 Hjort, P., Krozer, A., and Kasemo, B. (1996) *J. Alloy. Compd.*, **237**, 74.
112 Bloch, J. and Mintz, M.H. (1997) *J. Alloy. Compd.*, **253–254**, 529.
113 Schweppe, F., Martin, M., and Fromm, E. (1997) *J. Alloy. Compd.*, **261**, 254.
114 Inomata, A., Aoki, H., and Miura, T. (1998) *J. Alloy. Compd.*, **278**, 103.
115 Bloch, J. (2000) *J. Alloy. Compd.*, **312**, 135.
116 Chou, K.-C., Li, Q., Lin, Q. *et al.* (2005) *Int. J. Hydrogen Energy*, **30**, 301.
117 Gabis, I.E., Voit, A.P., Evard, E.A. *et al.* (2005) *J. Alloy. Compd.*, **404–406**, 312.
118 Christian, J.W. (2002) *The Theory of Transformations in Metals and Alloys. Part 1*, Pergamon, Oxford.
119 Cohen, R.L., West, K.W., and Wernick, J.H. (1980) *J. Less-Common Met.*, **73**, 272.
120 Cohen, R.L., West, K.W., and Wernick, J.H. (1980) *J. Less-Common Met.*, **70**, 229.
121 Schroder Pedersen, A., Kjoller, J., Larsen, B. *et al.* (1984) *Int. J. Hydrogen Energy*, **9** (9), 799.
122 Bogdanovic, B. and Spliethoff, B. (1987) *Int. J. Hydrogen Energy*, **12** (12), 863.
123 Bogdanovic, B., Spliethoff, B., and Ritter, A. (1989) *Z. Phys. Chem. N. F.*, **164**, 1497.
124 Uchida, H., Terao, K., and Huang, Y.C. (1989) *Z. Phys. Chem. N. F.*, **164**, 1275.
125 Wang, Q.-D. and Wu, J. (1989) *Z. Phys. Chem. N. F.*, **164**, 1305.
126 Bowman, R.C., Lynch, F.E., Marmaro, R.W. *et al.* (1993) *Z. Phys. Chem.*, **181**, 269.
127 Pedersen, A.S. and Larsen, B. (1993) *Int. J. Hydrogen Energy*, **18** (4), 297.
128 Dehouche, Z., Goyette, J., Bose, T.K. *et al.* (2003) *Int. J. Hydrogen Energy*, **28**, 983.
129 Bouaricha, S., Huot, J., Guay, D. *et al.* (2002) *Int. J. Hydrogen Energy*, **27**, 909.
130 Goodell, P.D. (1983) *J. Less-Common Met.*, **89**, 45.
131 Sandrock, G.D. and Goodell, P.D. (1984) *J. Less-Common Met.*, **104**, 159.
132 Song, M.Y., Bobet, J.-L., and Darriet, B. (2002) *J. Alloy. Compd.*, **340**, 256.
133 Dehouche, Z., Klassen, T., Oelerich, W. *et al.* (2002) *J. Alloy. Compd.*, **347**, 319.
134 Gross, K.J., Spatz, P., Zuttel, A. *et al.* (1996) *J. Alloy. Compd.*, **240**, 206.
135 Dehouche, Z., Djaozandry, R., Goyette, J. *et al.* (1999) *J. Alloy. Compd.*, **288** (1–2), 312.
136 Reiser, A., Bogdanovic, B., and Schlichte, K. (2000) *Int. J. Hydrogen Energy*, **25**, 425.

137 Dehouche, Z., Djaozandry, R., Huot, J. *et al.* (2000) *J. Alloy. Compd.*, **305**, 264.
138 Friedlmeier, G., Manthey, A., Wanner, M. *et al.* (1995) *J. Alloy. Compd.*, **231**, 880.
139 Bowman, R.C., Luo, C.H., Ahn, C.C. *et al.* (1995) *J. Alloy. Compd.*, **217**, 185.
140 Kabutomori, T., Takeda, H., Wakisaka, Y. *et al.* (1995) *J. Alloy. Compd.*, **231**, 528.
141 Nishimura, K., Sato, K., Nakamura, Y. *et al.* (1998) *J. Alloy. Compd.*, **268**, 207.
142 Suzuki, K., Ishikawa, K., and Aoki, K. (2000) *Mater. T. JIM*, **41** (5), 581.
143 Joubert, J.-M., Latroche, M., Cerny, R. *et al.* (2002) *J. Alloy. Compd.*, **330–332**, 208.
144 Shen, C.-C., Lee, S.-M., Tang, J.-C. *et al.* (2003) *J. Alloy. Compd.*, **356–357**, 800.
145 Laurencelle, F., Dehouche, Z., and Goyette, J. (2006) *J. Alloy. Compd.*, **424**, 266.
146 Sandrock, G., Gross, K., Thomas, G. *et al.* (2000) presented at the International Symposium on Metal-Hydrogen Systems, Noosa, Australia (eds E. Gray and C. Sholl).
147 Liang, G., Huot, J., and Schulz, R. (2001) *J. Alloy. Compd.*, **320**, 133.
148 Smardz, L., Smardz, K., Jurczyk, M. *et al.* (2000) *J. Alloy. Compd.*, **313**, 192.
149 Sakaguchi, H., Sugioka, T., and Adachi, G. (1996) *Eur. J. Solid State Inorg. Chem.*, **33**, 101.
150 Sakaguchi, H., Sugioka, T., and Adachi, G.-Y. (1995) *Chem. Lett. (Jpn)*, **1995**, 561.
151 Aoyagi, H., Aoki, K., and Masumoto, T. (1995) *J. Alloy. Compd.*, **231**, 804.
152 Corre, S., Bououdina, M., Kuriyama, N. *et al.* (1999) *J. Alloy. Compd.*, **292**, 166.
153 Koch, C.C. and Whittenberger, J.D. (1996) *Intermetallics*, **4**, 339.
154 Jurczyk, M., Smardz, K., Rajewski, W. *et al.* (2001) *Mater. Sci. Eng. A-Struct.*, **303**, 70.
155 Fujii, H., Munehiro, S., Fujii, K. *et al.* (2002) *J. Alloy. Compd.*, **330–332**, 747.
156 Ares, J.R., Cuevas, F., and Percheron-Guégan, A. (2005) *Acta Mater.*, **53**, 2157.
157 Schlapbach, L. (1992) *Surface Properties and Activation in Hydrogen in Intermetallic Compounds II*, vol. 67 (ed. L. Schlapbach), Springer-Verlag, Berlin, p. 15.
158 Psoma, A. and Sattler, G. (2002) *J. Power Sources*, **106**, 381.
159 Tessier, P., Zaluski, L., Yan, Z.-H. *et al.* (1993) *Mater. Res. Soc. Symp. Proc.*, **286**, 209.
160 Chiang, C.-H., Chin, Z.-H., and Perng, T.-P. (2000) *J. Alloy. Compd.*, **307**, 259.
161 Zaluski, L., Tessier, P., Ryan, D.H. *et al.* (1993) *J. Mater. Res.*, **8** (12), 3059.
162 Sun, L., Liu, H., Bradhurst, D.H. *et al.* (1998) *J. Mater. Sci. Lett.*, **17**, 1825.
163 Trudeau, M.L., Dignard-Bailey, L., Schulz, R. *et al.* (1992) *Nanostruct. Mater.*, **1**, 457.
164 Trudeau, M.L., Schulz, R., Zaluski, L. *et al.* (1992) *Mater. Sci. Forum*, **88–90**, 537.
165 Bououdina, M., Fruchart, D., Jacquet, S. *et al.* (1999) *Int. J. Hydrogen Energy*, **24** (9), 885.
166 Zaluski, L., Hosatte, S., Tessier, P. *et al.* (1994) *Z. Phys. Chem.*, **183**, 45.
167 Zaluski, L., Zaluska, A., Tessier, P. *et al.* (1995) *J. Alloy. Compd.*, **227** (1), 53.
168 Zaluski, L., Zaluska, A., Tessier, P. *et al.* (1996) *J. Mater Sci.*, **31**, 695.
169 Zaluski, L., Zaluska, A., Tessier, P. *et al.* (1996) *Mater. Sci. Forum*, **225–227**, 875.
170 Tessier, P., Schulz, R., and Ström-Olsen, J.O. (1998) *J. Mater. Res.*, **13**, 1538.
171 Abe, M. and Kuji, T. (2007) *J. Alloy. Compd.*, **446–447**, 200.
172 Stein, F., Palm, M., and Sauthoff, G. (2004) *Intermetallics*, **12**, 713.
173 Shaltiel, D., Jacob, I., and Davidov, D. (1977) *J. Less-Common Met.*, **53**, 117.
174 Shoemaker, D.P. and Shoemaker, C.B. (1979) *J. Less-Common Met.*, **68**, 43.
175 Semenenko, K.N., Verbetskii, V.N., Mitrokhin, S.V. *et al.* (1980) *Zh. Neorganich. Khim.*, **25** (7), 1731.
176 Ivey, D.G. and Northwood, D.O. (1986) *Z. Phys. Chem. N. F.*, **147**, 191.
177 Bououdina, M., Soubeyroux, J.L., de Rango, P. *et al.* (2000) *Int. J. Hydrogen Energy*, **25**, 1059.
178 Bououdina, M., Grant, D., and Walker, G. (2006) *Int. J. Hydrogen Energy*, **31** (2), 177.
179 Bououdina, M., Enoki, H., and Akiba, E. (1998) *J. Alloy. Compd.*, **281**, 290.
180 Cracco, D. and Percheron-Guegan, A. (1998) *J. Alloy. Compd.*, **268**, 248.
181 Iba, H. and Akiba, E. (1995) *J. Alloy. Compd.*, **231**, 508.
182 Huot, J., Akiba, E., and Iba, H. (1995) *J. Alloy. Compd.*, **228**, 181.
183 Huot, J., Akiba, E., Ogura, T. *et al.* (1995) *J. Alloy. Compd.*, **218**, 101.
184 Akiba, E., Huot, J., and Iba, H. (1994) presented at the The Electrochemical

Society Conference, Pennington, NJ, The Electrochemical Society (eds P.D. Bannet and T. Sakai), **94-27**, 165.

185 Akiba, E. (1999) *Curr. Opin. Solid St. Mater.*, **4**, 267.
186 Iba, H. and Akiba, E. (1997) *J. Alloy. Compd.*, **253–254**, 21.
187 Iba, H., Shionoya, M., and Akiba, E. (1996) *Toyota Tech. Rev.*, **45** (2), 111.
188 Akiba, E. and Iba, H. (1998) *Intermetallics*, **6** (6), 461.
189 Tamura, T., Kazumi, T., Kamegawa, A. *et al.* (2003) *J. Alloy. Compd.*, **356–357**, 505.
190 Tamura, T., Kamegawa, A., Takamura, H. *et al.* (2002) *Mater. T. JIM*, **43** (3), 410.
191 Challet, S., Latroche, M., and Heurtaux, F. (2007) *J. Alloy. Compd.*, **439**, 294.
192 Yan, Yigang, Chen, Yungui, Liang, Hao *et al.* (2008) *J. Alloy. Compd.*, **454** (1–2), 427.
193 Cho, Sung-Wook, Shim, Gunchoo, Choi, Good-Sun *et al.* (2007) *J. Alloy. Compd.*, **430** (1–2), 136.
194 Yan, Yigang, Chen, Yungui, Zhou, Xiaoxiao *et al.* (2008) *J. Alloy. Compd.*, **453** (1–2), 428.
195 Yu, X.B., Yang, Z.Y., Feng, S.L. *et al.* (2006) *Int. J. Hydrogen Energy*, **31** (9), 1176.
196 Taizhong, H., Zhu, W., Baojia, X. *et al.* (2003) *Sci. Technol. Adv. Mater.*, **4** (6), 491.
197 Taizhong, H., Zhu, W., Jinzhou, C. *et al.* (2004) *Mat. Sci. Eng. A-Struct.*, **385**, 17.
198 Taizhong, H., Zhu, W., Baojia, X. *et al.* (2005) *Intermetallics*, **13** (10), 1075.
199 Ellinger, F.H., Holley, C.E., McInteer, B.B. *et al.* (1955) *J. Am. Chem. Soc*, **77**, 2647.
200 Percheron-Guégan, A. and Welter, J.-M. (1988) Preparation of intermetallics and hydrides, in *Hydrogen in Intermetallic Compounds I* (ed. L. Schlapbach), Springer-Verlag, Berlin, p. 11.
201 Dornheim, M., Doppiu, S., Barkhordarian, G. *et al.* (2007) *Scr. Mater.*, **56**, 841.
202 Guo, Z.X., Shang, C., and Aguey-Zinsou, K.F. (2008) *J. Eur. Ceram. Soc.*, **28** (7), 1467.
203 Zaluska, A., Zaluski, L., and Ström-Olsen, J.O. (1999) *J. Alloy. Compd.*, **288**, 217.
204 Zaluska, A., Zaluski, L., and Ström-Olsen, J.O. (2001) *Appl. Phys. A*, **72** (2), 157.
205 Huot, J., Liang, G., Boily, S. *et al.* (1999) *J. Alloy. Compd.*, **293–295**, 495.
206 Hanada, N., Ichikawa, T., Orimo, S. *et al.* (2004) *J. Alloy. Compd.*, **366**, 269.
207 Varin, R.A., Czujko, T., Chiu, Ch. *et al.* (2006) *J. Alloy. Compd.*, **424**, 356.
208 Varin, R.A., Czujko, T., and Wronski, Z. (2006) *Nanotechnology*, **17**, 3856.
209 Wronski, Z., Varin, R.A., Czujko, T. *et al.* (2007) *J. Alloy. Compd.*, **434–435**, 743.
210 Huhn, P.A., Dornheim, M., Klassen, T. *et al.* (2005) *J. Alloy. Compd.*, **404–406**, 499.
211 Dornheim, M., Eigen, N., Barkhordarian, G. *et al.* (2006) *Adv. Eng. Mater.*, **8** (5), 377.
212 Aguey-Zinsou, K.F., Ares Fernandez, J.R., Klassen, T. *et al.* (2007) *Int. J. Hydrogen Energy*, **32** (13), 2400.
213 Fátay, D., Révész, Á., and Spassov, T. (2005) *J. Alloy. Compd.*, **399** (1–2), 237.
214 Stepanov, A., Ivanov, E., Konstanchuk, I.G. *et al.* (1987) *J. Less-Common Met.*, **131**, 89.
215 Liang, G., Huot, J., Boily, S. *et al.* (1999) *J. Alloy. Compd.*, **291**, 295.
216 Gutfleisch, O., Schlorke-de Boer, N., Ismail, N. *et al.* (2003) *J. Alloy. Compd.*, **356–357**, 598.
217 Liang, G., Huot, J., Boily, S. *et al.* (1999) *J. Alloy. Compd.*, **292**, 247.
218 Huot, J., Pelletier, J.F., Liang, G. *et al.* (2002) *J. Alloy. Compd.*, **330–332**, 727.
219 Bobet, J.-L., Pechev, S., Chevalier, B. *et al.* (1999) *J. Mater. Chem.*, **9** (1), 315.
220 Liang, G. (2004) *J. Alloy. Compd.*, **370**, 123.
221 Terzieva, M., Khrussanova, M., and Peshev, P. (1998) *J. Alloy. Compd.*, **267**, 235.
222 Liang, G., Wang, E., and Fang, S. (1995) *J. Alloy. Compd.*, **223**, 111.
223 Gross, K.J., Chartouni, D., Leroy, E. *et al.* (1998) *J. Alloy. Compd.*, **269**, 259.
224 Khrussanova, M., Terzieva, M., Peshev, P. *et al.* (1989) *Z. Phys. Chem. N. F.*, **164**, 1261.
225 Oelerich, W., Klassen, T., and Bormann, R. (2001) *J. Alloy. Compd.*, **315**, 237.
226 Bobet, J.-L., Desmoulins-Krawiec, S., Grigorova, E. *et al.* (2003) *J. Alloy. Compd.*, **351**, 217.
227 Jung, Kyung Sub, Kim, Dong Hyun, Lee, Eun Young *et al.* (2007) *Catal. Today*, **120** (3–4), 270.

228 Varin, R.A., Czujko, T., Wasmund, E.B. *et al.* (2007) *J. Alloy. Compd.*, **432**, 217.

229 Friedrichs, O., Sanchez-Lopez, J.C., Lopez-Cartes, C. *et al.* (2006) *J. Appl. Chem. B*, **110**, 7845.

230 Hanada, Nobuko, Hirotoshi, Enoki, Ichikawa, Takayuki *et al.* (2008) *J. Alloy. Compd.*, **450** (1–2), 395.

231 Porcu, M., Petford-Long, A.K., and Sykes, J.M. (2008) *J. Alloy. Compd.*, **453** (1–2), 341.

232 Schimmel, H.G., Huot, J., Chapon, L.C. *et al.* (2005) *J. Am. Chem. Soc.*, **127**, 14348.

233 Didisheim, J.-J., Zolliker, P., Yvon, K. *et al.* (1984) *Inorg. Chem.*, **23**, 1953.

234 Kyoi, D., Sato, T., Rönnebro, E. *et al.* (2004) *J. Alloy. Compd.*, **372**, 213.

235 Vermeulen, P., Niessen, R.A.H., and Notten, P.H.L. (2006) *Electrochem. Comm.*, **8**, 27.

236 Li, S., Varin, R.A., Morozova, O. *et al.* (2004) *J. Alloy. Compd.*, **384**, 231.

237 Herrich, M., Ismail, N., Lyubina, J. *et al.* (2004) *Mater. Sci. Eng. B-Solid*, **108**, 28.

238 Gennari, F.C., Castro, F.J., and Andrade Gamboa, J.J. (2002) *J. Alloy. Compd.*, **339**, 261.

239 Huot, J., Boily, S., Akiba, E. *et al.* (1998) *J. Alloy. Compd.*, **280**, 306.

240 Dornheim, M., Pundt, A., Kirchheim, R. *et al.* (2003) *J. Appl. Phys.*, **93** (11), 8958.

241 Johnson, W.C. and Chiang, C.S. (1988) *J. Appl. Phys.*, **64** (3), 1155.

242 Vajo, J.J., Mertens, F., Ahn, C.C. *et al.* (2004) *J. Phys. Chem. B*, **108**, 13977.

243 Vajo, John J. and Olson, Gregory L. (2007) *Scr. Mater.*, **56** (10), 829.

244 Crivello, J.C., Nobuki, T., Kato, S. *et al.* (2007) *J. Alloy. Compd.*, **446–447**, 157.

245 Zhang, Q.A. and Wu, H.Y. (2005) *Mater. Chem. Phys.*, **94** (1), 69.

246 Bouaricha, S., Dodelet, J.P., Guay, D. *et al.* (2000) *J. Alloy. Compd.*, **297**, 282.

247 Dufour, J. and Huot, J. (2007) *J. Alloy. Compd.*, **446–447**, 147.

248 Takeichi, N., Tanaka, K., Tanaka, H. *et al.* (2007) *J. Alloy. Compd.*, **446–447**, 543.

249 Dufour, J. and Huot, J. (2007) *J. Alloy. Compd.*, **439**, L5.

250 Yvon, K. and Bertheville, B. (2006) *J. Alloy. Compd.*, **425**, 101.

251 Valiev, R. (2000) *Prog. Mater. Sci.*, **45**, 103.

252 Altan, B.S., ed. (2006) *Severe plastic deformation toward bulk production of nanostructured materials.* Nova Science, New York, p. 612.

253 Langdon, T.G. (2006) *Rev. Adv. Mater. Sci.*, **13** (1), 6.

254 Skripnyuk, V., Rabkin, E., Estrin, Y. *et al.* (2004) *Acta Mater.*, **52** (2), 405.

255 Skripnyuk, V., Buchman, E., Rabkin, E. *et al.* (2007) *J. Alloy. Compd.*, **436**, 99.

256 Kusadome, Yuichiro, Ikeda, Kazutaka, Nakamori, Yuko *et al.* (2007) *Scr. Mater.*, **57** (8), 751.

257 Zhang, L.T., Ito, K., Vasudevan, V.K. *et al.* (2001) *Acta Mater.*, **49**, 751.

258 Zhang, L.T., Ito, K., Vasudevan, V.K. *et al.* (2002) *Mater. Sci. Eng. A-Struct.*, **329–331**, 362.

259 Ueda, T.T., Tsukahara, M., Kamiya, Y. *et al.* (2004) *J. Alloy. Compd.*, **386**, 253.

260 Mori, R., Miyamura, H., Kikuchi, S. *et al.* (2007) *Mater. Sci. Forum*, **561–565**, 1609.

261 Saganuma, K., Miyamura, H., Kikuchi, S. *et al.* (2007) *Adv. Mater. Res.*, **26–28**, 857.

5
Complex Hydrides

Claudia Weidenthaler and Michael Felderhoff

5.1
Introduction

Complex hydrides are salt-like materials in which hydrogen is covalently bound to the central atoms, in this way a crystal structure consisting of so-called "complex anions" is formed. In general, complex hydrides have the chemical formula $A_xMe_yH_z$. Compounds where position A is preferentially occupied by elements of the first and second groups of the periodic table and Me is occupied either by boron or aluminum are well known and have been intensively investigated. However, another possibility is that complex hydrides are built by transition metal cations. Complex metal hydrides have been known for more than 50 years, but for many years they were not considered for reversible hydrogen storage due to the high kinetic barriers of the decomposition requiring high temperatures. The situation changed in 1997 when Bogdanović and Schwickardi discovered that the kinetic barrier of the decomposition of the complex metal hydride, $NaAlH_4$, can be lowered by the addition of Ti catalysts and that the material becomes reversible close to acceptable technical conditions [1]. They further showed not only that catalysts can enhance the kinetics of dehydrogenation but also that rehydrogenation under moderate conditions becomes feasible. This breakthrough induced the publication of a tremendous number of papers focusing on the synthesis and the properties of complex hydrides. The major goal is to understand the basic steps of dehydrogenation and rehydrogenation and the role of catalysts. Many complex metal hydrides have high hydrogen gravimetric storage capacities and some of them are commercially available. Some complex hydride systems, such as Mg_2FeH_6 and $Al(BH_4)_3$, have an extremely high volumetric hydrogen density of up to 150 kg m^{-3}. Compared to liquid hydrogen with a volumetric density of 70 kg m^{-3}, the amount of hydrogen in such metal hydrides is much higher. This makes complex transition metal hydrides interesting as potential storage materials.

Handbook of Hydrogen Storage. Edited by Michael Hirscher

ISBN: 978-3-527-32273-2

5.2
Complex Borohydrides

5.2.1
Introduction

In 1939 Schlesinger *et al.* prepared $Al(BH_4)_3$, the first example of a complex metal borohydride [2]. Starting from aluminum trimethyl and diborane, they tried to synthesize AlH_3, but the complex borohydride $Al(BH_4)_3$ was produced as in Eq. (5.1).

$$Al_2(CH_3)_6 + 4B_2H_6 \rightarrow 2Al(BH_4)_3 + 2B(CH_3)_3 \quad (5.1)$$

In the following years the Schlesinger group prepared different new complex borohydride compounds (e.g., $NaBH_4$ was first synthesized in 1943), but for secrecy during World War II, most of the results were not published before 1953 [3]. Nowadays some of the complex borohydrides ($NaBH_4$, KBH_4) are widely used in organic synthesis for the reduction of aldehydes or ketones to alcohols [4]. One important property is the high gravimetric H_2 density of complex borohydrides. The low atomic weight of boron and the high amount of bound hydrogen makes this complex system interesting for hydrogen storage. $Be(BH_4)_2$ is the compound with the highest hydrogen content (more than 20 wt.%), but its high toxicity excludes this material from any use as a hydrogen storage material. Most of the metal borohydrides show high thermal stability, making the liberation of hydrogen difficult. They cannot be decomposed with the waste heat of a PEM fuel cell, which makes their use impractical. Another important disadvantage is the poor reversibility under moderate technical conditions. However, even though metal borohydride compounds are not feasible for reversible hydrogen storage, they can be used as hydrogen delivering materials in aqueous solutions. Boron forms not only the simple $[BH_4]^-$ anion, it can also form a dimeric $[B_2H_7]^-$ anion with a bridging hydrogen atom. So far, dimeric anions have not been considered for hydrogen storage applications. Some physical properties of the most important complex borohydrides are listed in Table 5.1.

5.2.2
Stability of Metal Borohydrides

The bonding characteristics and the properties of metal borohydride systems strongly depend on the electronegativity difference between the metal atom and boron ($\chi = 2.01$). While for the compounds $NaBH_4$ ($\chi_{Na} = 0.93$) and KBH_4 ($\chi_K = 0.82$) the electronegativity difference is large, the difference between Al ($\chi = 1.47$) and B in covalently bound aluminum borohydride is much smaller. Therefore $Al(BH_4)_3$ is a liquid at room temperature. The same behavior is observed for $Be(AlH_4)_2$ with a small electronegativity difference between Be ($\chi = 1.47$) and B. In this compound the $[BH_4]^-$ units are covalently bonded via hydrogen bridges to beryllium. Theoretical calculations of a series of different metal borohydrides $Me(BH_4)_n$ (M = Li, Na, K, Mg, Sc, Cu, Zn, Zr, Hf, n = 1–4) have shown that in all

Table 5.1 Physical constants of selected boron hydrides.

	Molecular weight ($g\ mol^{-1}$)	Melting point (°C)	Start of decomposition[a] (°C)	Hydrogen content (wt.%)
$LiBH_4$	21.8	275	320 [13]	18.4
$NaBH_4$	37.8	505	450 [23]	10.6
KBH_4	53.9	585	584 [25]	7.4
$Be(BH_4)_2$	38.6	–	–	20.7
$Mg(BH_4)_2$	53.9	–	320 [33]	14.8
$Ca(BH_4)_2$	69.8	–	360 [44]	11.5
$Al(BH_4)_3$	71.4	-64	~40 [49]	16.8

a) Experimental data.

cases the bond between the metal cation and the $[BH_4]^-$ anion is an ionic bond and that charge transfer from the metal cation to the anion governs the stability of the metal borohydrides [5]. Figure 5.1 shows the linear relation between the heat of formation of $Me(BH_4)_n$ and the Pauling electronegativity χ_p of the metal cation Me, as predicted by first-principles calculations [5, 6].

5.2.3 Decomposition of Complex Borohydrides

Considering the decomposition of alkali metal borohydrides to form metal hydrides, the reaction can be described as follows (Eq. (5.2)):

$$MeBH_4 \rightarrow MeH + B + 1.5H_2 \tag{5.2}$$

with the liberation of 1.5 mol H_2. An alternative pathway releases hydrogen completely and metal boride is produced (Eq. (5.3)).

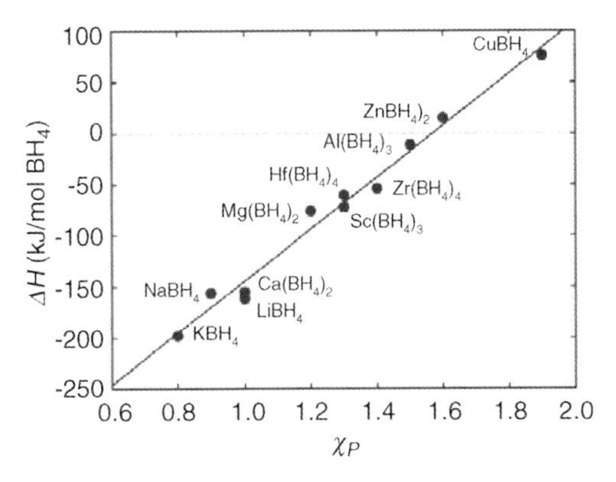

Figure 5.1 Relation between the heat of formation ΔH and the Pauling electronegativity χ_p of the cation. The straight line indicates the results of the least square fitting ($\Delta H = 253.6\ \chi_p - 398.0$). (Reprinted from [6], with permission from Elsevier.)

$$MeBH_4 \rightarrow MeB + 2H_2 \quad (5.3)$$

For alkaline earth borohydrides the following decomposition reactions are possible (Eqs. (5.4)–(5.6)):

$$Me(BH_4)_2 \rightarrow MeH_2 + 2B + 3H_2 \quad (5.4)$$

$$Me(BH_4)_2 \rightarrow MeB_2 + 4H_2 \quad (5.5)$$

$$Me(BH_4)_2 \rightarrow 2/3MeH_2 + 1/3MeB_6 + 10/3H_2 \quad (5.6)$$

with the production of alkaline earth metal hydride and boron metal, the formation of metal boride or a combination of the formation of the alkaline earth hydride and a metal boride compound. An alternative reaction to the release of hydrogen during the decomposition of metal borohydride compounds is the liberation of diborane or higher homologues. These volatile boranes have the disadvantage of damaging the catalyst or the membrane of a fuel cell. In the case of a potential reversible reaction, the storage capacity of the system is reduced over time. Most importantly, diborane and other volatile boranes are highly toxic.

Thermal desorption studies revealed for Li-, Na, and K-borohydrides the formation of alkali hydride as decomposition products above 430 °C. While Mg-, Sc, and Zr-borohydrides decompose through the formation of intermediate phases and/or borides, $Zn(BH_4)_2$ decomposes directly to elemental Zn [5].

5.2.4 Lithium Borohydride, $LiBH_4$

5.2.4.1 Synthesis and Crystal Structure

With a hydrogen content of 18.4 wt.% H_2, $LiBH_4$ could be attractive as a hydrogen storage material. $LiBH_4$ was first prepared by the reaction of ethyl lithium with diborane [7]. Under rigorous conditions of temperature between 550 and 700 °C and hydrogen pressures between 3 and 15 MPa, $LiBH_4$ can be prepared in a direct synthesis starting from lithium (or lithium hydride) and boron [8]. The first crystal structure analysis of $LiBH_4$ was performed in 1947, with the conclusion that the space group of the room temperature structure is Pcmn [9]. After re-investigation of the structure using synchrotron X-ray powder diffraction data, orthorhombic symmetry was confirmed. The space group is Pnma with cell dimensions $a = 7.17858$ Å, $b = 4.43686$ Å and $c = 6.80321$ Å at 25 °C [10]. The tetrahedral $[BH_4]^-$ anions are strongly distorted with respect to bond lengths and bonds angles and are aligned along two orthogonal directions. $LiBH_4$ shows a phase transition at 118 °C to a hexagonal structure (space group $P6_3mc$) with $a = 4.27631$ Å and $c = 6.94844$ Å at 135 °C. The phase transition is endothermic with a transition enthalpy of 4.18 kJ mol^{-1}.

5.2.4.2 Decomposition of $LiBH_4$

In the first thermal desorption experiments under ambient pressure and in a hydrogen atmosphere up to 1 MPa several endothermic peaks were observed

[11, 12]. The first two peaks at 110 and 280 °C describe a structural phase transformation and the melting of the material. The main amount of hydrogen (80 wt.%) is liberated from the molten $LiBH_4$ in a temperature range between 320 and 380 °C. Under these conditions, one hydrogen atom remains under in the structure of LiH. More advanced experiments have shown different stages of hydrogen liberation [13, 14]. In the range between 100 and 200 °C a small amount of hydrogen (0.3 wt.%) is released. After melting at 280 °C (without hydrogen release) an additional 1 wt.% hydrogen is released around 320 °C. After this decomposition reaction, a second hydrogen release step is observed starting at 400 °C. Up to 600 °C an overall amount of 9 wt.% hydrogen is produced. This is half of the total hydrogen in $LiBH_4$, resulting in a product with the general composition "$LiBH_2$". The addition of SiO_2 powder lowers the decomposition temperatures of all hydrogen release steps and 9 wt.% of hydrogen is liberated below 400 °C. Figure 5.2 shows the desorbed amount of hydrogen as a function of temperature and heating rates of such a mixed sample ($LiBH_4$: SiO_2 = 25: 75) [15]. The hydrogen flow as a function of temperature is shown for the very low heating rate of 0.5 °C min^{-1}. Three distinct peaks were observed, indicating that several intermediates are involved during the decomposition.

It was shown by Raman spectroscopy, X-ray diffraction studies, and NMR investigations that, during the decomposition of $LiBH_4$, the boron hydride cluster $[B_{12}H_{12}]^{2-}$ is formed as an intermediate [16, 17]. The dodecahydro-closo-dodecaborate-anion $[B_{12}H_{12}]^{2-}$ with an icosahedra structure is quite stable and only one example of the structural variety of boron hydride compounds. Therefore, it can be expected that several other boron hydride compounds might be discovered as intermediate products during the decomposition of metal borohydride systems. Rehydrogenation of the decomposition products LiH and B to $LiBH_4$ is, in principle, possible. However, even after more than 12 h under rigorous conditions (690 °C, 20 MPa H_2) the reaction is still not complete [18].

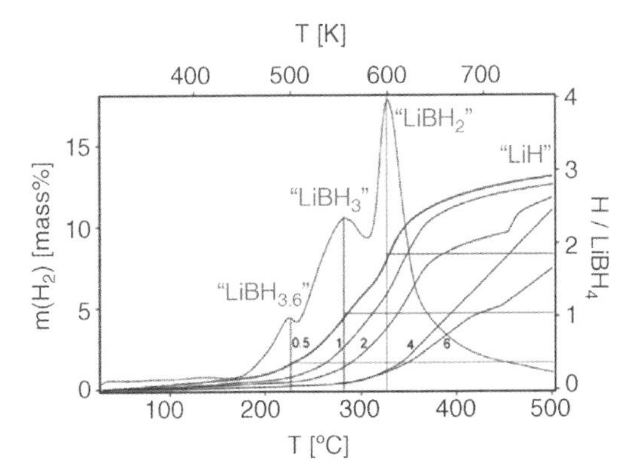

Figure 5.2 Integrated thermal desorption spectra for $LiBH_4/SiO_2$ mixtures obtained for various heating rates (from 0.5 to 6 °C min^{-1}) (Reprinted from [15], with permission from Elsevier.)

5.2.5 Sodium Borohydride, $NaBH_4$

5.2.5.1 Synthesis and Crystal Structure

Sodium borohydride contains 10.8 wt.% of hydrogen, but it shows unfavorable thermodynamics for use as reversible hydrogen storage material. Over 100 synthesis methods for the preparation of $NaBH_4$ have been described, but only two have reached practical significance. In the Schlesinger process (Eq. (5.7)), trimethyl borate is boiled together with NaH in hydrocarbon oil at 250 °C [19]:

$$4NaH + B(OCH_3)_3 \rightarrow NaBH_4 + 3NaOCH_3 \quad (5.7)$$

The addition of water hydrolyses $NaOCH_3$ to sodium hydroxide and methanol and causes separation from the hydrocarbon oil. Methanol is recovered by distillation and recycled to form trimethyl borate. $NaBH_4$ left in the NaOH solution is extracted with isopropyl amine.

The Bayer process is based on borosilicate $Na_2B_4O_7{\cdot}7\,SiO_2$, produced by fusion of borax and silica. The borosilicate is reacted with Na in an atmosphere of 0.3 MPa of hydrogen at 400–500 °C. Extraction with liquid ammonia under pressure yields $NaBH_4$. The use of $NaBH_4$ as a reducing agent in organic synthesis is well known and represents the major use of this compound. Under ambient conditions, α-$NaBH_4$ crystallizes in a cubic Fm3m structure with boron in tetrahedral coordination [20]. At lower temperatures of about −83 °C the compound shows a phase transformation to a tetragonal phase (β-$NaBH_4$). Additionally, pressure-induced transformations to a tetragonal (at about 6 GPa) and an orthorhombic phase ($>$ 9 GPa) have been reported [21, 22].

5.2.5.2 Decomposition of $NaBH_4$

Different to $LiBH_4$, with hydrogen liberation from the liquid state, the sodium compound releases most of its hydrogen from the solid material [23]. Hydrogen evolution starts at about 240 °C and proceeds in several steps during the heating process. However, only 1 wt.% of the total hydrogen content of 10.4 wt.% is liberated at lower temperatures whereas most of the hydrogen is released at about 450 °C.

5.2.6 Potassium Borohydride KBH_4

Potassium borohydride is a commercial product, but it can be prepared in a simple reaction from an aqueous solution of $NaBH_4$ with potassium hydroxide (Eq. (5.8)) [24].

$$NaBH_4 + KOH \rightarrow KBH_4 + NaOH \quad (5.8)$$

Compared to lithium and sodium borohydride, potassium borohydride has the highest hydrogen desorption temperature of 584 °C and melting temperature of 607 °C [25]. The hydrogen content is 7.7 wt.%. Potassium borohydride crystallizes in the cubic NaCl-type structure (space group Fm3m) as do sodium, rubidium, and caesium borohydrides. $[BH_4]^-$ units are octahedrally coordinated by K atoms.

At higher temperatures the α-form transforms into the closely related tetragonal structure [26]. Very similar to $NaBH_4$, KBH_4 exhibits phase transformations from the cubic phase, stable at ambient conditions, to a tetragonal phase at about 4 GPa and an orthorhombic phase at 7 GPa [27].

5.2.7 Beryllium Borohydride $Be(BH_4)_2$

With more than 20 wt.% H_2, $Be(BH_4)_2$ represents the complex borohydride with the highest hydrogen content. However, the high toxicity of Be metal excludes its use in hydrogen storage applications. $Be(BH_4)_2$ can be synthesized according to Eq. (5.9), starting from $BeCl_2$ and $LiBH_4$ at 145 °C [28]. The produced $Be(BH_4)_2$ sublimes under these conditions and can be continuously removed by pumping.

$$BeCl_2 + 2LiBH_4 \xrightarrow{145\,^{\circ}C} Be(BH_4)_2 + 2LiCl \qquad (5.9)$$

The crystal structure (space group $I4_1cd$) consists of a helical polymer of BH_4Be and BH_4 units with two hydrogen bridges between B and Be [29]. So far, experimental data are not available but first principle calculations suggest a decomposition temperature of −111 °C at 0.1 MPa [30].

5.2.8 Magnesium Borohydride $Mg(BH_4)_2$

5.2.8.1 Synthesis and Crystal Structure

With a hydrogen content of 14.8 wt.% $Mg(BH_4)_2$ seems to be interesting for hydrogen storage but the high thermal stability (decomposition temperature around 300 °C) makes it less suitable for reversible hydrogen storage. Starting from $MgCl_2$ and $LiBH_4$, $Mg(BH_4)_2$ can be prepared in ether solution. After drying at 145 °C a solvent-free crystalline material was obtained, which was used for high-resolution X-ray diffraction studies [31, 32]. The evaluation of several mechanochemical and wet chemical preparation methods showed that solvent-free and pure α-$Mg(BH_4)_2$ can only be obtained by direct wet chemical synthesis [33].

Qualitatively, the crystal structure of $Mg(BH_4)_2$ is similar to that of other boranates, but a closer look in more detail shows a very complex crystal structure. Each Mg^{2+} atom in the hexagonal structure (space group $P6_1$) is coordinated by eight H atoms from four $[BH_4]^-$ tetrahedra (Figure 5.3). Each Mg dodecahedra is linked by $[BH_4]^-$ bridges to four neighboring dodecahedra forming a network [31, 32].

5.2.8.2 Decomposition

During thermal treatment, α-$Mg(BH_4)_2$ first undergoes a phase transformation at about 190 °C before it further decomposes in several steps to MgH_2, Mg, and MgB_2 (Eqs. (5.10) and (5.11)) [33]. The weight loss between 290 and 500 °C is 13 wt.% H_2 with a maximum release between 300 and 400 °C. Small amounts of B_2H_6 were detected by mass spectrometric measurements. The presence of $TiCl_3$ decreases the

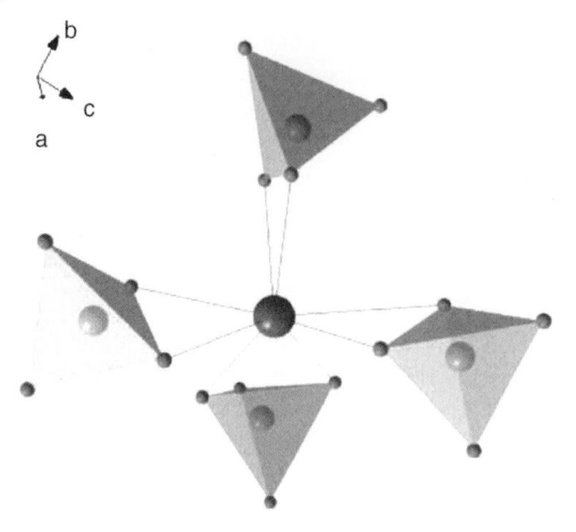

Figure 5.3 $Mg(BH_4)_2$: Coordination of the Mg atom (large circle) by four $[BH_4]^-$ units resulting in an eightfold coordination of Mg.

decomposition temperature to about 100 °C [34]. The decomposition mechanism can be described as follows:

$$Mg(BH_4)_2 \rightarrow MgH_2 + 2B + 3H_2 \quad 11.1\ \text{wt.\%} \tag{5.10}$$

$$MgH_2 + 2B \rightarrow MgB_2 + H_2 \quad 4.2\ \text{wt.\%} \tag{5.11}$$

For the first decomposition step the formation of $[B_{12}H_{12}]^{2-}$ anion as a possible intermediate phase is discussed [35]. Different enthalpy values in the range −40 to 57 kJ mol^{-1} H_2 for the first decomposition step at about 277 °C have been determined from experimental data [35, 36]. From density functional theory (DFT) calculations a reaction enthalpy of 38 kJ mol^{-1} H_2 at 27 °C was confirmed [37]. Cycling experiments performed at 350 °C under 10 MPa hydrogen lead to re-absorption of more than 3 wt.% hydrogen, showing that the second decomposition step is reversible [33, 36].

5.2.9 Calcium Borohydride $Ca(BH_4)_2$

5.2.9.1 Synthesis and Crystal Structure

Calcium borohydride, $Ca(BH_4)_2$, contains 11.5 wt.% H_2. It can be prepared by reacting calcium hydride or alkoxides with diborane or by reaction in THF [38–40]. Starting from ball-milled CaH_2 added to triethylamine borazane complex, $Ca(BH_4)_2$ and CaH_2 were obtained as by-products after heating to 140 °C, cooling and washing with n-hexane [41]. Without any solvent, $Ca(BH_4)_2$ can be obtained by ball milling of CaH_2 and MgB_2 and subsequent hydrogenation at elevated temperatures [42]. Mechanochemical synthesis from a mixture of $CaCl_2$ and $LiBH_4/NaBH_4$ under an argon pressure of 0.1 MPa has been reported to produce calcium borohydride and chlorides as by-phases [43]. Recently, calcium borohydride was prepared

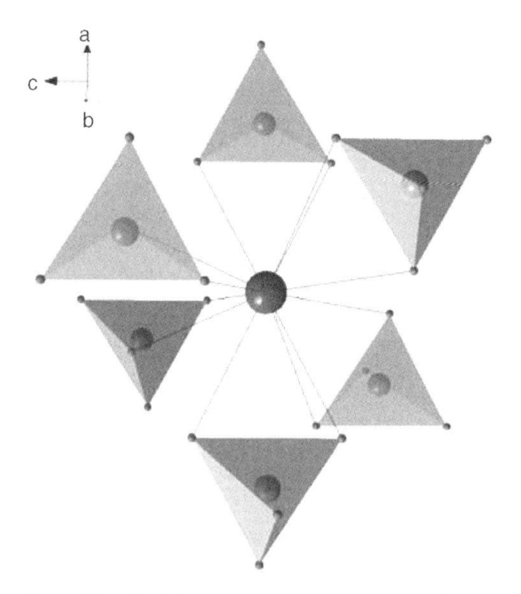

Figure 5.4 $Ca(BH_4)_2$: $Ca(BH_4)_2$ structure: Coordination of the Ca atom by six $[BH_4]^-$ tetrahedra.

under rather harsh conditions by reacting ball-milled CaB_6 and CaH_2 at 70 MPa hydrogen pressure and heating to 400–440 °C [44]. The crystal structure of α-$Ca(BH_4)_2$ (space group Fddd) consists of Ca^{2+} atoms each coordinated octahedrally by six $[BH_4]^-$ anions (Figure 5.4). Each $[BH_4]^-$ has three Ca^{2+} neighbors [45]. While the bond between Ca and $[BH_4]^-$ has ionic character, the internal bonding of $[BH_4]^-$ is covalent.

5.2.9.2 **Decomposition**

The decomposition reaction with a hydrogen release of 9.6 wt.% and an enthalpy change of 32 kJ mol^{-1} H_2 was predicted to be (Eq. (5.12)) [45]:

$$Ca(BH_4)_2 \rightarrow 2/3CaH_2 + 1/3CaB_6 + 10/3H_2 \tag{5.12}$$

Starting from THF educt and removing the solvent in vacuum at elevated temperatures leads to the formation of the solvent-free low temperature α-phase (orthorhombic). Riktor *et al.* proposed the existence of another orthorhombic low-temperature form called the γ-form [41]. However, since the given lattice parameters are very similar to the values of the α-form and no further structure details were provided, the existence of such a phase is questionable. Further heating of the α-phase leads to transformation into the β-form (tetragonal) at about 170 °C [46, 47]. The β-form decomposes in two endothermic steps between 360 and 500 °C into an unidentified intermediate phase and CaH_2. Interestingly, Rönnebro and Majzoub observed CaH_2 and CaB_6 as final products of the dehydrogenation at 400 °C, corresponding to the educts. From this they conclude a reversible system with 9.6 wt.% H_2 capacity [44]. However, the synthesis conditions and therewith the conditions for hydrogen loading of 70 MPa H_2 and 400 °C might be too challenging for any application. Even the positive influence of Ti and Nb catalysts on the desorption temperatures and the re-adsorption properties cannot hide the fact that the reaction temperature of 350 °C held for 24 h is unreasonable for reloading a hydrogen storage tank in mobile applications [48].

5.2.10 Aluminum Borohydride $Al(BH_4)_3$

5.2.10.1 Synthesis and Crystal Structure

Compared to the ionic alkali boron hydrides, the $[BH_4]^-$ units in $Al(BH_4)_3$ are covalently bonded by two hydrogen bridges to the Al metal center. With a hydrogen content of 16.8 wt.% H_2 the $Al(BH_4)_3$ systems seem to be attractive for hydrogen storage but, from a thermodynamic point of view, the material is too unstable for reversible storage of hydrogen. $Al(BH_4)_3$ is a liquid at room temperature with a melting point of −64 °C. It can be easily prepared in a metathesis reaction according to Eq. (5.13) starting from aluminum trichloride and sodium borohydride [49]:

$$AlCl_3 + 3NaBlH_4 \rightarrow Al(BH_4)_3 + 3NaCl \quad (5.13)$$

5.2.10.2 Decomposition

The decomposition and the release of hydrogen start at around 40 °C. However, the material has not been investigated in detail because of its extreme sensitivity to water and air. The solid $Al(BH_4)_3$ exists in two modifications with a phase transition temperature between −93 and −78 °C [50]. The heat of formation of $Al(BH_4)_3$ was calculated to be −132 and −131 kJ mol^{-1} for the α- and β-forms without zero point energy correction. Weak interactions between molecules in the solid state explain the low melting point of the material [6].

5.2.11 Zinc Borohydride $Zn(BH_4)_2$

Zinc borohydride as an example of a transition metal compound contains 8.5 wt.% H_2. The decomposition temperature of about 85 °C is quite low and is associated with the simultaneous melting of the compound. Between 85 and 140 °C an endothermic decomposition starts which is accompanied by the liberation of hydrogen and significant amounts of diborane. This excludes the use of $Zn(BH_4)_2$ for solid hydrogen storage [51].

5.2.12 $NaBH_4$ as a Hydrogen Storage Material in Solution

The high stability of $NaBH_4$ in alkaline solutions offers an alternative for the delivery of hydrogen for PEM fuel cells. Half of the four hydrogen molecules released originate from water which increases the storage capacity of the system.

Accordingly to Eq. (5.14), four moles of hydrogen are produced from one mole of $NaBH_4$ [52].

$$NaBH_4 + 2H_2O \rightarrow NaBO_2 + 4H_2 \qquad (5.14)$$

The major drawback of the system is that the amount of storable hydrogen is limited due to the solubility of the boron hydride and of the decomposition product in the alkaline solution. $NaBO_2$ has a solubility in water of 26 g/100 ml at 20 °C, whereas $NaBH_4$ is more soluble with an amount of 55 g/100 ml at 25 °C. To prevent the precipitation of sodium metaborate in the solution, the amount of $NaBH_4$ used must be lower than the maximum solubility of the metaborate. Millennium Cell has commercialized a hydrogen-producing system based on $NaBH_4$ in combination with a fuel cell for electricity on-demand. For a 20 wt.% $NaBH_4$ solution, stabilized with 1% of NaOH, the storage capacity of the hydride system is reduced to 4 wt.%. This low concentration is necessary to prevent blocking of the active sites of the Ru catalyst essential for fast decomposition of the hydride in alkaline solutions. Figure 5.5 shows a schematic view of the commercialized system from Millennium Cell.

Alternatively, $NaBH_4$ could be used in borohydride fuel cell systems, where the borohydride is directly oxidized [53]. Such systems have a slightly higher potential of $E^\circ = 1.64$ V, compared to the PEM fuel cell with hydrogen gas as energy source. The reaction at the cathode is described as follows (Eq. (5.15)):

$$2O_2 + 4H_2O + 8e^- \rightarrow 8OH^- \quad E^\circ = 0.40\ \text{V} \qquad (5.15)$$

and the reaction at the anode is (Eq. (5.16)):

$$NaBH_4 + 8OH^- \rightarrow NaBO_2 + 6H_2O + 8e^- \quad E^\circ = -1.24\ \text{V} \qquad (5.16)$$

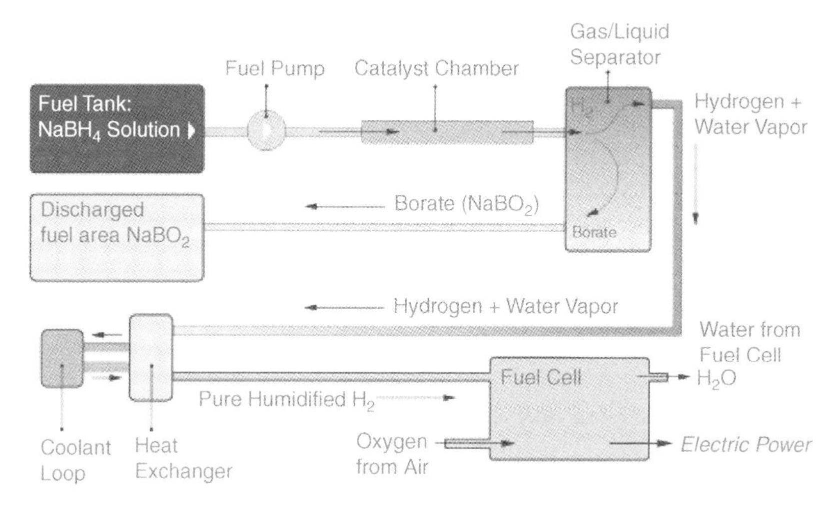

Figure 5.5 Operating scheme of a commercial $NaBH_4$–PEMFC system (With permission from Millennium Cell.)

The product of the decomposition is, like in the Millennium Cell system, sodium metaborate. The system shows the same problems for the regeneration of $NaBH_4$ from an aqueous solution of metaborate as described below. One advantage of the direct borohydride fuel cell systems is that platinum as catalyst is not needed. Unfortunately, depending on the temperature of the solution, some hydrogen gas is produced in a side reaction. However, this hydrogen can be piped out or can be used as additional fuel in a subsequent PEM fuel cell.

5.2.12.1 Regeneration of Decomposed $NaBH_4$ in Solution

For application of $NaBH_4$ as a hydrogen storage material, a cheap process for the mass production and/or recovery of $NaBH_4$ in solution has to be developed. Unfortunately, the produced sodium metaborate, $NaBO_2$, is a thermodynamic sink and makes every process quite unfavorable. In addition to the thermodynamic problem, $NaBO_2$ is dissolved in alkaline aqueous solution. A high amount of energy is necessary to boil off the water and dry the metaborate. Regeneration of $NaBH_4$ requires dry conditions. Starting from dry $NaBO_2$, two different processes have been developed. In the first, $NaBO_2$ is converted into $NaBH_4$ using MgH_2 during a ball-milling process (Eq. (5.17)) [54, 55].

$$NaBO_2 + 2MgH_2 \rightarrow NaBH_4 + 2MgO \tag{5.17}$$

The second process proceeds as a dynamic hydrogenation/dehydrogenation reaction. It starts from $NaBO_2$ and Mg and uses the ball-milling process in a hydrogen atmosphere [56]. Under these conditions highly reactive MgH_2 is produced, which reduces $NaBO_2$ directly to $NaBH_4$. Small particle size and a high specific surface area increase the reaction rate and the yield of the reaction.

5.3 Complex Aluminum Hydrides

5.3.1 Introduction

During the last decade complex aluminum hydrides have received increasing attention as potential candidates for hydrogen storage. The advantageous properties of complex aluminum hydrides are the low price of the compounds, their low weight, their nontoxicity, and the fact that no volatile gas products, except hydrogen, are formed. Most of the compounds possess high gravimetric hydrogen densities but the thermodynamics of most compounds is the limiting factor for their use as a reversible onboard hydrogen carrier. High kinetic barriers were the reason that complex aluminum hydrides were not considered as hydrogen carriers for many years. On the other hand, the thermodynamic properties of complex hydrides are most important for their application in combination with fuel cells. The thermodynamic

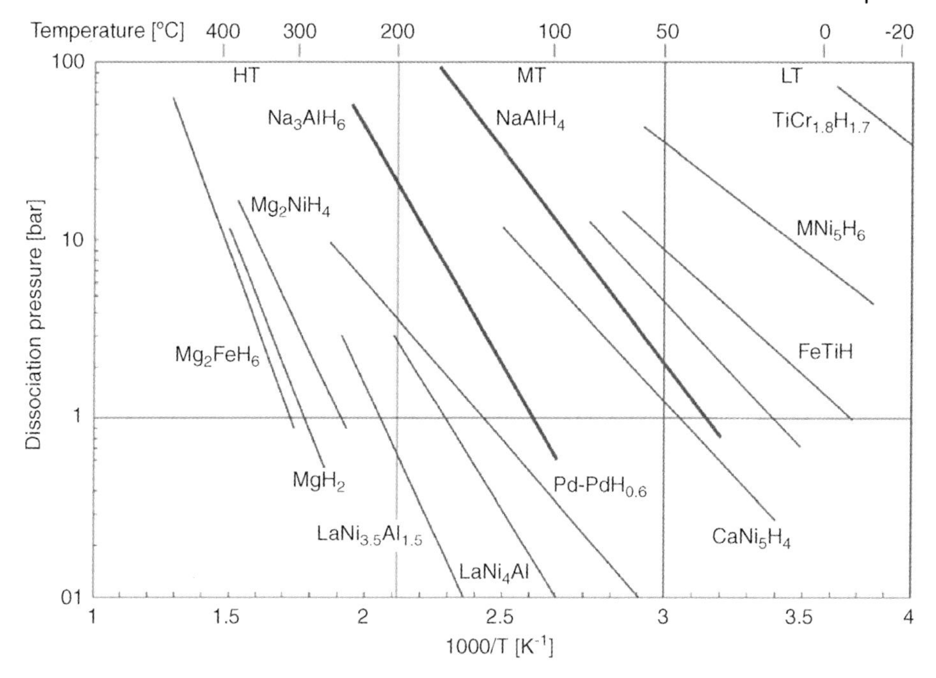

Figure 5.6 Van't Hoff diagram of selected complex hydrides and metal hydrides (Reproduced from [94], with permission from Elsevier).

properties of different compounds are shown by the van't Hoff plot describing the equilibrium pressure depending on the temperature (Figure 5.6).

The van't Hoff plot shows the reason why $NaAlH_4$ is one of the most promising candidates for reversible hydrogen storage. $NaAlH_4$ as a typical low-temperature metal hydride exhibits an equilibrium pressure of 0.1 MPa at 35 °C and a storage capacity of 3.7 wt.% for the first decomposition step. This is twice the amount stored in interstitial metal hydrides.

One detail which has to be taken into account is the heat management of a propulsion system. Considering a tank system containing 6 kg H_2 and exhibiting an enthalpy of formation of about 20 MJ kg^{-1} H_2, typical for many hydrides, 120 MJ thermal load would have to be compensated during refueling [57]. Another disadvantage of complex aluminum hydrides is that many members of the family cannot be rehydrogenated under reasonable physical conditions.

The crystal structures of all complex aluminum hydrides are built up by $[AlH_4]^-$ tetrahedra or $[AlH_6]^{3-}$ octahedral units. These building units can be either isolated, as for example in $NaAlH_4$, or they can form more complex structures such as chain-like structures, as for $CaAlH_5$. The decomposition of alkaline earth aluminum hydrides proceeds via hydrides to intermetallic compounds whereas alkali metal alanates decompose via an intermediate hexahydride structure to the corresponding hydride. Table 5.2 summarizes the physical data of selected complex aluminum hydrides.

Table 5.2 Physical constants of selected complex aluminum hydrides.

	Molecular weight (g mol^{-1})	Melting point (°C)	Start of decomposition[a)] (°C)	Hydrogen content (wt.%)
$LiAlH_4$	40.0		170 [69–73]	10.6
$NaAlH_4$	54.0	183	230 [92]	7.5
$KAlH_4$	70.1	—	>300 [153,154]	5.8
$Mg(AH_4)_2$	86.3	—	110–130 [76, 172–174]	9.3
$Ca(AlH_4)_2$	102.1	—	80 [76]	7.9
$LiMg(AlH_4)_3$	124.3	—	120 [76]	9.7
Li_3AlH_6	53.9	—	165–230 [69–73]	11.1
Na_3AlH_6	102.0	—	265 [94]	5.9
$LiMgAlH_6$	64.3	—	170 [76]	9.4

a) Experimental data.

5.3.2 $LiAlH_4$

5.3.2.1 Synthesis and Crystal Structure

Beside its use as a reducing agent in organic chemistry, $LiAlH_4$ belongs to the complex light metal aluminum hydrides with a very high hydrogen content of 10.5 wt.% H_2. Different routes exist for the preparation of lithium aluminum hydride. The first synthesis was reported in 1947 starting from LiH and an ether solution of $AlCl_3$ [58]. Alternatively, it can be obtained by direct synthesis from LiH and Al using $TiCl_3$ as catalyst precursor and tetrahydrofuran (THF) as complexing agent [59, 60]. The first single crystal structure investigations of $LiAlH_4$ were performed by Sklar and Post [61]. The monoclinic structure (space group $P2_1/c$) was described to consist of almost regular $[AlH_4]^-$ tetrahedra and lithium as bridging ions between the tetrahedra. Shortly after publication of the crystal structure data, other authors proposed different cells [62, 63]. More than 30 years later, Hauback *et al.* re-investigated the early structure analysis by combined neutron and X-ray powder diffraction studies and confirmed the results of Sklar and Post [64]. The re-investigation yielded more reasonable Al–D and Li–D distances than the very short distances reported in the early work [61]. The Li ions are coordinated by 5 D atoms forming LiD_5 bipyramids sharing edges with neighboring LiD_5 bipyramids. In the following years, several theoretical calculations on the crystal structure, the thermodynamic stability of $LiAlH_4$, and mechanistic investigations of its decomposition were published [65–68].

5.3.2.2 Decomposition of $LiAlH_4$

The decomposition of $LiAlH_4$ involves structural phase transformations and several decomposition steps and has been intensively investigated [69–73]. First, the endothermic melting of $LiAlH_4$ between 165 and 175 °C is observed, followed by the exothermic decomposition ($\Delta H = -10\,\mathrm{kJ\,mol^{-1}}\,H_2$) and recrystallization of Li_3AlH_6 between 175 and 220 °C (Eq. (5.18)). During this first step the hexahydride structure, consisting of isolated $[AlH_6]^{3-}$ octahedral, is formed from the tetrahydride structure [62, 74]. In a further endothermic reaction step ($\Delta H = 25\,kJ\,mol^{-1}\,H_2$), the hexahydride decomposes and LiH and Al metal are formed (Eq. (5.19)). The final dehydrogenation step is associated with the decomposition of LiH between 370 and 483 °C (Eq. (5.20)).

$$3LiAlH_4 \rightarrow Li_3AlH_6 + 2Al + 3H_2 \quad (5.3\,wt.\%\,H_2) \tag{5.18}$$

$$Li_3AlH_6 \rightarrow 3LiH + Al + 1.5H_2 \quad (2.65\,wt.\%\,H_2) \tag{5.19}$$

$$3LiH \rightarrow 3Li + 1.5H_2 \quad (2.65\,wt.\%\,H_2) \tag{5.20}$$

The exothermic decomposition of $LiAlH_4$ to Li_3AlH_6 makes rehydrogenation impossible and $LiAlH_4$ can therefore not be used as a reversible hydrogen storage material. The stability of $LiAlH_4$ at room temperature is caused by the slow kinetics of the solid state transformation to the hexahydride and not by the thermodynamics of the system. In addition to the melting and decomposition processes, a phase transformation of a metastable phase, $LiAlH_{4-\alpha'}$, was reported, which transforms to $LiAlH_{4\text{-}\alpha}$ at 95 °C [75]. The powder pattern of the metastable phase is different to that of $LiAlH_{4\text{-}\alpha}$ with a larger unit cell and with monoclinic symmetry ($a = 9.42(2)$, $b = 11.22(2)$, $c = 11.05(2)$ Å and $\beta = 108.9(4)°$). The existence of the metastable $LiAlH_{4-\alpha'}$ phase was also confirmed by *in situ* X-ray diffraction experiments showing a phase transformation to the α-form at about 70 °C [76].

5.3.2.3 **Role of Catalysts**

Most of the publications summarized so far deal only with the decomposition of $LiAlH_4$. However, a solid-state hydrogen storage material should reversibly re-adsorb hydrogen. For $NaAlH_4$, Bogdanović and Schwickardi [1] could show that the addition of a catalyst enhances the kinetics, lowers the decomposition temperature, and, most importantly, makes the rehydrogenation of the compound feasible. The influence of both high energy ball milling and the presence of catalysts on the stability of $LiAlH_4$ has been the topic of several publications [77–84]. The presence of catalysts accelerates the decomposition of $LiAlH_4$ and complete decomposition to the hexahydride is observed already after 5 min of ball milling with $TiCl_4$ (3 wt.%) as catalyst [81]. Other Ti compounds are less active. The catalyst does not maintain its original structure but transforms into an intermetallic alloy of composition $TiAl_3$, as verified by X-ray diffraction measurements [81]. Nevertheless, the presence of $TiAl_3$ does not answer the question whether the alloy is the catalytic active species. It describes only the final state of the system after ball-milling and decomposition. A more detailed description of the doping processes of complex aluminum hydrides with transition metal compounds follows in the section on $NaAlH_4$.

The theoretical work of Løvvik was driven by the question to which position Ti moves to when added to complex aluminum hydride surfaces [85]. As intensively discussed for $NaAlH_4$, without considering energetic or crystallographic issues, there are different sites at which Ti could anchor. It could first react with surface atoms and then remain in the surface regions. It might also form interstitial species in the bulk lithium aluminum hydride or replace one or more atoms in the crystal structure of the lithium aluminum hydride. There are different adsorption sites possible: (i) on-top adsorption above Li and Al ions (ii) bridge sites, and (iii) hollow sites. Even though there are some trends, the calculations do not allow an unambiguous statement as to whether Al or Li is the nearest metal atom to the Ti atom. The most notable trend is that the most stable sites are hollow or bridge sites below the surface.

The thermodynamics of the decomposition are important for a prospective application as a hydrogen storage material. For $LiAlH_4$ the first decomposition step is exothermic, which means the decomposition is irreversible under ambient conditions. The second step is slightly endothermic and, in principle, reversible under restrictive conditions. So far, reversibility in the solid state has been reported only by Chen *et al.* under 4 MPa of hydrogen pressure [82]. However, the results presented have not been reproduced [83]. Additionally, the reaction enthalpies of 9.8 kJ mol^{-1} for the first step and 15.7 kJ mol^{-1} for the second step, as predicted by Løvvik et al. [67], could not be confirmed by other calculations [86]. A combined study of DFT calculations and powder diffraction investigations underlined that the rehydrogenation of Li_3AlH_6 to $LiAlH_4$ is endothermic (9 kJ mol^{-1} H_2) [66]. This means that direct hydrogenation is not possible at ambient temperature [87]. The weakly exothermic hydrogenation of LiH and Al to form Li_3AlH_6 might enable a rehydrogenation under very high pressures. Recently Wang *et al.* confirmed the previous results of Clasen [59] and Ashby *et al.* [60] for the rehydrogenation of Li_3AlH_6, LiH, and Al in the presence of THF [88, 89]. Dehydrogenation takes place below 100 °C and provides about 4 wt.% H_2. Rehydrogenation takes place at room temperature at pressures up to about 10 MPa. However, the presence of the organic adduct leads to additional problems, that is, the organic compound has to be removed prior to dehydrogenation.

Summarizing the studies on $LiAlH_4$ as a potential candidate for reversible hydrogen storage shows that $LiAlH_4$ is thermodynamically very unstable. Only the slow kinetics prevent a spontaneous decomposition of $LiAlH_4$ to the more stable Li_3AlH_6. For thermodynamic reasons, rehydrogenation to $LiAlH_4$ under reasonable physical conditions is not possible.

5.3.3 Li_3AlH_6

5.3.3.1 **Synthesis and Crystal Structure**

A very convenient way for the direct preparation of Li_3AlH_6 avoiding the use of any solvent is ball milling (Eq. (5.21)) [90]:

$$LiAlH_4 + 2LiH \rightarrow Li_3AlH_6 \tag{5.21}$$

Another way to obtain Li_3AlH_6 is by thermal decomposition of $LiAlH_4$ producing Al as the by-phase. After the first report on lithium aluminum hexahydride in 1966 [90], it took quite a long time before the crystal structure could be determined from combined X-ray and neutron diffraction data [91]. The trigonal structure (space group R-3) of Li_3AlH_6 consists of isolated $[AlH_6]^{3-}$ octahedra. The Li atoms are connected to two corners and two edges of the $[AlH_6]^{3-}$ octahedra. Due to the smaller size of the Li cations in comparison to the Na cations in the Na_3AlH_6 structure, the coordination number of Li is 6 whereas the coordination number of Na in Na_3AlH_6 is both 6 and 8. Each $[LiH_6]^{3-}$ octahedron shares edges with four other Li octahedra and additionally corners with four other Li octahedra. Pure Li_3AlH_6 obtained by direct synthesis starts to decompose at the same temperatures as Li_3AlH_6 obtained by thermolysis.

5.3.4 $NaAlH_4$

5.3.4.1 Synthesis and Crystal Structure

Sodium aluminum hydride can be synthesized by many different routes. The first preparation was performed by the reaction of NaH and $AlBr_3$ in the presence of dimethyl ether as a solvent [92]. The direct synthesis from NaH and Al under elevated hydrogen pressures and in various solvents is possible [59, 60]. However, sodium aluminum hydride can also be synthesized by melting the elements in the absence of any solvents (Eq. (5.22)) [93].

$$Na + Al + 2H_2 \rightarrow NaAlH_4 \quad (T < 270\text{–}280\,°C,\ p > 17.5\,MPa) \qquad (5.22)$$

Metal-doped sodium aluminum hydride can be obtained by ball milling or by wet chemical reactions of pre-synthesized $NaAlH_4$ with a catalyst [94–97] or by "direct synthesis" of NaH/Al powders with the doping agent in a ball mill or wet chemical reaction. This step is followed by hydrogenation under elevated hydrogen pressure [98]. Highly reactive $NaAlH_4$ can be synthesized from NaH and Al with the addition of $TiCl_3$ in a "one-step direct synthesis" using the ball milling method under hydrogen pressures [99]. The properties of sodium aluminum hydride prepared by this procedure are much improved compared to doping of pre-synthesized sodium aluminum hydride with $TiCl_3$. The use of nanosized Ti catalysts also leads to improved kinetics [100, 101].

The tetragonal crystal structure (space group $I4_1/a$) of $NaAlH_4$ was determined by Lauher *et al.* from single crystal data [102]. The structure consists of isolated $[AlH_4]^-$ tetrahedra and distorted triangular NaH_8 dodecahedra (Figure 5.7). While the interatomic Al–H distance of 1.532 Å of the original structure was too short, the Al–H bond for the re-determined structure was reported to be reasonably large at 1.61 Å [103]. The localization of the hydrogen/deuterium atoms was obtained from neutron diffraction data [104].

5.3.4.2 Decomposition and Thermodynamics of $NaAlH_4$

Finholt *et al.* described the decomposition of sodium aluminum hydride to be a very slow process starting between 145 and 183 °C where, at the latter temperature, the

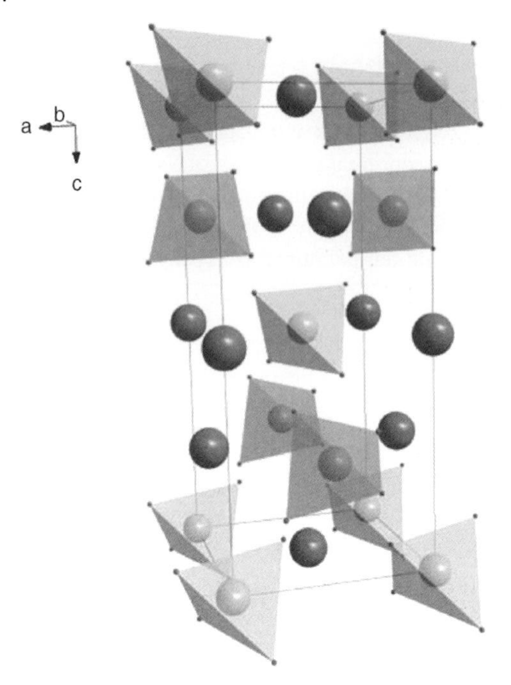

Figure 5.7 Crystal structure of $NaAlH_4$ showing the presence of isolated $[AlH_4]^-$ tetrahedra and the coordinating Na atoms in one unit cell.

solid starts to melt. At about 230–240 °C the following decomposition reaction was proposed (Eq. (5.23)) [92]:

$$NaAlH_4 \rightarrow NaH + Al + 1.5H_2 \quad (5.23)$$

For many years, the hydrogen content seemed to be sufficiently high to make $NaAlH_4$ a promising material for solid hydrogen storage. The disadvantages of $NaAlH_4$ were the high dehydrogenation temperature and the unfavorable kinetics limiting the reversibility. The kinetics of dehydrogenation as well as of rehydrogenation are way too slow for any application [93]. However, the real break-through came with the discovery of Bogdanović and Schwickardi that doping sodium aluminum hydride with a catalyst improves the kinetics of dehydrogenation and makes the reaction reversible. The decomposition and re-adsorption of hydrogen in the $NaAlH_4$ system is one of the most intensively investigated reactions in the field of solid hydrogen storage materials. The process of reversible hydrogenation, the influence of different catalysts and the doping process itself have been the topic of numerous publications [105–116]. The question of how sodium aluminum hydride behaves during cycling is an important issue for an applicable storage system. It could be shown that Ti-doped $NaAlH_4$ is very stable during cycling (Figure 5.8) maintaining the storage capacity of about 3 wt.% over 100 cycles [117].

In situ diffraction studies during the release of hydrogen allowed a comprehensive insight into the phase transformation processes taking place during the reversible storage of hydrogen [113, 114]. The decomposition mechanism for sodium aluminum

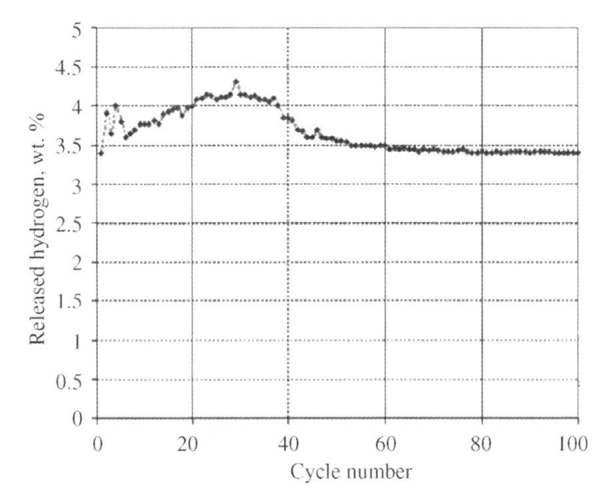

Figure 5.8 The release of hydrogen (wt.%) vs. 100 dehydrogenation cycles. (Reproduced from [117], with permission from Elsevier.)

hydride can be summarized by the following two steps (Eqs. (5.24) and (5.25)):

$$3NaAlH_4 \rightarrow Na_3AlH_6 + 2Al + 3H_2 \tag{5.24}$$

$$Na_3AlH_6 \rightarrow 3NaH + Al + 1.5H_2 \tag{5.25}$$

During the first endothermic step 3.7 wt.% hydrogen is released, the enthalpy of the reaction was measured to be $\Delta H = 37\,kJ\,mol^{-1}\,H_2$. The second step is associated with a hydrogen release of 1.8 wt.% with $\Delta H = 47\,kJ\,mol^{-1}\,H_2$. A further decomposition of NaH would increase the overall capacity to 7.4 wt.%. However, NaH is very stable and releases hydrogen only at temperatures above 450 °C.

5.3.4.3 **Role of Catalysts**

The slow hydrogen release rate and the fact that rehydrogenation cannot be achieved under moderate conditions can be influenced by using a catalyst. At first, titanium compounds such as titanium butoxide were under consideration but the catalyst precursor preferred nowadays is $TiCl_3$ which is added already during synthesis of $NaAlH_4$ by ball milling. Not only the Ti system was intensively investigated, other compounds were also considered as catalysts [118]. Even though the positive effect of the catalysts was obvious, the state of the catalyst and its role was unclear and was the topic of numerous publications. First, more descriptive studies were performed. Anton observed a dependence on the ionic size of the catalyst atom and the hydrogen discharge rate [118]. Very low rates were observed for dopants with radii significantly larger or smaller than 0.76 Å. For significant influence, the ionic radius of the catalyst atom should be between the radii of Al^{3+} and Na^{+}. Several research groups tried to localize the catalyst as part of the crystal structure. Sun *et al.* investigated the dependence of the unit cell parameters of $NaAlH_4$ on the amount of dopant [119]. For both Ti-and Zr-doped sodium aluminum hydride they noted an increase in the lattice parameters upon doping, indicating the substitution of Ti or Zr into the bulk lattice of sodium aluminum hydride. These results could not be confirmed by other groups. Crystal structure refinements did not show any changes in the lattice parameters by doping [120–122]. Weidenthaler *et al.* doped sodium aluminum hydride with $TiCl_3$, removed the excess sodium aluminum hydride and investigated the residue representing a high concentration of the catalyst [120]. The X-ray powder patterns showed the presence of crystalline alumina but no crystalline Ti-containing phase could be observed. This pointed to highly dispersed Ti in an X-ray amorphous phase along with metallic aluminum. During heating of the sample, crystalline Al_3Ti formed while metallic aluminum was consumed. This is an indication that, directly after doping, Ti is closely associated with Al without forming a crystallographically defined alloy. XANES measurements confirmed the reduction of Ti^{3+} to the zerovalent state after doping [123–125]. However, the local environment of Ti was discussed controversially. Most investigations pointed to Al as the neighboring atom to Ti [123, 124] but from XPS investigations it was concluded that the next neighbors of Ti are rather Ti than Al [126]. However, it might be daring to deduce such detailed information about the coordination of Ti from X-ray photoelectron data rather than from X-ray absorption studies.

By means of theoretical calculations much effort was made to localize the catalyst on the surface or in the sub-surface of sodium aluminum hydride. The first theoretical calculations showed that it could be energetically possible that Ti substitutes Na in the structure [127], but from a second paper it became evident that it is energetically more stable for Ti to stay on the surface rather than to migrate into the bulk [128]. Chaudhuri *et al.* tried to go one step further and computed the reverse formation of $NaAlH_4$ in the presence of Ti [129]. The aluminum metal surface has a very low affinity for hydrogen but there is a strong indication that a particular local environment of two Ti atoms substituting surface Al is essential for the rehydrogenation reaction. The diffusion of hydride on the metallic aluminum phase and the formation of alane species are proposed to play the key role in the rehydrogenation reaction. Experiments on hydrogen/deuterium scrambling and exchange reactions demonstrated that Ti dissociates dihydrogen on the surface of sodium aluminum hydride [130]. This became evident by the appearance of HD after mixing D_2 with the gas phase of H_2. This does not happen for undoped $NaAlH_4$. About 30% of bulk H atoms are replaced by D atoms over a period of one week. Thus, Ti very probably catalyzes the H–H bond breaking and formation. The dissociation of H_2 on the surface of Ti-doped sodium aluminum hydride has been modeled by DFT calculations [129, 131] Besides Ti, many different metals, such as Zr, Fe, or V, have been tested as possible catalysts. However, $TiCl_3$ is still one of the best catalysts in terms of accelerated dissociation and rehydrogenation. Excellent performance was observed for Ti nanoparticles but the preparation of the nanoparticles is difficult [100, 101]. The use of Sc, Ce, or Pr can substantially decrease the hydrogenation times by a factor of ten at low pressures [132]. As an example, the cycle stability is shown for Ce-doped $NaAlH_4$ (Figure 5.9). At 110 °C and 9.5 MPa, the storage capacity of 4 wt.% H_2 for Ce-doped alanate remains unchanged for 80 cycles. No memory effect with variation of temperature is observed during cycling.

The discovery that $NaAlH_4$ can be reversibly hydrogenated was a breakthrough on the way towards potential solid hydrogen storage systems. The most promising

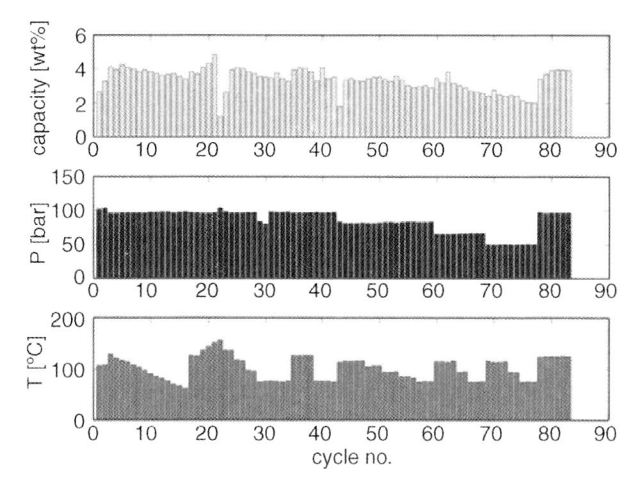

Figure 5.9 Cycle stability test of Ce-doped $NaAlH_4$ (4 mol.%) over 83 cycles (Reproduced from [133], with permission from Elsevier.)

candidate to date, MgH_2 with 7.6 wt.% hydrogen, exhibits very unfavorable thermodynamic properties, such as very high desorption temperatures between 250 and 300 °C. Transition metal hydrides possess only low gravimetric hydrogen contents ($LaNi_5H_6$ 1.5 wt.%). $NaAlH_4$ belongs to the low/middle-temperature reversible hydrides with an equilibrium pressure of 0.1 MPa at ambient temperatures. The hydrogen amount of 3.7 wt.% released during the low temperature decomposition step was by far higher than for all other systems. Also, the second decomposition step with an equilibrium pressure of 0.1 MPa at 130 °C makes the hexahydride a middle-temperature hydride. The rehydrogenation properties of doped $NaAlH_4$ are outstanding compared to any other complex aluminum hydride. Rehydrogenation times of 3–5 min have been achieved [133]. However, certain problems have to be faced with respect to the requirements of low temperature (LT) fuel cells for transportation systems. The decomposition temperature for the second decomposition step is too high to be acceptable for low-temperature PEM fuel cells. On the other hand, using only the hydrogen released during the first decomposition step would not be sufficient to run the overall system. Nevertheless, the development of high-temperature (HT) PEM fuel cells with operating temperatures up to 200 °C opens new prospects for $NaAlH_4$. With the excess heat from the fuel cell both decomposition steps of $NaAlH_4$ can be utilized for hydrogen liberation. At working temperatures of 150 °C for the $NaAlH_4$ tank and 200 °C for the fuel cell system, a temperature difference of 50 °C is available as driving force for the heat transfer from the fuel cell to the tank. At this temperature, the equilibrium pressure of the first decomposition step is roughly 6 MPa ($NaAlH_4$) and 0.3 MPa for the second step (Na_3AlH_6), which means that the hydrogen pressure is high enough to operate the fuel cell systems [134]. Important for a well-operating tank system and fast decomposition of $NaAlH_4$ is the thermal conductivity of the storage material. For sodium aluminum hydride values of 0.2 and 0.35–0.50 $W\,m^{-1}\,K^{-1}$ were reported [135, 136]. It has been shown that the thermal properties of doped $NaAlH_4$ vary significantly with the phase composition and the gas pressure. The thermal conductivity is enhanced with increasing hydrogen pressure [137]. Prototype tanks of doped $NaAlH_4$ containing up to 3 kg of storage material are now under development. Such tank systems are comprised of carbon fiber composite vessels. Heat exchange is carried out with finned tubes for extensive and fast heat transfer [138].

5.3.5 Na_3AlH_6

5.3.5.1 Synthesis and Crystal Structure

Sodium hexaaluminum hydride can be directly synthesized (i) from the elements [139] (Eq. (5.26)) or (ii) by heating $NaAlH_4$ with NaH in heptane (Eq. (5.27)) [140]:

$$3Na + Al + 3H_2 \rightarrow Na_3AlH_6 \quad (\text{toluene}/165\,^{\circ}C,\ AlEt_3,\ 35\,MPa) \tag{5.26}$$

$$NaAlH_4 + 2NaH \rightarrow Na_3AlH_6 \quad (\text{heptane}/160\,^{\circ}C,\ 14\,MPa) \tag{5.27}$$

Another possibility is the direct synthesis by mechanical alloying of $NaAlH_4$ + 2NaH [141].

The powder pattern of Na_3AlH_6 shows significant similarities to the pattern of cryolite, Na_3AlF_6 [142, 143]. Rönnebro *et al.* performed structure refinements and confirmed the isotypic structures of sodium hexaaluminate and cryolite [144]. The crystals structure (space group $P2_1/n$) consists of isolated distorted $[AlH_6]^{3-}$ octahedra (Figure 5.10).

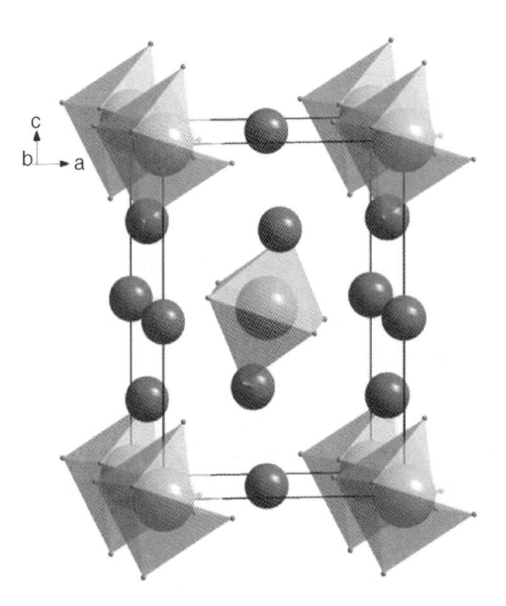

Figure 5.10 Crystal structure of Na_3AlH_6: distorted, isolated $[AlH_6]^{3-}$ octahedra forming a body centered unit cell. Sodium atoms are represented by large spheres.

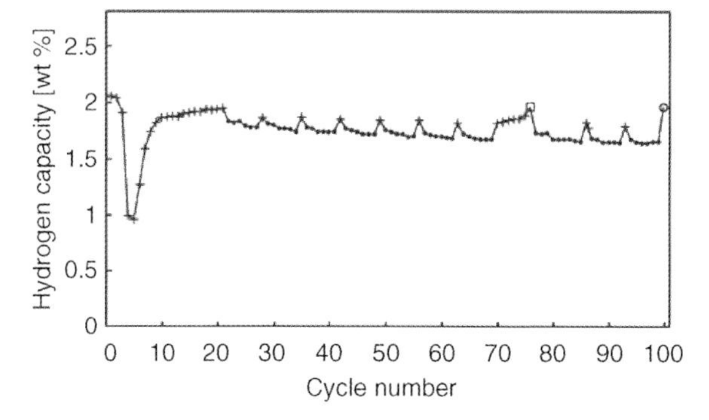

Figure 5.11 Cycle test for Ti-doped Na_3AlH_6 over 100 cycles in a closed system. Hydrogenation conditions: 170 °C/4.2–3.0 MPaH_2 (Reproduced from [1], with permission from Elsevier.)

The formation of a high-temperature stable modification of Na_3AlH_6 was assumed to occur at 222 °C [105]. Bastide *et al.* were the first to report the existence of a cubic β-Na_3AlH_6 phase resulting from a transformation of the monoclinic α-Na_3AlH_6 at 252 °C, as observed by X-ray diffraction [145]. The authors also reported the formation of β-Na_3AlH_6 under high pressure (3.5 GPa) and high temperatures (900 °C). Recently, *in situ* X-ray powder diffraction studies confirmed the reversible phase transformation of α-Na_3AlH_6 to the cubic β-Na_3AlH_6 form [146]. Due to the very narrow stability window of the cubic phase, no structure refinements could be performed. Starting from Na_3AlH_6 the storage capacity is slightly increased to 3 wt.% compared to the second release step when starting from $NaAlH_4$. This is due to the absence of additional Al in the hexahydride synthesized directly. Figure 5.11 shows the stability of the storage capacity over 100 cycles.

5.3.6 $KAlH_4$

5.3.6.1 Synthesis and Crystal Structure

$KAlH_4$ contains 7.4 wt.% hydrogen in total and 4.9 wt.% if only the decomposition to KH is considered. The hydride can be synthesized by different methods: (i) starting from the elements K and Al treated at 250 °C under elevated H_2 pressures [59], (ii) directly from the reaction of KH and Al at H_2 pressures $>$ 17.5 MPa and high temperatures ($>$270 °C) [93], (iii) reaction of either $NaAlH_4$ or $LiAlH_4$ with KCl in THF, followed by extraction in diglyme and subsequent evaporation [147].

The first structural investigations from powder diffraction data were contradictory in terms of the symmetries of the unit cell [148, 149]. However, NMR investigations showed a statistical disorder of the $[AlH_4]^-$ anion [150]. Crystal structure predictions were published with unit cell parameters very close to the orthorhombic unit cell originally proposed by Chini *et al.* [148] almost 40 years ago [66, 151]. Shortly after the theoretical structure calculations, the structure was refined from experimental data [152]. The orthorhombic crystal structure (space group Pbnm) is referred to the $BaSO_4$ structure with isolated $[AlH_4]^-$ tetrahedra. Seven $[AlH_4]^-$ tetrahedra coordinate the potassium atom, resulting in an overall coordination with ten deuterium atoms.

5.3.6.2 Decomposition of $KAlH_4$

Dymova *et al.* investigated the decomposition of $KAlH_4$ up to 750 °C by DSC and X-ray diffraction methods [153, 154]. The amount of hydrogen released was determined by volumetric methods. Three endothermic thermal effects associated with the release of hydrogen are assigned to the following decomposition steps (Eqs. (5.28)–(5.30)):

$$3KAlH_4 \rightarrow K_3AlH_6 + 2Al + 3H_2 \quad (270 - 317\,^\circ C) \tag{5.28}$$

$$K_3AlH_6 \rightarrow 3KH + Al + 1.5H_2 \quad (324 - 360\,^\circ C) \tag{5.29}$$

$$3KH \rightarrow 3K + 1.5H_2 \quad (418 - 438\,^\circ C) \tag{5.30}$$

The thermal decomposition was monitored by ^{27}Al NMR investigations showing for the initial $KAlH_4$ a major signal at 107 ppm assigned to $[AlH_4]^-$ units and a signal at about 20 ppm assigned to $[AlH_3]_n$ [147]. However, $[AlH_3]_n$ is supposed to be unstable and to decompose to Al and H_2. The signal at −40 ppm appearing during the thermolysis of $KAlH_4$, is assigned to $[AlH_6]^{3-}$ units.

The possibility of reversible rehydrogenation was investigated by Morioka *et al.* [155]. Heating $KAlH_4$ in a TPD system at a heating rate of 2 °C min^{-1} leads to a first decomposition to K_3AlH_6, Al, and H_2 (2.4 wt.%) at 300 °C. The second step occurs at 340 °C with a hydrogen release of 1.3 wt.% and the third step at 430 °C with an additional 1.2 wt.% of hydrogen. For the rehydrogenation experiment, KH and Al were exposed to 0.9 MPa H_2 while the temperature was increased to 500 °C. The gradual rehydrogenation was observed to start below 200 °C and to stop at 340 °C. The authors conclude a reversible hydrogenation of $KAlH_4$ at 1 MPa and 250–330 °C, even without the presence of a catalyst. They report a cycle-stable hydrogen capacity between 3 and 4 wt.%. Compared to $NaAlH_4$, $KAlH_4$ is thermodynamically more stable, resulting in a lower enthalpy of hydrogenation. The pressure required for hydrogenation is thus lower. This also means that much higher decomposition temperatures of about 300 °C for the first step are required. First-principles investigations of the crystal structure of the potassium hexahydride revealed α-K_3AlH_6 to have a structure similar to α-Na_3AlF_6. The structure consists of isolated $[AlH_6]^{3-}$ octahedral with an fcc sublattice formed by Al [156].

5.3.7 $Mg(AlH_4)_2$

5.3.7.1 Synthesis and Crystal Structure

$Mg(AlH_4)_2$ contains 9.3 wt.% hydrogen and seemed to be an interesting candidate for solid hydrogen storage provided dehydrogenation is possible.

Three different synthesis routes (Eqs. (5.31)–(5.33)) for $Mg(AlH_4)_2$ in diethyl ether were described by Wiberg and Bauer [157–159]:

$$4MgH_2 + 2AlCl_3 \rightarrow Mg(AlH_4)_2 + 3MgCl_2 \tag{5.31}$$

$$MgH_2 + 2AlH_3 \rightarrow Mg(AlH_4)_2 \quad (5.32)$$

$$MgBr_2 + 2LiAlH_4 \rightarrow Mg(AlH_4)_2 + 2LiBr \quad (5.33)$$

$Mg(AlH_4)_2$ was reported to be soluble in Et_2O and to decompose at 140 °C. The preparation of $Mg(AlH_4)_2$ by hydrogenolysis of a Grignard reagent in Et_2O followed by the addition of $AlCl_3$ was reported by Hartwig [160]. Another synthesis route also using solvents is the reaction of $NaAlH_4$ or $LiAlH_4$ and $MgCl_2$, $MgBr_2$, and MgI_2 in diethyl ether and THF [161–163]. Crystalline $Mg(AlH_4)_2{\cdot}Et_2O$ is further evacuated to remove the solvent. The preparation by mechanochemical activation in a high energy ball mill is an alternative way to prepare $Mg(AlH_4)_2$ [164]. To do this, MgH_2 and AlH_3 are ball milled for several hours. The evidence for the formation of the magnesium aluminum hydride was deduced from DTA experiments by comparing the DTA data of the products with those of the educts. Solvate-free preparations of $Mg(AlH_4)_2$ and of $MgAlH_5$ by mechanochemical activation were reported by Dymova *et al.* [165]. However, the intermediate of the decomposition, $MgAlH_5$, was never observed by others. Ball milling of either $NaAlH_4$ or $LiAlH_4$ with $MgCl_2$ for several hours produces $Mg(AlH_4)_2$ and NaCl or LiCl as by-phases in a metathesis reaction (Eq. (5.34)) [166]:

$$MgCl_2 + 2NaAlH_4/LiAlH_4 \rightarrow Mg(AlH_4)_2 + 2NaCl/LiCl \quad (5.34)$$

The orthorhombic crystal structure of $Mg(AlH_4)_2 \cdot 4THF$ consists of Mg octahedrally coordinated by four THF molecules and two H atoms, bridging two Al atoms. The coordination of the $[AlH_4]^-$ tetrahedra is completed by three terminal hydrogen atoms [167, 168]. The monoclinic crystal structure of the diethyl ether adduct is formed by a polymeric, ribbon-like structure [168]. The Mg atoms are coordinated by five H atoms and one O atom from dimethyl ether. The Al atoms are tetrahedrally coordinated by four H atoms. In contrast to the THF adduct, the $[AlH_4]^-$ groups act as bridging units between the Mg octahedra forming ribbons. For solvent-free $Mg(AlH_4)_2$ the crystal structure could not be solved from early X-ray powder patterns. The application of modern DFT calculations enabled the prediction of a structure model [169]. Finally, the structure was refined from experimental data [170]. The $[AlH_4]^-$ tetrahedra are surrounded by six Mg atoms in a distorted $[MgH_6]^{4-}$ octahedron. $[AlH_4]^-$ tetrahedra and $[MgH_6]^{4-}$ octahedra form a sheet-like structure.

5.3.7.2 Decompositon

Claudy *et al.* investigated the decomposition of solvent-free $Mg(AlH_4)_2$ prepared from $NaAlH_4$, $MgCl_2$ and THF [171]. The decomposition starts between 130 and 160 °C (endothermic, $\Delta H \sim 0\,kJ\,mol^{-1}$, reaction 1) with the formation of MgH_2 and Al. The second step proceeds at about 300 °C (endothermic reaction 2) which was originally assigned to the formation of Mg and Al as crystalline compounds. However, more recent X-ray diffraction studies show the formation of Mg–Al alloy [76, 164, 166, 172–174]. The decomposition mechanism can be formulated as follows (Eqs. (5.35) and (5.36)) [76]:

$$Mg(AlH_4)_2 \rightarrow MgH_2 + 2Al + 3H_2 \quad (110-130\,^{\circ}C) \quad (5.35)$$

$$MgH_2 + 2Al \rightarrow 0.5Al_3Mg_2 + 0.5Al + H_2 \quad (250\,°C) \tag{5.36}$$

Both, ball milling and adding a Ti catalyst have a positive effect on the kinetics of the decomposition [172, 174, 175]. Magnesium aluminum hydride is more unstable than $NaAlH_4$ or $KAlH_4$ [174]. Even after doping with Ti only 0.3 wt.% H_2 can be absorbed at long reaction times at 30 MPa and 80 °C. Magnesium aluminum hydride cannot be reversibly rehydrogenated due to the instability of the compound. With the known formation enthalpy of MgH_2, Claudy *et al.* calculated the formation enthalpy ΔH of $Mg(AlH_4)_2$ to be $-80.2\,kJ\,mol^{-1}$ [171]. Experimentally determined reaction enthalpies for the first step of the decomposition from $Mg(AlH_4)_2$ to MgH_2 of $1.7\,kJ\,mol^{-1}$, confirmed by DFT calculations, show that a reversible rehydrogenation of magnesium aluminum hydride under acceptable physical conditions is not possible [166, 176, 177]. The enthalpy measured for the second decomposition step of MgH_2 to Mg_3Al_2 was $48.8\,kJ\,mol^{-1}$. For the hydrogenation of the Mg_3Al_2 alloy with formation of MgH_2 and Al, an enthalpy of $-63.4\,kJ\,mol^{-1}$ is reported which in consideration of different Mg/Al ratios for both cases is a satisfactory agreement [178, 179]. The decomposition enthalpy of $Mg(AlH_4)_2$ is thus far below that required for reversible hydrogen storage at room temperature (30–40 $kJ\,mol^{-1}$ H_2) [180]. Dymova *et al.* proposed the existence of $MgAlH_5$ formed by two different pathways: (i) during the mechanochemical activation where $Mg(AlH_4)_2$ and MgH_2 react at elevated temperature or (ii) during the decomposition of $Mg(AlH_4)_2$ between 120 and 155 °C [165]. However, the existence of this phase has never been shown by X-ray diffraction. Neither the formation of an intermediate hexahydride, $Mg_3(AlH_6)_2$, as observed for the alkali metal aluminum hydrides, nor the formation of $MgAlH_5$ could be confirmed by other studies. Klaveness *et al.* investigated the theoretical stability of the proposed $MgAlH_5$ compound and postulated the structure to be stable only at −273 °C at ambient pressure [176].

5.3.8 $Ca(AlH_4)_2$

5.3.8.1 Synthesis and Crystal Structure

For complete desorption of $Ca(AlH_4)_2$ 7.7 wt.% hydrogen can be released.

$Ca(AlH_4)_2$ can be synthesized from CaH_2 and $AlCl_3$ in non-aqueous solvents such as THF [181] or dimethyl ether [92]. Unfortunately, most of the products were either not free of solvent or the yield of pure calcium aluminum hydride was very low. Mechanochemical activation of a mixture of $AlCl_3$ and CaH_2 was reported to successfully provide $Ca(AlH_4)_2$ [182]. Another synthesis path would be a metathesis reaction of $NaAlH_4$ or $LiAlH_4$ with $CaCl_2$ by ball milling, forming $Ca(AlH_4)_2$ and NaCl or LiCl as by-phases [166]. Fichtner *et al.* investigated several ways to prepare solvated and desolvated calcium aluminum hydride by wet-chemical and mechanically assisted methods [183]. The crystal structure of the THF adduct, solved from single crystal data, was shown to be similar to that of the $Mg(AlH_4)_2{\cdot}4THF$ adduct. Shortly after, the crystal structure of the solvent-free calcium aluminum hydride was predicted from DFT calculations [184]. The proposed crystal structure is formed by

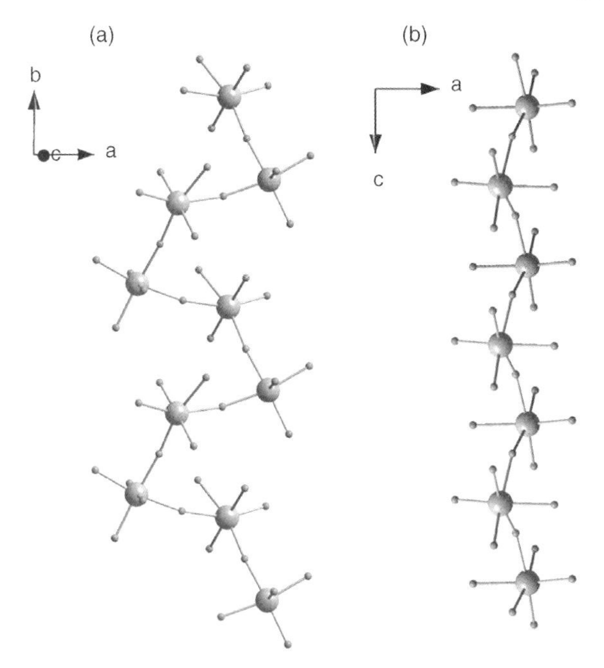

Figure 5.12 (a) Crystal structure of $CaAlH_5$ with helical arrangement of $[AlH_6]^{3-}$ octahedra, (b) crystal structure of $BaAlH_5$ showing the zizag arrangement of corner sharing $[AlH_6]^{3-}$ octahedra.

isolated $[AlH_4]^-$ tetrahedra, of which eight tetrahedra coordinate one calcium atom. The formation enthalpy of $Ca(AlH_4)_2$ was calculated to be around -12.5 kJ mol^{-1} H_2. The crystal structure of the intermediate phase $CaAlH_5$ formed during the first step of decomposition has been predicted by DFT calculations [185]. The structure is based on a α'-$SrAlF_5$ type structure in $P2_1/n$ symmetry. The very complex structure consists of corner sharing $[AlH_6]^{3-}$ octahedra (Figure 5.12a). The helical arrangement is different to the zigzag chains of $[AlH_6]^{3-}$ octahedra in the crystal structure of $BaAlH_5$ which are formed by corner-sharing $[AlH_6]^{3-}$ octahedra (Figure 5.12b) [186]. Each of the calcium atoms of the $CaAlH_5$ structure is coordinated by 9 hydrogen atoms with a distance between 2.3 and 2.6 Å and an additional hydrogen atom with a distance of more than 3 Å.

5.3.8.2 Decomposition of $Ca(AlH_4)_2$

Mal'tseva *et al.* investigated the decomposition of $Ca(AlH_4)_2$ by DTA experiments which showed several exothermic and endothermic reactions associated with hydrogen evolution [182]. From *in situ* X-ray diffraction studies and DSC analysis of solvent-free $Ca(AlH_4)_2$, it is evident that the decomposition proceeds in three steps (Eqs. (5.37)–(5.39)) [76]:

$$Ca(AlH_4)_2 \rightarrow CaAlH_5 + Al + 1.5H_2 \quad (80-100\,^\circ C) \tag{5.37}$$

$$CaAlH_5 \rightarrow CaH_2 + Al + 1.5H_2 \quad (180\text{–}220\,^\circ C) \tag{5.38}$$

$$CaH_2 + 2Al \rightarrow CaAl_2 + H_2 \quad (350\text{–}400\,^\circ C) \tag{5.39}$$

The first step was confirmed to be exothermic ($\Delta H \sim -7.5\,kJ\,mol^{-1}$) which makes a reversible rehydrogenation under reasonable physical conditions almost impossible. The reaction enthalpy for the second step was determined to be about 32 kJ mol^{-1} H_2. The experimental data have been confirmed by first-principles DFT calculations of the dehydrogenation enthalpies, showing that the first step of dehydrogenation is indeed exothermic [187, 188].

5.3.9 Na_2LiAlH_6

Claudy *et al.* obtained Na_2LiAlH_6 by the reaction of $NaAlH_4$ and LiH at 200 °C and 37 MPa H_2 pressure. A mixture of complex $AlEt_3$–NaH heated to 160 °C and pressurized at 30 MPa H_2 for 10 h produced $LiNa_2AlH_6$ [189]. The synthesis of the doped mixed aluminum hydride from $NaAlH_4$, LiH, and NaH in heptane under H_2 pressure was reported first by Bogdanović and Schwickardi. Synthesis by ball milling of a mixture of NaH, LiH, and $NaAlH_4$ avoids the use of any organic solvent [141, 190]. Mechanical alloying of $LiAlH_4$ and NaH also leads to the formation of Na_2LiAlH_6 [191].

The mixed aluminum hydride decomposes in one endothermic step between 220 and 230 °C (Eq. (5.40)):

$$Na_2LiAlH_6 \rightarrow 2NaH + LiH + Al + 1.5H_2 \tag{5.40}$$

The crystal structure was solved from synchrotron powder diffraction data and is in good agreement with the predicted theoretical structure [192, 193]. The structure crystallizes in the perovskite structure type with Li and Al in tetrahedral and octahedral coordination. $[LiH_6]^{5-}$ and $[AlH_6]^{3-}$ octahedra form a very complex network of corner-shared octahedra. Each octahedra is surrounded by six other octahedra of different type. The structure can also be described as a cubic close packing of isolated $[AlH_6]^{3-}$ octahedra with Li filling the octahedral sites and Na filling the tetrahedral sites. Sodium is coordinated to three H atoms from each of the four surrounding $[AlH_6]^{3-}$ units, resulting in a coordination number of 12.

The thermodynamic and kinetic properties of the complex aluminum hydride were analyzed by Ma *et al.* [194]. Different metals salts ($TiCl_3$, TiF_3, $CrCl_3$, and CeO_2) were used as potential catalysts and the influence of ball milling time and type of catalyst on the educts and product formation were analyzed. The strongest effect on the kinetics of rehydrogenation is observed for $TiCl_3$ but the highest capacity for rehydrogenation (2.6 wt.%) is achieved with CeO_2. However, without any catalyst the capacity is about 2.3 wt.% raising the question whether the dopants act as a catalyst at all. The measured capacity of the catalyzed complex is 3.0 wt.% for the first cycle and 2.6 wt.% for the following cycles [191]. The measured decomposition enthalpy of

$\Delta H = 53.5\,kJ\,mol^{-1}\,H_2$ is slightly higher than that of Na_3AlH_6. *In situ* X-ray diffraction experiments followed the decomposition of Na_2LiAlH_6, starting at 160 °C with the appearance of weak reflections belonging to NaH which gain intensity during further heating. Between 250 and 320 °C very weak reflections assigned to LiH are detected. Finally, the decomposition of NaH starts at about 350 °C [76]. All decomposition steps are endothermic. For the mixed aluminum hydride, the decomposition temperatures are lowered compared to the pure aluminum hydrides $NaAlH_4$ and $LiAlH_4$. Na_2LiAlH_6 is a medium temperature material with plateau pressures below 2.5 MPa in the temperature range 180–230 °C [194].

5.3.10 K_2LiAlH_6

Graetz *et al.* report the synthesis of K_2LiAlH_6 via the reaction (Eq. (5.41)) [191]:

$$LiAlH_4 + 2KH \rightarrow K_2LiAlH_6 \tag{5.41}$$

The symmetry of the structure, different to that of pure Li_3AlH_6 and K_3AlH_6 structures, was first predicted to be Fm3m [191, 192]. Rönnebro and Majzoub solved the crystal structure from powder data combined with DFT calculations to be isostructural with the rhombohedral high-temperature structure of K_2LiAlF_6 [195].

The decomposition of the mixed complex aluminum hydride takes place at about 230 °C. The hydrogen absorption without catalysts proceeds slowly at 300 °C. The lack of a clearly defined pressure plateau indicates that the phase does not exhibit a two phase region at about 300 °C. The measured H_2 storage capacity is about 2.3 wt.% [191].

5.3.11 K_2NaAlH_6

K_2NaAlH_6 can be synthesized by ball milling from different educts (Eq. (5.42) and (5.43)) [191]:

$$NaAlH_4 + 2KH \rightarrow K_2NaAlH_6 \tag{5.42}$$

$$KAlH_4 + KH + NaH \rightarrow K_2NaAlH_6 \tag{5.43}$$

Another possibility is the synthesis in an organic medium under a high H_2 pressure of 25 GPa [196]. The crystal structure was predicted by DFT calculations and confirmed by Rietveld refinements [191]. The cubic structure (space group Fm3m) consists of $[AlH_6]^{3-}$ units, octahedrally coordinated Na atoms and K atoms in 12-fold coordination. The structure is very similar to that of β-Na_3AlH_6. Desorption starts at about 260 °C with a hydrogen release of 2.0 wt.% and the re-adsorption of hydrogen is very slow, needing more than 1000 h to reach equilibrium pressure. The estimated decomposition enthalpy is about $97\,kJ\,mol^{-1}\,H_2$ [191].

5.3.12 $LiMg(AlH_4)_3$, $LiMgAlH_6$

5.3.12.1 Synthesis and Crystal Structure

The first report on the preparation and properties of lithium magnesium aluminum hydride, $LiMg(AlH_4)_3$, goes back to the work of Bulychev *et al.* who prepared the compound from $MgCl_2$, $LiAlH_4$, and $NaAlH_4$ in ether [197]. As shown later, $LiMg(AlH_4)_3$ in mixture with LiCl can also be prepared by ball milling of $MgCl_2$ with $LiAlH_4$ in a molar ratio of 1:3 [166]. The crystal structure of $LiMgAlH_6$, the intermediate formed during the decomposition of $LiMg(AlH_4)_3$, was solved by a combination of DFT predictions and Rietveld refinements. The structure is related to Na_2SiF_6, consisting of isolated $[AlH_6]^{3-}$ octahedra connected by Li and Mg atoms. The structure can also be described as alternating $AlMg_3$ and Al_2Li_3 layers [198].

5.3.12.2 Decomposition

$LiMg(AlH_4)_3$ decomposes in a two-step reaction according to (Eqs. (5.44) and (5.45)) [76]:

$$LiMg(AlH_4)_3 \rightarrow LiMgAlH_6 + 2Al + 3H_2 \quad (5.44)$$

$$LiMgAlH_6 \rightarrow LiH + MgH_2 + Al + 1.5H_2 \quad (5.45)$$

The decomposition products are LiH and MgH_2 which could be further decomposed to the elements at higher temperatures. The first step between 100 and 130 °C is exothermic which makes the materials not applicable for hydrogen storage. The second step between 150 and 180 °C is endothermic. The intermediate $LiMgAlH_6$ contains 9.4 wt.% H_2 of which 4.8 wt.% are released in the second step. When heating to about 250 °C, another 3.6 wt.% can be released. From the peak areas of DSC measurements dissociation enthalpies of about $-15\,kJ\,mol^{-1}$ for the first decomposition step (exothermic) and $13\,kJ\,mol^{-1}$ for the second step (endothermic) were calculated [166].

5.3.13 Sr_2AlH_7

Several routes have been published for the synthesis of Sr_2AlH_7. Using $SrAl_2$ as educt, the alloy can be hydrogenated to $SrAl_2H_2$ at 463 K and to Sr_2AlH_7 at 240 °C [199–201]. Another possibility for the synthesis of Sr_2AlH_7 is to ball mill Sr_2Al alloy under hydrogen followed by additional hydrogenation [202, 203]. Ball milling of Sr_2Al under 0.6 MPa hydrogen for 10 h leads first to the formation of SrH_2 and Al. In a second step, hydrogenation at pressures above 3 MPa and temperatures of 260°C for 2 days results in the formation of Sr_2AlH_7. However, the sample still contains about 30% SrH_2. The sample was shown to slowly release hydrogen at 1.3 MPa and 260 °C forming SrH_2 and Al, indicating that the reaction is reversible [203]. Keeping Sr_2AlH_7 under lower pressures (below 1.76 MPa) leads to the formation of $SrAl_4$. A sample containing the Sr–Al alloy cannot be rehydrogenated under reasonable conditions. The formation enthalpy of Sr_2AlH_7 from SrH_2 and Al and H_2 was determined to be $-51\,kJ\,mol^{-1}$ H_2 whereas the enthalpy change for the formation of $SrAl_4$ from the reaction of SrH_2 with Al was determined to be $48\,kJ\,mol^{-1}$ H_2. The reason for the irreversibility of the second reaction might be related to kinetic problems. Sr_2AlH_7 decomposes directly into SrH_2, $SrAl_4$, and H_2 when the hydrogen pressure is below the equilibrium pressure of the irreversible reaction. Zhang *et al.* have reported that Sr_2AlH_7 cannot be obtained successfully by ball milling of a mixture of commercial SrH_2 and Al powders under hydrogen but by ball milling of Sr_2Al under hydrogen [204]. A possible reason for this might be that large particles of Al form (100 nm) during milling while ball milling of Sr_2Al alloy leads to the formation of much smaller Al crystals ($\sim$10 nm). Lower specific surface area of the large particles and enhanced kinetics due to short diffusion pathways for the small particles are very likely the reasons for the distinct hydrogenation characteristics.

Sr_2AlH_7 possesses a very interesting and unusual crystal structure. The monoclinic structure consists of isolated $[AlH_6]^{3-}$ octahedra and of $[HSr_4]^{7+}$ units. The $[HSr_4]^{7+}$ tetrahedra are connected via shared edges and form infinite one-dimensional twisted chains [200, 202]. The presence of $[HSr_4]^{7+}$ units enables an easy rearrangement to SrH_2 during decomposition.

5.3.14 $BaAlH_5$

5.3.14.1 Synthesis and Crystal Structure

The synthesis and the crystal structure of $BaAlH_5$ were reported by Zhang *et al.* [186]. First Ba_7Al_{13} was prepared by arc melting and then the hydrogenation was performed under a hydrogen pressure of 7 MPa and at 240 °C for 5 days. A new hydride was formed together with Al. The orthorhombic crystal structure of $BaAlH_5$, solved from powder data, consists of corner-sharing $[AlH_6]^{3-}$ octahedra, forming one-dimensional zigzag chains along the crystallographic *c*-axis. Due to the size of the Ba atoms, the octahedra are rather distorted. The crystal structure of $BaAlH_5$ was reproduced by DFT calculations [176]. Additionally, the formation enthalpy was calculated to be $-224\,kJ\,mol^{-1}$. Estimating the reaction enthalpy of the decomposition of $BaAlH_5$ to BaH_2 to be $61\,kJ\,mol^{-1}$ H_2, $BaAlH_5$ should be stable. The compound could be used as a reversible storage material but higher temperatures are needed. The measured hydrogen content of 2.56 wt.% is in good agreement with the theoretical hydrogen content of 2.6 wt.%.

While $BaAlH_5$ is formed between 100 and 280 °C, Ba_2AlH_7 was reported to form together with Al during the hydrogenation of Ba_7Al_{13} at temperatures higher than 280 °C [205]. Starting from Ba_7Al_{13} at about 60 °C, $BaAlH_5$ and Al are formed. With increasing temperature, $BaAlH_5$ becomes unstable at about 280 °C and decomposes into Ba_2AlH_7, Al, and H_2. If the temperature is raised above 327 °C, $BaAl_4$ is formed as the most stable decomposition product. The crystal structure of Ba_2AlH_7 is isostructural to Sr_2AlH_7 with isolated $[AlH_6]^{3-}$ octahedra and edge-sharing $[Ba_4H]^{7+}$ tetrahedra forming infinite one-dimensional chains.

5.4
Complex Transition Metal Hydrides

5.4.1
Introduction

Complex metal hydrides are hydrides with very specific properties in terms of synthesis conditions (high pressures up 500 MPa [206]), high decomposition temperatures, high costs for the metal precursor, and often very low storage capacities. However, for completeness, a brief overview of this type of hydride will be given. Most of the work concerning preparation and structure determination of these compounds was reported by the groups of Bronger [206], Noreus [207], and Yvon [208]. For further reading these articles are highly recommended.

Complex metal hydrides can be distinguished by the type of central atom of the anionic complex. Metal borohydrides, complex aluminum hydrides, and aluminum hexahydrides are the most important examples for main group elements. These systems have been described in the previous chapters. Complex hydrides are known for most of the transition metals with the exception of Ti, V, Cr, Ag, Au, and Hg. Depending on the oxidation state and the type of metal, different structural units of the complex anions are observed. Typical examples for building units are octahedral $[FeH_6]^{4-}$, square-pyramidal $[CoH_5]^{4-}$, and square-planar $[PtH_4]^{2-}$ units and linear $[PdH_2]^{2-}$ systems [208]. A summary of the known examples of transition metal complex anions is given in Table 5.3. It can be seen from the table, that not all of the transition metals produce complex hydride anions. Most of the complex anions fulfill the 18 electron rule. It is obvious that the systems with light metal cations are the most promising for hydrogen storage.

5.4.2
Properties

The gravimetric hydrogen content of complex transition metal hydrides can reach values of more than 5 wt.% ($Mg_2FeH_6 = 5.5$ wt%), but most of these systems have a

Table 5.3 Anions of complex transition metal hydrides (from Ref. [208]).

Mn	Fe	Co	Ni	Cu	Zn
$[MnH_4]^{2-}$	$[FeH_6]^{4-}$	$[CoH_4]^{5-}$	$[NiH_4]^{4-}$	$[CuH_4]^{3-}$	$[ZnH_4]^{2-}$
$[MnH_6]^{5-}$		$[CoH_5]^{4-}$			
Tc	Ru	Rh	Pd	Ag	Cd
$[TcH_9]^{2-}$	$[RuH_3]^{6-}$	$[RhH_4]^{3-}$	$[PdH_2]^{2-}$		$[CdH_4]^{2-}$
	$[RuH_4]_n^{4n-}$	$[RhH_6]^{3-}$	$[PdH_3]^{3-}$		
	$[RuH_5]^{5-}$		$[PdH_4]^{2-}$		
	$[RuH_6]^{4-}$		$[PdH_4]^{4-}$		
Re	Os	Ir	Pt	Au	Hg
$[ReH_6]^{5-}$	$[OsH_6]^{4-}$	$[IrH_5]^{4-}$	$[PtH_4]^{2-}$		
$[ReH_9]^{2-}$		$[IrH_6]^{3-}$	$[PtH_6]^{2-}$		
			$[Pt_2H_9]^{5-}$		

hydrogen content lower than 4 wt.%. The volumetric hydrogen storage capacities of many of these systems are similar or higher than for liquid hydrogen and reach 150 g L^{-1} for Mg_2FeH_6. The desorption temperatures at 0.1 MPa vary from $< 100\,°C$ ($BaReH_9$) to more than 400 °C (Mg_3RuH_3). The desorption enthalpies lie in the range 64–130 kJ mol^{-1} H_2. For several reasons these materials are not suitable for application as hydrogen storage materials in combination with fuel cells. One reason is the high price in the case of noble metals (Mg_2RuH_4). Another reason is that the storage capacity is often too low and/or the desorption temperatures are too high. An overview of the composition, the structure of the anions and the properties of these systems is given in Table 5.4.

5.4.3
Synthesis

A first example of a complex transition metal hydride, K_2ReH_9, was prepared in 1961 [209]. Since then numerous new systems have been developed. Starting from potassium perrhenate in ethylenediamine, the hydride was prepared by reduction with potassium metal. In the original work the complex hydride was described as K_2ReH_8 but X-ray analysis showed that discrete $[ReH_9]^{2-}$ units are present [210]. Usual methods for the preparation of complex transition metal hydrides are synthesis in solution, preparation under high pressure, and the ball milling method. One example for the synthesis in solution is the preparation of Li_2ZnH_4 according to the following reaction scheme (Eqs. (5.46) and (5.47)) [211]:

$$2LiCH_3 + Zn(CH_3)_2 \rightarrow Li_2Zn(CH_3)_4 \tag{5.46}$$

$$Li_2Zn(CH_3)_4 + 2LiAlH_4 \xrightarrow{Et_2O} Li_2ZnH_4 + 2LiAl(CH_3)_2H_2 \tag{5.47}$$

In a first step $Li_2Zn(CH_3)_4$ was prepared from methyl lithium and dimethyl zinc. Reduction of this compound with $LiAlH_4$ produces the complex zinc hydride in

Table 5.4 Selected examples of complex transition metal hydrides.

	Molecular weight (g mol^{-1})	Type of anion	Start of decomposition[a)] (°C)	Hydrogen content (wt.%)
$BaReH_9$	332.6	$[ReH_9]^{2-}$	<100	2.7
Mg_3MnH_7	134.9	$[MnH_6]^{5-}$, H^-	280	5.2
Mg_2FeH_6	110.5	$[FeH_6]^{4-}$	320	5.5
Mg_2CoH_5	112.6	$[CoH_5]^{4-}$	280	4.5
Mg_2NiH_4	111.3	$[NiH_4]^{4-}$	280	3.6
$Ba_7Cu_3H_{17}$	1169.2	$[CuH_4]^{3-}$, 5H-	20	1.5
K_2ZnH_4	147.6	$[ZnH_4]^{2-}$	310	2.7

a) Experimental data.

high yields because the product is insoluble in ether and precipitates from the solution. Similar reaction sequences are used for the preparation of Na_2ZnH_4 and K_2ZnH_4. From elemental powders mixtures, Mg_2FeH_6 was prepared at 500 °C under 2–12 MPa hydrogen pressure [212]. The same procedure was used for the synthesis of Mg_2CoH_5 at 350–500 °C and hydrogen pressures between 4 and 6 MPa [213]. In recent years, high energy ball milling has been used for the preparation of different complex transition metal hydrides. Advantages of this method are shorter reaction times and a simple preparation technique. In the case of reactive milling, which means that milling is carried out under a hydrogen atmosphere, the reaction mixture is directly hydrogenated during the milling process. With this technique, Mg_2FeH_6 was prepared starting from a mixture of 2 mole Mg and 1 mole Fe under a pressure of 0.5 MPa and reaction times of 100 h [214, 215]. In a similar manner, Mg_2CoH_5 was synthesized from a stoichiometric mixture of 2 mole MgH_2 and 1 mole Co under a hydrogen atmosphere of 0.1 MPa H_2 [216].

5.4.4 Examples of Complex Transition Metal Hydrides

Mg_2FeH_6, containing octahedral $[FeH_6]^{4-}$ units, is one of the most investigated complex transition metal hydrides. Mg_2FeH_6 is a material with a very high volumetric hydrogen storage capacity of 150 g L^{-1} and a gravimetric hydrogen capacity of 5.5 wt.%. With a decomposition temperature of 320 °C at 0.1 MPa hydrogen, this material is not suitable for room temperature or medium-temperature fuel cell applications. The compound exhibits a high decomposition enthalpy of $\Delta H = 98$ kJ mol^{-1}. Advantageous are the low price of the precursor metals and excellent cycle stability over hundred cycles. These properties could make the material useful for heat storage application in the temperature range around 450–500 °C, for example, for solar heat storage applications [217].

5.5 Summary

It can be summarized that of all the complex aluminum hydrides presented in this chapter only $NaAlH_4$ up till now partially fulfils the requirements for a storage material for mobile fuel cell applications. Over the last decade, the kinetics as well as the cycle stability of doped sodium alanates has been improved. Nevertheless, for application in a low-temperature fuel cell the storage capacity seems to be too low because only the first decomposition step (3.6 wt.% hydrogen) can be used.

Due to their thermodynamic properties some complex aluminum hydrides are not reversible under acceptable technical conditions. Others have too low hydrogen storage capacities which exclude these materials from broad industrial applications.

Boron hydrides do have high hydrogen contents. However, the thermodynamics is in most cases unfavorable, the kinetics or the dehydrogenation and rehydrogenation are slow and the possible evolution of diborane is an additional problem. Therefore boron hydrides do not at present not allow any use as reversible hydrogen storage material.

The complex transition metal hydrides known so far show too low hydrogen contents and can therefore not be considered for hydrogen storage.

A critical view of the properties of complex hydrides shows that it might not be recommended to focus on a single material for a global solution of the hydrogen storage problem. The unfavorable thermodynamics of most systems combined with high material costs and reaction conditions which are not applicable in mobile applications underline that a "single solution" is unrealistic. At present, no system is available which could allow commercial use. Nevertheless, the use of compounds with suitable properties for a particular application cannot be excluded. As an example, solutions of $NaBH_4$ can serve as a hydrogen carrier for PEM fuel cells.

Fundamental research is necessary to understand all the mechanisms involved in the dehydrogenation and rehydrogenation in complex hydrides. Additionally to the understanding of the kinetics it is exceedingly important to develop new systems with much more advantageous thermodynamic properties.

References

1 Bogdanović, B. and Schwickardi, M. (1997) *J. Alloys Compd.*, **253–254**, 1.

2 Schlesinger, H.I., Sanderson, R.T., and Burg, A.B. (1939) *J. Am. Chem. Soc.*, **61**, 536.

3 Schlesinger, H.I., Brown, H.C., Abraham, B., Bond, A.C., Davidson, N., Finholt, A.E., Gilbreath, J.R., Hoekstra, H., Horvitz, L., Hyde, E.K., Katz, J.J., Knight, J., Lad, R.A., Mayfield, D.L., Rapp, L., Ritter, D.M., Schwartz, A.M., Sheft, I., Tuck, L.D., and Walker, A.O. (1953) *J. Am. Chem. Soc.*, **75**, 186.

4 Brown, H.C. (1975) *Organic Syntheses via Boranes*, John Wiley & Sons, Inc., New York, pp. 260.

5 Nakamori, Y., Miwa, K., Ninomiya, A., Li, H., Ohba, N., Towata, S., Züttel, A., and Orimo, S. (2006) *Phys. Rev. B*, **74**, 045126.

6 Miwa, K., Ohba, N., Towata, S., Nakamori, Y., Züttel, A., and Orimo, S. (2007) *J. Alloys Compd.*, **446–447**, 310.

7 Schlesinger, H.I. and Brown, H.C. (1940) *J. Am. Chem. Soc.*, **62**, 3429.

8 Goerrig, D. (27 December 1958) Ger. Pat. 1077644.

9 Harris, P.M. and Meibohm, E.P. (1947) *J. Am. Chem. Soc.*, **69**, 1231.

10 Soulié, J.-Ph., Renaudin, G., Černý, R., and Yvon, K. (2002) *J. Alloys Compd.*, **346**, 200.

11 Fedneva, E.M., Alpatova, V.L., and Mikheeva, V.I. (1964) *Russ. J. Inorg. Chem.*, **9**, 826.

12 Staninevich, D.S. and Egorenko, G.A. (1988) *Russ. J. Inorg. Chem.*, **13**, 341.

13 Züttel, A., Wenger, P., Rentsch, S., Sudan, P., Mauron, Ph., and Emmenegger, Ch. (2003) *J. Power Sources*, **118**, 1.

14 Züttel, A., Rentsch, S., Fischer, P., Wenger, P., Sudan, P., Mauron, Ph., and Emmenegger, Ch. (2003) *J. Alloys Compd.*, **356–357**, 515.

15 Züttel, A., Borgschulte, A., and Orimo, S.I. (2007) *Scr. Mater.*, **56**, 823.

16 Orimo, S., Nakamori, Y., Ohba, N., Miwa, K., Aoki, M., Towata, S., and Züttel, A. (2006) *Appl. Phys. Lett.*, **89**, 021920.

17 Hwang, S.J., Bowman, R.C. Jr., Reiter, J.W., Rijssenbeck, J., Soloveichik, G.L., Zhao, J.C., Kabbour, H., and Ahn, C.C. (2008) *J. Phys. Chem. Lett.*, **112**, 3164.

18 Orimo, S., Nakamori, Y., Kitahara, G., Miwa, K., Ohba, N., Towata, S., and Züttel, A. (2005) *J. Alloys Compd.*, **404–406**, 427.
19 Schlesinger, H.J. and Brown, H.C. (1953) *J. Am. Chem. Soc.*, **75**, 186.
20 Damian, G.A. and Hudson, B.S. (2004) *Chem. Phys. Lett.*, **385**, 166.
21 Kumar, R.S. and Cornelius, A.L. (2005) *Appl. Phys. Lett.*, **87**, 261916.
22 Filinchuk, Y., Talyzin, A.V., Chernyshov, D., and Dmitriev, V. (2007) *Phys. Rev. B*, **76**, 092104-1.
23 Urgnanai, J., Torres, F.J., Palumbo, M., and Baricco, M. (2008) *Int. J. Hydrogen Energy*, **33**, 3111.
24 Banus, M.D., Bragdon, R.W., and Hinckley, A.A. (1954) *J. Am. Chem. Soc.*, **78**, 3848.
25 Orimo, S., Nakamori, Y., and Züttel, A. (2004) *Mater. Sci. Eng. B*, **108**, 51.
26 Vajeeston, P., Ravindran, P., Kjekshus, A., and Fjellvåg, H. (2005) *J. Alloys Compd.*, **387**, 97.
27 Kumar, R.S., Kim, E., and Cornelius, A.L. (2008) *J. Phys. Chem. C*, **112**, 8452.
28 Schlesinger, H.I., Brown, H.C., and Hyde, E.K. (1953) *J. Am. Chem. Soc.*, **75**, 209.
29 Marynick, D.S. and Lipscomb, W.N. (1972) *Inorg. Chem.*, **11**, 820.
30 van Setten, M.J., de Wijs, G.A., and Brocks, G. (2008) *Phys. Rev. B*, **77**, 165115.
31 Černý, R., Filinchuk, Y., Hagemann, H., and Yvon, K. (2007) *Angew. Chem.*, **119**, 5867.
32 Her, J.H., Stephens, P.W., Gao, Y., Soloveichik, G.L., Rijssenbeek, J., Andrus, M., and Zhao, J.C. (2007) *Acta Crystallogr. B*, **63**, 561.
33 Chlopek, K., Frommen, C., Leon, A., Zabara, O., and Fichtner, M. (2007) *J. Mater. Chem.*, **17**, 3496.
34 Li, H.W., Kikuchi, K., Nakamori, Y., Miwa, K., Towata, S., and Orimo, S. (2007) *Scr. Mater.*, **57**, 679.
35 Li, H.W., Kikuchi, K., Nakamori, Y., Ohba, N., Miwa, K., Towata, S., and Orimo, S. (2008) *Acta Mater.*, **56**, 1342.
36 Matsunaga, T., Buchter, F., Mauron, P., Bielman, M., Nakamori, Y., Orimo, S., Ohba, N., Miwa, K., Towata, S., and Züttel, A. (2008) *J. Alloys Compd.*, **459**, 583.
37 van Setten, M.J., de Wijs, G.A., Fichtner, M., and Brocks, G. (2008) *Chem. Mater.*, **20**, 4952.
38 Wiberg, E. and Hartwimmer, R.Z. (1955) *Z. Naturforsch. B: Chem. Sci.*, **10**, 295.
39 Wiberg, E., Noth, H., and Hartwimmer, R.Z. (1955) *Z. Naturforsch. B: Chem. Sci.*, **10**, 292.
40 Mikheeva, V.I. and Titov, L.V. (1964) *Zh. Neorg. Khim.*, **9**, 789.
41 Riktor, M.D., Sørby, M.H., Chlopek, K., Fichtner, M., Buchter, F., Züttel, A., and Hauback, B.C. (2007) *J. Mater. Chem.*, **17**, 4939.
42 Barkhordarian, G., Klassen, T., Bornheim, M., and Bormann, R. (2007) *J. Alloys Compd.*, **440**, L18.
43 Nakamori, Y., Li, H.W., Kikuchi, K., Aoki, M., Miwa, K., Towata, S., and Orimo, S. (2007) *J. Alloys Compd.*, **446–447**, 296.
44 Rönnebro, E. and Majzoub, E.H. (2007) *J. Phys. Chem. B*, **111**, 12045.
45 Miwa, K., Aoki, M., Noritake, T., Ohba, N., Nakamori, Y., Towata, S., Züttel, A., and Orimo, S. (2006) *Phys. Rev. B*, **74**, 155122.
46 Aoki, M., Miwa, K., Noritake, T., Ohba, N., Matsumoto, M., Li, H.W., Nakamori, Y., Towata, S., and Orimo, S. (2008) *Appl. Phys. A*, **92**, 601.
47 Kim, J.H., Jin, S.A., Shim, J.H., and Cho, Y.W. (2008) *J. Alloys Compd.*, **461**, L20.
48 Kim, J.H., Shim, J.H., and Cho, Y.W. (2008) *J. Power Sources*, **181**, 140.
49 Schlesinger, H.I., Brown, H.C., and Hyde, E.K. (1953) *J. Am. Chem. Soc.*, **75**, 209.
50 Aldridge, S., Blake, A.J., Downs, A.J., Gould, R.O., Parsons, S., and Pulham, C.R. (1997) *J. Chem. Soc., Dalton Trans.*, 1007.
51 Jeon, E. and Cho, Y.W. (2006) *J. Alloys Compd.*, **422**, 273.
52 Amendola, S.C., Sharp-Goldman, S.L., Saleem Janjua, M., Spencer, N.C., Kelly, M.T., Petillo, P.J., and Binder, M. (2000) *Int. J. Hydrogen Energy*, **25**, 969.
53 Li, Z.P., Liu, B.H., Arai, K., and Suda, S. (2005) *J. Alloys Compd.*, **404–406**, 648.
54 Kojima, Y. and Haga, T. (2003) *Int. J. Hydrogen Energy*, **28**, 989.
55 Li, Z.P., Morigasaki, N., Liu, B.H., and Suda, S. (2003) *J. Alloys Compd.*, **349**, 232.

56 Suda, S., Morigasaki, M., Iwase, Y., and Li, Z.P. (2005) *J. Alloys Compd.*, **404–406**, 643.
57 Felderhoff, M., Weidenthaler, C., von Helmolt, R., and Eberle, U. (2007) *Phys. Chem. Chem. Phys.*, **9**, 2643.
58 Finholt, A.E., Bond, A., and Schlesinger, H. (1947) *J. Am. Chem. Soc.*, **69**, 1199.
59 Clasen, H. (1961) *Angew. Chem.*, **73**, 322.
60 Ashby, E.C., Brendel, G.J., and Redman, H.E. (1963) *Inorg. Chem.*, **2**, 499.
61 Sklar, N. and Post, B. (1967) *Inorg. Chem.*, **6**, 669.
62 Mikheeva, V.I. and Arkhipov, S.M. (1967) *Russ. J. Inorg. Chem.*, **12**, 1066.
63 Gorin, P., Marchon, J.C., Tranchant, J., Kovacevic, S., and Marsault, J.P. (1970) *Bull. Soc. Chim. Fr.*, **11**, 3790.
64 Hauback, B.C., Brinks, H.W., and Fjellvåg, H.J. (2001) *J. Alloys Compd.*, **346**, 184.
65 Vajeeston, P., Ravindran, P., Vidya, R., Fjellvåg, H., and Kjekshus, A. (2004) *Cryst. Growth Des.*, **4**, 471.
66 Chung, S.C. and Morioka, H. (2004) *J. Alloys Compd.*, **372**, 92.
67 Kang, J.K., Lee, J.Y., Muller, R.P., and Goddard, W.A. III (2004) *J. Chem. Phys.*, **121**, 10623.
68 Løvvik, O.M., Opalka, S.M., Brinks, H.W., and Hauback, B.C. (2004) *Phys. Rev. B*, **69**, 134117.
69 Garner, W.E. and Haycock, E.W. (1952) *Proc. R. Soc. A*, **211**, 335.
70 Dymova, T.N., Konoplev, V.N., Aleksandrov, D.P., Sizareva, A.S., and Silina, T.A. (1995) *Russ. J. Coord. Chem.*, **21** (3), 163.
71 Mikheeva, V.I., Selivokhina, M.S., and Kryukova, O.N. (1956) *Dokl. Akad. Nauk SSSR*, **109**, 439.
72 Bloch, J. and Gray, A.P. (1965) *Inorg. Chem.*, **4**, 304.
73 Brinks, H.W., Hauback, B.C., Blanchard, D., Jensen, C.M., Fichtner, M., and Fjellvåg, H. (2004) *Advanced Materials for Energy Conversion II*, (eds. D. Chandra, R.G. Bautista, L. Schlapbach), TMS, Warrendale, PA, p. 197.
74 Dilts, J.A. and Ashby, E.C. (1972) *Inorg. Chem.*, **11**, 1230.
75 Bastide, J.P., Bonnetot, B., and Létoffé, J.M. (1985) *Mater. Res. Bull.*, **20**, 999.
76 Mamatha, M., Weidenthaler, C., Pommerin, A., Felderhoff, M., and Schüth, F. (2006) *J. Alloys Compd.*, **416**, 303.
77 Dymova, T.N., Aleksandrov, D.P., Konoplev, V.N., Siliana, T.A., and Sizareva, A.S. (1994) *Russ. J. Coord. Chem.*, **20**, 263.
78 Zaluski, L., Zaluska, A., and Ström-Olsen, J.O. (1999) *J. Alloys Compd.*, **290**, 71.
79 Balema, V.P., Pecharsky, V.K., and Dennis, K.W. (2000) *J. Alloys Compd.*, **313**, 69.
80 Balema, V.P., Dennis, K.W., and Pecharsky, V.K. (2000) *Chem. Commun.*, 1665.
81 Balema, V.P., Wiench, J.W., Dennis, K.W., Pruski, M., and Pecharsky, V.K. (2001) *J. Alloys Compd.*, **329**, 108.
82 Chen, J., Kuriyama, N., Xu, Q., Takeshita, H.T., and Sakai, T. (2001) *J. Phys. Chem. B.*, **105**, 11214.
83 Blanchard, D., Brinks, H.W., Hauback, B.C., Norby, P., and Muller, J. (2005) *J. Alloys Compd.*, **404**, 743.
84 Suttisawat, Y., Rangsunvigit, P., Kitiyanan, B., Muangsin, N., and Kulprathipanja, S. (2007) *Int. J. Hydrogen Energy*, **32**, 1277.
85 Løvvik, O.M. (2003) *J. Alloys Compd.*, **178**, 356.
86 Frankcombe, T.J. and Kroes, G. (2006) *J. Chem. Phys. Lett.*, **423**, 102.
87 Orimo, S., Nakamori, Y., Eliseo, J.R., Züttel, A., and Jensen, C.M. (2007) *Chem. Rev.*, **107**, 4111.
88 Wang, J., Ebner, A.D., and Ritter, J.A. (2006) *J. Am. Chem. Soc.*, **128**, 5949.
89 Wang, J., Ebner, A.D., and Ritter, J.A. (2007) *J. Phys.Chem. C*, **111**, 14917.
90 Ehrlich, R., Young, A., Rice, G., Dvorak, J., Shapiro, P., and Smith, H. (1966) *J. Am. Chem. Soc.*, **88**, 858.
91 Brinks, H.W. and Hauback, B.C. (2003) *J. Alloys Compd.*, **354**, 143.
92 Finholt, A.E., Barbaras, G.D., Barbaras, G.K., Urry, G., Wartik, T., and Schlesinger, H.I. (1955) *J. Inorg. Nucl. Chem.*, **1**, 317.
93 Dymova, T.N., Eliseeva, N.G., Bakum, S.I., and Dergachev, Y.M. (1974) *Dokl. Akad. Nauk. SSSR*, **215**, 1369.
94 Bogdanović, B., Brand, R.A., Marjanović, A., Schwickardi, M., and Tölle, J. (2000) *J. Alloys Compd.*, **302**, 36.

95 Zidan, R.A., Takara, S., Hee, A.G., and Jensen, C.M. (1999) *J. Alloys Compd.*, **285**, 119.

96 Jensen, C.M., Zidan, R.A., Mariels, W., Hee, A.G., and Hagen, C. (1999) *Int. J. Hydrogen Energy*, **24**, 461.

97 Sandrock, G., Gross, K., and Thomas, G. (2002) *J. Alloys Compd.*, **339**, 299.

98 Zaluska, A., Zaluski, L., and Ström-Olsen, J.O. (2001) *Appl. Phys. A*, **72**, 157.

99 Bellosta von Colbe, J.M., Felderhoff, M., Bogdanović, B., Schüth, F., and Weidenthaler, C. (2005) *Chem. Commun.*, **37**, 4732.

100 Bogdanović, B., Felderhoff, M., Kaskel, S., Pommerin, A., Schlichte, K., and Schüth, F. (2003) *Adv. Mater.*, **15**, 1012.

101 Fichtner, M., Fuhr, O., Kircher, O., and Rothe, J. (2003) *Nanotechnlogogy*, **14**, 778.

102 Lauher, J.W., Dougherty, D., and Herley, P.J. (1979) *Acta Crystallogr. B*, **35**, 1454.

103 Bel'skii, V.K., Bulychev, B.M., and Golubeva, A.V. (1983) *Russ. J. Inorg. Chem.*, **28**, 1528.

104 Hauback, B.C., Brinks, H.W., Jensen, C.M., Murphy, K., and Maeland, A.J. (2003) *J. Alloys Compd.*, **358**, 142.

105 Claudy, P., Bonnetot, B., Chahine, G., and Létoffé, J.M. (1980) *Thermochim. Acta*, **38**, 75.

106 Gross, K.J., Thomas, G.J., and Jensen, C.M. (2002) *J. Alloys Compd.*, **330**, 683.

107 Gross, K.J., Sandrock, G., and Thomas, G.J. (2002) *J. Alloys Compd.*, **330**, 691.

108 Gross, K.J., Guthrie, S., Takara, S., and Thomas, G. (2000) *J. Alloys Compd.*, **297**, 270.

109 Gross, K.J., Sandrock, G., and Thomas, G.J. (2002) *J. Alloys Compd.*, **330–322**, 691.

110 Balogh, M.P., Tibbetts, G.G., Pinkerton, F.E., Meisner, G.P., and Olk, C.H. (2003) *J. Alloys Compd.*, **337**, 254.

111 Bogdanović, B., Felderhoff, M., Germann, M., Härtl, M., Pommerin, A., Schüth, F., Weidenthaler, C., and Zibrowius, B. (2003) *J. Alloys Compd.*, **350**, 246.

112 Meisner, G.P., Tibbetts, G.G., Pinkerton, F.E., Olk, C.H., and Balogh, M.P. (2002) *J. Alloys Compd.*, **337**, 254.

113 Sandrock, G., Gross, K., Thomas, G., Jensen, C., Meeker, D., and Takara, S. (2002) *J. Alloys Compd.*, **330–332**, 696.

114 Gross, K.J., Majzoub, E.H., and Spangler, S.W. (2003) *J. Alloys Compd.*, **356**, 423.

115 Majzoub, E.H. and Gross, K.J. (2003) *J. Alloys Compd.*, **356**, 363.

116 Bellosta von Colbe, J.M., Bogdanović, B., Felderhoff, M., Pommerin, A., and Schüth, F. (2004) *J. Alloys Compd.*, **370**, 104.

117 Srinivasan, S.S., Brinks, H.W., Hauback, B.C., Sun, D., and Jensen, C.M. (2004) *J. Alloys Compd.*, **337**, 283.

118 Anton, D.L. (2003) *J. Alloys Compd.*, **356–357**, 400.

119 Sun, D., Kiyobayashi, T., Takeshita, H.T., Kuriyama, N., and Jensen, C.M. (2008) *J. Alloys Compd.*, **337**, L8–L11.

120 Weidenthaler, C., Pommerin, A., Felderhoff, M., Bogdanović, B., and Schüth, F. (2003) *Phys. Chem. Chem. Phys.*, **5**, 5149.

121 Brinks, H.W., Jensen, C.M., Srinivasan, S.S., Hauback, B.C., Blanchard, D., and Murphy, K. (2004) *J. Alloys Compd.*, **376**, 215.

122 Majzoub, E.H., Herberg, J.L., Stumpf, R., Spangler, R., and Maxwell, R.S. (2005) *J. Alloys. Compd.*, **394**, 265.

123 Felderhoff, M., Klementiev, K., Grünert, W., Spliethoff, B., Tesche, B., Bellosta von Colbe, J.M., Bogdanović, B., Härtel, M., Pommerin, A., Schüth, F., and Weidenthaler, C. (2004) *Phys. Chem. Chem. Phys.*, **6**, 4369.

124 Graetz, J., Reilly, J.J., Johnson, J., Ignatov, A.Y., and Tyson, T.A. (2004) *Appl. Phys. Lett.*, **85**, 500.

125 Léon, A., Kircher, O., Rothe, J., and Fichtner, M. (2004) *J. Phys. Chem. B*, **108**, 16372.

126 Léon, A., Schild, D., and Fichtner, M. (2005) *J. Alloys Compd.*, **404–406**, 766.

127 Íñiguez, J., Yildirim, T., Udovic, T.J., Sulic, M., and Jensen, C.M. (2004) *Phys. Rev. B*, **70**, 060101. 1.

128 Íñiguez, J. and Yildirim, T. (2005) *Appl. Phys. Lett.*, **86**, 103109.

129 Chaudhuri, S. and Muckerman, J.T. (2005) *J. Phys. Chem. B*, **109**, 6952.

130 Bellosta von Colbe, J.M., Schmidt, W., Felderhoff, M., Bogdanović, B., and

Schüth, F. (2006) *Angew. Chem. Int. Ed.*, **45**, 3663.
131 Chaudhuri, S., Graetz, J., Ignatov, A., Reilly, J.J., and Muckermann, J.T. (2006) *J. Am. Chem. Soc.*, **128**, 11404.
132 Bogdanović, B., Felderhoff, M., Pommerin, A., Schüth, F., and Spielkamp, N. (2006) *Adv. Mater.*, **18**, 1198.
133 Bogdanović, B., Felderhoff, M., Pommerin, A., Schüth, F., Spielkamp, N., and Stark, A. (2009) *J. Alloys Compd.*, **471**, 383.
134 Jensen, J.O., Li, Q., He, R., Pan, C., and Bjerrum, N.J. (2005) *J. Alloys Compd.*, **404–406**, 653.
135 Gross, K.J., Majzoub, E., Thomas, G.J., and Sandrock, G. (2002) Hydride development for Hydrogen Storage. Proceedings of the 2002 U.S. DOE Hydrogen Program Review, NREL/CP-610-32405, Golden, CO.
136 Anton, D.L., Mosher, D.A., and Opalka, S.M. (2003) High density H2 storage demonstration using NaAlH4-based complex compound hydrides 2003. Proceedings of the U.S. Hydrogen Program Review, May 18–22, Berkeley, CA.
137 Dedrick, D.E., Kanouff, M.P., Replogle, B.C., and Gross, K.J. (2005) *J. Alloys Compd.*, **389**, 299.
138 Mosher, D.A., Arsenault, S., Tang, X., and Anton, D.L. (2007) *J. Alloys Compd.*, **446–447**, 707.
139 Ashby, E.C. and Kobetz, P. (1966) *Inorg. Chem.*, **5**, 1615.
140 Zakharkin, L.I. and Gavrilenko, V.V. (1962) *Dokl. Akad. Nauk SSSR*, **145**, 793.
141 Huot, J., Boily, S., Güther, V., and Schulz, R. (1999) *J. Alloys Compd.*, **383**, 304.
142 Subertova, V. (1966) *Collect. Czech. Chem. Commun.*, **31**, 4455.
143 Narray-Szabo, S. and Sasvari, K. (1938) *Z. Kristallogr. A*, **99**, 27.
144 Rönnebro, E., Noréus, D., Kadir, K., Reiser, A., and Bogdanović, B. (2000) *J. Alloys Compd.*, **299**, 101.
145 Bastide, J.P., Bonnetot, B.B., Létoffé, J.M., and Claudy, P. (1981) *Mater. Res. Bull.*, **16**, 91.
146 Weidenthaler, C., Pommerin, A., Felderhoff, M., Schmidt, W., Bogdanović, B., and Schüth, F. (2005) *J. Alloys Compd.*, **398**, 228.
147 Tarasov, V.P., Bakum, S.I., and Novikov, A.V. (2001) *Russ. J. Inorg. Chem.*, **46**, 409.
148 Chini, P., Baradel, A., and Vacca, C. (1966) *Chem. Ind.*, **48**, 596.
149 Bakulina, V.M., Bakum, S.I., and Dymova, T.N. (1968) *Zh. Neorg. Khim.*, **13**, 1288.
150 Tarasov, V.P., Bakum, S.I., and Novikov, A.V. (2000) *Russ. J. Inorg.Chem.*, **45**, 1890.
151 Vajeeston, P., Ravindran, P., Fjellvåg, H., and Kjekshus, A.J. (2003) *J. Alloys Compd.*, **363**, L7.
152 Hauback, B.C., Brinks, H.W., Heyn, R.H., Blom, R., and Fjellvåg, H. (2005) *J. Alloys Compd.*, **394**, 35.
153 Dymova, T.N., Selivokhina, M.S., and Eliseeva, N.G. (1963) *Dokl. Akad. Nauk SSSR*, **153**, 133.
154 Dymova, T.N., Bakum, S.I., and Mirsaidov, U. (1974) *Dokl. Akad. Nauk SSSR*, **216**, 87.
155 Morioka, H., Kakizaki, K., Chung, S.C., and Yamada, A. (2003) *J. Alloys Compd.*, **352**, 310.
156 Vajeeston, P., Ravindran, P., Kjekshus, A., and Fjellvåg, H. (2005) *Phys. Rev. B*, **71**, 092103.
157 Wiberg, E. and Bauer, R. (1950) *Z. Naturforsch.*, **5b**, 397.
158 Wiberg, E. (1953) *Angew. Chem.*, **65**, 16.
159 Wiberg, E. and Bauer, R. (1952) *Z. Naturforsch.*, **7b**, 131.
160 Hartwig, A. (1955) German Patent 921986
161 Ethyl Corp . (1962) British Patent 905985.
162 Ashby, E.C., Schwartz, R.D., and James, B.D. (1970) *Inorg. Chem.*, **9**, 325.
163 Srivastava, S.C. and Ashby, E.C. (1971) *Inorg. Chem.*, **10**, 186.
164 Dymova, T.N., Knoplev, V.N., Sizareva, A.S., and Aleksandrov, D.P. (1999) *Russ. J. Coord. Chem.*, **25**, 312.
165 Dymova, T.N., Mal'tseva, N.N., Konoplev, V.N., Golovanova, A.I., Aleksandrov, D.P., and Sizareva, A.S. (2003) *Russ. J. Coord. Chem.*, **29**, 385.
166 Mamatha, M., Bogdanović, B., Felderhoff, M., Pommerin, A., Schmidt, W., Schüth, F., and Weidenthaler, C. (2006) *J. Alloys Compd.*, **407**, 78.
167 Noeth, H., Schmidt, M., and Treitl, A. (1995) *Chem. Ber.*, **128**, 999.

168 Fichtner, M. and Fuhr, O. (2002) *J. Alloys Compd.*, **345**, 286.

169 Fichtner, M., Engel, J., Fuhr, O., Glöss, A., Rubner, O., and Ahlrichs, R. (2003) *Inorg. Chem.*, **42**, 7060.

170 Fossdal, A., Brinks, H.W., Fichtner, M., and Hauback, B.C. (2005) *J. Alloys Compd.*, **387**, 47.

171 Claudy, P., Bonnetot, B., and Létoffé, J.M. (1979) *J. Therm. Anal.*, **15**, 119.

172 Fichtner, M., Fuhr, O., and Kirchner, O. (2003) *J. Alloys Compd.*, **356**, 418.

173 Fossdal, A., Brinks, H.W., Fichtner, M., and Hauback, B.C. (2005) *J. Alloys Compd.*, **404**, 752.

174 Komiya, K., Morisaku, N., Shinzato, Y., Ikeda, K., Orimo, S., Ohki, Y., Tasumi, K., Yukawa, H., and Morinaga, M. (2007) *J. Alloys Compd.*, **446**, 237.

175 Wang, J., Ebner, A.D., and Ritter, J.A. (2005) *Adsorption*, **11**, 811.

176 Klaveness, A., Vajeeston, P., Ravindran, P., Fjellvåg, H., and Kjekshus, A. (2006) *Phys. Rev. B*, **73**, 0941221.

177 Varin, R.A., Chiu, Ch., Czujko, T., and Wronski, Z. (2007) *J. Alloys Compd.*, **439**, 302.

178 Mintz, M.H., Gavra, Z., Kimmel, G., and Hadari, Z. (1980) *J. Less-Common Met.*, **74**, 263.

179 Gavra, Z., Hadari, Z., and Mintz, M.H. (1981) *J. Inorg. Nucl. Chem.*, **43**, 1763.

180 Schlapbach, L. and Züttel, A. (2001) *Nature*, **414**, 353.

181 Schwab, W. and Wintersberger, K.Z. (1953) *Z. Naturforsch.*, **8b**, 690.

182 Mal'tseva, N., Golovanova, A.I., Dymova, T.N., and Aleksandrov, D.P. (2001) *Russ. J. Inorg. Chem.*, **46**, 1793.

183 Fichtner, M., Frommen, C., and Fuhr, O. (2005) *Inorg. Chem.*, **44**, 3479.

184 Løvvik, O.M. (2005) *Phys. Rev. B*, **71**, 14411-1.

185 Weidenthaler, C., Frankcombe, T.J., and Felderhoff, M. (2006) *Inorg. Chem.*, **45**, 3849.

186 Zhang, Q.A., Nakamura, Y., Oikawa, K.I., Kamiyama, T., and Akiba, E. (2002) *Inorg. Chem.*, **41**, 6941.

187 Wolverton, O. and Ozolins, V. (2007) *Phys. Rev. B.*, **75**, 064101.

188 Marasheda, A. and Frankcombe, T.J. (2008) *J. Chem. Phys.*, **128**, 234505.

189 Claudy, P., Bonnetot, B., Bastide, J.P., and Létoffé, J.M. (1982) *Mat. Res. Bull.*, **17**, 1499.

190 Ma, X.Z., Klassen, T., Martinez-Franco, E., Bormann, R., Mao, Z.Q., **and** Laufs, R. (eds) (2004) Clean Energy for the 21th Century. Proceedings of HYFORUM 2004, EFO Energy Forum GmbH, Berlin, p. 225.

191 Graetz, J.J., Lee, Y., Reilly, J.J., Park, S., and Vogt, T. (2005) *Phys. Rev. B*, **71**, 184115.

192 Løvvik, O.M. and Swang, O. (2004) *Europhys. Lett.*, **67**, 607.

193 Brinks, H.W., Hauback, B.C., Jensen, C.M., and Zidan, R. (2005) *J. Alloys Compd.*, **392**, 27.

194 Ma, X.Z., Martinez-Franco, E., Dornheim, M., Klassen, T., and Bormann, R. (2005) *J. Alloys. Compd.*, **404**, 771.

195 Rönnebro, E. and Majzoub, E.H. (2006) *J. Phys. Chem. B.*, **110**, 25686.

196 Bastide, J.P., Claudy, P., Létoffé, J.M., and El Hajri, J. (1987) *Rev. Chim. Miner.*, **24**, 248.

197 Bulychev, B.M., Semenenko, K.N., and Bitcoev, K.B. (1978) *Koord. Khim.*, **4**, 374.

198 Grove, H., Brinks, H.W., Løvvik, O.M., Heyn, R.H., and Hauback, B.C. (2008) *J. Alloys Compd.*, **460**, 64.

199 Gingl, F., Vogt, T., and Akiba, E. (2000) *J. Alloys Compd.*, **306**, 127.

200 Zhang, Q.A., Nakamura, Y., Oikawa, K.I., Kamiyama, T., and Akiba, E. (2002) *Inorg. Chem.*, **41**, 6547.

201 A: Zhang, Q., Nakamura, Y., Oikawa, K.I., Kamiyama, T., and Akiba, E. (2003) *Inorg. Chem.*, **42**, 3152.

202 Zhang, Q.A. and Akiba, E. (2005) *J. Alloys Compd.*, **394**, 308.

203 Zhang, Q.A. (2008) *, E. Akiba, J. Alloys Compd.*, **460**, 272.

204 Zhang, Q.A., Enoki, H., and Akiba, E. (2007) *J. Alloys Compd.*, **427**, 153.

205 Zhang, Q.A., Nakamura, Y., Oikawa, K.I., Kamiyama, T., and Akiba, E. (2003) *J. Alloys Compd.*, **361**, 180.

206 Bronger, W. and Auffermann, G. (1998) *Chem. Mater.*, **10**, 2723.

207 Olofsson-Mårtensson, M., Häussermann, U., Tomkinson, J., and Noreus, D. (2000) *J. Am. Chem. Soc.*, **122**, 6969.

208 Yvon, K. (1998) *Chimia*, **52**, 613.
209 Ginsberg, A.P., Miller, J.M., and Koubek, E. (1961) *J. Am. Chem. Soc.*, **83**, 4909.
210 Knox, K. and Ginsberg, A.P. (1964) *Inorg. Chem.*, **3**, 555.
211 Ashby, E.C. and Watson, J.J. (1973) *Inorg. Chem.*, **12**, 2493.
212 Didisheim, J.-J., Zolliker, P., Yvon, K., Fischer, P., Schefer, J., Gubelmann, M., and Williams, A.F. (1984) *Inorg. Chem.*, **23**, 1953.
213 Zolliker, P., Yvon, K., Fisher, P., and Schefer, J. (1985) *Inorg. Chem.*, **24**, 4177.
214 Gennari, F.C., Castro, F.J., and Andrade Gamboa, J.J. (2002) *J. Alloys Compd.*, **339**, 261.
215 Varin, R.A., Li, S., Calka, A., and Wexler, D. (2004) *J. Alloys Compd.*, **373**, 270.
216 Chen, J., Takeshita, T., Chartouni, D., Kuriyama, N., and Sakai, T. (2001) *J. Mater.Sci.*, **36**, 5829.
217 Bogdanović, B., Reiser, A., Schlichte, K., Spliethoff, B., and Tesche, B. (2002) *J. Alloys Compd.*, **345**, 77.

Further Reading

(a) Schlapbach, L. and Züttel, A. (2001) *Nature*, **414**, 353; (b) Schüth, F., Bogdanović, B., and Felderhoff, M. (2004) *Chem. Commun.*, 2249; (c) Felderhoff, M., Weidenthaler, C., von Helmolt, R., and Eberle, U. (2007) *Phys. Chem. Chem. Phys.*, **9**, 2643; (d) Züttel, A., Borgschulte, A., and Schlapbach, L. (eds) (2008) *Hydrogen as a Future Energy Carrier*, Wiley-VCH, Weinheim, (e) Züttel, A. (2004) *Naturwissenschaften*, **91**, 157; (f) Orimo, S., Nakamori, Y., Eliseo, J.R., Züttel, A., and Jensen, C.M. (2007) *Chem. Rev.*, **107**, 4111; (g) Wang, P. and Zhang, D. (2008) *Dalton Trans.*, 5400; (h) Leon, A. (ed.) (2008) *Hydrogen Technology: Mobile and Portable Applications*, Springer, Berlin.

6
Amides, Imides and Mixtures

Takayuki Ichikawa

6.1
Introduction

As early as 1910, Dafert and Miklauz [1] reported that the reaction between lithium nitride (Li_3N) and hydrogen (H_2) generated Li_3NH_4. However, the product was proved by Ruff and Goeres [2] to be a mixture of lithium amide ($LiNH_2$) and lithium hydride (LiH):

$$Li_3N + 2H_2 \rightarrow LiNH_2 + 2LiH \tag{6.1}$$

Furthermore, the mixture was decomposed by heating, releasing H_2:

$$LiNH_2 + 2LiH \rightarrow Li_2NH + LiH + H_2 \tag{6.2}$$

The chemistry of alkali amides was well investigated in the early twentieth century [3, 4], especially, sodium amide ($NaNH_2$) has been used as a reagent in synthetic organic chemistry because of its ability to promote condensation reactions, to introduce amino groups into a molecule, and to remove the elements of water or of a hydrohalide acid. Lithium nitride has also been investigated for more than 50 years [5].

In 2002, Chen *et al.* [6] reported the system Li_3N to be a high capacity hydrogen storage material, where the hydrogenation and dehydrogenation of Li_3N were performed by the following two-step reversible reactions;

$$Li_3N + 2H_2 \leftrightarrow Li_2NH_2 + LiH + H_2 \leftrightarrow LiNH_2 + 2LiH \tag{6.3}$$

They claimed that a high amount of hydrogen (10.4 mass%) could be reversibly stored in this Li–N–H system. Since then, the hydrogen storage properties of metal–N–H systems have also been investigated all over the world. In this chapter, our current understanding of the thermodynamics, kinetics, crystal structure, electronic structure, characterization techniques, practical properties and reaction mechanism of the metal–N–H systems will be summarized from the viewpoint of hydrogen storage materials.

Handbook of Hydrogen Storage. Edited by Michael Hirscher

ISBN: 978-3-527-32273-2

6.2
Hydrogen Storage Properties of Amide and Imide Systems

6.2.1
Li–N–H System

In the first report of the hydrogen storage properties of the Li–N–H system [6], Li_3N was reported to absorb hydrogen in a two-step reaction to form $LiNH_2$ and 2LiH, with a theoretical hydrogen capacity of 10.4 mass%. However, only the second step in the reaction (6.3) is reversible under practical conditions of temperature and pressure [6, 7], and it releases only 5.2 mass% hydrogen. The pressure–composition (PC) isothermal properties are shown in Figure 6.1. By eliminating an extra LiH in the second step, the hydrogen capacity increases to 6.5 mass% for the following reaction [8, 9],

$$LiNH_2 + LiH \leftrightarrow Li_2NH + H_2. \quad (6.4)$$

The enthalpy change of the reaction (6.4) was deduced to be –44.5 kJ $(mol\ H_2)^{-1}$ from the database indicated in Chen's report [6]. However, Chen *et al.* [6] and Kojima and Kawai [10] have carefully measured the PC isothermal curve for the Li–N–H system at different temperatures and evaluated ΔH from the van't Hoff plot to be −66 kJ $(mol\ H_2)^{-1}$. Recently, Isobe *et al.* evaluated ΔH for hydrogen desorption on the Li–N–H system by the direct measurement of DSC, indicating ΔH to be –67 kJ $(mol\ H_2)^{-1}$ [11].

Because the first reaction step in the reaction (6.3) showed quite low equilibrium pressure (Figure 6.1) for the hydrogen desorption due to a large enthalpy change, a number of research groups have focused on the "amide–imide" reaction instead of the "imide–nitride" as indicated in reaction (6.4).

Concerning the mechanism of reaction (6.4), after the report on the Li–N–H system, some researchers have indicated the reaction mechanism on the hydrogen desorption from two solid phases of LiH and $LiNH_2$. Chen *et al.* [12] have claimed that H in $LiNH_2$ is positively charged while H in LiH is negatively charged, so that the strong affinity between $H^{\delta+}$ and $H^{\delta-}$ gives rise to a hydrogen molecule. This model suggests hydrogen gas desorption due to direct molecule–molecule interaction, in which the LiH and $LiNH_2$ molecules should be liberated from two solid phases. On the other hand, Hu and Ruckenstein [13] and Ichikawa *et al.* [14] have proposed that the hydrogen desorption reaction (6.4) proceeds through the following two-step elementary reaction mediated by ammonia (NH_3):

$$2LiNH_2 \rightarrow Li_2NH + NH_3 \quad (6.5)$$

and

$$LiH + NH_3 \rightarrow LiNH_2 + H_2 \quad (6.6)$$

The detailed mechanism will be mentioned later in this chapter.

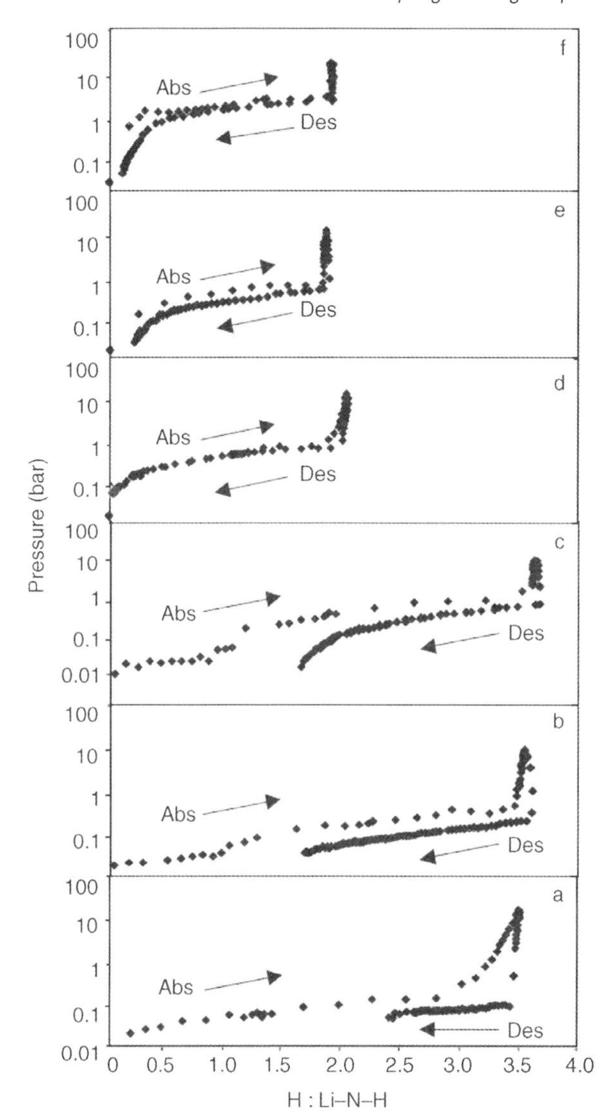

Figure 6.1 Pressure–composition (PC) isotherms of Li_3N and Li_2NH samples. The pressure was increased step by step to 2 MPa then gradually reduced to 0.004 MPa. The *x* axis represents the molar ratio of H atom to Li–N–H molecule. (a) Li_3N at 195 °C; (b) Li_3N at 230 °C; (c) Li_3N at 255 °C; (d) Li_3N re-PCI at 255 °C; (e) Li_2NH at 255 °C and (f) Li_2NH at 285°C [6].

6.2.2 **Li–Mg–N-H Systems**

As mentioned in the previous section, the thermodynamic properties of the Li–N–H system are thought to be unsuitable for practical application, because the temperature required to release hydrogen at usable pressures is too high due to the large change in enthalpy. Therefore, after Chen's report [6], some researchers tried to improve the thermodynamic properties. Orimo's group [15–17] started from a magnesium nitride (Mg_3N_2) and $4Li_3N$ mixture, because their concept to improve this Li–N–H system is based on the replacement of Li atom by Mg. In their papers, as an effective method for destabilization of $LiNH_2$, partial substitution of Li by elements with larger eletronegativity, such as Mg, was attempted. A mixture of Li_3N–20 mol% Mg_3N_2 was investigated in detail. The reversible reaction reported in this paper is as follows (9.1 mass%):

$$3Mg(NH_2)_2 + 12LiH \leftrightarrow Mg_3N_2 + 4Li_3N + 12H_2 \tag{6.7}$$

where, $Mg(NH_2)_2$ is magnesium amide. Luo [9] and Xiong *et al.* [18] started from a mixture of MgH_2 and $LiNH_2$ and obtained the Li–Mg–N–H system as well, because their concept to develop this system was based on the replacement of LiH by MgH_2 to destabilize the Li–N–H system. Luo expected the partial replacement of Li by Mg in the $LiNH_2$–LiH system to improve the sorption characteristics since MgH_2 is less stable than LiH [9]. Xiong *et al.* synthesized the ternary imides by mixing, heating, hydrogenating, and dehydrogenating lithium amide with the corresponding hydrides of alkaline earth metals because considerable changes in the thermodynamic properties of a binary hydride were expected by forming ternary or multinary hydrides [18]. After the dehydrogenation and rehydrogenation treatments, they finally obtained the following reaction (5.5 mass%).;

$$Mg(NH_2)_2 + 2LiH \leftrightarrow MgLi_2(NH)_2 + 2H_2 \tag{6.8}$$

On the other hand, Leng *et al.* investigated a composite material produced by ball milling a 3 : 8 molar mixture of $Mg(NH_2)_2$ and LiH under 1.0 MPa H_2 atmosphere [19] as follows (6.9 mass%).,

$$3Mg(NH_2)_2 + 8LiH \leftrightarrow Mg_3N_2 + 4Li_2NH + 8H_2 \tag{6.9}$$

Because their group investigated the reaction mechanism [14] of the Li–N–H system, they thought that the elementary step shown in reaction (6.5) needed to be improved. Therefore, $Mg(NH_2)_2$ was used instead of $LiNH_2$ because $Mg(NH_2)_2$ was expected to be decomposed at a much lower temperature than $LiNH_2$.

Thus, all these groups have investigated and developed the Li–Mg–N–H systems with different strategies. These systems, which are quite hot topics in the H-storage field because of having thermodynamically suitable characteristics for on-board H-storage, are composed of LiH and $Mg(NH_2)_2$ with different mixing ratios in the hydrogenated states. The difference between them is only in their mixing ratios, resulting in different dehydrogenated states.

The PC isothermal properties on the above Li–Mg–N–H system have been reported by a few groups for desorption [9, 20–22], absorption [23] and both absorption and desorption reactions [24–26], as shown in Figure 6.2 (desorption) and Figure 6.3 (absorption). However, thermodynamically accurate results have not yet been reported, because the kinetic properties are significantly worse, even around

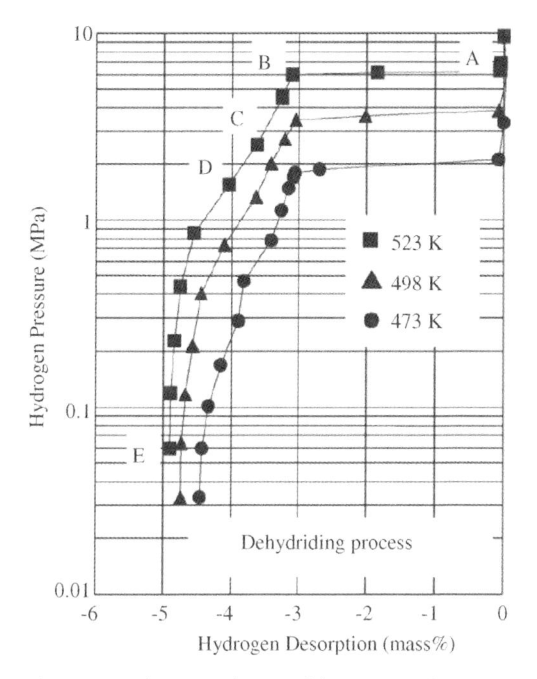

Figure 6.2 The PC isotherms of the mixture of $Mg_3N_2 + 4Li_3N$ after hydrogenation during dehydrogenating at 473, 498, and 523 K. Points A, B, C, D, and E mark the amounts of the desorbed hydrogen of 0, 3.1, 3.5, 4.0, 4.9 mass% on the isotherm at 523 K, respectively [110].

200 °C, than for conventional hydrogen storage alloys. As shown in Figures 6.2 and 6.3, the reactions (6.7)–(6.9) have a quite narrow plateau region and possess a slope region. Therefore, the thermodynamic quantity is quite difficult to estimate from these PC isothermal experiments. Akbarzadeh *et al.* [27] and Araujo [28]

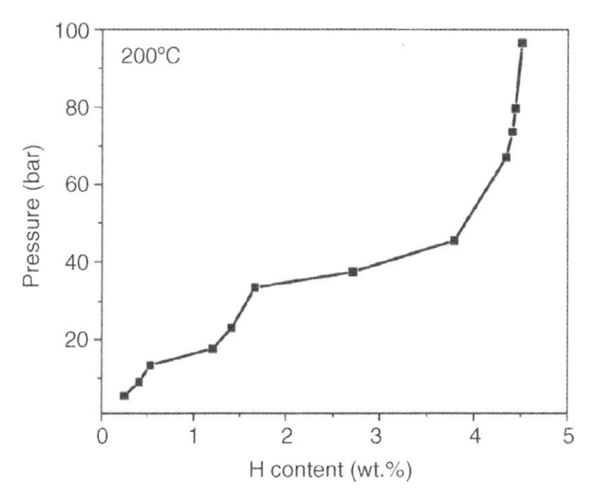

Figure 6.3 Absorption isotherm at 200 °C of $Li_2Mg(NH)_2$ [26].

Table 6.1 Calculated enthalpy changes for the corresponding reactions.

Reaction	ΔH (kJ (mol H_2)$^{-1}$)	Reference
$Mg(NH_2)_2 + 2LiH \leftrightarrow MgLi_2(NH)_2 + 2H_2$	40	[27]
	46.1	[28]
$MgLi_2(NH)_2 + 2Li \leftrightarrow 4Li_2NH + Mg_3N_2 + 2H_2$	84.1	[28]
$4Li_2NH + Mg_3N_2 + 4LiH \leftrightarrow 3Li_3N + 3LiMgN + 4H_2$	103.9	[28]
$Li_2Mg(NH)_2 + Mg_3N_2 + 2LiH \leftrightarrow 4LiMgN + 2H_2$	86	[27]

performed a first-principles study on the thermodynamics of hydrogen desorption for the Li–Mg–N–H systems. The results are shown in Table 6.1.

As mentioned above, four groups independently reported on Li–Mg–N–H systems with different mixing ratio of $Mg(NH_2)_2$ and LiH. Xiong *et al.* [20] and Leng *et al.* [29] investigated the hydrogen desorption and structural properties of the systems with different mixing ratios. In both reports, an equal amount (1 : 1) of LiH mixed with $Mg(NH_2)_2$ generated a considerable amount of ammonia gas upon heating, as shown in Figure 6.4. This has been the case with a slightly less amount of LiH mixed with

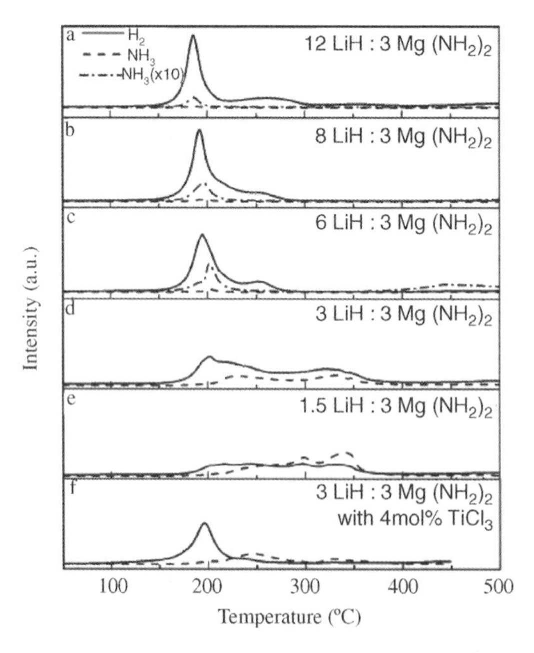

Figure 6.4 (a)–(e) TDMS profiles of the mixture of LiH and $Mg(NH_2)_2$ with the LiH/Mg $(NH_2)_2$ ratios of 12/3, 8/3, 6/3, 3/3, and 1.5/3, respectively, with temperatures increasing up to 500 °C under a helium flow at a ramp rate of 5 °C min^{-1}. (f) TDMS profiles of the mixture of LiH and $Mg(NH_2)_2$ with a 3/3 ratio and 4 mol % $TiCl_3$ as catalyst. The "-••-•-" line represents the NH_3 signals enlarged 10 times [29].

$Mg(NH_2)_2$. From these results, one may assume that if enough amount LiH is mixed with $Mg(NH_2)_2$, mainly hydrogen gas can be desorbed (see also Figure 6.4). However, lack of LiH or a poor mixing led to NH_3 emission due to the decomposition of $Mg(NH_2)_2$.

6.2.3 Other Metal–N–H Systems

After the discovery of the Li–Mg–N–H systems, many combinations of metal hydride (MH_x) and metal amide ($M(NH_2)_x$) were investigated to obtain suitable hydrogen storage properties.

The mixture of MgH_2 and $Mg(NH_2)_2$, the so-called the Mg–N–H system, was studied by Nakamori *et al.* [30], Leng *et al.* [31], Hu *et al.* [32] and Xie *et al.* [33], who showed a quite low starting temperature of hydrogen desorption compared with the Li–N–H system. The corresponding reaction can be written as (7.3 mass%):

$$Mg(NH_2)_2 + 2MgH_2 \rightarrow Mg_3N_2 + 4H_2 \tag{6.10}$$

However, the ball-milling treatment to make a good mixture induced hydrogen desorption even at room temperature. During the treatment, H_2 pressure in the milling vessel increased to about 1.2 MPa [32], meaning that the enthalpy change of this reaction was estimated to have a small value. Therefore, complete hydrogenation of Mg_3N_2 has not been successful so far.

The Ca–N–H system possesses even more complicated properties. Hino *et al.* [34] reported the thermal desorption properties of the mixture of calcium amide ($Ca(NH_2)_2$) and calcium hydride (CaH_2) with two different ratios (1 : 1 and 1 : 3) and the mixture of CaNH and CaH_2 as shown in Figure 6.5. From this result, $Ca(NH_2)_2$ and CaH_2 with the 1 : 3 ratio desorbs H_2 gas by the following two-step reaction (3.5 and 2.1 mass%, respectively):

$$Ca(NH_2)_2 + CaH_2 \rightarrow 2CaNH + 2H_2 \tag{6.11}$$

and

$$2CaNH + 2CaH_2 \leftrightarrow 2Ca_2NH + 2H_2 \tag{6.12}$$

Before that, Chen's group had already focused on the reaction (6.12) as a similar hydrogen storage system to the Li–N–H system [6, 35], which showed reversible hydrogen storage properties at a temperature of more than 500 °C, as shown in Figure 6.6.

The Li–Ca–N–H systems were composed of $Ca(NH_2)_2$ and LiH. Xiong *et al.* [18] first synthesized the single phase of Li–Ca imide, then succeeded in the hydrogenation. Tokoyoda *et al.* [36] started from mixtures of CaH_2 + $LiNH_2$ and LiH + $Ca(NH_2)_2$ in the suitable mixed ratio, and confirmed that the reversible hydrogen storage reaction can be understood by the following reactions (4.5 mass%):

$$Ca(NH_2)_2 + 2LiH \leftrightarrow CaNH + Li_2NH + 2H_2 \tag{6.13}$$

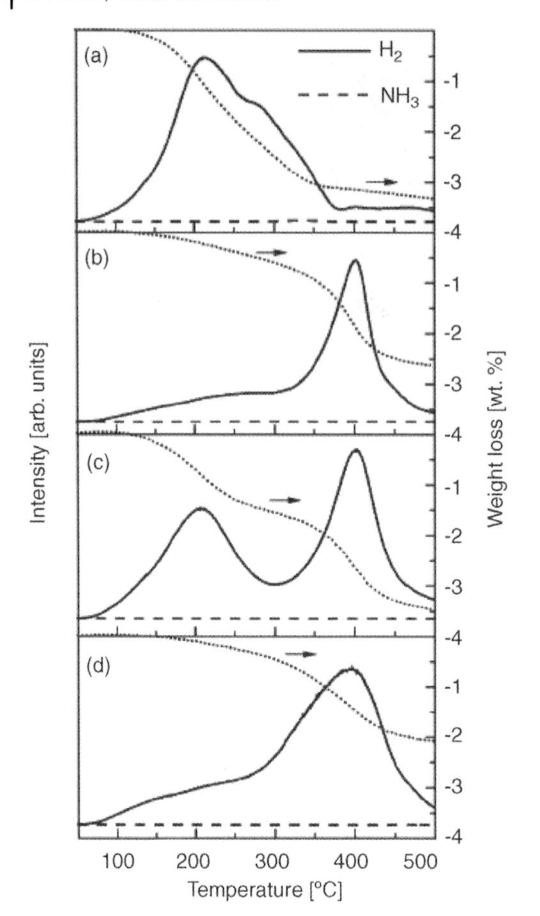

Figure 6.5 Thermal desorption mass spectra (TDMS) and thermogravimetry (TG) analysis for the mixtures of (a) $Ca(NH_2)_2$ and CaH_2 (molar ratio 1 : 1), (b) CaNH and CaH_2 and (c) $Ca(NH_2)_2$ and CaH_2 (molar ratio 1 : 3) milled for 2 h. Sample (d) was the sample (c) milled for 20 h. The solid line indicates H_2 desorption, the dashed line shows NH_3 emission and the dotted line shows weight loss percent [34].

and

$$CaH_2 + 2LiNH_2 \rightarrow CaNH + Li_2NH + 2H_2 \leftrightarrow Ca(NH_2)_2 + 2LiH \qquad (6.14)$$

Recently, Wu [37] reported the crystal structure of the ternary imide $Li_2Ca(NH)_2$ which was determined using neutron powder diffraction data on a deuterated sample. In his paper, the reaction, $Li_2Ca(NH)_2 + H_2 \rightarrow LiNH_2 + LiH + CaNH$, was indicated. However, with respect to the reaction mechanism of the Li–Mg–N–H system [38], the reaction process in this report was thought to be stopped halfway.

The Na–Li–N–H [14], Mg–Na–N–H [39], Ca–Na–N–H [40], Mg–Ca–N–H [41, 42] and Mg–Ca–Li–N–H [43] systems have also been investigated, where the metal elements used in these works were mainly limited to the alkali and alkaline earth

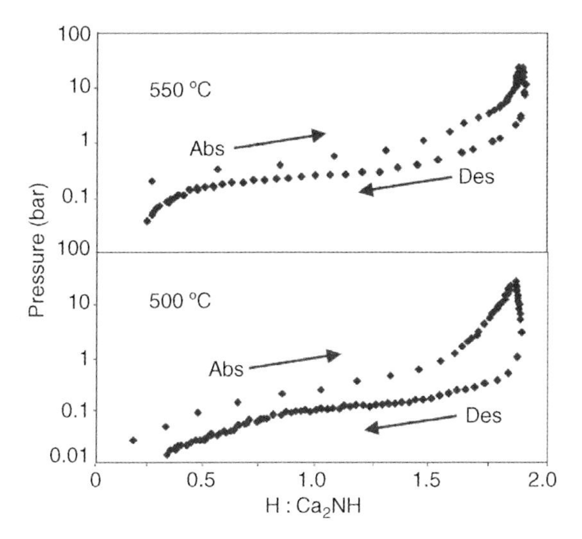

Figure 6.6 PC isotherms of the Ca_2NH sample measured at 500 and 550 °C, respectively. For details of measurement, please refer to the Methods [6].

elements. On the other hand, Yamane *et al.* [44] focused on the elements of Group 13 for the ternary nitride as the dehydrogenated state of metal–N–H systems. They synthesized Li_3BN_2, Li_3AlN_2 and Li_3GaN_2, and then, systematically investigated the hydrogenation and dehydrogenation properties. However, the results showed that the $LiNH_2$, LiH and GaN were generated after the hydrogenation, indicating that the reversible step should be equivalent to that in the Li–N–H system.

Kojima *et al.* [45] and Xiong *et al.* [46, 47] tried to modify the Li–N–H system. Instead of LiH possessing the negatively charged hydrogen $H^{\delta-}$, the Li_3AlH_6 + $2LiNH_2$ and $LiAlH_4$ + $2LiNH_2$ mixtures were investigated. Finally, Xiong *et al.* [47] concluded that the 5.17 mass% of hydrogen was able to be reversibly stored by the following reaction,

$$Li_3AlN_2 + 2H_2 \leftrightarrow LiNH_2 + 2LiH + AlN \tag{6.15}$$

in the temperature range 50–500 °C.

Janot *et al.* [48] prepared lithium aluminum amide ($LiAl(NH_2)_4$) and used it instead of $LiNH_2$ in the Li–N–H system. The mixture of $LiAl(NH_2)_4$ and 4LiH released more than 5 mass % hydrogen at 130 °C. However, the experimental result of the hydrogenation indicated that this combination would have a poor reversibility because of the existence of AlN after the dehydrogenation.

6.3
Structural Properties of Amide and Imide

Sodium and potassium amides ($NaNH_2$ and KNH_2) were discovered in the early part of the nineteenth century [3]. At that time, the metal amides were synthesized from

KNH_2 by the exchange of cation atoms as follows,

$$KNH_2 + MOH \rightarrow KOH + MNH_2 \quad (6.16)$$

in which KNH_2 was synthesized by the following reaction,

$$2K + 2NH_3 \rightarrow 2KNH_2 + H_2 \quad (6.17)$$

In this section, recent techniques to characterize the metal amides and imides will be summarized from the viewpoint of their use as hydrogen storage materials. Recently, many metal amides have become available commercially. The corresponding imides or nitrides can be synthesized from these commercial amides.

6.3.1 Lithium Amide and Imide

$LiNH_2$ is a white ionic substance which was first synthesized in 1894 [49]. It has been synthesized by heating molten Li at 200 °C under an ammonia gas flow and recently by ball milling LiH under an ammonia gas atmosphere at room temperature [50], based on the reaction (6.6). $LiNH_2$ contains 8.7 mass% hydrogen. The crystal structure is a tetragonal structure (space group I-4, $a = 5.03164$ Å, $c = 10.2560$ Å [51]), where the Li atoms are tetrahedrally coordinated by four NH_2 species (Figure 6.7) [51–55]. Chellappa *et al.* [56] reported on the pressure-induced phase transitions in $LiNH_2$. The pressure-induced changes in the Raman spectra of $LiNH_2$ indicate a phase transition from ambient-pressure α-$LiNH_2$ (tetragonal, I-4) to a high-pressure phase β-$LiNH_2$ that begins at ~12 GPa and is completed at ~14 GPa. The two N–H distances are between 0.967 and 0.978 Å and the H–N–H angle is ~104 °, which is very close to the angle in the H_2O molecule. The electronic structure of $LiNH_2$ was investigated by first-principles calculations [16, 55, 57–60], indicating strong ionic characteristics between the ionic Li^+ cation and the covalent bonded $(NH_2)^-$ anion. In the IR absorption spectrum, the peaks at 3312, 3258 cm^{-1} and 1561, 1539 cm^{-1} corresponding to N–H stretching vibration modes and the H–N–H deformation vibration mode are observed, respectively [10, 61]. Solid-state NMR measurements using ^1H and ^6Li elements were performed for $LiNH_2$ [62–64], indicating that the LiOH and $LiOH{\cdot}H_2O$ resonant peaks were separated from the LiH and $LiNH_2$ resonant peaks. Palumbo *et al.* [65] reported that the inelastic spectroscopy in $LiNH_2$ during its high temperature decomposition into Li_2NH can be used to monitor the time and temperature evolution of decomposition Figure 6.7.

Li_2NH has been synthesized according to reaction (6.5) by heat treating $LiNH_2$ in vacuum at 350–450 °C. It desorbs 36.9 mass% of NH_3 but no H_2 on decomposing to Li_2NH. The $LiNH_2$ and LiH mixed by ball milling $LiNH_2$ with LiH under an Ar atmosphere at room temperature, was transformed into Li_2NH by heating at 250–300 °C evolving hydrogen, as shown in the reaction (6.4). Hu *et al.* [66] reported a preparation of Li_2NH from Li_3N and $LiNH_2$, which reveals the ultrafast solid reaction. Li_2NH contains 3.5 mass% of hydrogen atom. Its crystal structure was studied by a number of researchers. However, the crystal structure of Li_2NH has not

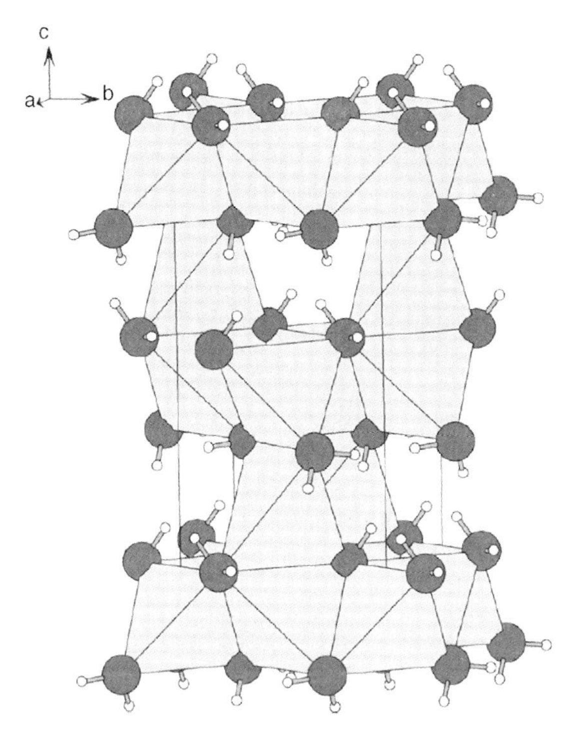

Figure 6.7 The crystal structure of $LiND_2$. Li: in centers of gray tetrahedra; N: dark gray spheres; D: white spheres. Note the tetrahedral coordination of amide (ND_2^-) units around Li [51].

yet been refined. Juza *et al.* [52] proposed by an X-ray diffraction study, primarily in the early 1950s, that Li_2NH has an anti-fluorite structure with Fm-3m symmetry but gave no information about the positions of the H atoms. By using neutron diffraction to study the hydride material of Li_2NH, Ohoyama *et al.* [67] reported the two possible models of a cubic structure with a lattice constant of $a = 5.0769$ Å. Their models were Fm-3m symmetry having hydrogen atoms at the 48h site, and F-43m with the hydrogen atoms at the 16*e* site, respectively. In both models, the sites were randomly occupied by hydrogen atoms. Similarly, Noritake *et al.* [68] performed synchrotron powder X-ray diffraction experiments and charge density analyses for the hydride material of Li_2NH, they reported that Li_2NH showed the cubic structure (Fm-3m) with a lattice constant of $a = 5.07\,402$ Å, as shown in Figure 6.8, in which the sites were randomly occupied by hydrogen atoms as well. Following this result, Balogh *et al.* [69] have carefully investigated the crystal structure of the deuterated material, Li_2ND, by means of neutron and X-ray diffraction and studied its temperature dependence. Li_2ND shows an order–disorder phase transition at 85 °C. The high-temperature phase is best characterized as disordered cubic (Fm-3m) with D atoms randomized over the 192*l* sites. Below that temperature, Li_2ND can be described, to the same level of accuracy, as a disordered cubic (Fd-3m) structure with partially occupied Li 32*e* sites or as a fully occupied orthorhombic

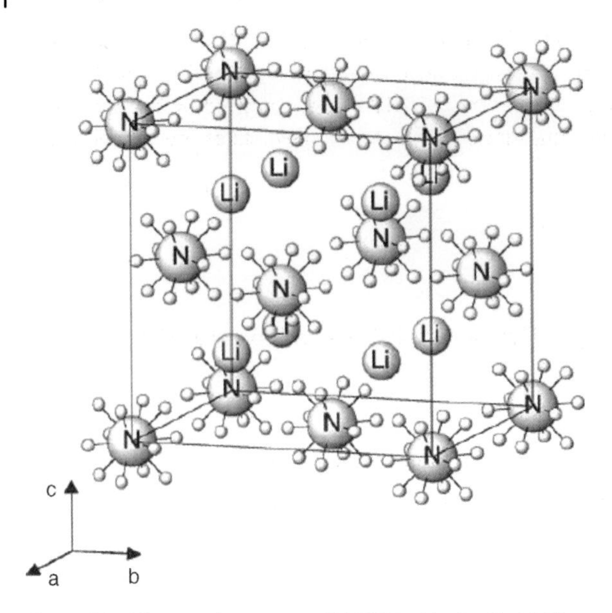

Figure 6.8 The crystal structure of Li_2NH (cubic, Fm-3m) [68].

(Ima2 or Imm2) structure, where density functional theory calculations complement and support the diffraction analyses in the latter case. Thus, Balogh *et al.* reported a feasible structure model [69], the corresponding cubic crystal system with $a \sim 10$ Å. However, the model is indistinguishable from other theoretical crystal structure models [58, 70]. Magyari-Köpe *et al.* [71] theoretically found several crystal structures (orthorhombic, Pnma and monoclinic, C2/m) with significantly lower calculated energies than that reported above, suggesting structures as shown in Figure 6.9.

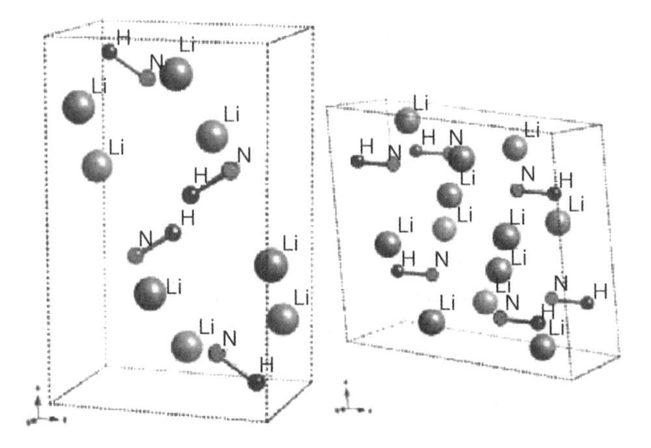

Figure 6.9 The predicted structures in the low-temperature phase of Li_2NH (orthorhombic Pnma and monoclinic C2/m) [71].

Thus, its structure has proven difficult to fully characterize. The key problem is to identify the location of the hydrogen and lithium positions and the N–H bond orientations. The electronic structure of Li_2NH was investigated by first-principles calculations [58–60], indicating that the highest occupied states are non-bonding, consisting of N p_π orbitals.

6.3.2
Sodium Amide

Gaseous NH_3 reacts readily with molten Na at 300 °C and forms $NaNH_2$ [49] by the following reaction:

$$2Na + 2NH_3 \rightarrow 2NaNH_2 + H_2 \tag{6.18}$$

Furthermore, $NaNH_2$ has been synthesized by ball milling NaH under an NH_3 atmosphere at room temperature [50] in the following reaction:

$$NaH + NH_3 \rightarrow NaNH_2 + H_2. \tag{6.19}$$

$NaNH_2$ contains 5.1 mass% hydrogen. The crystal structure is orthorhombic (Fddd) [72, 73]. On heating, unlike $LiNH_2$, molten $NaNH_2$ does not decompose into the imide and ammonia or nitride and ammonia, but seems to decompose into N_2, H_2 and Na between 400 and 500 °C through some intermediate reactions.

6.3.3
Magnesium Amide and Imide

$Mg(NH_2)_2$ has been synthesized by a direct reaction between Mg and gaseous NH_3 at 300 °C [74]:

$$Mg + 2NH_3 \rightarrow Mg(NH_2)_2 + H_2 \tag{6.20}$$

$Mg(NH_2)_2$ has also been prepared by ball milling MgH_2 under a gaseous NH_3 atmosphere at room temperature [50]:

$$MgH_2 + 2NH_3 \rightarrow Mg(NH_2)_2 + 2H_2 \tag{6.21}$$

$Mg(NH_2)_2$ contains 7.2 mass% hydrogen and is crystallized in a tetragonal structure where the Mg^{2+} cation is tetrahedrally coordinated by four amide $(NH_2)^-$ ions [51], as shown in Figure 6.10 [75]. The absorption peaks at 3332, 3277 and 1577 cm^{-1} corresponding to, respectively, N–H stretching modes and H–N–H deformation modes are observed in the IR spectrum [76]. On heating, $3Mg(NH_2)_2$ decomposes into 3MgNH and $3NH_3$ at 180–300 °C, and 3MgNH further decomposes into Mg_3N_2 and NH3 at 300–500 °C [50] according to the following reaction:

$$3Mg(NH_2)_2 \rightarrow 3MgNH + 3NH_3 \rightarrow Mg_3N_2 + 4NH_3 \tag{6.22}$$

$Mg(NH_2)_2$ is less stable than $LiNH_2$ because of its higher electronegativity and a lower temperature of ammonia release than that in $LiNH_2$. For the first time,

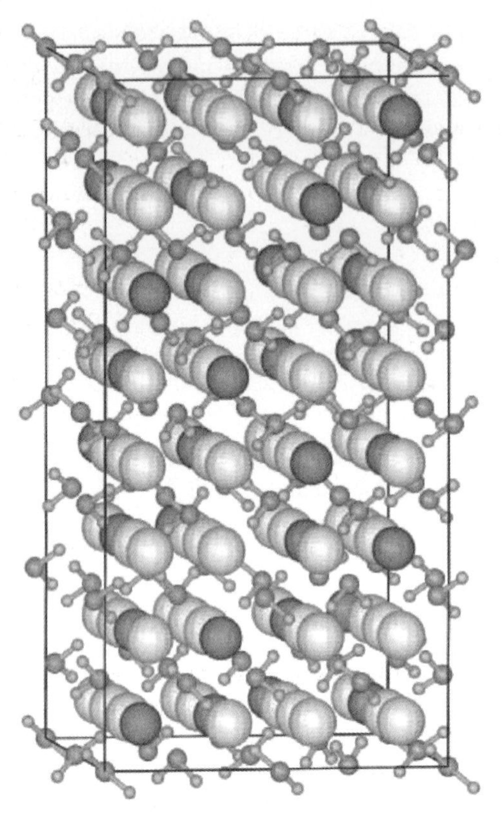

Figure 6.10 Crystal structures of $Mg(NH_2)_2$. The small dark and large white spheres represent the Mg and vacancies, respectively. Nitrogen atoms form approximate fcc lattices, while Mg cations and vacancies are located at the tetrahedral interstices. N–H bonds point to the vacancy sites [75].

Velikokhantnyi *et al.* showed the electronic structure, density of states and the vibrational properties of $Mg(NH_2)_2$ by first-principles calculations [77]. Magnesium imide (MgNH) is a hexagonal structure (P6322) [74], but it is quite unstable and easily decomposes into Mg_3N_2 with emission of ammonia during heating.

6.3.4
Other Amides and Imides

$Ca(NH_2)_2$ contains 5.5 mass% hydrogen and possesses a tetragonally distorted structure with ionic character [78]. The observed IR absorption peaks at 3318, 3295, 3257, 3228 cm^{-1} and 1580, 1520 cm^{-1} correspond to N–H stretching modes and H–N–H deformation modes, respectively [79]. CaNH is a cubic structure (Fm-3m) but distorts into a tetragonal one on heating [80]. At higher temperature, CaNH further reacts with CaH_2, then transforms into Ca_2NH by releasing H_2 [6, 34, 35].

$Li_2Mg(NH)_2$ was investigated by Rijssenbeek *et al.* [81] using synchrotron X-ray diffraction and neutron diffraction. $Li_2Mg(NH)_2$ undergoes two structural transitions from an orthorhombic structure (α-$Li_2Mg(NH)_2$, Iba2) to a primitive cubic structure (β-$Li_2Mg(NH)_2$, P-4$_3$m) at intermediate temperature (350 °C) followed by a face-centered cubic crystal structure (γ-$Li_2Mg(NH)_2$, Fm-3m) at high temperature (500 °C). The electronic structure of α-$Li_2Mg(NH)_2$ was investigated by first-principles calculations [77], showing the heat of reaction (6.8) to be 53 kJ (mol H_2)$^{-1}$ in contrast to the experimental value of 44 kJ (mol H_2)$^{-1}$. Wang and Chou [75] examined possible ordered configurations of $Li_2Mg(NH)_2$ by total-energy density-functional calculations, then concluded that the obtained configurations were consistent with those of a disordered α-$Li_2Mg(NH)_2$ obtained from a recent diffraction experiment at room temperature. $Li_2Ca(NH)_2$ was synthesized and its structural properties were investigated by Wu *et al.* [82] and Wu [37]. They claimed that the crystal structure was a trigonal structure (P-3m1) with lattice constants of $a = 3.5664$ Å and $c = 5.9540$ Å.

$LiAl(NH_2)_4$ has been synthesized from the reaction of Li and Al with liquid ammonia at temperatures from 80 to100 °C [83] or $LiAlH_4$ in liquid ammonia at room temperature [48]. The X-ray diffraction was attributed to a monoclinic structure ($P2_1/n$) and strong ammonia desorption was observed at about 150 °C by thermal decomposition.

6.4 Prospects of Amide and Imide Systems

6.4.1 Kinetic Analysis and Improvement

In order to improve the kinetics of reaction (6.4), Ichikawa *et al.* [8] showed superior catalytic effects of $TiCl_3$ compared to transition metals. For the hydrogen desorption reaction, the activation energy was estimated, by applying the Kissinger method, to be 110 kJ. Matsumoto *et al.* [84] also studied the hydrogen desorption reaction (6.4) to estimate the activation energy of the reaction by means of the temperature programmed desorption technique. Although they obtained comparable results with that of Ichikawa *et al.* for the sample with the catalyst, the activation energy of the pristine sample showed a much smaller value of 54 kJ despite a slower kinetic response for the pristine sample, which is thought to be an important pending issue. Lohstroh and Fichtner [85] investigated the $TiCl_3$ catalytic effect for the $2LiNH_2 + MgH_2$ mixture, and concluded that the addition of 2 mol% $TiCl_3$ to the initial $2LiNH_2 + MgH_2$ mixture caused the hydrogen release to shift to lower temperature. However, this catalytic effect disappeared after two hydrogenation cycles. Figure 6.11 Isobe *et al.* [86] investigated the catalytic effect of some different types of Ti additives on hydrogen desorption properties for the reaction (6.4). They found that Ti^{nano}, $TiCl_3$ and TiO_2^{nano} have a superior catalytic effect, while Ti^{micro} and TiO_2^{micro} did not lead to a lower desorption temperature. In the XRD profiles, there are traces of Ti and TiO_2 phases in the Ti^{micro} and TiO_2^{micro}, whereas no trace of Ti, $TiCl_3$ and TiO_2 was found in the Ti^{nano}, $TiCl_3$ and TiO_2^{nano} samples. From these results, it appears that the particle

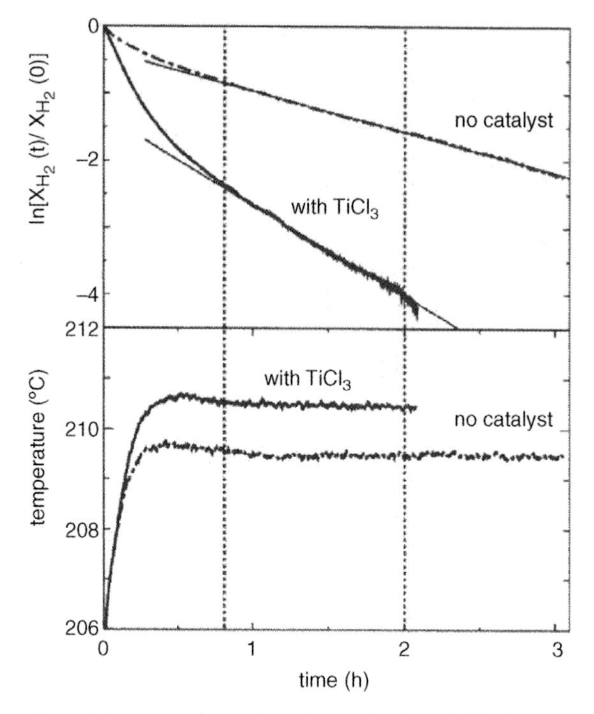

Figure 6.11 Logarithmic plots of normalized residual hydrogen amount, $X_{H_2}(t)/X_{H_2}(t)$, and linear plots of measuring temperature as a function of time for the 1 : 1 ball-milled powders with and without 1 mass% $TiCl_3$ catalyst [14].

size of the additive might be the important factor for improving the kinetics of hydrogen desorption, and metallic Ti highly dispersed on the nanometer scale samples acts as an effective catalyst for the Li–N–H system. These invisible catalysts were characterized by X-ray absorption spectroscopy [87], indicating that the Ti atoms in the titanium compounds, which have a catalytic effect on the kinetics of hydrogen desorption, have a common chemical bonding state. Moreover, it was clarified that this common state of Ti atom was generated by a reaction with $LiNH_2$, not with LiH. Tsumuraya *et al.* [88] focused on this effect and analyzed the X-ray absorption spectra of Ti K-edge in the several kinds of Ti compounds by first-principles calculations. The spectrum shape of Li_3TiN_2 has a prominent pre-edge structure, as seen in the common spectrum shape in the catalytically-active Ti state (Figure 6.12). The common structural feature in Li_3TiN_2, Li_2NH, and $LiNH_2$ is that Ti in Li_3TiN_2 and, Li in Li_2NH and $LiNH_2$ are tetrahedrally coordinated by N. The enhancement in the pre-edge structure is attributed to the on-site hybridization between Ti-p and Ti-d due to the lack of centro-symmetry realized in the tetrahedrally coordinated structure.

Hu and Ruckenstein [89] tried to improve the kinetics of the Li–N–H system by partial oxidation of Li_3N and concluded that this oxidation is highly effective for hydrogen storage with fast kinetics and high reversible hydrogen capacity as well as

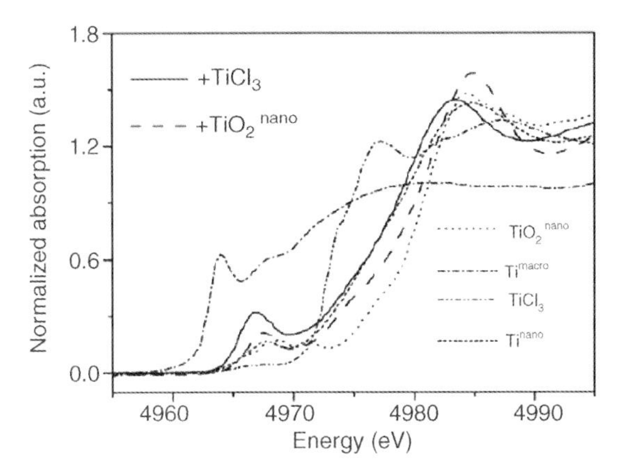

Figure 6.12 X-ray absorption near edge structure (XANES) profiles of Ti K-edges for the composites of LiH + $LiNH_2$ catalyzed by $TiCl_3$ and TiO_2^{nano} and XANES profiles for TiO_2^{nano}, Ti^{micro}, $TiCl_3$ and Ti^{nano} themselves as references [87].

excellent stability. Yao *et al.* [90] confirmed that Mn, V and their oxides, enhanced the release of NH_3 from $LiNH_2$, but posed little influence on the production of H_2 from the $LiNH_2$ + LiH mixture. From these results, they concluded the rate-limiting step for dehydrogenation from the mixture was the reaction between LiH and NH_3. Chen *et al.* [91] added different carbon materials, such as single-walled carbon nanotubes (SWNTs), multi-walled carbon nanotubes (MWNTs), graphite and activated carbon (AC) to the $LiNH_2/MgH_2$ mixture. Among them, small amount of SWNTs markedly improved the dehydrogenating kinetics because SWNTs act as tiny milling balls to facilitate particle pulverization. Sudik *et al.* [92] tested a seeding technique, that is, the $Mg(NH_2)_2$–2LiH mixture contains various amounts of its decomposition product, $Li_2Mg(NH)_2$ as seed. From this technique, as an optimum condition, they found that 10 mass% seed in the $Mg(NH_2)_2$–2LiH mixture caused a 40 °C reduction in the peak temperature in the hydrogen desorption.

Pinkerton [93] found that the reaction rate of the Li–N–H system depended on the sample weight as a consequence of the very low NH_3–$LiNH_2$ equilibrium vapor pressure at temperatures below 300 °C. Larger sample weight produced a local concentration of NH_3 high enough to inhibit further reaction and direct isothermal measurements of the initial reaction rate at temperatures between 200 and 300 °C agreed well with the values calculated from the heating rate-derived kinetic parameters. They claimed that the activation energy for the decomposition reaction was virtually independent of the source and purity of the $LiNH_2$, its stoichiometry, ball milling time, and TGA sample weight. However, Markmaitree *et al.* [94] and Shaw *et al.* [95] indicated an opposite conclusion.

In order to improve the kinetics of the Li–N–H system, Xie *et al.* [96] prepared Li_2NH hollow nanospheres by plasma metal reaction based on the Kirkendall effect. The special nanostructure showed significantly improved hydrogen storage kinetics compared to that of the Li_2NH micrometer particles. The absorption temperature decreased markedly, and the absorption rate was enhanced dramatically because of the larger specific surface area and nanometer diffusion distance in the special hollow nanosphere structure. Meanwhile, the desorption temperature decreased dramatically due to the hydrogenated product $LiNH_2$ and LiH maintaining the hollow nanosphere structure, in which the two phases are limited in the nanometer range. Furthermore, this synthetic method is expected to extend to the preparation of other inorganic compounds with hollow nanosphere structure and can be scaled up for industrial needs.

6.4.2
NH_3 Amount Desorbed from Metal–N–H Systems

For the application of the metal–N–H system in hydrogen storage technologies ammonia release should be avoided or minimized. Therefore, the existence of a competing reaction in which metal amides decompose by releasing ammonia, as indicated in reaction (6.5) is of critical importance. Pinkerton [93] pointed out that slow but significant decomposition by NH_3 release occurred in a hydrogen atmosphere, even under moderate pressure (0.13 MPa) and temperature (175 °C). Hino *et al.* [97, 98] measured NH_3 and H_2 partial pressures in the desorbed gas from LiH + $LiNH_2$ + 1 mol% $TiCl_3$ in a closed system by means of infrared spectroscopy and gas chromatography. The amounts of desorbed H_2 and NH_3 were drastically increased at high temperatures of 300 °C, and the NH_3/H_2 ratio was of the order of 0.1% at 300–400 °C, as shown in Figure 6.13. Luo and Stewart [99] performed an accurate determination of the amounts of NH_3 in the H_2 with a Draeger Tube with an accuracy of ±15% for the Li–Mg–N–H system. The NH_3 concentration in the desorbed hydrogen increased with the temperature of the storage material bed, from 180 ppm at 180 °C to 720 ppm at 240 °C. Liu *et al.* [100] also measured NH_3 concentration with a Metrohm 781 pH/ion meter for the Li–Mg–N–H system. The results indicated that the equilibrium concentrations of NH_3 were 240 ppm at 167 °C

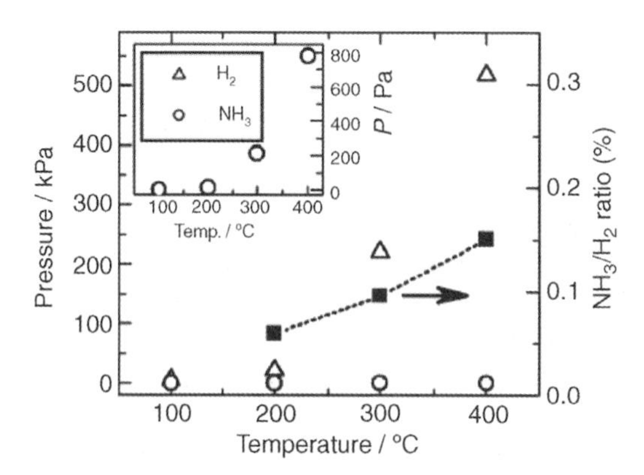

Figure 6.13 Calculated pressures of the emitted gases from the mixtures of LiH + $LiNH_2$ + 1 mol% $TiCl_3$ while heating to 400 °C. The NH_3/H_2 ratios are also represented. Inset shows NH_3 emission on an enlarged scale [98].

and 498 ppm at 198 °C, which are comparable levels to other estimations [98, 99]. Moreover, the equilibrium concentrations of NH_3 were 30 and 17 ppm, respectively, at 90 and 25 °C, as estimated by thermodynamic calculation based on extrapolation from the experimental data [100].

6.4.3
Practical Properties

To obtain the Li–Mg–N–H system in an easy and economical way, Okamoto *et al.* [101] tried to establish a new method to synthesize the $Mg(NH_2)_2$ and LiH mixture. They focused on Mg and $LiNH_2$ as initial materials because of the cost effectiveness. A homogeneous mixture of Mg and $LiNH_2$ under an inert gas atmosphere was heat-treated at 250 °C for 16 h under vacuum conditions when it was transformed into Mg_3N_2 and Li_2NH. Then, the target materials, $Mg(NH_2)_2$ and LiH, were synthesized by a further heat-treatment of the Mg_3N_2 and Li_2NH at 200 °C under 10 MPa hydrogen atmosphere for 12 h. Finally, they obtained quite similar hydrogen desorbing properties to those of the directly ball-milled mixture of $Mg(NH_2)_2$ and LiH.

Luo *et al.* [102] investigated the effect of exposure to water-saturated air on the hydrogen storage properties of the Li–Mg–N–H system, because it is the most likely gaseous contaminant during routine handling. The effect of H_2O-saturated air exposure is nearly the same whether the sample is exposed in its hydrided or dehydrided forms. The "final" mass% of hydrogen obtained was either unchanged or slightly enhanced after exposure of the sample to H_2O-saturated air at 25 °C. There was a decrease in the reaction times for reaching the same H mass% after exposure to the H_2O-saturated air for both absorption and desorption, which offered the possibility of the use of impure H_2.

The hydrogen produced during the steam reforming of natural gas ($CH_4 + 2H_2O \rightarrow 4H_2 + CO_2$) will be contaminated by a small amount of CO_2 gas. Hu and Ruckenstein [103] reported that a CO_2/H_2 mixture can be used directly as a hydrogen source for hydrogen storage in Li_3N. The results indicated that the kinetics and capacity of hydrogen absorption of Li_3N were not affected by CO_2, and the kinetics was only slightly affected. $LiNH_2/Li_3N$ can be employed for hydrogen storage by using a CO_2/H_2 mixture as hydrogen source without pre-separation. Furthermore, the presence of CO_2 in H_2 did not affect the high reversibility of the hydrogen capacity (about 6.7 mass%) and the fast hydrogen absorption kinetics of the $LiNH_2/Li_3N$ material, leading to the $LiNH_2$ and LiH mixture after hydrogenation. This can tremendously reduce the cost of hydrogen fuel for transportation and also provides a novel process to obtain pure hydrogen [104].

In order to obtain bulk morphologies, even after repeated hydrogenation and dehydrogenation reactions, for the Li–N–H system, Hao *et al.* [105] demonstrated the feasibility of an impregnation method for the synthesis of Li_2NH (Li-imide) and $LiNH_2$ (Li-amide) for application in practical hydrogen storage. As shown in Figure 6.14, a metallic Li foil with 99.9% purity was placed on top of the Ni foam installed in a steel reaction cell with diameter 15 mm, and then heated to 500 °C for

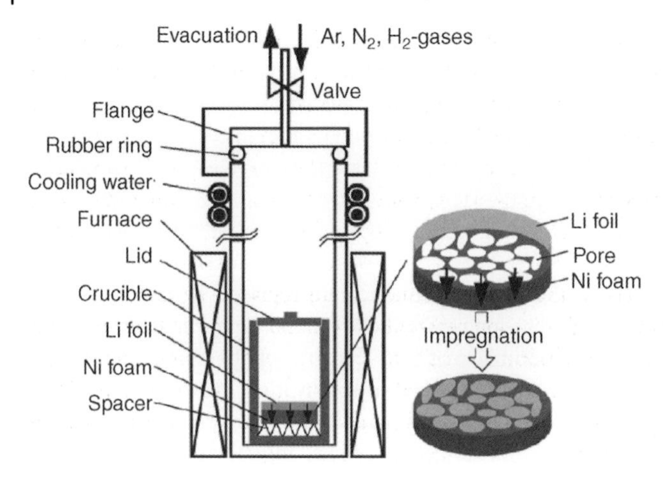

Figure 6.14 Schematic drawing of the set-up for the impregnation method. A metallic Li foil was placed on top of the Ni foam for the impregnation. A bellows-shaped Ni foil was used as a "spacer" to avoid any direct contact between the Ni foam and the bottom of the crucible [105].

4 h in an Ar atmosphere. During the process, the Li foil melts and impregnates into the Ni foam, the element Ni being chosen due to the high stability with lithium, hydrogen and nitrogen. When the samples were placed inside the cell, the metallic Li impregnated into the Ni foam was nitrogenated under 0.2 MPa of nitrogen with 99.999 95% purity at 220 °C for 3 h and then hydrogenated under 1.5 MPa of hydrogen with 99.999 99% purity at 300 °C for 10 h, in order to obtain Li_2NH and $LiNH_2$. The bulk morphology of the Ni foam was retained after impregnation, nitrogenation, and hydrogenation and dehydrogenation, even after ten cycles. In this manner, the impregnation method is thought to be useful to handle the Li–N–H systems as hydrogen storage materials since the bulk morphology was retained.

6.5 Proposed Mechanism of the Hydrogen Storage Reaction in the Metal–N–H Systems

As indicated above, for metal–N–H systems the disproportionation takes place during the hydrogenation, whereas the hydrogen desorbing reaction is carried out by a solid–solid reaction. After Chen's report [6], a number of studies contributed to our understanding of these reactions. In this section, the proposed mechanisms of hydrogen storage reactions in the metal–N–H systems will be introduced.

6.5.1 Ammonia-Mediated Model for Hydrogen Desorption

Usually, it is very difficult to understand the appearance of the first order solid–solid reaction [14], because the reaction rate can be controlled by the diffusion process of elements between the solid phases and/or the growth of the nuclei of the Li_2NH

phase. Hu and Ruckenstein [13] claimed that the ultrafast reaction between LiH and NH_3 generated by the decomposition of $LiNH_2$ is very important for the solid–solid reaction during the hydrogen absorbing and desorbing processes. Ichikawa *et al.* [14] also proposed a NH_3-mediated model for understanding the reaction mechanism of the Li–N–H system, and reported that the reaction (6.4) is composed of two kinds of elementary reactions, as expressed by reactions (6.5) and (6.6). Later, Isobe *et al.* [106] verified the above mechanism by analysis with thermal desorption mass spectroscopy, thermogravimetry and Fourier transform infrared spectroscopy, in which LiH or $LiNH_2$ in the mixed products were replaced by LiD or $LiND_2$, respectively. The hydrogen desorption reaction corresponding to the latter reaction occurs as soon as $LiNH_2$ has decomposed into Li_2NH and NH_3, since the former and the latter reactions are endothermic and exothermic, respectively. Therefore, the hydrogen desorption reaction can be understood by the following processes: In the first step, $LiNH_2$ decomposes into $Li_2NH/2$ and $NH_3/2$, and then the generated $NH_3/2$ quickly reacts with LiH/2, transforming into $LiNH_2/2$ and $H_2/2$. In the second step, the produced $LiNH_2/2$ decomposes into $Li_2NH/4 + NH_3/4$ and then $NH_3/4 +$ LiH/4 transform to $LiNH_2/4 + H_2/4$, and such successive steps continue until $LiNH_2$ and LiH completely transform into Li_2NH and H_2. From the microscopic point of view, the reaction of LiH and NH_3 generates H_2 gas at the boundary between LiH and $LiNH_2$, then the LiH phase shrinks and the decomposition of $LiNH_2$ into Li_2NH proceeds by exchanging Li and H atoms. This idea can be followed by the evidence of Hu's result [2008 Hu JPS], which showed that the NH_3 release from $LiNH_2$ was dramatically increased at about 75 °C, detected by ^1H NMR. David *et al.* [107] demonstrated that the transformation between $LiNH_2$ and Li_2NH during hydrogen cycling in the Li–N–H hydrogen storage system is a bulk reversible reaction that occurs in a non-stoichiometric manner within the cubic anti-fluorite-like Li–N–H structure, as obtained by the structural refinement from synchrotron X-ray diffraction data. These results could be understood by the above microscopic feature as well. Based on the NH_3-mediated reaction model, Shaw *et al.* [108] investigated the reaction pathway and rate-limiting step of dehydrogenation of the $LiNH_2$ and LiH mixture. A reaction pathway was proposed to explain the ultrafast reaction between LiH and NH_3 as well as the slow decomposition of $LiNH_2$ to Li_2NH and NH_3. The model proposed is in good agreement with experimental observations including (i) the change in the volume of the solid before and after reaction, (ii) the limited increase in particle agglomeration over 35 h at very high homologous temperatures, and (iii) the substantially different reaction rates between reactions (6.2) and (6.3) although both reactions generate solid and gaseous products.

Using this NH_3-mediated model, the reaction between LiH and $Mg(NH_2)_2$ can also be explained, with the Li^+ cation being supplied from LiH after the reaction between LiH and NH_3. Here, we discuss the transformation of the phases during the dehydrogenating reaction processes in the Li–Mg–N–H systems. Rijssenbeek *et al.* [81], for the case of "Li/Mg = 6/3", determined three new imide crystal structures by high-resolution X-ray and neutron diffraction, identifying a new family of imides with formula $Li_{4-2x}Mg_x(NH)_2$ (up to 6 mass% of H_2, at $\sim$220 °C).

Luo *et al.* [25] studied the case of "Li/Mg = 6/3" and characterized the products at several points on the PC isotherm and proposed the reactions as follows:

$$Mg(NH_2)_2 + 2LiH \leftrightarrow Li_2MgN_2H_{3.2} + 1.4H_2 \quad (6.23)$$

$$Li_2MgN_2H_{3.2} \leftrightarrow Li_2Mg(NH)_2 + 0.6H_2 \quad (6.24)$$

For the case of "Li/Mg = 8/3", Nakamura *et al.* determined the phase of $Li_{2.6}MgN_2D_{1.4}$ as a dehydrogenated state (in vacuum conditions, at 200 °C) by using synchrotron radiation (SR)-X-ray and neutron diffraction [109]. For "Li/Mg = 6/3, 8/3, 12/3" systems, Aoki *et al.* performed PC-isotherm measurements only for the dehydrogenation at 250 °C and observed the complex imide $Li_2Mg(NH)_2$ as a dehydrogented state by SR-XRD [110]. Considering all these experimental results, the reaction steps of the Li–Mg–N–H system can be stated as below,

$$nLiH + 3Mg(NH_2)_2$$

$$\rightarrow (n-3)LiH + 3LiMgN_2H_3 + 3H_2 \quad (6.25)$$

$$\rightarrow (n-6)LiH + 3Li_2MgN_2H_2 + 6H_2 \quad (6.26)$$

$$\rightarrow (n-8)LiH + 3Li_{2.7}MgN_2H_{1.3} + 8H_2 \quad (6.27)$$

$$\rightarrow (n-12)LiH + 3Li_3N + 3LiMgN + 12H_2 \quad (6.28)$$

In first step (6.25), 3LiH reacts with $3Mg(NH_2)_2$, generating $3H_2$ and $3LiMgN_2H_3$, which can be regarded as a complex of Li-amide and Mg-imide. Although this single phase has not yet been reported, Aoki *et al.* reported the existence of $Li_{1.3}MgN_2H_{2.7}$ which is a quite similar phase to $LiMgN_2H_3$ [110]. In the next step (6.26), 3LiH reacts with $3LiMgN_2H_3$, generating $3H_2$ and $3Li_2MgN_2H_2$, which can be regarded as a complex of Li-imide and Mg-imide and has already been reported by many researchers. In the third step (6.27), 2LiH reacts with $3Li_2MgN_2H_2$ to generate $2H_2$ and $3Li_{2.7}MgN_2H_{1.3}$, which can be regarded as a complex of Li-imide and Mg-nitride and an equivalent phase of $Li_{2.6}MgN_2H_{1.4}$ reported by Nakamura *et al.* [109]. In fact, this Li-imide and Mg-nitride single phase should be obtained below 200 °C under vacuum conditions. At a much higher temperature than ~450 °C, 4LiH reacts with $3Li_{2.6}MgN_2H_{1.3}$, generating $4H_2$, $3Li_3N$ and LiMgN, which is a complex of Li-nitride and Mg-nitride. In brief, the phase transformation proceeds in the order of "Mg-amide to Mg-imide", "Li-amide to Li-imide", "Mg-imide to Mg-nitride" and "Li-imide to Li-nitride".

6.5.2
Direct Solid–Solid Reaction Model for Hydrogen Desorption

Chen *et al.* [12] examined the dehydrogenation reaction from the mixture of $LiNH_2$ and 2LiD, and confirmed that the desorbing gas was a mixture of H_2, HD and D_2, all

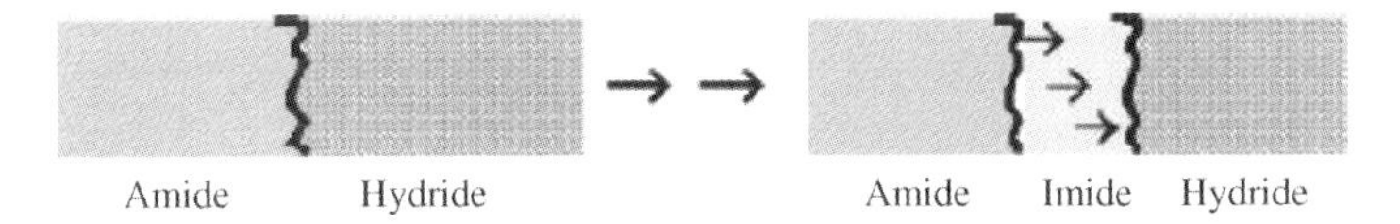

Figure 6.15 Chemical reactions at the interfaces of amide/hydride, amide/imide, and imide/hydride and mass transport through the imide layer [113].

of which were desorbed in the temperature range from 150 to 400 °C. These results cannot be understood by a simple solid–solid reaction model. Xiong *et al.* [20] pointed out that the TPD profiles of hydrogen desorption from $Mg(NH_2)_2$ + 2LiH mixture and ammonia desorption from $Mg(NH_2)_2$ were totally different, indicating the unlikely involvement of ammonia desorption in the reaction path of hydrogen desorption. Moreover, Weidner *et al.* [111] offered a mechanism for direct solid–solid reaction. Aguey-Zinsou *et al.* [112] denied the ammonia mediated reaction mechanism based on their experimental results. Chen *et al.* [113] measured the apparent activation energies to be 88.1 and 130 kJ for the isothermal and non-isothermal kinetics, respectively. From these results, they concluded that the thermal decomposition of $Mg(NH_2)_2$ was unlikely to be an elementary step in the chemical reaction of $Mg(NH_2)_2$ and 2LiH. Finally, they proposed that the reaction rate is controlled by the interface reaction in the early stage of the reaction and by mass transport through the imide layer in the later stage, as shown in Figure 6.15.

6.5.3 Hydrogenating Mechanism of the Li-Mg-N-H System

Leng *et al.* [114] proposed that $4Li_2NH$ was first hydrogenated to 4LiH and $4LiNH_2$. In the next step, $4LiNH_2$ decomposes to $2Li_2NH$ and $2NH_3$, and the generated $2NH_3$ reacts with $(1/2)Mg_3N_2$ to produce the $(3/2)Mg(NH_2)_2$ phase, while $2Li_2NH$ is hydrogenated back to 2LiH and $2LiNH_2$ Such successive steps continue until all the $4Li_2NH$ and Mg_3N_2 are completely transform into 8LiH and $3Mg(NH_2)_2$ (6.29) by hydrogenation. In this paper, they confirmed the reaction between Mg_3N_2 and $4NH_3$. The remaining reaction between $LiNH_2$ and H_2 was confirmed later [115]. The complete hydrogenating reaction can be written as follows:

$$
\begin{aligned}
&4Li_2NH + Mg_3N_2 + 8H_2 \\
&\rightarrow 4LiH + 4LiNH_2 + Mg_3N_2 + 4H_2 \\
&\rightarrow 4LiH + (2Li_2NH + 2H_2) + (2NH_3 + 1/2Mg_3N_2) + 1/2Mg_3N_2 + 2H_2 \\
&\rightarrow 4LiH + 2LiH + 3/2Mg(NH_2)_2 + 2LiNH_2 + 1/2Mg_3N_2 + 2H_2 \\
&\rightarrow (4+2)LiH + 3/2Mg(NH_2)_2 + (Li_2NH + H_2) + (NH_3 + 1/4Mg_3N_2) \\
&\quad + 1/4Mg_3N_2 + H_2 \\
&\rightarrow (4+2)LiH + LiH + (3/2 + 3/4)Mg(NH_2)_2 + LiNH_2 + 1/4Mg_3N_2 + H_2 \\
&\rightarrow (4+2+1)LiH + (3/2 + 3/4)Mg(NH_2)_2 + (1/2Li_2NH + 1/2H_2) \\
&\quad + (1/2NH_3 + 1/8Mg_3N_2) + 1/8Mg_3N_2 + 1/2H_2
\end{aligned}
$$

$$
\begin{aligned}
&\rightarrow (4+2+1)\mathrm{LiH}+1/2\mathrm{LiH}+(3/2+3/4+3/8)\mathrm{Mg(NH_2)_2}\\
&\quad +1/2\mathrm{LiNH_2}+1/8\mathrm{Mg_3N_2}+1/2\mathrm{H_2}\\
&\rightarrow \ldots\\
&\rightarrow 4\left(\sum_{k=0}^{n}\frac{1}{2^k}\right)\mathrm{LiH}+3\left(\sum_{k=1}^{n}\frac{1}{2^k}\right)\mathrm{Mg(NH_2)_2}+\frac{1}{2^{n-2}}\mathrm{LiNH_2}\\
&\quad +\frac{1}{2^n}\mathrm{Mg_3N_2}+\frac{1}{2^{n-2}}\mathrm{H_2}\\
&\rightarrow 8\mathrm{LiH}+3\mathrm{Mg(NH_2)_2}.
\end{aligned}
\tag{6.29}
$$

6.6 Summary

In this chapter, amides, imides and mixtures were introduced from the viewpoint of hydrogen storage properties with respect to various combinations, the crystal structures of amides and imides, their electronic structures, catalytic properties and the reaction mechanism of hydrogen absorption and desorption. In the section on the hydrogen storage properties of amide and imide systems, almost all the combinations of amide and hydride have been reviewed, which suggests that the Li–Mg–N–H systems could be the most promising because the hydrogen storage capacity can reach up to 5.5 mass% and 4 kg/100 L. The hydrogen absorption and desorption can be operated in the temperature range 150 to 200 °C, at which the absorption pressure should be less than 10 MPa, and the desorption pressure can be partially obtained at more than 1 MPa and the remaining hydrogen can be obtained below 200 °C under vacuum conditions. On the other hand, several hundred ppm of NH_3 is essentially emitted from the amide component in the $Mg(NH_2)_2$ and LiH mixture during hydrogen desorption, indicating that a special filter or absorbent for an NH_3 trap is necessary to use this system for a fuel cell. The kinetics of this system becomes worse after a few cycles, indicating that drastic improvement in the cycle properties and measures to decrease the amount of NH_3 emission should be addressed for the practical application of hydrogen storage.

Because the structures of some imides have not yet been determined, a great effort should be directed towards this. With respect to hydrogen desorption from the metal–N–H systems, two representative reaction models have been proposed – the direct solid–solid reaction model and the ammonia-mediated model. Understanding the reaction mechanism is still a pending issue. Therefore, research on the metal–N–H system should focus on these fundamental aspects.

References

1 Dafert, F.W. and Miklauz, R. (1910) *Monatsh. Chem.*, **31**, 981.

2 Ruff, O. and Goeres, H. (1911) *Chem. Ber.*, **44**, 502.

3 Bergstrom, F.W. and Fernelius, W.C. (1933) *Chem. Rev.*, **12**, 43.
4 Bergstrom, F.W. and Fernelius, W.C. (1937) *Chem. Rev.*, **20**, 413.
5 Gregory, D.H. (2001) *Coord. Chem. Rev.*, **215**, 301.
6 Chen, P., Xiong, Z., Luo, J., Lin, J., and Tan, L. (2002) *Nature*, **420**, 302.
7 Hu, Y.H. and Ruckenstein, E. (2003) *Ind. Eng. Chem. Res.*, **42**, 5135.
8 Ichikawa, T., Isobe, S., Hanada, N., and Fujii, H. (2004) *J. Alloys Compd.*, **365**, 271.
9 Luo, W. (2004) *J. Alloys Compd.*, **381**, 284.
10 Kojima, Y. and Kawai, Y. (2005) *J. Alloys Compd.*, **395**, 236.
11 Isobe, S., Ichikawa, T., Tokoyoda, K., Hanada, N., Leng, H., Fujii, H., and Kojima, Y. (2008) *Thermochem. Acta*, **468**, 35.
12 Chen, P., Xiong, Z., Luo, J., Lin, J., and Tan, K.L. (2003) *J. Phys. Chem. B*, **107**, 10967.
13 Hu, Y.H. and Ruckenstein, E. (2003) *J. Phys. Chem. A*, **107**, 9737.
14 Ichikawa, T., Hanada, N., Isobe, S., Leng, H.Y., and Fujii, H. (2004) *J. Phys. Chem. B*, **108**, 7887.
15 Nakamori, Y. and Orimo, S. (2004) *J. Alloys Compd.*, **370**, 271.
16 Orimo, S., Nakamori, Y., Kitahara, G., Miwa, K., Ohba, N., Noritake, T., and Towata, S. (2004) *Appl. Phys. A*, **79**, 1765.
17 Nakamori, Y., Kitahara, G., Miwa, K., Towata, S., and Orimo, S. (2005) *Appl. Phys. A*, **80**, 1.
18 Xiong, Z., Wu, G., Hu, J., and Chen, P. (2004) *Adv. Mater.*, **16**, 1522.
19 Leng, H.Y., Ichikawa, T., Hino, S., Hanada, N., Isobe, S., and Fujii, H. (2004) *J. Phys. Chem. B*, **108**, 8763.
20 Xiong, Z., Hu, J., Wu, G., Chen, P., Luo, W., Gross, K., and Wang, J. (2005) *J. Alloys Compd.*, **398**, 235.
21 Aoki, M., Noritake, T., Kitahara, G., Nakamori, Y., Towata, S., and Orimo, S. (2007) *J. Alloys Compd.*, **428**, 307.
22 Yang, J., Sudik, A., and Wolverton, C. (2007) *J. Alloys Compd.*, **430**, 334.
23 Ichikawa, T., Tokoyoda, K., Leng, H., and Fujii, H. (2005) *J. Alloys Compd.*, **400**, 245.
24 Xiong, Z., Wu, G., Hu, J., Chen, P., Luo, W., and Wang, J. (2006) *J. Alloys Compd.*, **417**, 190.
25 Luo, W. and Sickafoose, S. (2006) *J. Alloys Compd.*, **407**, 274.
26 Janot, R., Eymery, J.B., and Tarascon, J.M. (2007) *J. Power Sources*, **164**, 496.
27 Akbarzadeh, A.R., Ozolins, V., and Wolverton, C. (2007) *Adv. Mater.*, **19**, 3233.
28 Araujo, C.M., Scheicher, R.H., and Ahuja, R. (2008) *Appl. Phys. Lett.*, **92**, 021907.
29 Leng, H.Y., Ichikawa, T., and Fujii, H. (2006) *J. Phys. Chem. B*, **110**, 12964.
30 Nakamori, Y., Kitahara, G., and Orimo, S. (2004) *J. Power Sources*, **138**, 309.
31 Leng, H.Y., Ichikawa, T., Isobe, S., Hino, S., Hanada, N., and Fujii, H. (2006) *J. Alloys Compd.*, **404–406**, 443.
32 Hu, J., Wu, G., Liu, Y., Xion, Z., Chen, P., Murata, K., Sakata, K., and Wolf, G. (2006) *J. Phys. Chem. B*, **110**, 14688.
33 Xie, L., Li, Y., Yang, R., Liu, Y., and Li, X. (2008) *Appl. Phys. Lett.*, **92**, 231910.
34 Hino, S., Ichikawa, T., Leng, H., and Fujii, H. (2005) *J. Alloys Compd.*, **398**, 62.
35 Xiong, Z., Chen, P., Wu, G., Lin, J., and Tan, K.L. (2003) *J. Mater. Chem.*, **13**, 1676.
36 Tokoyoda, K., Hino, S., Ichikawa, T., Okamoto, K., and Fujii, H. (2007) *J. Alloys Compd.*, **439**, 337.
37 Wu, H. (2008) *J. Am. Chem. Soc.*, **130**, 6515.
38 Leng, H.Y., Ichikawa, T., Hino, S., Nakagawa, T., and Fuii, H. (2005) *J. Phys. Chem. B*, **109**, 10744.
39 Xiong, Z., Hu, J., Wu, G., and Wu, P. (2005) *J. Alloys Compd.*, **395**, 209.
40 Xiong, Z., Wu, G., Hu, J., and Chen, P. (2007) *J. Alloys Compd.*, **441**, 152.
41 Hu, J., Xiong, Z., Wu, G., Chen, P., Murata, K., and Sakata, K. (2006) *J. Power Sources*, **159**, 116.
42 Liu, Y., Hu, J., Xiong, Z., Wu, G., Chen, P., Murata, K., and Sakata, K. (2007) *J. Alloys Compd.*, **432**, 298.
43 Liu, Y., Xiong, Z., Hu, J., Wu, G., Chen, P., Murata, K., and Sakata, K. (2006) *J. Power Sources*, **159**, 135.
44 Yamane, H., Kano, T., Kamegawa, A., Shibata, M., Yamada, T., Okada, M., and Shimada, M. (2005) *J. Alloys Compd.*, **402**, L1.
45 Kojima, Y., Matsumoto, M., Kawai, Y., Haga, T., Ohba, N., Miwa, K., Towata, S.,

Nakamori, Y., and Orimo, S. (2006) *J. Phys. Chem. B*, **110**, 9632.
46 Xiong, Z., Wu, G., Hu, J., and Chen, P. (2006) *J. Power Sources*, **159**, 167.
47 Xiong, Z., Wu, G., Liu, Y., Chen, P., Luo, W., and Wang, J. (2007) *Adv. Funct. Mater.*, **17**, 1137.
48 Janot, R., Eymery, J.B., and Tarascon, J.M. (2007) *J. Phys. Chem. C*, **111**, 2335.
49 Titherley, A.W. (1894) *J. Chem. Soc.*, **65**, 504.
50 Leng, H.Y., Ichikawa, T., Hino, S., Hanada, N., Isobe, S., and Fujii, H. (2006) *J. Power Sources*, **156**, 166.
51 Sørby, M.H., Nakamura, Y., Brinks, H.W., Ichikawa, T., Hino, S., Fujii, H., and Hauback, B.C. (2007) *J. Alloys Compd.*, **428**, 297.
52 Juza, R. and Opp, K. (1951) *Z. Anorg. Allg. Chem.*, **266**, 325.
53 Jacobs, H. and Juza, R. (1972) *Z. Anorg. Allg. Chem.*, **391**, 271.
54 Grotjahn, D.B. and Sheridan, P.M. (2001) *J. Am. Chem. Soc.*, **123**, 5489.
55 Yang, J.B., Zhou, X.D., Cai, Q., James, W.J., and Yelon, W.B. (2006) *Appl. Phys. Lett.*, **88**, 041914.
56 Chellappa, R.S., Chandra, D., Somayazulu, M., Gramsch, S.A., and Hermley, R.J. (2007) *J. Phys. Chem. B*, **111**, 10785.
57 Miwa, K., Ohba, N., Towata, S., Nakamori, Y., and Orimo, S. (2005) *Phys. Chem. B*, **71**, 195109.
58 Hebst, J.F. and Hector, L.G. Jr., (2005) *Phys. Rev. B*, **72**, 125120.
59 Song, Y. and Guo, Z.X. (2006) *Phys. Rev. B*, **74**, 195120.
60 Tsumuraya, T., Shishidou, T., and Oguchi, T. (2007) *J. Alloys Compd.*, **446–447**, 323.
61 Bohger, J.-P.O., Eßmann, R.R., and Jacobs, H. (1995) *J. Mol. Struct.*, **348**, 325.
62 Lu, C., Hu, J., Kwak, J.H., Yang, Z., Ren, R., Markmaitree, T., and Shaw, L.L. (2007) *J. Power Sources*, **170**, 419.
63 Hu, J.Z., Kwak, J.H., Yang, Z., Osborn, W., Markmaitree, T., and Shaw, L.L. (2008) *J. Power Sources*, **181**, 116.
64 Hu, J.Z., Kwak, J.H., Yang, Z., Osborn, W., Markmaitree, T., and Shaw, L.L. (2008) *J. Power Sources*, **182**, 278.
65 Palumbo, O., Paolone, A., Cantelli, R., and Chandra, D. (2008) *Int. J. Hydrogen Energy*, **33**, 3107.
66 Hu, Y.H. and Ruckenstein, E. (2006) *Ind. Eng. Chem. Res.*, **45**, 4993.
67 Ohoyama, K., Nakamori, Y., Orimo, S., and Yamada, K. (2005) *J. Phys. Soc. Jpn.*, **74**, 483.
68 Noritake, T., Hozaki, H., Aoki, M., Towata, S., Kitahara, G., Nakamori, Y., and Orimo, S. (2005) *J. Alloys Compd.*, **393**, 264.
69 Balogh, M.P., Jones, C.Y., Herbst, J.F., Hector, L.G., Jr., and Kundrat, M. (2006) *J. Alloys Compd.*, **420**, 326.
70 Mueller, T. and Ceder, G. (2006) *Phys. Rev. B*, **74**, 134104.
71 Maagyari-Köpe, B., Ozolins, V., and Wolverton, C. (2006) *Phys. Rev. B*, **73**, 220101.
72 Juza, R., Weber, H.H., and Opp, K. (1956) *Z. Anorg. Allg. Chem.*, **284**, 73.
73 Zalkin, A. and Templeton, D.H. (1956) *J. Phys. Chem.*, **60**, 821.
74 Jacobs, H. and Juza, R. (1969) *Z. Anorg. Allg. Chem.*, **370**, 254.
75 Wang, Y. and Chou, M.Y. (2007) *Phys. Rev. B*, **76**, 014116.
76 Linde, G. and Juza, R. (1974) *Z. Anorg. Allg. Chem.*, **409**, 199.
77 Velikokhatnyi, O.I. and Kumta, P.N. (2007) *Mater. Sci. Eng. B*, **140**, 114.
78 Juza, R. and Schumacher, H. (1963) *Z. Anorg. Allg. Chem.*, **324**, 278.
79 Bouclier, P., Portier, J., and Turrell, G. (1969) *J. Mol. Struct.*, **4**, 1.
80 Sichla, T. and Jacobs, H. (1996) *Z. Anorg. Allg. Chem.*, **622**, 2079.
81 Rijssenbeek, J., Gao, Y., Hnson, J., Huang, Q., Jones, C., and Toby, B. (2008) *J. Alloys Compd.*, **454**, 233.
82 Wu, G., Xiong, Z., Liu, T., Liu, Y., Hu, J., Chen, P., Feng, Y., and Wee, A.T.S. (2007) *Inorg. Chem.*, **46**, 517.
83 Jacobs, H., Jänichen, K., Hadenfeldt, C., and Juza, R. (1985) *Z. Anorg. Allg. Chem.*, **531**, 125.
84 Matsumoto, M., Haga, T., Kawai, Y., and Kojima, Y. (2007) *J. Alloys Compd.*, **439**, 358.
85 Lohstroh, W. and Fichtner, M. (2007) *J. Alloys Compd.*, **446–447**, 332.

86 Isobe, S., Ichikawa, T., Hanada, N., Leng, H.Y., Fichtner, M., Fuhr, O., and Fujii, H. (2005) *J. Alloys Compd.*, **404–406**, 439.
87 Isobe, S., Ichikawa, T., Kojima, Y., and Fujii, H. (2007) *J. Alloys Compd.*, **446–447**, 360.
88 Tsumuraya, T., Shishidou, T., and Oguchi, T. (2008) *Phys. Rev. B*, **77**, 235114.
89 Hu, Y.H. and Ruckenstein, E. (2004) *Ind. Eng. Chem. Res.*, **43**, 2464.
90 Yao, J.H., Shang, C., Aguey-Zinsou, K.F., and Guo, Z.X. (2007) *J. Alloys Compd.*, **432**, 277.
91 Chen, Y., Wang, P., Liu, C., and Cheng, H.M. (2007) *Int. J. Hydrogen Energy*, **32**, 1262.
92 Sudik, A., Yang, J., Halliday, D., and Wolverton, C. (2007) *J. Phys. Chem. C*, **111**, 6568.
93 Pinkerton, F.E. (2005) *J. Alloys Compd.*, **400**, 76.
94 Markmaitree, T., Ren, R., and Shaw, L.L. (2006) *J. Phys. Chem. B*, **110**, 20710.
95 Shaw, L.L., Ren, R., Markmaitree, T., and Osborn, W. (2008) *J. Alloys Compd.*, **448**, 263.
96 Xie, L., Zheng, J., Liu, Y., Li, Y., and Li, X. (2008) *Chem. Mater.*, **20**, 282.
97 Hino, S., Ichikawa, T., Ogita, N., Udagawa, M., and Fujii, H. (2005) *Chem. Comm.*, **24**, 3038.
98 Hino, S., Ichikawa, T., Tokoyoda, K., Kojima, Y., and Fujii, H. (2007) *J. Alloys Compd.*, **446–447**, 342.
99 Luo, W. and Stewart, K. (2007) *J. Alloys. Compd.*, **440**, 357.
100 Liu, Y., Hu, J., Wu, G., Xiong, Z., and Chen, P. (2008) *J. Phys. Chem. C*, **112**, 1293.
101 Okamoto, K., Tokoyoda, K., Ichikawa, T., and Fujii, H. (2007) *J. Alloys Compd.*, **432**, 289.
102 Luo, S., Flanagan, T.B., and Luo, W. (2007) *J. Alloys Compd.*, **440**, L13.
103 Hu, Y.H. and Ruckenstein, E. (2007) *Ind. Eng. Chem. Res.*, **46**, 5940.
104 Hu, Y.H. and Ruckenstein, E. (2008) *Ind. Eng. Chem. Res.*, **47**, 48.
105 Hao, T., Matsuo, M., Nakamori, Y., and Orimo, S. (2008) *J. Alloys Compd.*, **458**, L1.
106 Isobe, S., Ichikawa, T., Hino, S., and Fujii, H. (2005) *J. Phys. Chem. B*, **109**, 14855.
107 David, W.I.F., Jones, M.O., Gregory, D.H., Jewell, C.M., Johnson, S.R., Walton, A., and Edwards, P.P. (2007) *J. Am. Chem. Soc.*, **129**, 1594.
108 Shaw, L.L., Osborn, W., Markmaitree, T., and Wan, X. (2008) *J. Power Sources*, **177**, 500.
109 Nakamura, Y., Hino, S., Ichikawa, T., Fujii, H., Brinks, H.W., and Hauback, B.C. (2008) *J. Alloys Compd.*, **457**, 362.
110 Aoki, M., Noritake, T., Kitahara, G., Nakamori, Y., Towata, S., and Orimo, S. (2007) *J. Alloys Compd.*, **428**, 307.
111 Weidner, E., Bull, D.J., Shabalin, I.L., Keens, S.G., Telling, M.T.F., and Ross, D.K. (2007) *Chem. Phys. Lett.*, **444**, 76.
112 Aguey-Zinsou, K.F., Yao, J.H., and Guo, Z.X. (2007) *J. Phys. Chem. B*, **111**, 12531.
113 Chen, P., Xiong, Z., Yang, L., Wu, G., and Luo, W. (2006) *J. Phys. Chem. B*, **110**, 14221.
114 Leng, H.Y., Ichikawa, T., Hino, S., Nakagawa, T., and Fuii, H. (2005) *J. Phys. Chem. B*, **109**, 10744.
115 Leng, H.Y., Ichikawa, T., Hino, S., and Fujii, H. (2008) *J. Alloys Compd.*, **463**, 462.

7
Tailoring Reaction Enthalpies of Hydrides

Martin Dornheim

7.1
Introduction

In view of the future use of hydrogen as a renewable fuel for mobile and stationary applications the development of cost and energy efficient, safe and reliable hydrogen storage technologies enabling high volumetric and gravimetric storage densities remains one of the most challenging tasks.

One major advantage of metal hydrides is the ability to store hydrogen in a very energy efficient way enabling hydrogen storage at rather low pressures without further need for liquefaction or compression. However, depending on the hydrogen reaction enthalpy of the specific storage material during hydrogen uptake, a huge amount of heat (equivalent to 15% or more of the energy stored in hydrogen) can be generated and has to be removed in a rather short time. On the other hand, during desorption the same amount of heat has to be applied to facilitate the endothermic hydrogen desorption process. If, in the case of stationary applications, during hydrogen uptake the generated amount of heat is stored and used for desorption again, or if, in the case of mobile applications, it is used as process heat for applications near to a filling station, the highest energy efficiencies of the whole hydrogen storage tank–utilization system can be accomplished, see Figure 7.1.

In most mobile applications fast recharging of a hydrogen storage tank ought to be possible within only a few minutes. If a 4 kg hydrogen storage tank based on conventional room temperature hydrides with reaction enthalpies of around $\Delta H = 30\,\text{kJ}\,(\text{mol H}_2)^{-1}$ is considered, like in the case of $FeTiH_2$ ($-28\,\text{kJ}\,(\text{mol H}_2)^{-1}$ for TiFeH and $-35\,\text{kJ}\,(\text{mol H}_2)^{-1}$ for $TiFeH_{2-x}$, Buchner [1]) or $LaNi_5$ ($\Delta H = -31\,\text{kJ}\,(\text{mol H}_2)^{-1}$) about 60 MJ heat is thereby produced. This value increases still further if more stable medium- or even high-temperature metal hydrides/complex hydrides are chosen. Therefore, it is desirable to develop materials with high storage capacities as well as moderate reaction heats. MgH_2 is an example of a high-temperature metal hydride with a high gravimetric storage capacity and a large value of reaction enthalpy. It has a formation enthalpy of $\Delta H = -75\,\text{kJ}\,(\text{mol H}_2)^{-1}$ and thus during

Handbook of Hydrogen Storage. Edited by Michael Hirscher

ISBN: 978-3-527-32273-2

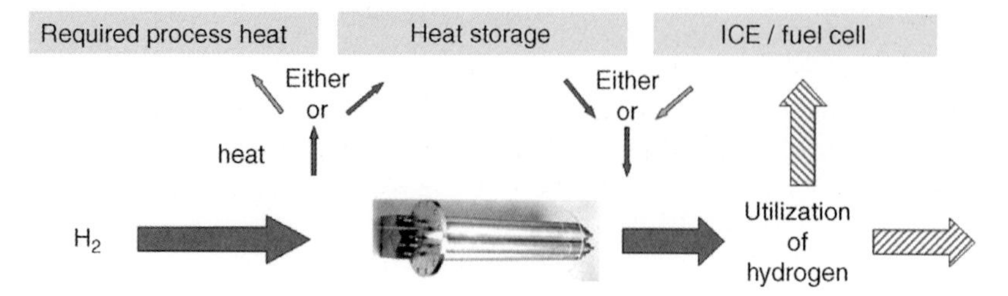

Figure 7.1 Heat management in a metal hydride hydrogen storage device.

charging of the respective 4 kg hydrogen storage tank about 150 MJ heat would be released which excludes the possibility of fast charging of such a tank, see Figure 7.2.

During hydrogen release the same amount of heat is required – however, generally on a longer time scale. On the one hand this enhances the inherent safety of such a tank system. Without an external heat supply hydrogen release would lead to cooling of the tank and finally hydrogen desorption will stop. On the other hand it implies further restrictions for the choice of suitable storage materials. The highest energy efficiencies of the whole tank to fuel combustion or fuel cell system can only be achieved if, in the case of desorption, the energy required for hydrogen release can be supplied by the waste heat generated, in the case of mobile applications, on-board by the hydrogen combustion process.

Furthermore, to realize certain hydrogen release pressures the storage tank needs to be kept at a certain temperature level which again depends on the exact alloy/composite composition. Therefore, there is a lower limit for the temperature of the generated heat so that it can be used directly without any additional heat pumps so on, for heat supply to the storage tank. This shows that the reaction enthalpy is a

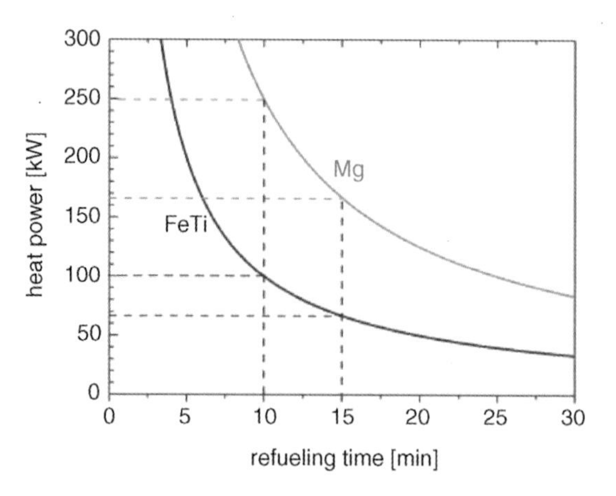

Figure 7.2 Heat generation of a 4 kg hydrogen storage tank based on a room-temperature hydride with $\Delta H \approx -30\,kJ\,(mol\,H_2)^{-1}$ and high-temperature hydride with $\Delta H \approx -75\,kJ\,(mol\,H_2)^{-1}$, respectively.

major parameter to be considered in the choice of specific metal hydrides for different applications.

Actually, there is only a limited number of materials having the required storage capacities in weight and volume as well as showing sufficient kinetics and being producible on a larger scale at affordable prices. This shows both the need for research for novel storage materials and the need for new strategies to tune the reaction enthalpies of already known storage materials.

Only a few binary hydrides like VH_2 or NbH_2 show reasonable hydrogen desorption pressures at temperatures suitable for most technological applications. Therefore, thermodynamic tuning of metal hydrides was, and still is, an important challenge to be faced.

The development of conventional room-temperature hydrides based on intermetallic compounds led to a large number of storage materials exhibiting very favorable sorption enthalpies with values of around 25 kJ (mol H_2)$^{-1}$, which can be operated in combination with conventional (80 °C operation temperature) polymer electrolyte membrane (PEM) fuel cells. However, their gravimetric storage capacity is limited to less than 3 wt.% H_2.

The remaining challenge, therefore, is the development of lightweight hydrides or hydride composites with suitable kinetics in an appropriate temperature range as well as suitable thermodynamic properties.

7.2 Thermodynamic Limitations of Lightweight Hydrides

In recent years several new lightweight hydrides with gravimetric storage capacities higher than 8 wt.% have been shown to react reversibly with hydrogen. Furthermore, it has been demonstrated that the reaction kinetics of such high capacity hydrides can be influenced significantly by altering the microstructure and/or adding suitable dopants to the system.

Magnesium has such a favorable storage capacity, 7.6 wt.%, is the eighth most abundant element on the earth and thus comparatively inexpensive. Consequently, it has been considered as a promising storage material if past kinetic restrictions could be successfully overcome. The breakthrough was achieved using nanocrystalline hydrides prepared by high-energy ball milling [2–4]. The development of suitable catalysts led to very short hydrogen absorption and desorption times of less than 2 min for MgH_2 [5–15].

In addition, MgH_2 with minor additions of transition metal oxides shows excellent cycling stability. This can partly be attributed to the fast kinetics that involves a high nucleation rate during phase transformation, thus limiting grain growth. In addition, the oxide catalysts stabilize the grain boundaries, as shown by Friedrichs *et al.* [16]. Dehouche *et al.* [17] have cycled Cr_2O_3 catalyzed MgH_2 1000 times. No decrease in sorption capacity and only a slight decrease in sorption kinetics were found.

However, the temperatures of operation and the reaction heat are too high for most technical applications, for example, >200 °C and 75 kJ (mol H_2)$^{-1}$, respectively.

Therefore, strong efforts have been undertaken to search for novel Mg-based hydrides with lowered hydrogen reaction enthalpy.

As will be shown in the following sections many attempts have been made to favorably modify the enthalpy of hydride formation, not only of MgH_2 but also of other hydrides, by alloy chemistry. One of the most prominent examples is Mg_2NiH_4 [18]. However, alloying in most cases leads to a substantial decrease in capacity.

Similarly, the demonstration by Bogdanovic *et al.* [12, 19] in 1997 of reversible hydrogen uptake from the gas phase and release of $NaAlH_4$ has been a great step forward in the development of high capacity storage materials with adapted reaction enthalpies. The operating temperatures are in the range 100–150 °C. Thus, combinations with, for example, combustion engines or high-temperature PEM fuel cells, which are currently under development, are feasible. However, the storage capacity is significantly reduced compared to MgH_2. $NaAlH_4$ exhibits a practical storage capacity of only 4–4.5 wt.% H_2. Unfortunately, alanates decompose in two reaction steps upon dehydrogenation which implies two different pressure plateaus instead of just one:

$$MeAlH_4 \rightarrow 1/3\ Me_3AlH_6 + 2/3\ Al + H_2(g) \rightarrow MeH + Al + 3/2\ H_2(g) \quad (7.1)$$

with Me being Li or Na.

The remaining hydrogen bonded to Li or Na is technically not exploitable due to the high stability of the respective hydride.

Meanwhile, a number of different promoters have been tested with respect to their effects on the kinetics. $NaAlH_4$ can be readily reformed within 2 to 10 min at around 100 °C, but at rather high pressures of 80 bar or more [20–22].

While the reaction kinetics could be optimized significantly, the desorption enthalpy of $NaAlH_4$ of 37 kJ (mol H_2)$^{-1}$ and Na_3AlH_6 of 47 kJ (mol H_2)$^{-1}$ is still too large for many applications.

$LiAlH_4$ and Li_3AlH_6 are much less stable than $NaAlH_4$ and Na_3AlH_6 and, therefore, for some time have been considered as potential storage materials.

However, it has been shown that the reaction

$$3\ LiAlH_4 \rightarrow Li_3AlH_6 + 2\ Al + 3\ H_2(g) \quad (7.2)$$

is irreversible and the value of the reaction enthalpy is too small [23]. Ke and Chen [24] calculated the Gibbs energy change for reaction (7.1) as a function of temperature at atmospheric pressure and found it to be negative at all temperatures. Concerning the second reaction step Brinks *et al.* [25] showed that the equilibrium pressure for reaction

$$Li_3AlH_6 + 3\ Al \rightarrow 3\ LiH + 3\ Al + 3/2\ H_2(g) \quad (7.3)$$

is at least 85 bar at 353 K.

Barkhordarian *et al.* (unpublished results) have succeeded in partially synthesizing Li_3AlH_6 from LiH and Al by reactive milling at 180 bar H_2 pressure.

Another high capacity single component hydride based on Al is $Mg(AlH_4)_2$. Because of its high storage capacity of up to 9.3 wt.% it had also been considered as a promising hydrogen storage material [26]. However, in this case, again, the reaction enthalpy has been found to be close to 0 kJ mol^{-1}, hence rehydrogenation is not possible by gas-phase loading [27, 28].

There are only a few stable hydrides showing an even higher gravimetric hydrogen storage density than MgH_2. Among these are the boron-hydrogen compounds which show gravimetric hydrogen storage densities up to 18 wt.% in the case of $LiBH_4$.

For the total thermal decomposition of $LiBH_4$, desorption temperatures higher than 450 °C are required [29]. Furthermore, rehydrogenation of LiH + B is very difficult. Despite the high thermodynamic stability of −66 kJ (mol $H_2)^{-1}$ the borohydride is not formed from a ball-milled (spex mill with ball to powder ratio of 10 for 24 h) LiH + B composite at 350 bar and 400 °C for 24 h [30]. Orimo *et al.* [31] succeeded in rehydrogenating LiH + B partially at 350 bar and 600 °C. This implies that there are both strong thermodynamic and kinetic restrictions for the use of such borohydrides as potential hydrogen storage materials.

Up to now no reversible lightweight hydride with a storage capacity of 5 wt.% or more has been identified which could meet the storage criteria concerning operation temperature or suitable reaction enthalpy.

Thus, new concepts have been, and are still to be, developed for the design of new lightweight hydrides or hydride composites with suitable reaction enthalpy and high capacities.

7.3 Strategies to Alter the Reaction Enthalpies of Hydrides

In this section the main strategies to manipulate the reaction enthalpies of hydrogen absorbing and releasing reactions are discussed.

7.3.1 Thermodynamic Tuning of Single Phase Hydrides by Substitution on the Metal Site

In 1958 Libowitz *et al.* [32] discovered that the intermetallic compound ZrNi reacts reversibly with H_2 to form the ternary hydride $ZrNiH_3$. It was determined that the stability of this metal hydride lies between the stable ZrH_2 and the unstable NiH. This opened up a new research field which led to the discovery of hundreds of new storage materials with different thermodynamic properties following Miedema's rule of reversed stability (see Section 4.4). Around 1970, the discovery of hydrides such as Mg_2Ni, $LaNi_5$ and FeTi followed. In the meantime, several hundred intermetallic hydrides have been reported and a number of interesting compositional types identified. Generally, these intermetallic alloys consist of a hydride-forming element A and a non-hydride-forming element B.

Some important ones are given in Table 7.1.

Table 7.1 Selected examples of intermetallic hydrides.

Composition	A	B	Compounds
A_2B	Mg, Zr	Ni, Fe, Co	Mg_2Ni, Mg_2Co, Zr_2Fe
AB	Ti, Zr	Ni, Fe	TiNi, TiFe, ZrNi
AB_2	Zr, Ti, Y, La	V, Cr, Mn, Fe, Ni	$LaNi_2$, YNi_2,YMn_2, $ZrCr_2$, $ZrMn_2$,ZrV_2, $TiMn_2$
AB_3	La, Y, Mg	Ni, Co	$LaCo_3$,YNi_3,$LaMg_2Ni_9$
AB_5	Ca, La, rare earth	Ni, Cu, Co, Pt, Fe	$CaNi_5$, $LaNi_5$, $CeNi_5$, $LaCu_5$, $LaPt_5$, $LaFe_5$

Such hydrides based on intermetallic compounds are not only attractive for stationary hydrogen storage applications, but also for electrochemical hydrogen storage in rechargeable metal hydride electrodes, reaching capacities of up to 400 mAh g^{-1}. They are produced and sold in more than a billion metal hydride batteries per year.

Because of their high volumetric density, intermetallic hydrides are also used as hydrogen storage materials in advanced fuel cell driven submarines, prototype passenger ships, forklifts and hydrogen automobiles as well as auxiliary power units for laptops.

As described earlier, conventional hydrides exhibit the clear advantage of an extremely high volumetric density of hydrogen and are a safe alternative to compressed gas or liquid hydrogen storage. However, the gravimetric hydrogen density of reversible hydrides working at room temperature is currently limited to less than 3 wt.%, which, for mobile applications, is often not satisfactory. Therefore, in the following sections the discussion is focused on novel lightweight hydrides with much higher gravimetric storage capacities.

MgH_2 is an example of such a material having a very high gravimetric storage capacity, however, it does not have suitable thermodynamics.

As shown above, one possible strategy to lower the absolute value of reaction enthalpies in lightweight hydrides is to look for elements, which exhibit negative heats of mixing with the hydride-forming elements/compounds and thus stabilize the dehydrogenated state.

In principle, there are two possibilities to alter hydrogen reaction enthalpies of Mg-based materials: Either (i) a ternary/multinary hydride is formed during hydrogenation of a Mg alloy, or (ii) only MgH_2 is formed and the alloying element precipitates in elemental form during hydrogenation. In the first case, in order to decrease the overall amount of reaction enthalpy, the change in the enthalpy of formation (per mol H_2) of the ternary hydride with respect to MgH_2 has to be smaller than the heat of formation of the intermetallic compound. In the second case, the value of the reaction enthalpy is directly lowered by the heat of formation of the intermetallic compound. Consequently, this approach is thermodynamically most successful in the case of very stable intermetallic compounds and when the alloying element forms a complex with hydrogen, which forms a weak ionic bond with Mg or phase separation occurs during hydrogenation.

Mg_2Ni is a prominent example for the first class of materials, $MgCu_2$ for the second case.

In this section we concentrate on the first case of single phase hydrides. The case of the formation of different phases/precipitates during hydrogenation is treated in Section 7.3.2.

7.3.1.1 **Lightweight Hydrides Forming Stable Compounds in the Dehydrogenated State**

The enthalpies of formation for the ternary hydride Mg_2NiH_4 and Mg_2Ni amount to $\Delta_f H^0(Mg_2NiH_4) = -176\,kJ\,mol^{-1}$ and $\Delta_f H^0(Mg_2Ni) = -42\,kJ\,mol^{-1}$, respectively. Thus, the enthalpy of reaction for the formation of the hydride Mg_2NiH_4 from the intermetallic compound is $\Delta_r H^0(Mg_2NiH_4) = -67\,kJ\,(mol\,H_2)^{-1}$ and is thus increased by $11\,kJ\,(mol\,H_2)^{-1}$ with respect to MgH_2. However, for many applications, the substantially reduced storage capacity of 3.6 wt.% is still far too low and, furthermore, Mg_2NiH_4 is too stable, thus the desorption temperatures and the amount of energy required to release the hydrogen are too high. One possible solution to overcome this drawback is to search for new ternary/multinary intermetallic phases to destabilize the hydride. Darnaudery *et al.* [33] studied $Mg_2Ni_{0.75}M_{0.25}$ alloys with M being a 3d element. During hydrogenation, they observed the formation of quaternary hydrides for M = V, Cr, Fe, Co and Zn. However, the thermal stabilities of these compounds were found to be very close to that of Mg_2NiH_4.

In addition, Mg is immiscible with V, Cr, and Fe, and, therefore, any ternary phases based on such elements tend to decompose into more stable binary intermetallics at elevated temperatures.

Nevertheless, Tsushio *et al.* [34] reported the hydrogenation properties of $MgNi_{0.86}M1_{0.03}$ (M1 = Cr, Fe, Co, Mn) alloys. Using the van't Hoff equation based on the desorption plateaus and the standard entropy of hydrogen gas $130.9\,J\,mol^{-1}\,K^{-1}$ they determined an enthalpy of hydride formation of the Cr-substituted alloy of $-50\,kJ\,(mol\,H_2)^{-1}$, which is, for Mg-based alloys, a very high value.

Klassen *et al.* investigated $Mg_2(Ni,Me)$ systems [35], Me being Al, Si, Ca, Co, or Cu, by milling MgH_2 together with the elemental metals. After milling, new phases were observed only in the systems Mg–Ni–Si and Mg–Ni–Cu. For a given temperature, $Mg_2Ni_{0.5}Cu_{0.5}H_4$ showed higher plateau pressures than Mg_2NiH_4, and thus lower thermodynamic stability, see Figure 7.3. While pure Mg–H reached a hydrogen plateau pressure of around 1 bar at around 300 °C, in case of the Mg_2Ni a plateau pressure of 1 bar is already reached at 240 °C and in the case of $Mg_2Ni_{0.5}Cu_{0.5}$ already at 230 °C.

However, the storage capacity is reduced to only 3.05 wt.%.

The new phase in the system Mg–Ni–Si did not show any hydrogen interaction as a ternary compound. Instead, it decomposed into Mg_2Ni and Mg_2Si. Upon thermal cycling, hydrogen reacted only with Mg_2Ni and formed a composite consisting of Mg_2NiH_4–Mg_2Si.

Terashita *et al.* [37] found Mg-based pseudo-binary alloys $(Mg_{1-x}Ca_x)Ni_2$ which desorb hydrogen at room temperature. Figure 7.4 shows the PC isotherms of annealed $(Mg_{0.68}Ca_{0.32})Ni_2$ at 278, 313 and 353 K. The observed values of enthalpy

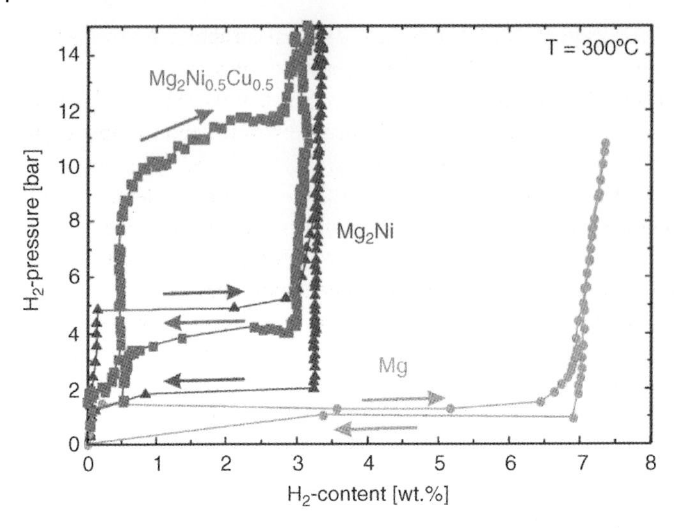

Figure 7.3 Pressure–composition diagrams of Mg–H, Mg_2Ni–H and $Mg_2Ni_{0.5}Cu_{0.5}$. Due to alloying the plateau pressure at $T = 300\,°C$ increases while the storage capacity decreases, taken from [35].

and entropy of hydride formation were $-37\,kJ\,(mol\,H_2)^{-1}$ and $-94\,J\,(mol\,H_2)^{-1}\,K^{-1}$, respectively. These values are already quite comparable to conventional alloys such as $LaNi_5$–H ($-30\,kJ\,(mol\,H_2)^{-1}$) and thus suitable for many applications. However, again, the gravimetric storage capacity is lowered significantly from 3.6 wt.% in the case of Mg_2Ni to only 1.4 wt.%.

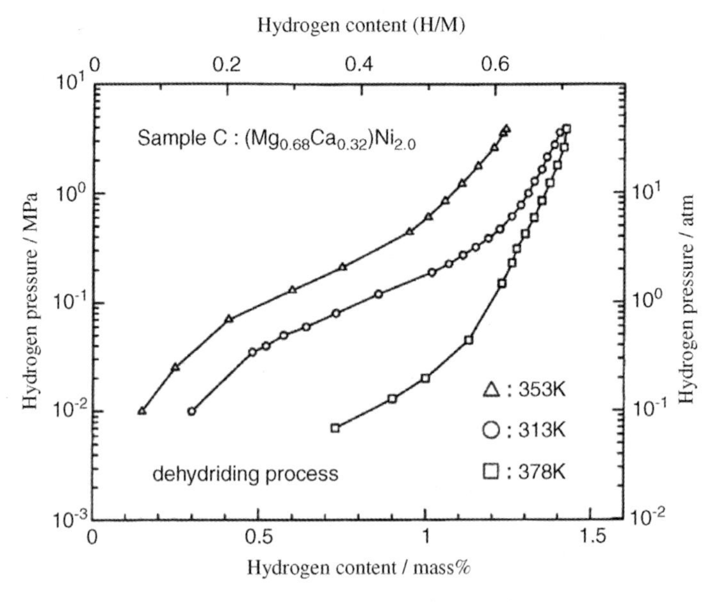

Figure 7.4 Pressure–composition isotherms of annealed $(Mg_{0.68}Ca_{0.32})Ni_2$-H [37].

Mg_2Ca itself exhibits a rather favorable Laves crystal structure (C14) and an enthalpy of formation of $\Delta_f H^0(Mg_2Ca) = -42\,kJ\ mol^{-1}$. However, during hydrogenation, formation of CaH_2 has to be avoided, as it is substantially more stable than $MgH_2 \Delta_f H^0(Mg_2CaH_2) = -175\,kJ\ (mol\ H_2)^{-1}$), and thus would unfavorably influence the reaction enthalpy. Another phase formed during hydrogenation of Ca–Mg compounds is $Ca_4Mg_3H_{14}$ [38]. However, again this phase is more stable than MgH_2 as desorption requires temperatures above 400 °C. So far, attempts to avoid the very stable Ca hydride phases and reversibly store hydrogen in Mg_2Ca by forming MgH_2 and CaH_2 have not been successful under moderate conditions.

Using high pressures in the GPa range as well as thin-film processing, a destabilization of MgH_2 has been reported [39] and new Mg-based hydrides with significantly lowered hydrogen bonding enthalpies have been reported, such as Mg–Sc–H [40] and Mg–Ti–H [41, 42]. Whereas Mg–Sc hydrides would be too expensive for most commercial applications, an $Mg_xTi_yH_z$ phase would be an interesting hydride, if it would form at ambient pressures in bulk materials. However, on the one hand reversibility and long term stability must be achieved and, therefore, the stability of the phase should not be much lower than the respective stability of the MgH_2–TiH_2 composite, while, on the other hand such an $Mg_xTi_yH_z$ phase would have to be substantially less stable than the respective competing composite, otherwise no decrease in the reaction enthalpy compared to MgH_2 would be achieved.

Kyoi *et al.* [41] first reported a ternary Mg–Ti hydride alloy with the composition $Mg_7TiH_{12.7}$. They synthesized this alloy using a high-pressure anvil cell by the reaction of MgH_2 and $TiH_{1.9}$ at 8 GPa and 873 K. However, by processing of MgH_2 and $TiH_{1.9}$ by high-energy ball milling and subsequent hydrogen cycling, the ternary hydride phase could not be observed [43]. This suggests a stabilization of the phase by the applied high pressures and a practical application is thus very questionable.

Vermeulen *et al.* [42] reported hydrogen reactions at low temperatures in thin Mg–Ti films, which could hint at a ternary hydride phase. In thin films, however, which are usually clamped onto hard substrates, huge biaxial stresses in the GPa range can be generated during hydrogen loading due to hydrogen-induced expansion of the absorbing material [44, 45]. Therefore, this result supports the observations of Kyoi *et al.* [41].

Such a film clamped onto a hard substrate ideally cannot expand during absorption in all three spatial dimensions and thus is laterally compressed. This causes, in isotropic materials, a change in the free energy by

$$\Delta F = \sum_i \int V\sigma_i \mathrm{d}\varepsilon_i = \sum_i \int V(M \cdot \varepsilon_i)\mathrm{d}\varepsilon_i$$

with V being the volume, s the biaxial stress, ε the expansion which would occur in a free film, M the biaxial modulus. Taking an isotropic expansion ε given by

$\varepsilon = \alpha_H \cdot X_H$, with α_H being the expansion coefficient and X_H being the atomic hydrogen concentration in the units mol(H)/mol(Me), it follows that

$$\Delta F \approx 2 \cdot \left(V\alpha_H^2 M\right) \cdot X_H^2.$$

Using typical values of Mg: $V_{mol} = 14\,cm^3$, $M = E/(1-\nu) = 63\,GPa$ and $\alpha_H = 0.05$, the contribution of the biaxial stress to the free energy is $\Delta F \approx 4.4\,kJ \cdot X_H^2$ per mol Mg or $0.1\,eV \cdot X_H^2$ per Mg atom, with $[0 \leq X_H \leq 2]$. According to this rough estimation, MgH_2 laterally fixed on a hard substrate could be destabilized by about 10 to 20 kJ (mol H_2)$^{-1}$. This value is in agreement with the experimentally determined destabilization of MgH_2 thin films observed by Krozer *et al.* [39]. It represents a very substantial shift in the stability of phases and can explain the differences in bulk and thin film samples.

The approach of altering the reaction enthalpy of single phase metal hydrides on the basis of stable dehydrogenated phases has been proven as a very successful method in the case of the intermetallic hydrides. For Mg-based hydrides, however, the chances of finding any new Mg-based intermetallic phase that forms a multi-component hydride are rather slim because of the limited number of elements exhibiting a negative heat of mixing with Mg. Up to now all successful attempts leading to novel Mg-based hydrides with reduced reaction enthalpies have led also to a significant reduction in gravimetric storage capacity.

Nevertheless, in general, there still exists the possibility of finding novel hydrides by this approach. For example, in 2004 Aoki *et al.* [46] discovered the reversible formation of $CaSiH_{1.3}$ by hydrogenation of CaSi.

7.3.1.2 Lightweight Hydrides with Positive Heat of Mixing in the Dehydrogenated State

As indicated above, Mg is often immiscible with late-transition metals. However, if a ternary hydride is formed and the hydrogen per magnesium atom ratio is enhanced by the addition, this can result in a clear influence on the reaction enthalpy per hydrogen molecule. Owing to the formation of a more stable hydride in comparison to the decomposed hydride/additive state and the absence of a stable intermetallic phase in this state, in general, an increase in the value of the reaction enthalpy is achieved by this concept. Mg_2FeH_6 is such an example. This material is interesting for thermo-chemical energy storage at around 500 °C [47]. Mg_2FeH_6 is among those materials with the highest known volumetric hydrogen density of 150 kg m^{-3} [48]. This is more than double that of liquid hydrogen. In spite of the addition of Fe the gravimetric hydrogen density is still high at 5.6 wt%. As Mg and Fe do not form any intermetallic compounds, the hydride phase is more difficult to synthesize than, for example, Mg_2NiH_4. However, hydrogen can act as a binding component to form Mg_2FeH_6 [49]. It can thus be produced by activation through hydrogenation at hydrogen pressures of at least 90 bar and temperatures of at least 450 °C [50]. Reactive milling of MgH_2 and Fe in a hydrogen atmosphere is another favorable option. According to Gutfleisch *et al.* the yield of the Mg_2FeH_6 phase is found to depend on the Fe concentration, and can reach very high values when proper initial starting mixtures are used [51]. The desorption enthalpy is found to be $\Delta H_{dis} = 98 \pm 3$ kJ (mol H_2)$^{-1}$ (648–723 K) [52], 86 $\pm$ 6 kJ (mol H_2)$^{-1}$ (623–698 K) [53], 77 kJ (mol H_2)$^{-1}$ (623–798 K) [54] and 87 $\pm$ 3 kJ (mol H_2)$^{-1}$ [55] as obtained from respective van't Hoff plots, indicating the increase in the absolute value of reaction enthalpy compared to single MgH_2. This increase in the amount of reaction enthalpy is, in the case of Mg-based systems, unfavorable for many applications.

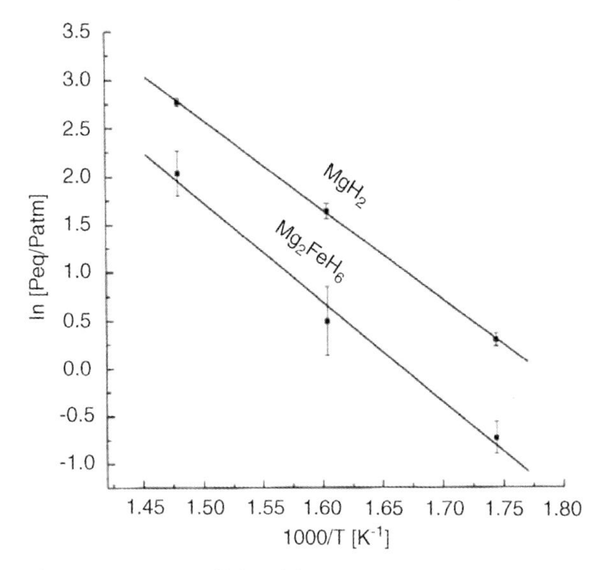

Figure 7.5 Van't Hoff plot of the 2Mg–Fe–H system constructed from desorption isotherms at 573, 623 and 673 K [55].

Figure 7.5 shows the van't Hoff plot of the 2Mg–Fe–H system constructed from desorption isotherms at 573, 623 and 673 K [55].

As described in Sections 7.1 and 7.2 the value of the reaction enthalpy in the Na–Al–H system is still too high for use in combination with low-temperature fuel cells. In contrast, the value of the reaction enthalpy in the Li–Al–H system is too low to reach complete reversibility under moderate conditions. To increase the storage capacity and tailor the reaction enthalpy of alanates it is, therefore, a reasonable approach to replace some of the Na in the Na–Al–H system by Li, and, thereby, try to achieve a higher storage capacity in combination with a reduced value of reaction enthalpy$|\Delta_r H|$.

Huot *et al.* [56] demonstrated the feasibility of the formation of Na_2LiAlH_6 without a solvent by ball-milling. As proven by XRD, hydrogen sorption in the Na–Li–Al–H system is reversible according to:

$$Na_2LiAlH_6 \leftrightarrow 2\,NaH + LiH + Al + 3/2\,H_2(g) \qquad (7.4)$$

Theoretically, the reversible hydrogen storage capacity amounts to 3.5 wt%.

For all temperatures, the addition of transition metal compounds enhances the hydrogenation and dehydrogenation kinetics. Metal oxide, chloride or fluoride additions lead to a significant enhancement of the reaction kinetics, especially during desorption [57].

The results of several research groups clearly demonstrate that partial replacement of Na by Li is feasible and thus higher hydrogen storage capacities are accessible for the hexahydride phase. The nanocrystalline microstructure ensures reversibility with respect to the decomposition/recombination reaction during the hydrogen desorption/absorption reaction.

As discussed above, due to the absence of any stable compound in the dehydrogenated state and the formation of a rather stable hydride the desorption enthalpy is not decreased compared to the original single alkali composed hydrides. Fossdal *et al.* [58] measured pressure–composition isotherms for TiF_3-enhanced material in the temperature range 170–250 °C. From the van't Hoff plot they determined a dissociation enthalpy of 56.4 ± 0.4 kJ (mol $H_2)^{-1}$ and a corresponding entropy of 137.9 ± 0.7 J K^{-1} (mol $H_2)^{-1}$. The respective value of the reaction enthalpy of reaction (7.4) is 47 kJ (mol $H_2)^{-1}$ [19].

Figure 7.6 shows the respective van't Hoff Plots of Na_3AlH_6 and Na_2LiAlH_6.

As for the Na–Li–Al–H system, different groups have searched for mixed alkaline or alkaline earth-based alanates. So far, however, none of the novel identified hydrides have shown promising properties. By ball milling of $MgCl_2$ and $CaCl_2$ with $NaAlH_4$ or $LiAlH_4$ Mamatha *et al.* [59] succeeded in the preparation of lithium–magnesium alanates [$LiMg(AlH_4)_3$] in a mixture with NaCl and LiCl. From a thermovolumetric measurement the authors assumed that, in analogy to the two-step decomposition of $NaAlH_4$, $LiMg(AlH_4)_3$ decomposes according to Eqs. (7.5) and (7.6).

$$LiMg(AlH_4)_3 \xrightarrow{100\,^\circ C} LiMgAlH_6 + 2\,Al + 3\,H_2 \tag{7.5}$$

$$LiMgAlH_6 + 2\,Al \xrightarrow{150^\circ C} MgH_2 + LiH + 3\,Al + 1.5\,H_2 \tag{7.6}$$

The respective DSC measurements of the decomposition reactions revealed that reaction (7.5) is exothermic with an enthalpy of about 15 kJ mol^{-1} and reaction (7.6) is endothermic with an enthalpy of 13 kJ mol^{-1}. Thus, this system again has rather unfavorable thermodynamic properties and is not reversible by gas-phase H_2 loading.

By investigating the thermodynamic stabilities for a series of metal borohydrides $M(BH_4)_n$ (M = Li, Na, K, Cu, Mg, Zn, Sc, Zr, and Hf; n = 1–4) Nakamori *et al.* [60] and

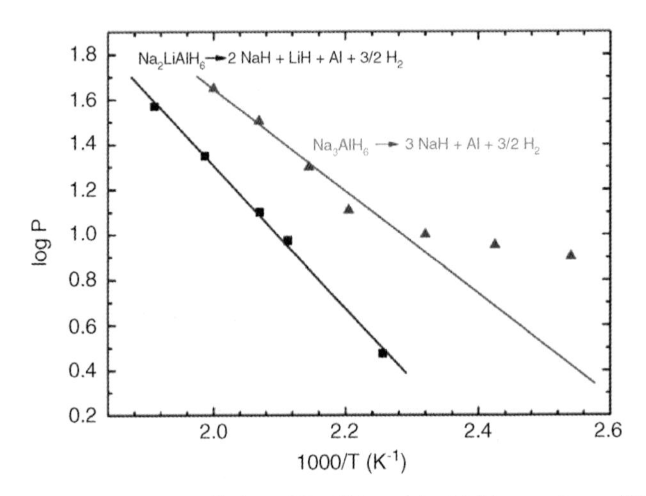

Figure 7.6 Van't Hoff plots of Na_3AlH_6 and Na_2LiAlH_6 according to [19, 58], respectively.

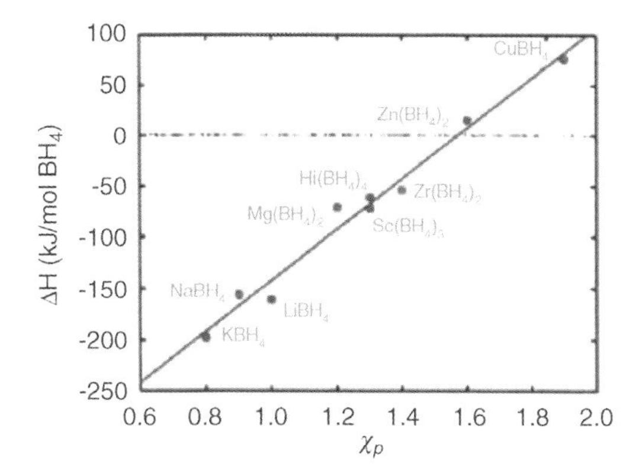

Figure 7.7 Relation between the heats of formation of $M(BH_4)_n$, ΔH_{boro} and the Pauling electronegativity of M, χ_P [60].

Grochala *et al.* [61] found a good correlation between the decomposition temperature/heat of formation H_{boro} of $M(BH_4)_n$ and the Pauling electronegativity of the cation χ_P, see Figure 7.7.

Therefore, the development of double-cation (multi-cation) borohydrides of the type $MM^*(BH_4)_n$ with M and M^* having different Pauling electronegativities is a valuable approach to fine-tuning the thermodynamic stabilities of borohydrides. Indeed, in analogy to mixed cation alanates, Orimo *et al.* [62, 63] discovered several double-cation borohydrides $MLi_{m-n}(BH_4)_m$ with M being Zn, Al, Zr and $ZrLi_{n-4}(BH_4)_n$ being sufficiently stable upon heating with regard to the possible disproportionation into the respective hydrides $Zr(BH_4)_4$ and $LiBH_4$. Furthermore, Jones and Nickels *et al.* [64, 65] found the new borohydride $LiK(BH_4)_2$. The decomposition temperature and enthalpy of formation of both $ZrLi_{n-4}(BH_4)_n$ and $LiK(BH_4)_2$ seem to agree very well with the Pauling electronegativity correlation model. Thus, their results indicate that the appropriate combination of cations might be an effective method to adjust the thermodynamic stability of metal borohydrides.

In summary, the search for systems with a positive heat of mixing in the dehydrogenated state is a valuable approach which can indeed lead to the discovery of novel hydrogen storage materials. In general, this approach enables the tuning of reaction enthalpy in both conventional hydrogen storage alloys and in lightweight complex hydrides.

7.3.2
Thermodynamic Tuning of Single Phase Hydrides by Substitution on the Hydrogen Sites: Functional Anion Concept

In recent years transition metal fluorides have attained more and more attention as effective dopant precursors for complex hydrides [66, 67]. It is well known that aluminum hydrides, as well as borohydrides, react with fluorine to form fluorides [68]. In addition, many fluorides and hydrides are isotypic. One example is

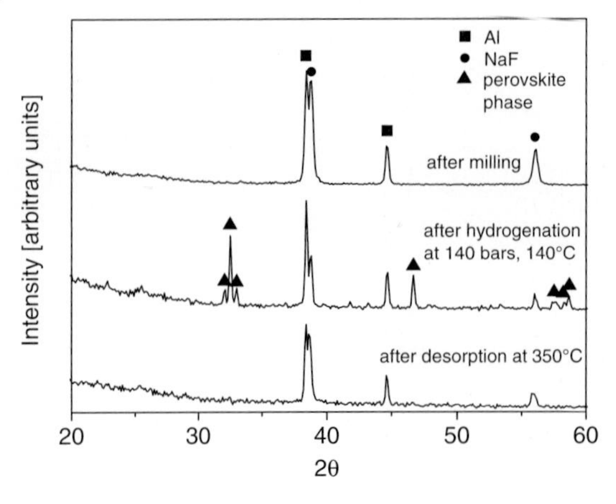

Figure 7.8 XRD analysis of a mixture of NaF and Al after milling, after subsequent hydrogenation at 145 bar and 140 °C for 8 h and after subsequent dehydrogenation at 380 °C for several hours [75].

Na_3AlH_6 and Na_3AlF_6 [69, 70]. Bouamrane *et al.* reported on the successful synthesis of the hydridofluorides $NaMgH_2F$ [71] and $KCaH_{3-x}F_x$ [72].

In 2007 Yin *et al.* [73] presented DFT calculations of TiF_3- and $TiCl_3$-doped Na_3AlH_6. In contrast to many other groups and investigations this group studied in detail the function of the anions of the dopant precursors and concluded that a substitution of F^- for H^- could render a favorable thermodynamic modification.

Indeed Brinks *et al.* [74] and Eigen *et al.* [75] demonstrated the reversible hydrogenation and dehydrogenation of ball-milled 3NaF–Al composites and NaF–Al composites, respectively.

Figure 7.8 shows an X-ray analysis of a hydrogenated and the re-dehydrogenated sample. As expected the X-ray analysis of NaF and Al after ball milling shows only the two fcc-phases of the initial components. After hydrogenation, the formation of a perovskite-like structure is visible, which is very similar to the structures for Na_3AlF_6 and Na_3AlH_6. After desorption at 380 °C into an argon stream, the original phases NaF and Al reappear, showing that the reaction is reversible.

The synchrotron XRD analysis (Figure 7.9) shows the respective diffraction peaks of NaF, Al and the perovskite-like phase, but also evidences the formation of $NaAlH_4$. The perovskite-like phase shows similarities to both Na_3AlH_6 and Na_3AlF_6. A thorough investigation of the pattern, however, shows the lack of several reflections of pure Na_3AlH_6 and pure Na_3AlF_6. Instead a single $Na_3AlH_{6-x}F_x$ compound has formed upon hydrogen absorption. On the other hand, in the case of the pure $NaAlH_4$-phase no peak shift in the reflections is observed, indicating that most likely no significant amounts of fluorine atoms are substituted for hydrogen in the $NaAlH_4$ phase.

Aiming at understanding the dehydrogenation reaction, *in situ* PXD analysis of a hydrogenated mixture of pure NaF and Al was carried out during heating from 20 to 350 °C in vacuum (Figure 7.10). At about 180 °C the reflections corresponding to $NaAlH_4$ disappear, confirming its melting. In correspondence to a second strong

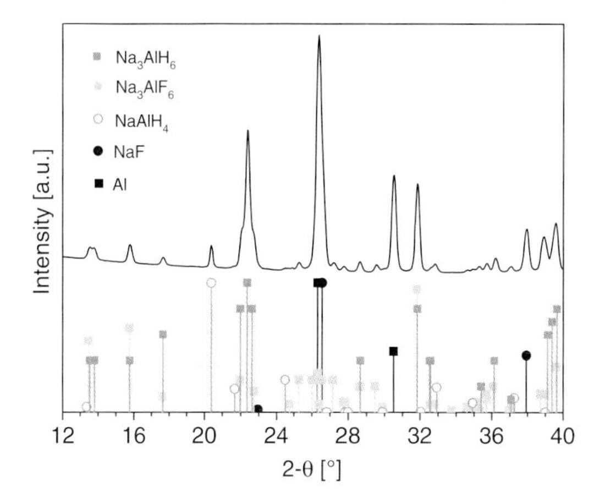

Figure 7.9 PXD analysis of a mixture of NaF and Al hydrogenation at 145 bar and 140 °C for 8 h.

peak of hydrogen release observed in mass spectrometry measurements, between 240 and 300 °C the content of the $Na_3AlH_{6-x}F_x$ phase decreases strongly and, correspondingly, the NaF and Al reflections increase. In addition to the continuous peak shift, observed in all phases and caused by the lattice expansion upon heating, the NaF reflection shows an additional small lattice widening at around 240–300 °C, and a reduced peak shift/stagnation at higher temperatures, despite further heating, indicating the formation of $NaF_{1-y}H_y$ with varying amount of hydrogen [76].

By performing different sets of measurements Eigen *et al.* [75, 76] developed the following reaction scheme:

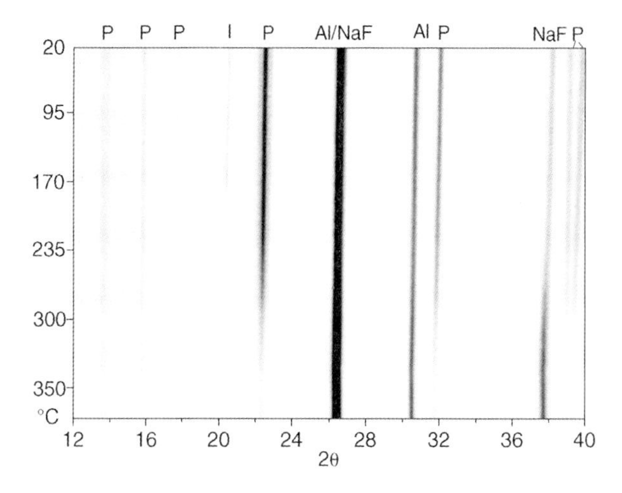

Figure 7.10 *In-situ* PXD analysis of a hydrogenated mixture of NaF and Al during heating from 20 to 350 °C in vacuum (P = $Na_3AlH_{6-x}F_x$, I = $NaAlH_4$) [75].

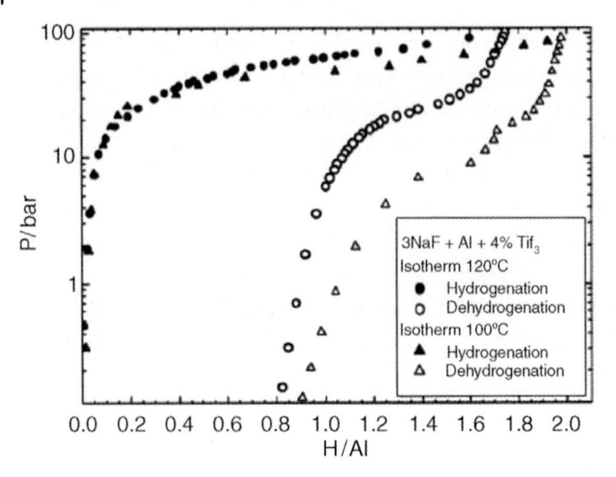

Figure 7.11 Pressure–composition isotherms of 3 NaF + Al + 4 mol% TiF_3 at 100 and 120 °C [74].

$$
\begin{aligned}
&1/2\ NaAlH_4 + 1/6\ Na_3AlF_6 \\
&\leftrightarrow 1/3\ Na_3\ AlH_3F_3 + 1/3\ Al + 1/2H_2 \\
&\leftrightarrow NaF + 2/3\ Al + H_2
\end{aligned}
\tag{7.7}
$$

DSC measurements revealed reduced reaction enthalpies for the two respective reaction steps.

Brinks *et al.* [74] measured the pressure composition isotherms of 3 NaF + Al 4 mol% TiF_3 at 100 and 120 °C. At 100 °C they observe a plateau pressure of about 50 bar for hydrogenation and 8 bar for dehydrogenation (to be compared with 0.7 bar for Na_3AlH_6 [19, 77]. At 120 °C the respective observed plateau pressures are 70 and 20 bar.

The result that both absorption and desorption isotherms are well above the plateau pressures for Na_3AlH_6 (Figure 7.11) proves clearly the successful reduction of desorption enthalpy in this system by the anion concept. For kinetic reasons in these measurements desorption stopped already at a concentration of 0.8 H/M. The corresponding diffraction data of Brinks *et al.* [74] show the formation of the $Na_3AlH_xF_{(1-x)}$ phase during hydrogenation. After dehydrogenation to 0.8 H/M the amount of $Na_3AlH_XF_{(1-x)}$ is reduced, and the amounts of NaF and Al are increased.

First-principle calculations based on a density functional theory approach revealed that, in borohydrides, the effect on reducing the reaction enthalpy is expected to be even more pronounced than in aluminohydrides. By using a $1 \times 2 \times 1$ supercell consisting of 48 atoms ($Li_8B_8H_{32}$), a $1 \times 2 \times 1$ LiH supercell containing 16 atoms (Li_8H_8), and a crystallographic unit cell containing 12 atoms to represent α-boron Yin *et al.* [78] calculated reaction enthalpies for different reaction products.

The most likely reactions are

$$Li_8B_8H_{32-x}F_x \rightarrow Li_8H_{8-x}F_x + 8B + 12H_2 \tag{7.8}$$

$$Li_8B_8H_{32-x}F_x \rightarrow Li_{8-x}H_{8-x} + Li_xF_x + 8B + 12H_2 \tag{7.9}$$

The respective decomposition enthalpies are dependent on the amount of hydrogen substituted by fluorine in the range 25–65 kJ (mol H_2)$^{-1}$ compared to 78 kJ (mol H_2)$^{-1}$ or, after taking zero-point energy corrections into account, 13–59 kJ (mol H_2)$^{-1}$ compared to 61 kJ (mol H_2)$^{-1}$ for the pure $LiBH_4$.

To conclude: The functional anion concept opens up new possibilities to tailor reaction enthalpies of high capacity hydrides towards potential applications. At present, however, not much is known about how successful this method will eventually be and which material classes can be successfully altered by it.

F anion-substituted systems are currently being investigated in several different national and international projects and the future will reveal the range of its usability for hydrogen storage applications.

7.3.3 Multicomponent Hydride Systems

Another attractive alternative to tailor the reaction enthalpies of already known single component hydrides MH_x is to mix them with suitable reactants (A) which react exothermically with the hydrides during heating whereby hydrogen is released and one or many stable compounds are formed according to the scheme:

$$MH_x + yA \leftrightarrow MA_y + \tfrac{x}{2}H_2 \qquad (7.10)$$

The stability of these compounds in the desorbed state has to be in the right range to allow reversibility under moderate conditions. Furthermore, the stability of these compounds formed upon desorption directly determines the change in reaction enthalpy of the multicomponent hydride system in comparison to the single-component hydride.

This approach goes back to Reilly and Wiswall [79] who were the first to prove that this concept indeed allows the modification of the thermodynamics of hydrogenation/dehydrogenation reactions.

In recent years, several systems applying this concept to light metal hydrides have been studied. Theoretical calculations of different groups like for example, Alapati *et al.* [80] hint at a huge number of possible very promising mixed systems with high gravimetric storage capacities and suitable thermodynamic properties. However, for kinetic reasons, only some of them follow the thermodynamically expected reaction pathway.

7.3.3.1 Mixtures of Hydrides and Reactive Additives

$MgH_2/MgCu_2 \leftrightarrow Mg_2Cu/xH_2$ is the earliest and, therefore, probably the most prominent example of such systems. During desorption the stable compound Mg_2Cu is formed. During absorption the reactant $MgCu_2$ precipitates and MgH_2 is formed [79]:

$$2\,Mg_2Cu + 3\,H_2 \leftrightarrow 3\,MgH_2 + MgCu_2 \qquad (7.11)$$

The value of the reaction enthalpy per mole H_2 is lowered compared to the value of the formation enthalpy of MgH_2. This can be understood by considering the following partial reactions for absorption:

$$2\,Mg_2Cu \rightarrow MgCu_2 + 3\,Mg \qquad (7.12)$$

$$3\,Mg + 3\,H_2 \rightarrow 3\,MgH_2 \qquad (7.13)$$

Here, reaction (7.12) is endothermic and thus counteracts the exothermic heat release for partial reaction (7.13). Therefore, the total amount of hydrogen reaction enthalpy according to reaction (7.11) is reduced to about 73 kJ (mol H_2)$^{-1}$ [81]. Consequently, the equilibrium temperature for 1 bar hydrogen pressure is reduced to about 240 °C. In addition, Mg_2Cu is much more easily hydrogenated than pure Mg. The lower dissociation temperatures and enthalpies make the design engineer's requirements for an integrated system less stringent than for Mg. However, the storage capacity is reduced significantly to only 2.6 wt.%.

Similarly, the *magnesium-aluminum* system shows several intermediate phases which react with hydrogen, such as Mg_2Al_3 [81] and $Mg_{17}Al_{12}$ [82].

Figure 7.12 shows the pressure vs. composition isotherms for the system Mg_2Al_3–H_2 according to the reaction (7.14)

$$Mg_2Al_3 + 2\,H_2 \rightarrow 2\,MgH_2 + 3\,Al \qquad (7.14)$$

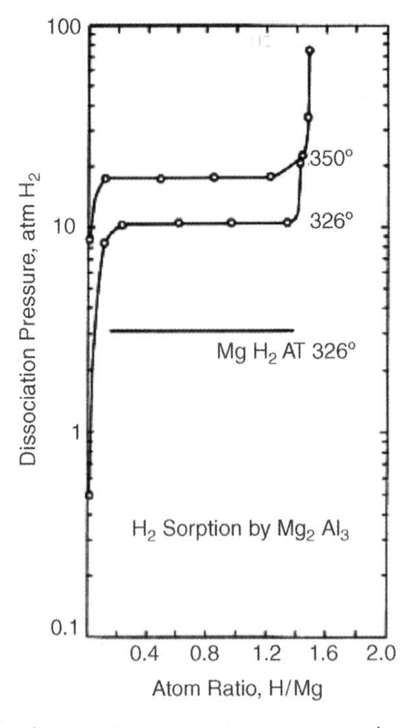

Figure 7.12 Pressure–composition isotherms for the system, Mg_2Al_3 [81].

While the dissociation pressure of pure MgH_2 at 326 °C is 3 bar the respective dissociation pressure of the system $2MgH_2 + 3\,Al$ is found to be increased to 10 bar.

According to Bouaricha *et al.* [82] $Mg_{17}Al_{12}$ reacts with hydrogen in two steps:

$$2\,MgH_2 + 3\,Al \leftrightarrow Mg_2Al_3 + 2H_2 \tag{7.15a}$$

$$4\,Mg_2Al_3 + 9\,MgH_2 \leftrightarrow Mg_{17}Al_{12} + 9\,H_2 \tag{7.15b}$$

While Mg_2Al_3 absorbs only about 3 wt.% H_2, $Mg_{17}Al_{12}$ can reversibly absorb 4.4 wt.% H_2 with Mg_2Al_3 being an intermediate.

$Mg_{17}Al_{12}$ has a formation enthalpy of about $-102\ kJ\,mol^{-1}$, and Mg_2Al_3 $-12.5\ kJ\,mol^{-1}$. Therefore, the addition of Al leads to a decrease in the value of the total reaction enthalpy of reactions (7.15a) and (7.15b) of about $6\,kJ\,(mol\ H_2)^{-1}$ each.

Similarly to Mg_2Cu the equilibrium pressure of 1 bar is reached at around 240 °C and, therefore, is significantly lowered compared to pure MgH_2.

Mg_2Si has an enthalpy of formation of $\Delta_f H^0(Mg_2Si) = -79\,kJ\,mol^{-1}$. Therefore, the formation of Mg_2Si reduces the standard enthalpy of dehydrogenation of MgH_2 by $37\,kJ\,(mol\ H_2)^{-1}$.

According to the reaction:

$$2\,MgH_2 + Si \leftrightarrow Mg_2Si + 4H_2 \tag{7.16}$$

Theoretically, 5 wt.% could be stored reversibly. The thermodynamic data indicate a quite high equilibrium pressure of approximately 1 bar at 20 °C, 50 bar at 120 °C and 100 bar at about 150 °C [36]. Vajo *et al.* [36] investigated the desorption behavior of MgH_2 and a $2MgH_2 + Si$ mixture. Both samples were milled under identical conditions. In both cases desorption started at 295 °C. In the case of the MgH_2 sample Vajo observed at 300 °C a constant equilibrium pressure of 1.65 bar. For the $2MgH_2 + Si$ mixture, desorption also began at 295 °C; however, hydrogen evolution did not stop at the equilibrium value for pure MgH_2 but continued to increase until the temperature was lowered after 5.4 h. At 300 °C, the equilibrium pressure is, therefore, >7.5 bar. This is at least a factor of 4 larger than the pressure for pure MgH_2.

However, Mg_2Si has a rather unfavorable, dense-packed crystal structure and shows no hydrogen interaction under moderate conditions (300 °C, 50 bar H_2; 150 °C and 100 bar [36]). Attempts have been made to modify the crystal structure or to partly substitute Si by ternary alloying additions, however, up to now, only with very limited success.

A number of other reactions between Mg-based intermetallics like $MgCd_x$ [83] have been proven to resemble this reaction pattern. However, all these systems suffer from a significantly reduced gravimetric storage capacity compared to MgH_2.

Due to its high reversible storage capacity MgH_2 is one of the most investigated lightweight hydrogen storage materials. The approach shown above, however, is not limited to MgH_2, as will be shown in the following.

LiH has not been considered a practical hydrogen storage material because of its high stability. Nevertheless, it is another very suitable system to show that the

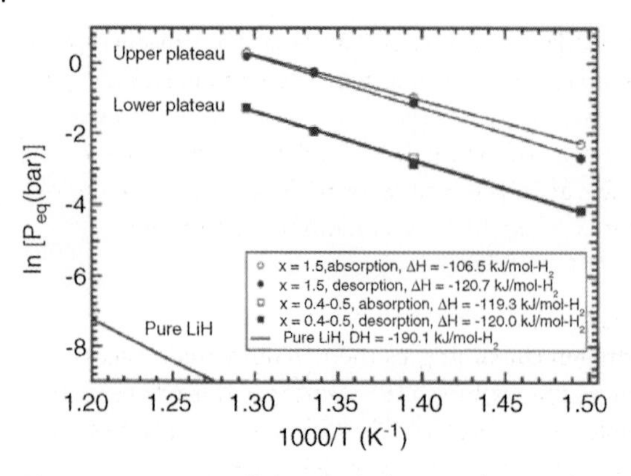

Figure 7.13 Van't Hoff plots for the lower and upper plateaus determined from the isotherms at 400–500 °C and, for comparison, pure LiH [36].

approach of using reactive additives to decrease the value of the reaction enthalpy is quite generally valid for many different hydrides. The enthalpy of dehydrogenation of LiH is reported to be 190 kJ (mol H_2)$^{-1}$ [84]. LiH, therefore, is a high-temperature hydride having an equilibrium pressure of 1 bar at 910 °C or alter 5×10^{-5} bar at 490 °C. According to Vajo *et al.* [36], using Si as an additive to LiH leads to the following dehydrogenation/rehydrogenation reactions:

$$2.35\,\text{LiH} + \text{Si} \leftrightarrow \text{Li}_{1.71}\text{Si} + 0.64\,\text{LiH} + 0.855\,\text{H}_2 \leftrightarrow \text{Li}_{2.35}\text{Si} + 1.175\,\text{H}_2. \quad (7.17)$$

The absorption and desorption isotherms for milled 2.5 LiH + Si mixtures measured by Vajo *et al.* show two plateaus. Figure 7.13 shows the Van't Hoff plots of the lower and upper plateaus from the LiH–Si–H system as well as the pure Li–H system.

Thus, the addition of Si, increases the equilibrium pressure by $>10^4$ times while lowering the dehydrogenation enthalpy from 190 to 120 kJ (mol H_2)$^{-1}$.

In addition, Yu *et al.* [85] demonstrated that the similar reaction

$$2\,\text{LiH} + 2\text{C} \leftrightarrow \text{Li}_2\text{C}_2 + \text{H}_2 \quad (7.18)$$

also shows reversibility.

Because of its high gravimetric storage capacity $LiBH_4$ has attracted, and still attracts, much attention from various research groups. Nevertheless, the thermodynamic properties are far from ideal for most applications. By use of the CALPHAD approach Cho *et al.* [86] determined the decomposition temperature of pure $LiBH_4$ at 1 bar H_2 pressure to be 403 °C.

However, using the same approach for the reaction

$$2\,\text{LiBH}_4 + \text{Al} \leftrightarrow 2\,\text{LiH} + \text{AlB}_2 + 3\,\text{H}_2. \quad (7.19)$$

they derived a corresponding equilibrium temperature of 188 °C at 1 bar.

Using TiF_3 as additive, Kang *et al.* [87] and Jin *et al.* [88] showed that reaction (7.19) indeed shows reversibility and under much more suitable conditions than the pure $LiBH_4$ system.

All these examples show that the use of such additives which react with the hydride upon desorption is a rather suitable approach to tune the reaction enthalpy. However, the total hydrogen storage capacity of the hydride is reduced significantly.

7.3.3.2 Mixed Hydrides/Reactive Hydride Composites

While the addition of non-hydrogen-containing elements/compounds necessarily reduces the total hydrogen storage capacity, the use of high-capacity hydrides such as MgH_2 as a reacting additive enables maintenance of the high storage capacities of the single hydrides and tuning of the reaction enthalpy at the same time.

In 2002 Chen *et al.* [89] discovered the reversible hydrogen release and uptake of $LiNH_2$ and LiH according to the reaction scheme:

$$LiNH_2 + 2\,LiH \leftrightarrow Li_2NH + LiH + H_2 \leftrightarrow Li_3N + 2H_2 \qquad (7.20)$$

If the different hydrides of such composites react upon desorption in an exothermic reaction to form a stable compound, they are denoted as reactive hydride composites (RHC) [90].

Unfortunately, the value of the overall reaction enthalpy of RHC is $|\Delta H| \sim 80$ kJ (mol $H_2)^{-1}$ and thus is still too high for many applications. However, a wide range of other amide-containing hydride systems could be identified having quite interesting properties. The chemical reaction of $Mg(NH_2)_2$ + 2LiH according to (7.21) is of special interest because of its rather low heat of desorption ($|\Delta H| \sim 40$ kJ (mol $H_2)^{-1}$), an equilibrium pressure of 1 bar H_2 at approximately 90 °C [91], comparatively low sorption temperatures and a theoretical storage capacity of 5.6 wt.%.

$$Mg(NH_2)_2 + 2\,LiH \leftrightarrow Li_2Mg(NH)_2 + 2\,H_2 \qquad (7.21)$$

In addition, altering the molar ratio of $Mg(NH_2)_2$ and LiH leads to different dehydrogenation products, such as nitrides [92], and thus alters thermodynamic and kinetic properties, and storage capacity.

In 2004 Vajo *et al.* [36, 93] and Barkhordarian *et al.* [30, 94] independently discovered the reversible hydrogenation of MgH_2/borohydrides composites such as $MgH_2 + NaBH_4$, $MgH_2 + LiBH_4$ and $MgH_2 + Ca(BH_4)_2$ forming magnesium diboride as well as alkaline or alkaline earth hydrides during desorption.

Because of the combination of comparatively low values of the reaction enthalpy $|\Delta_r H|$ in the range of around 20–60 kJ (mol $H_2)^{-1}$ for the different borohydrides – magnesium hydride composites with high theoretic reversible hydrogen storage capacities of 7.8, 11.4 and 8.3 wt.%, these RHCs are very promising candidates for hydrogen storage materials. Furthermore, MgB_2 has been shown to facilitate reversible hydrogenation of borohydrides from the gas phase by providing a pathway for the borohydride formation [15].

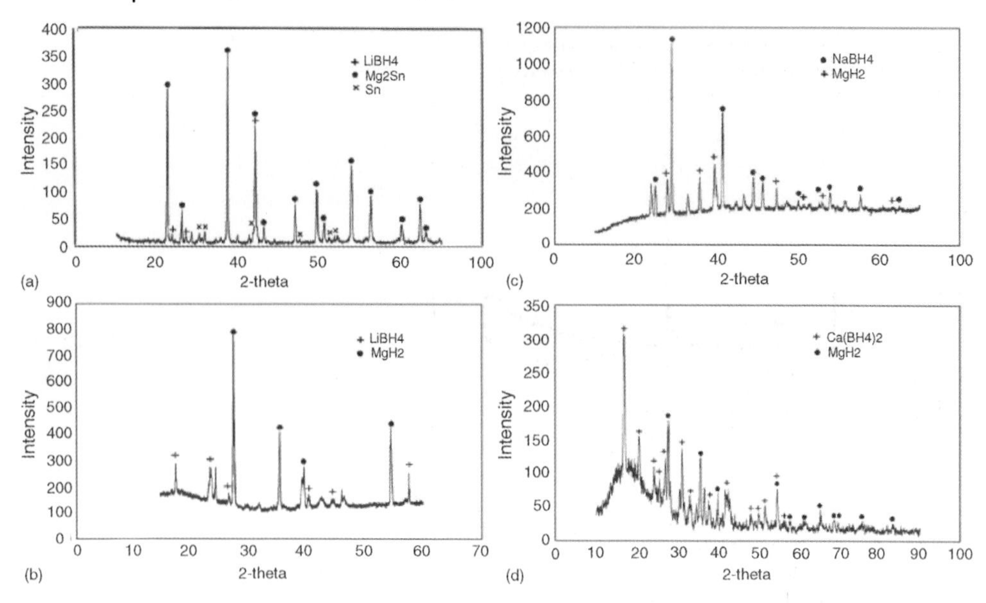

Figure 7.14 X-ray patterns of the products of hydriding (a) LiH + MgB_2 (b) Li_7Sn_2 + MgB_2 (c) NaH + MgB_2 (d) CaH_2 + MgB_2 at 400 °C, 350 bar of hydrogen pressure, measured at room temperature, taken from [30].

Figure 7.14a–d show the XRD patterns of the hydrogenation products of 2LiH MgB_2, Li_7Sn_2 + 3.5MgB_2, 2NaH + MgB_2 and CaH_2 + MgB_2 after high-pressure synthesis performed by Barkhordarian *et al.* [30]. While it was not possible to synthesize the borohydrides from the respective alkaline and alkaline earth metal hydrides and boron under the applied conditions, formation of the borohydrides occurred in these reactive systems in spite of the reduced driving force for hydrogenation.

Further measurements performed by Vajo *et al.* [36], Barkhordarian *et al.* [95], and Bösenberg *et al.* [96] confirmed that much lower pressures and temperatures are sufficient for hydrogenation of these compounds. So far, all measurements hint at a one-step absorption process (7.22), while desorption in the case of 2$LiBH_4$ + MgH_2 for instance, appears to be at least a two-step process under the normally applied conditions (7.23), see Figure 7.15.

$$2/x\,AH_x + MgB_2 \rightarrow A(BH_4)_x, A = \{Li, Na, Ca, \ldots\} \quad (7.22)$$

$$2\,LiBH_4 + MgH_2 \rightarrow 2\,LiBH_4 + Mg + H_2 \rightarrow 2\,LiH + MgB_2 + 4\,H_2 \quad (7.23)$$

Figure 7.15 shows DSC curves of MgH_2– $LiBH_4$ composites measured at 3 and 50 bar H_2 pressure and a heating/cooling rate of 5 K min^{-1} [15]. The gray curve shows the first dehydrogenation upon heating, the black curve the subsequent first hydrogen absorption reaction. There are two sharp endothermic peaks at 120 and 270 °C which are due to a structural transformation and the melting of $LiBH_4$, respectively. Hydrogen desorption starts at around 350 °C if no catalysts are used. Experiments show that MgB_2 formation is facilitated if a residual hydrogen pressure

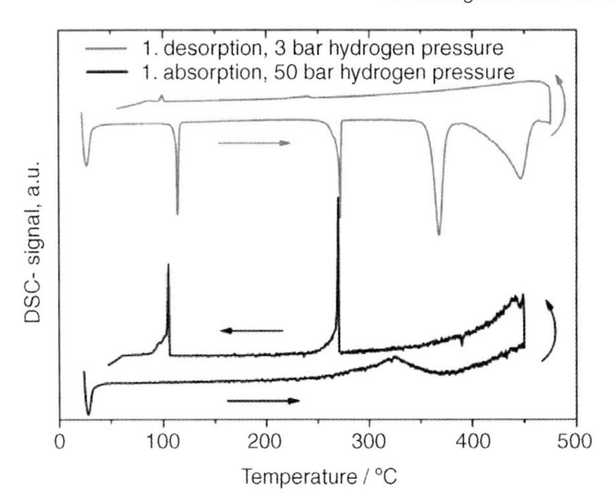

Figure 7.15 High-pressure DSC curves of $MgH_2 + LiBH_4$. The gray curve represents the first desorption reaction at 3 bar hydrogen pressure of pure $MgH_2 + LiBH_4$ without any additives after ball-milling. It shows a two-step desorption behavior of the pure RHC. The black curve represents the first absorption performed at 50 bar hydrogen pressure taken from [15].

of some bar is applied. At 50 bar hydrogen pressure hydrogen absorption starts at 250 °C if no catalysts are used.

In situ XRD-patterns taken during the hydrogenation of pure $CaH_2 + MgB_2$ performed by Barkhordarian *et al.* [95] indicate the simultaneous formation of $Ca(BH_4)_2$ as well as MgH_2 at 125 bar H_2 and 230 °C, as shown in Figure 7.16.

By changing the stoichiometries of the $NaBH_4$–MgH_2 composites Pistidda *et al.* [97] discovered that, similarly to the $Mg(NH_2)_2$–LiH system, the reaction path taken during desorption differs strongly. In the case of MgH_2–$LiBH_4$ composites MgB_2 is formed directly during desorption only if a residual pressure of several bar of hydrogen is present. Likewise, in MgH_2–$Ca(BH_4)_2$ composites, different reaction paths leading to the formation of B, CaB_6 (Y. W. Cho, personal communication) and MgB_2 are possible. The reaction path can be controlled by additives. The formation of MgB_2 during desorption, for example, can be triggered by a few mol% of Ti-isopropoxide addition, which also lowers the required desorption temperature significantly [95, 96].

In addition to MgH_2/borohydride RHCs other promising composites have been identified.

Pinkerton *et al.* [98] and Jin *et al.* [99] demonstrated the reversibility of $TiCl_3$-doped $LiBH_4$–CaH_2 composites and $LiBH_4$–MH_2 (M = Ce, Ca) composites, respectively, according to the reactions:

$$6\,LiBH_4 + CaH_2 \xleftrightarrow{Ti^*} 6\,LiH + CaB_6 + 10\,H_2 \tag{7.24}$$

$$6\,LiBH_4 + CeH_2 \xleftrightarrow{Ti^*} 6\,LiH + CeB_6 + 10\,H_2 \tag{7.25}$$

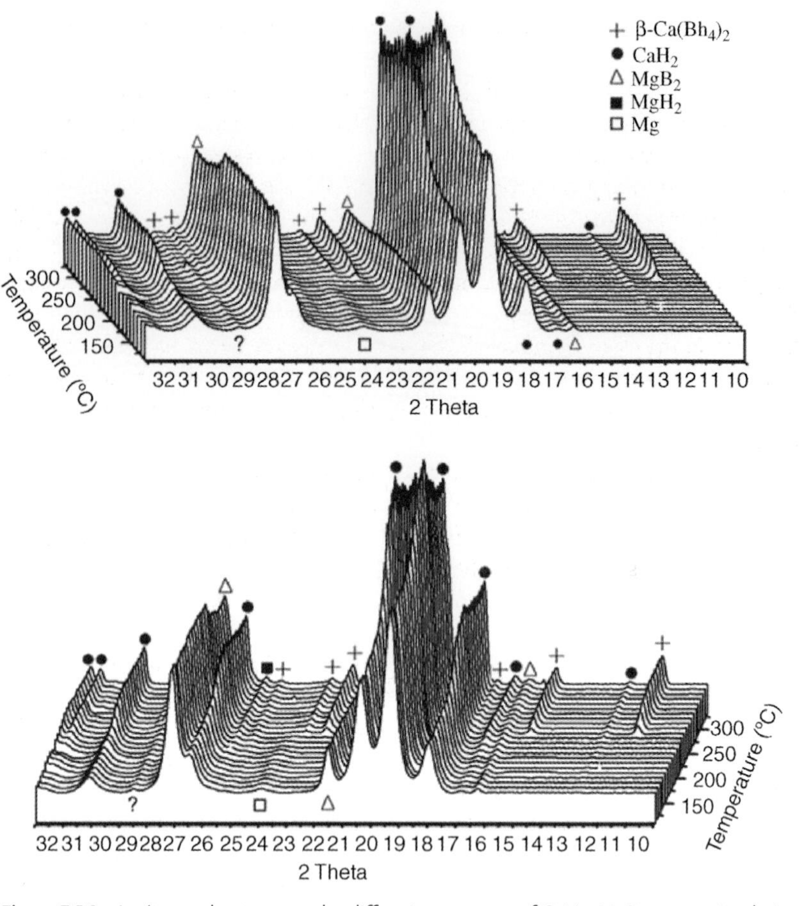

Figure 7.16 *In situ* synchrotron powder diffraction patterns of CaH_2–MgB_2 composite during absorption at 125 bar hydrogen taken from [95].

These composites again combine high theoretic gravimetric storage capacities of 11.7 and 7.4 wt.% with considerably reduced values of reaction enthalpy $|\Delta_r H| = 65$ and 44 kJ $(\text{mol } H_2)^{-1}$ [99], respectively.

7.4 Summary and Conclusion

The development of novel lightweight hydrides for hydrogen storage applications is a very challenging task. Many lightweight hydrides show neither suitable kinetic nor thermodynamic properties. While, in the past, kinetic limitations could be solved very successfully in some cases like MgH_2 and $NaAlH_4$, thermodynamic restrictions still remained. In this chapter some of the most important strategies to alter reaction enthalpies of hydrides or hydride composites and recent developments have been reviewed.

The classical strategy of destabilizing hydrides is substitution on the metal lattice, thereby forming new intermetallic hydrides. This has proven to be a very powerful approach in the case of conventional metal hydrides, leading to a huge number of hydrides with different thermodynamic properties but with gravimetric storage capacities of less than 3 wt.% H/M. There are two possibilities: either a stable compound is formed upon desorption or, if there is a positive heat of mixing, segregation occurs. This approach has recently been extended towards lightweight hydrides and complex hydrides.

Contrary to the classical approach the recently applied functional anion concept considers the substitution on H-sites by F^- anions. First results indicate that this approach again is very promising for altering reaction enthalpies of complex hydrides.

In addition to these single-phase hydrides multiphase mixed hydrides or so-called reactive composites attract increasing attention. By addition of another element or compound the reaction enthalpy of the hydride can be tailored very efficiently if a stable compound is formed from the hydride and the additive(s) upon desorption. Unfortunately, this approach leads to an unwanted decrease in hydrogen storage capacity.

By extension of this approach – of using additives which react reversibly with the hydride during desorption and thereby lower the value of the reaction enthalpy – towards mixed hydrides or Reactive Hydride Composites, new possibilities for substantially reducing the values of the reaction enthalpy of hydride systems are opened up. By this extension, alteration of thermodynamic properties is possible without any losses in the gravimetric storage capacity. It overcomes the limitations of past monolithic systems, where high storage capacities are often linked to high enthalpies of reaction. The remaining key issue for such novel hydrogen-releasing reactions is to enhance the reaction kinetics significantly to allow lower temperature applications.

References

1 Buchner, H., (1982) Energiespeicherung in metallhydriden, in *Innovative Energietechnik*, Springer-Verlag Wien, New York.

2 Schulz, R., Boily, S., Zaluski, L., Zaluska, A., Tessier, P., and Ström-Olsen, J.O. (1995) *Proceedings of the International Conference on Composite Materials and Energy*, Montreal, 529–535.

3 Schulz, R., Huot, J., Liang, G., Boily, S., Lalande, G., Denis, M.C., and Dodelet, J.P. (1999) *Mater. Sci. Eng.*, **A267**, 240–245.

4 Zaluska, A., Zaluski, L., and Ström-Olsen, J.O. (1999) *J. Alloys Compd.*, **288**, 217–225.

5 Suryanarayana, C. (2001) *Prog. Mater. Sci.*, **46**, 1.

6 Liang, G., Huot, J., Boily, S., and Schulz, R. (2000) *J. Alloys Compd.*, **305**, 239–245.

7 Huot, J., Liang, G., and Schulz, R. (2001) *Appl. Phys. A*, **72**, 187–195.

8 Grigorova, E., Khristov, M., Khrussanova, M., Bobet, J.-L., and Peshev, P. (2005) *Int. J. Hydrogen Energy*, **30**, 1099–1105.

9 Oelerich, W., Klassen, T., and Bormann, R. (2001) *J. Alloys Compd.*, **315**, 237–242.

10 Barkhordarian, G., Klassen, T., and Bormann, R. (2003) *Scr. Mater.*, **49**, 213.

11 Barkhordarian, G., Klassen, T., and Bormann, R. (2004) *J. Alloys Compd.*, **364**, 242.

12 Bogdanovic, B., and Schwickardi, M. (1997) *J. Alloys Compd.*, **253**, 1–9.

13 Gutfleisch, O., Dal Toe, S., Herrich, M., Handstein, A., and Pratt, A., (2005) *J. Alloys Compd.*, **404–406**, 413–416.

14 Hanada, N., Ichikawa, T., Hino, S., and Fujii, H. (2006) *J. Alloys Compd.*, **420**, 46–49.

15 Dornheim, M., Doppiu, S., Barkhordarian, G., Boesenberg, U., Klassen, T., Gutfleisch, O., and Bormann, R. (2007) Viewpoint Paper, *Scr. Mater.*, **56**, 841–846.

16 Friedrichs, O., Aguey-Zinsou, F., Ares Fernandez, J.R., Sanchez-Lopez, J.C., Justo, A., Klassen, T., Bormann, R., and Fernandez, A. (2006) *Acta Mater.*, **54**, 105–110.

17 Dehouche, Z., Klassen, T., Oelerich, W., Goyette, J., Bose, T.K., and Schulz, R. (2002) *J. Alloys Compd.*, **347**, 319–323.

18 Reilly, J.J., and Wiswall, R.H.Jr., (1968) *Inorg. Chem.*, **7**, 2254–2256.

19 Bogdanovic, B., Brand, R.A., Marjanovic, Ankica, Schwickardi, M., and Tölle, J. (2000) *J. Alloys Compd.*, **302**, 36–58.

20 Eigen, N., Keller, C., Dornheim, M., Klassen, T., and Bormann, R. (2007) Viewpoint Paper, *Scr. Mater.*, **56**, 847–851.

21 Zidan, R.A., Takara, S., Hee, A.G., and Jensen, C.M. (1999) *J. Alloys Compd.*, **285**, 119.

22 Eigen, N., Gosch, F., Dornheim, M., Klassen, T., and Bormann, R. (2008) *J. Alloys Compd.*, **465**, 310–316.

23 Orimo, S., Nakamori, Y., Eliseo, J.R., Züttel, A., and Jensen, C.M. (2007) *Chem. Rev.*, **107**, 4111–4132.

24 Ke, Xuezhi, and Chen, Changfeng (2007) *Phys. Rev. B*, **76**, 024112-1 to -11.

25 Brinks, H.W., Fossdal, A., Fonneløp, J.E., and Hauback, B.C. (2005) *J. Alloys Compd.*, **397**, 291–295.

26 Fichtner, M., Fuhr, O., and Kircher, O. (2003) *J. Alloys Compd.*, **356–357**, 418–422.

27 Schüth, F., Bogdanovic, B., and Felderhoff, M. (2004) *Chem. Comm.*, 2249–2258.

28 Komiya, K., Morisaku, N., Shinzato, Y., Ikeda, K., Orimo, S., Ohki, Y., Tatsumi, K., Yukawa, H., and Morinaga, M. (2007) *J. Alloys Compd.*, **446–447**, 237–241.

29 Züttel, A., Rentsch, S., Fischer, P., Wenger, P., Sudan, P., Mauron, Ph., and Emmenegger, Ch. (2003) Hydrogen storage properties of LiBH4. *J. Alloys Compd.*, **356–357**, 515–520.

30 Barkhordarian, G., Klassen, T., Dornheim, M., and Bormann, R. (2007) *J. Alloys Compd.*, **440**, L18–L21.

31 Orimo, S., Nakamori, Y., Kitahara, G., Miwa, K., Ohba, N., Towata, S., and Züttel, A. (2005) *J. Alloys Compd.*, **404–406**, 427–430.

32 Libowitz, G.G., Hayes, H.F., and Gibb, T.R.G.Jr., (1958) *J. Phys. Chem.*, **62**, 76.

33 Darnaudery, J.P., Darriet, B., and Pezat, M. (1983) *Int. J. Hydrogen Energy*, **8**, 705–708.

34 Tsushio, Y., Enoki, H., and Akiba, E. (1998) *J. Alloys Compd.*, **281**, 301–305.

35 Klassen, T., Oelerich, W., Zeng, K., Bormann, R., and Huot, J. (1998) *Magnesium Alloys and their Applications* (eds B.L., Mordike and K.U., Kainer), Werkstoff-Informationsgesellschaft mbH, Frankfurt, pp. 307–311.

36 Vajo, J.J., Mertens, F., Ahn, C.C., Bowman, R.C.Jr., and Fultz, B. (2004) *J. Phys. Chem. B*, **108**, 13977–13983.

37 Terashita, N., Kobayashi, K., Sasai, T., and Akiba, E. (2001) *J. Alloys Compd.*, **327**, 275–280.

38 Gingl, F., Bonhomme, F., Yvon, K., and Fischer, P. (1992) *J. Alloys Compd.*, **185**, 273–278.

39 Krozer, A., and Kasemo, B. (1990) *J. Less-Common Met.*, **160**, 323–342.

40 Kalisvaart, W.P., Niessen, R.A.H., and Notten, P.H.L. (2006) *J. Alloys Compd.*, **417**, 280–291.

41 Kyoi, D., Sato, T., Rönnebro, E., Kitamura, N., Ueda, A., Ito, M., Katsuyama, S., Hara, S., Noreus, D., and Sakai, T. (2004) *J. Alloys Compd.*, **372**, 213–217.

42 Vermeulen, P., Niessen, R.A.H., and Notten, P.H.L. (2006) *Electrochem. Commun.*, **8**, 27–32.

43 Liang, G., Huot, J., Boily, S., van Neste, A., and Schulz, R. (1999) *J. Alloys Compd.*, **292**, 247–252.

44 Dornheim, M., Pundt, A., Kirchheim, R., Molen, S.J.v.d., Kooij, E.S., Kerssemakers, J., Griessen, R., Harms, H., and Geyer, U. (2003) *J. Appl. Phys.*, **93**, 8958–8965.

45 Laudahn, U., Pundt, A., Bicker, M., Hülsen, U.v., Geyer, U., Wagner, T., and Kirchheim, R. (1999) *J. Alloys Compd.*, **293–295**, 490–494.

46 Aoki, M., Ohba, N., Noritake, T., and Towata, S. (2004) *Appl. Phys. Lett.*, **85**, 387–388.

47 Reiser, A., Bogdanovic, B., and Schlichte, K. (2000) *Int. J. Hydrogen Energy*, **25**, 425.

48 Züttel, A., Wenger, P., Rentsch, S., Sudan, P., Mauron, Ph., and Emmenegger, Ch. (2003) *J. Power Sources*, **118**, 1.

49 Porutsky, S.G., Zhurakovsky, E.A., Mogilevsky, S.A., Verbetsky, V.N., Bakuma, O.S., and Semenenko, K.N. (1990) *Solid State Commum.*, **74**, 551.

50 Selvam, P., and Yvon, K. (1991) *Int. J. Hydrogen Energy*, **16**, 615–617.

51 Herrich, M., Ismail, N., Lyubina, J., Handstein, A., Pratt, A., and Gutfleisch, O. (2004) *Mater. Sci. Eng. B*, **108**, 28–32.

52 Didisheim, J.-J., Zolliker, P., Yvon, K., Fischer, P., Schefer, J., Gubelmann, M., and Williams, A.F. (1984) *Inorg. Chem.*, **23**, 1952.

53 Konstanchuk, I.G., Ivanov, E.Y., Pezat, M., Darriet, B., Boldyrev, V.V., and Hagenmüller, P. (1987) *J. Less-Common Met.*, **131**, 181.

54 Bogdanovic, B., Reiser, A., Schlichte, K., Spliethoff, B., and Tesche, B. (2002) *J. Alloys Compd.*, **345**, 77.

55 Puszkiel, J.A., Arneodo larochette, P., and Gennari, F.C. (2008) *Int. J. Hydrogen Energy*, **33**, 3555–3560.

56 Huot, J., Boily, S., Güther, V., and Schulz, R. (1999) *J. Alloys Compd.*, **283**, 304–306.

57 Ma, X.Z., Martinez-Franco, E., Dornheim, M., Klassen, T., and Bormann, R. (2005) *J. Alloys Compd.*, **404–406**, 771–774.

58 Fossdal, A., Bringks, H.W., Fonnelop, J.E., and Hauback, B.C. (2005) *J. Alloys Compd.*, **397**, 135–139.

59 Mamatha, M., Bogdanovic, B., Felderhoff, M., Pommerin, A., Schmidt, W., Schüth, F., and Weidenthaler, C. (2006) *J. Alloys Compd.*, **407**, 78–86.

60 Nakamori, Y., Miwa, K., Ninomiya, A., Li, H., Ohba, N., Towata, S., Züttel, A., and Orimo, S. (2006) *Phys. Rev. B*, **74**, 045126.

61 Grochala, W., and Edwards, P.P. (2004) *Chem. Rev.*, **104**, 1283–1316.

62 Li, H.-W., Orimo, S., Nakamori, Y., Miwa, K., Ohba, N., Towata, S., and Züttel, A. (2007) *J. Alloys Compd.*, **446–447**, 315–318.

63 Nakamori, Y., Miwa, K., Li, H.W., Ohba, N., Towata, S., and Orimo, S. (2007) *Mater. Res. Soc. Symp. Proc.*, **971**, 0971-Z02-01.

64 Jones, M.O., David, W.I.F., Johnson, S.R., Sommariva, M., Lowton, R.L., Nickels, E.A., and Edwards, P.P. (2008) *Mater. Res. Soc. Symp. Proc.*, **1098**, 1098-HH02-05.

65 Nickels, E.A., Jones, M.O., Davikd, W.I.F., Johnson, S.R., Lowton, R.L., Sommariva, M., and Edwards, P.P. (2008) *Angew. Chem. Int. Ed.*, **47**, 2817–2819.

66 Kim, J.-H., Shim, J.-H., and Cho, Y.W. (2008) *J. Power Sources*, **181**, 140–143.

67 Wang, P., Kang, X.D., and Cheng, H.M. (2005) *Chem. Phys. Chem.*, **6**, 2488.

68 Lagow, R.J., and Margrave, J.L. (1974) *Inorg. Chim. Acta*, **10**, 9–11.

69 Rönnebro, E., Noreus, D., Kadir, A., Reiser, A., and Bogdanovic, B. (2000) *J. Alloys Compd.*, **299**, 101.

70 Yang, H.X., Ghose, X., and Hatch, D.M. (1993) *Phys. Chem. Miner.*, **19**, 528.

71 Bouamrane, A., Laval, J.P., Soulie, J.-P., and Bastide, J.P. (2000) *Mater. Res. Bull.*, **35**, 545–549.

72 Soulie, J.-P., Laval, J.-P., and Bouamrane, A. (2003) *Solid State Sci.*, **5**, 273–276.

73 Yin, L.-C., Wang, P., Kang, X.-D., Sun, C.-H., and Cheng, H.-M. (2007) *Phys. Chem. Chem. Phys.*, **9**, 1499–1502.

74 Brinks, H., Fossdal, A., and Hauback, B.C. (2008) *J. Phys. Chem. C*, **112**, 5658–5661.

75 Eigen, N., Boesenberg, U., Bellosta, J., Colbe, v., Jensen, T., Dornheim, M., Klassen, T., and Bormann, R. (2008) *J. Alloys Compd.* doi: 10.1016/j.jallcom.2008.10.002

76 Eigen *et al.* in preparation.

77 Gross, K.J., Thomas, G.J., and Jensen, C.M. (2002) *J. Alloys Compd.*, **330–332**, 683.

78 Yin, L., Wang, P., Fang, Z., and Cheng, H. (2008) *Chem. Phys. Lett.*, **450**, 318–321.

79 Reilly, J.J., and Wiswall, R.H. (1967) *Inorg Chem*, **6**, 2220–2223.

80 Alapati, S.V., Johnson, J.K., and Sholl, D.S. (2008) *J. Phys. Chem. C Lett.*, **112**, 5258–5262.

81 Wiswall, R. (1978) Hydrogen storage in metals, in *Hydrogen in Metals II, vol.* **29** (eds G., Alefeld and J., Völkl), Springer-Verlag, Berlin, Heidelberg, New York.

82 Bouaricha, S., Dodelet, J.P., Guay, D., Huot, J., Boily, S., and Schulz, R. (2000) *J. Alloys Compd.*, **297**, 282–293.

83 Liang, G., and Schulz, R. (2004) *J. Mater. Sci.*, **39**, 1557–1562.

84 Sangster, J.J., and Pelteon, A.D. (2000) *Phase Diagrams of Binary Hydrogen Alloys* (ed. F.D., Manchester), ASM International, Materials Park, OH, p. 74.Veleckis, E. (1979) *J. Nucl. Mater.*, **79**, 20.

85 Yu, X.B., Wu, Z., Chen, Q.R., Li, Z.L., Weng, B.C., and Huang, T.S. (2007) *Appl. Phys. Lett.*, **90**, 034106-1–3.

86 Cho, Y.W., Shim, J.-H., and Lee, B.-J. (2006) *CALPHAD*, **30**, 65–69.

87 Kang, X.-D., Wang, P., Ma, L.-P., and Cheng, H.-M. (2007) *Appl. Phys. A*, **89**, 963–966.

88 Hin, S.-A., Shim, J.-H., Cho, Y.W., Yi, K.-W., Zabara, O., and Fichtner, M. (2008) *Scr. Mater.*, **58**, 963–965.

89 Chen, P., Xiong, Z.T., Luo, J.Z., Lin, J., and Tan, K.L. (2002) *Nature*, **420**, 302–304.

90 Dornheim, M., Eigen, N., Barkhordarian, G., Klassen, T., and Bormann, R. (2006) *Adv. Eng. Mater.*, **8**, 377–385.

91 Xiong, Z.T., Hu, J.J., Wu, G.T., Chen, P., Luo, W., Gross, K., and Wang, J. (2005) *J. Alloys Compd.*, **398**, 235–239.

92 Chen, P., Xiong, Z., Wu, G., Liu, Y., Hu, J., and Luo, W. (2007) *Scr. Mater.*, **56**, 817–822.

93 Vajo, J.J., Mertens, F.O., Skeith, S., and Balogh, M.P. (2004) Patent 375 pending, Int. Pub. No: WO 2005/097671 A2 (priority date 2004)

94 Barkhordarian, G., Klassen, T., and Bormann, R. (2004) int. patent application, Publ. No. WO2006063627 (priority date 2004).

95 Barkhordarian, G., Jensen, T.R., Doppiu, S., Bösenberg, U., Borgschulte, A., Gremaud, R., Cerenius, Y., Dornheim, M., Klassen, T., and Bormann, R. (2008) *J. Phys. Chem. C*, **112**, 2743–2749.

96 Bösenberg, U., Doppiu, S., Mosegaard, L., Barkhordarian, G., Eigen, N., Borgschulte, A., Jensen, T.R., Cerenius, Y., Gutfleisch, O., Klassen, T., Dornheim, M., and Bormann, R. (2007) *Acta Mater.*, **55**, 3951–3958.

97 Pistidda, C., Barkhordarian, G., Garroni, S., Monatto Minella, C., Jensen, T., Dornheim, M., Lohstroh, W., Fichtner, M., and Bormann, R. (2009) presentation in the Symposium Hydrogen and Energy, Braunwald, Switzerland.

98 Pinkerton, F.E., and Meyer, M.S. (2008) *J. Alloys Compd.*, **464**, L1–L4.

99 Jin, S.-A., Lee, Y.-S., Shim, J.-H., and Cho, Y.W. (2008) *J. Phys. Chem. C*, **112**, 9520–9524.

8
Ammonia Borane and Related Compounds as Hydrogen Source Materials

Florian Mertens, Gert Wolf, and Felix Baitalow

8.1
Introduction

Mainly driven by the efforts of the automotive industry, a strong push towards new concepts to store hydrogen has occurred over recent years. Generally, in this framework, high hydrogen content materials have been of great interest for chemical storage concepts.

From weight considerations, the most natural choice to develop hydrogen storage and source materials would be the use of first row element hydrides. Since a high valency allows more hydrogen to bond per atom, the mid-first row hydrides borane, methane, and ammonia with 21, 25, and 17.5 wt% are amongst the materials with the highest hydrogen contents. Although the thermodynamics is not favorable, due to the strong bond between carbon and hydrogen, the utilization of methane as a hydrogen source material has been widely investigated and small natural gas reformers for a fueling station to supply polymer electrolyte membrane (PEM) fuel cell vehicles with hydrogen have been developed [1].

However, ammonia and boranes alone are also not very attractive as hydrogen source compounds because of thermodynamic considerations and the harm they potentially bring to fuel cell catalysts (PEM). The interesting fact that nitrogen and boron are on opposite sides of the electronegative–electropositive border line that runs through the periodic table, offers attractive new features to the ammonia borane adduct BH_3NH_3 (AB). Especially, their ability to release hydrogen at low temperatures, combined with the very high hydrogen content, has brought these materials into the focus of recent research efforts.

Besides the fact that the known world boron resources are insufficient to permanently fuel a significant portion of the world vehicle fleet, it is very clear from the energy expense needed for AB regeneration and the cost aspects that ammonia borane-based materials are not likely to become a means of mass hydrogen storage. These materials, however, can be of interest for small scale applications demanding very high hydrogen storage density, such as autonomously operated small electrical

Handbook of Hydrogen Storage. Edited by Michael Hirscher

ISBN: 978-3-527-32273-2

devices. Research efforts to put AB materials to use primarily encompass the control of the hydrogen release kinetics, the production of AB materials and the rehydrogenation of the spent materials.

Since spent ammonia borane, or borazane as it is also called, and most of its derivatives are not easily rechargeable, for thermodynamic reasons, new recycling strategies need to be developed. An essential part of possible recycling schemes is the generation of hydridic, that is, negatively charged, hydrogen located at the boron centers. Because of the importance of borohydrides in organic synthesis and the recent developments in nonmetal-center-based hydrogen activation [2], the relevance of these research efforts is outside the area of hydrogen source and storage materials.

8.2 Materials Description and Characterization

Ammonia borane is a white crystalline material that is relatively stable in air, although reports on its stability vary slightly [1–5]. Stored under vacuum at room temperature no significant hydrogen loss was reported for a period of over 2 months and its solubility in organic solvents remained unchanged [6]. In contrast, reports of its storage in air indicated a slow hydrogen loss, presumably induced by humidity [3, 4]. AB melts with hydrogen release, leading to a strong volume expansion by foaming [7]. After some time the product solidifies. Reports on the melting point vary strongly from 104 °C [8, 9], via 110–116 °C [5, 8, 12, 13] up to 124 °C [10, 11]. Crystal structure investigations of the original AB material encompassed powder and single crystal -ray diffraction (XRPD, XRSD) [3, 10, 16–18], and neutron diffraction studies [12]. At ambient temperature the structure is tetragonal with a BN distance unusually short for this type of electron donor–acceptor complex. In contrast to AB in the solid phase, the B–N distance of gaseous AB determined by microwave spectroscopy is 1.67 Å, significantly longer [13, 14]. It is assumed that the shortening of the B–N distance is caused by the influence of a dipolar field in the crystal [18] or by cooperative dipole–dipole interactions [22, 23]. The conclusion from self-consistent reaction field (SCRF) calculations that the B−N bond in the AB molecule, if dissolved in organic solvents, becomes shorter with increased solvent polarity is in accordance with this observation [18]. At temperatures below −48 °C, AB possesses an orthorhombic structure with an even shorter B−N distance, that is, 1.56–1.58 Å [18, 24, 25]. The phase transition, which is accompanied by a rotational order–disorder transition, was investigated by ^{11}B-NMR [26], ^{15}N-NMR [27], ^{2}H-NMR [28, 29], powder neutron diffraction [25], Raman microscopy [30], and heat capacity measurements [31]. The phase transition enthalpy $\Delta_P H$ was determined to be 1.3 kJ mol^{-1} [31] Table 8.1.

A particularly important feature of AB materials is the existence of the so-called dihydrogen bond. Guided by the unusually short distance of 2.02 Å compared to the sum of the van der Waal's radii of 2.4 Å between the nearest hydridic and protic hydrogen atoms located at the boron and nitrogen atom of different AB molecules, an unknown type of bond, the dihydrogen bond, was postulated [19, 32]. Its detailed

Table 8.1 Physical properties of AB.

Property	Value
Molar mass, [$g\,mol^{-1}$]	31[a)]
Density, [$g\,cm^{-3}$]	0.73[b)]
Melting point, [°C]	104[c)], 110–116[d)], 125[e)]
Phase transition point, [°C]	−48[f)]
Vapor pressure at 25 °C, [bar]	$<1.3 \times 10^{-6}$[g)]
Dipole moment, [D]	4.88[h)], 5.05[i)], 5.22[j)]

a) In solution, from cryoscopy [3, 10].
b) For polycrystalline pellets [10].
c) From [10, 11].
d) From [5, 12, 13].
e) From [14, 15].
f) From [18, 24, 25, 31].
g) Upper limit [43].
h) In solution [44].
i) In solution [45].
j) In gas phase, from microwave spectroscopy [20, 21].

investigation was carried out with neutron diffraction [19], Raman spectroscopy [33, 34], and *ab initio* calculations [32, 35–38]. Kullkarni showed with perturbation calculations at the MP2 level that a high tendency to form dihydrogen bonds, indicated by the energy of formation for a AB dimer, facilitates the release of hydrogen [39]. Calculations of the H–H bond strength in AB-dimers and -tetramers yield values of −12 to −25 $kJ\,mol^{-1}$ for the dihydrogen bonds [23, 32, 36, 38, 40, 41]. These values are comparable to typical N−H···Base H-bonds (−12 to −33 $kJ\,mol^{-1}$) and demonstrate that boron-bonded hydrogen possesses strong Lewis base character [32]. The existence of the dihydrogen bond largely explains the very different melting points of BH_3NH_3 (110 °C) and the isoelectronic compound ethane (−181 °C) [32, 42], as well as the higher volumetric hydrogen storage density of AB compared to ethane.

Although the vapor pressure of AB at room temperature was measured to be low [43], it is still possible to purify the material by sublimation at 60–80 °C under vacuum (10^{-2} mbar) [3–5, 8]. It is not completely clear though, whether or not in these cases BH_3NH_3 exists in the gas phase in molecular form or in the form of its constituents (di)borane and ammonia. If the latter is true, it implies that the precipitation onto the cooling finger creates AB instead of the ionic dimer diammoniate of diborane (DADB), $[BH_2(NH_3)]BH_4$. The special meaning of DADB for the AB decomposition and generation will be discussed in later sections. In spectroscopic measurements AB was detected in the gas phase but the amount of diborane exceeded the amount of AB significantly, which was explained by the instability of the AB molecule with respect to wall collisions [46].

AB is soluble in water and in many organic solvents. It possesses high solubility, that is, >5 wt%, in H_2O, EtOH, THF, Diglyme, Triglyme; significant solubility, that is, 0.5–1 wt%, in Et_2O, and dioxane, low solubility, that is, <0.1 wt%, in toluene and

Table 8.2 Solubility of AB[a].

Solvent	Solubility [wt.%]
NH_3 (l)	72.2[b]
Water	25.1[b]
Ethanol	6.1[b]
Isopropanol	3.8[b]
Tetrahydrofuran	20.0[b]
Dioxane	0.50[c]
Diethyl ether	0.74[b]
Diglyme	27.2[d]
Triglyme	25.6[d]
Benzene	0.05[c]
Toluene	0.03[c]
CCl_4	0.02[c]

a) In NH_3(l) at −78 °C, in other solvents at room temperature.
b) From [14].
c) From [10].
d) Data measured by the authors, the concentration of the saturated solution was determined with ICP-OES.

CCl_4. More detailed values can be found in Table 8.2. Since the solubility of AB in Et_2O is larger at −78 °C than at 25 °C, the system possesses a negative temperature dependence of the solubility [3, 4]. This effect can be exploited for the purification via recrystallization with Et_2O.

The stability of the ether-based solutions is comparatively high if the solvent is well dried [4, 5]. Adding small amounts of water leads to a fast decomposition of AB with hydrogen release [4]. In contrast to the last observation, an AB – water solution using deionized water is rather stable as well, proceeding with a degree of hydrolyzation of 0.5–1% a day [5, 10, 47]. From the determination of the molecular weight [3, 4, 10] dimerization and the formation of associates of AB in solution (Et_2O, benzene, dioxane, water) can be excluded. Since AB is a simple donor-acceptor complex, it was expected that it could be directly synthesized from diborane and NH_3 rather easily. However, early investigations generally carried out at low temperatures (−78 °C) did not produce AB material but instead the ionic isomer diammoniate of diborane (DADB) [48–50]. The structure of this compound was the object of debate for a significant period of time and was widely investigated. Shore and co-workers in a early series of articles [4, 51–54] as well as subsequent work by other authors using ^{11}B-NMR and Raman studies [55, 56] proved its ionic nature and clarified its structure.

DADB is usually obtained as an X-ray amorphous white powder, but reports of its presence as a microcrystalline material with a characteristic XRPD diffractogram also exist [8, 13]. It appears in two modifications. Modification I is stable at ambient temperature [13]. Modification II is only formed at higher temperatures [8] and changes at ambient temperature to modification I within a few days. DADB is stable up to 80 °C and decomposes at 90 °C with melting [4]. It is soluble in NH_3 and is, in contrast to AB, not soluble in Et_2O [4, 8]. From vapor pressure measurements it was concluded that DADB stays intact if dissolved in NH_3 [57].

8.3 Production

Besides necessary technical improvement, the most important prerequisite for the widespread use of ammonia borane as a hydrogen source is its availability at low cost. From a chemical perspective, however, hydridic hydrogen is an expensive species. Currently, AB material is generated by two main processes. The first, the standard laboratory synthesis, is based on borohydrides and ammonium salts dissolved in organic solvents [3–5, 58]. The second, potentially the one that may allow an up-scaling to lower production costs, uses the direct reaction of ammonia with diborane [8, 10], $BH_3{\cdot}O(CH_3)_2$ [4], $BH_3{\cdot}S(CH_3)_2$ [15], or $BH_3{\cdot}THF$ [13]. Because of simplicity with respect to handling and safety, the first method is preferred for laboratory scale production. The reaction proceeds according to $BH_4^- + NH_4^+ \rightarrow BH_3NH_3 + H_2$.

$LiBH_4$, $NaBH_4$ and the ammonium salts all possess low solubility in the organic solvents in question, with slow dissolution kinetics. The slow dissolution of the reactants is the rate-limiting step and prohibits, together with the production of large amounts of by-products, the economic production of AB materials.

Of all available studies on the standard laboratory synthesis based on ammonium salts, Ramachandran *et al.* [58] have conducted the most extensive survey, screening 33 reactant variations at different temperatures (the study included two hydrides, seven salts, and eight solvents). A selection of all available results, including the variants with the highest yields, can be found in Table 8.3.

The low solubility of the salts and the by-products helps to gain product of sufficient purity (>95%) after evaporation of the solvent. The procedure generates stoichometric amounts of by-product which prohibits the scale-up of the procedure. Given the cost of boranes according to [58], the development and use of an optimized direct synthesis from ammonia and diborane seems to be mandatory.

Table 8.3 Ammonia borane synthesis data.

Ammonium salt	Borohydride	Solvent	Yield	Reference
NH_4Cl	LiBH4	Et_2O	33%	[4]
$(NH_4)_2SO_4$	LiBH4	Et_2O	47%	[4]
NH_4HCOO	$NaBH_4$	THF	95%	[58]
$(NH_4)_2SO_4$	$NaBH_4$	THF	95%	[58]
NH_4CH_3COO	$NaBH_4$	THF	81%	[58]
$(NH_4)_2CO_3$	$NaBH_4$	THF	80%	[5]
NH_4Cl	NaBH4	THF	45%	[58]

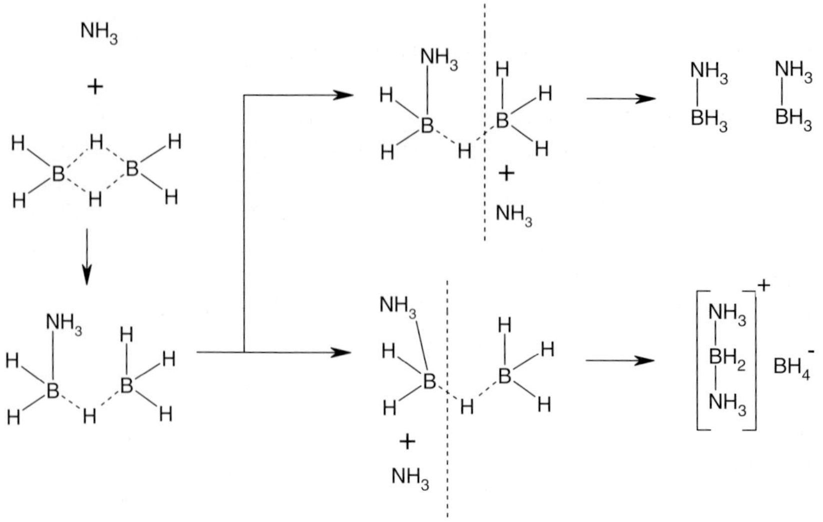

Scheme 8.1 Symmetric and asymmetric splitting of diborane by ammonia.

Besides the problem of the use of borane, which is a toxic inflammable gas, and its generation, which conventionally utilizes boranates as well, the synthesis of AB materials is not as straightforward as it seems. A particularly interesting feature of the direct synthesis of BN-hydrogen materials is the symmetric or asymmetric diborane splitting, depending on the reaction conditions used. If the solvent is a sufficiently strong electron donor then simple adduct formation to AB occurs, according to $1/2\ B_2H_6 + NH_3 \rightarrow BH_3NH_3$ [8]. At temperatures below $-78\,^\circ C$ and in the absence of solvent, an asymmetric splitting of the diborane molecule takes place, resulting in the formation of diammoniate of diborane [13, 49, 50]. Since the diborane reacts fairly vigorously with ammonia the reaction needs to be controlled in such a way that the formation of by-products is avoided [13]. The two splitting mechanisms described above can be summarized in Scheme 8.1:

AB and DADB production directly from (di)borane and ammonia can be put into practice as follows: AB material can be simply produced by first splitting diborane with an electron pair donating solvent, such as dimethylsulfide [15], ethers [8], monoglyme [8] or water [10], to form a borane solvent adduct, and then replacing the solvent part in the adduct by bubbling ammonia through the solution. In the case of diethyl ether, the procedure can be carried out at room temperature, resulting in a yield of 75%[1)] and in the case of monoglyme a yield of 76% was reported at a temperature of $-63\,^\circ C$ [8]. For the synthesis of DADB, gaseous diborane diluted in N_2 is bubbled through liquid NH_3 at $-78\,^\circ C$ [13]. After the reaction, ammonia is distilled off and a white, microcrystalline powder is obtained which contains almost exclusively DADB. The method results in a yield better than 97%. Since DADB is a stable compound it could serve as a hydrogen source material in the same way as AB.

1) U. Wietelmann, Chemetall GmbH, 2009, personal communication.

8.4 Thermally Induced Decomposition of Pure Ammonia Borane

In comparison to other boranates and complex metal hydrides, ammonia borane possesses a rather low decomposition temperature. Since the dehydrogenation reaction is exothermic the applied temperature is only needed to start the activated dehydrogenation process. The exothermicity of ammonia borane can be viewed, with respect to technical applications, as a positive feature if a fast hydrogen release of the complete hydrogen content is desired. With large storage containers where only a small portion of the hydrogen is supposed to be extracted, the exothermicity poses a control problem. If one does not want to run the risk of a thermal run-away from an unintentional emission of the complete hydrogen content in the case of a malfunction of the control mechanism, the storage of the AB material in small compartments is inevitable. A possible alternative to the pyrolytic decomposition of solid ammonia borane is decomposition in solution. The decomposition of AB material in solution has the advantage of separating the decomposition zone from the storage area because of the improved transport of the stored material in liquid form. Better AB transport properties also facilitate the application of heterogeneous catalysts for the decomposition of AB material, since the reactants and products can be easily moved to or away from the catalyst. Although the solvent needed for the preparation of an ammonia borane solution lowers the effective storage capacity of the hydrogen generator, concepts are conceivable where the solvent acts solely as the transportation medium and is retrieved from the reaction products after the ammonia borane decomposition, which then allows its reuse. By applying such a separation technique, the amount of solvent needed does not scale with the amount of ammonia borane stored because of the solvent reuse and, thus, could be potentially excluded to a certain degree from the weight percentage consideration in large systems.

8.4.1 Pyrolysis

Solid ammonia borane is thermally decomposed in several reaction steps, each releasing one molecule of hydrogen per formula unit at different temperature levels. The thermal decomposition path is complicated and still not fully understood. Detailed knowledge of its underlying processes may help to control the H_2 release kinetics and help to avoid a significant contamination of the released hydrogen gas by volatile products. DSC, TG, and volumetric measurements have been carried out by several groups [9, 14, 59–63]. The early decomposition experiments performed by Wendlandt and coworkers revealed a two-step process in the temperature range 30–200 °C, each step releasing approximately 1 mol of hydrogen per mol AB or 6.5 wt% [9, 59, 60]. If the temperature increase is conducted at a comparably high heating rate of 10 K min^{-1} the hydrogen release sets in shortly after the melting point at 112 °C and reaches a maximum at 130 °C. The second step, releasing another equivalent of molecular hydrogen, occurs in the range 150–200 °C. The corresponding signals of both decomposition steps displayed a strong overlap. At much higher temperatures the decomposition chain can be even carried all the way through to boron nitride with further hydrogen release, according to the scheme: $BH_3NH_3 \rightarrow BH_2NH_2 \rightarrow BHNH \rightarrow BN$.

In several papers the Freiberg group have demonstrated that the appearance of AB decomposition or melting is strongly kinetically controlled [47, 61–63]. The first decomposition step occurs below the melting point if low heating rates ($\leq 0.1\,K\,min^{-1}$) are applied. Due to the slow decomposition kinetics of the first AB decomposition step a faster temperature ramp not only causes the signals of the individual decompositions steps to overlap but also allows the well known melting of ammonia borane at 112 °C. Completion of the first dehydrogenation step below the melting temperature is possible at isothermal conditions of approximately 70–90 °C. The hydrogen release rate is strongly temperature dependent, at 90 °C the release is completed after 4 h whereas at 70 °C it takes 40 h for completion [61]. The decomposition product, with the approximate composition of BNH_4, is a white, nonvolatile, amorphous solid, presumably of polymeric nature $(BH_2NH_2)_n$ [61, 64, 65], which would naturally explain why melting is not possible after dehydrogenation. The exact structure of the obtained amino borane type and that obtained from other dehydrogenation procedures is still unknown. Attempts to determine the molecular weight distribution of the polymers failed because of their low solubility in common organic solvents [64, 66]. Investigations using ^{11}B-NMR, ^{15}N-NMR and mass spectrometry indicate that the material is rather inhomogeneous and not well-defined, consisting of linear, branched, and ring systems as well as, potentially, of cyclic oligomers of various length [61, 67]. Some of the cyclic compounds were separately synthesized and identified by Shore and coworkers using ^{11}B-NMR, infrared spectroscopy, and molecular weight investigations [65].

A significant problem for the practical application of AB materials as a hydrogen source may be the existence of an induction period for the H_2 production. If the material is heated to 90 °C, almost no hydrogen release is initially observed. Subsequently, the release rate increases slowly and then reaches then its maximum after 40 min [47, 61]. At 70 °C this interval, the induction period, is extended to 24 h. The mechanistic reason for this phenomenon will be discussed further below. The resulting polymeric $(BH_2NH_2)_n$ does not release hydrogen below 110 °C. Above this temperature the next step in the dominant decomposition chain occurs, which is also strongly temperature and heating rate dependent. In total, up to 13 wt.% of hydrogen can be released if the temperature is raised to 180 °C. The decomposition product of the second step with a net composition of BNH_2 is also a white, amorphous solid of presumably polymeric nature $(BHNH)_n$ [63]. It is known that the corresponding monomer BHNH (BN analog of acetylene) is very reactive and can only be characterized using matrix isolation techniques [68, 69]. The cyclic trimer $(BHNH)_3$ is the well known BN analog of benzene, called borazine.

Both decomposition steps are exothermic. For the first reaction step a consistent value of $\Delta_R H = -21.7 \pm 1.2\,kJ\,(mol\ AB)^{-1}$ was obtained from isothermal DSC measurements [61]. For the second decomposition step, depending on the heating rate, values from -15 to $-24\,kJ\,(mol\ AB)^{-1}$ were obtained [63]. The variation may be explained by the rate-dependent product spectrum generated. Figure 8.1a displays

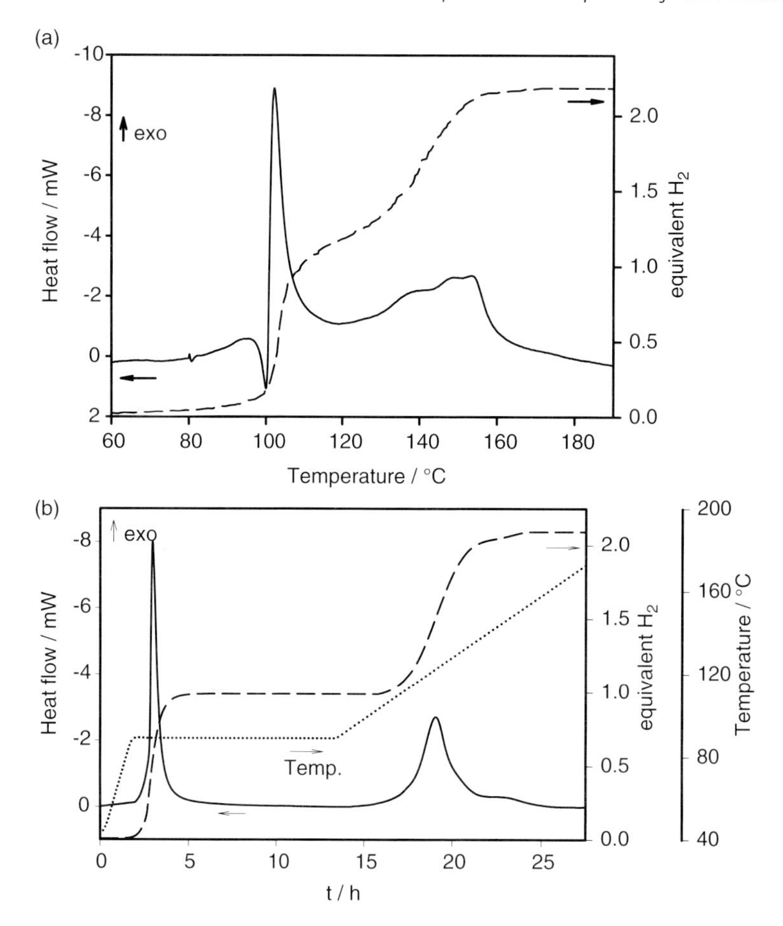

Figure 8.1 Heat flow and H_2 release during the thermal decomposition of ammonia borane (a) heating rate 1 K min^{-1} (b) isothermal conditions followed by a heating rate of 0.1 K min^{-1}.

the DSC signal obtained from temperature ramp experiments. Figure 8.1b shows the isolated signal of the first decomposition step from a combined isothermal–temperature ramp experiment.

There is not enough thermodynamic data for calculation of the reaction entropy. Essentially, for all relevant BN compounds the entropy values are missing. Only for AB itself was it determined to be 96 J K^{-1} mol^{-1} [31]. If one assumes that the reaction entropy is almost exclusively determined by the contribution of the released hydrogen gas an equilibrium pressure of approximately 10^5 bar at 25 °C is calculated, excluding AB regeneration by simple exposure to high pressure hydrogen gas by all practical means just from thermodynamic considerations. Indeed measurements up to 600 bar did not reveal any influence on the H_2 release from solid AB [70].

The nominal very high hydrogen content of AB needs to be seen in the light of energy efficient hydrogen release. That is why the last dehydrogenation step, that is, the hydrogen release from polymeric $(BHNH)_x$ is usually discarded for hydrogen source purposes, although AB was taken into consideration as a hydrogen delivering rocket fuel and the thermolysis of it at very high temperatures was investigated. Above 500 °C conversion to BN sets in. FTIR measurements showed that with increasing temperature first the signals of the hydrogen atoms bonded to boron and then those of the hydrogen atoms bonded to nitrogen disappear. At 900 °C the conversion to amorphous boron nitride is complete [71, 72].

From thermogravimetric measurements simultaneously carried out during the dehydrogenation occurring in the first step, a mass loss of approx. $\Delta m = (10.0 \pm 0.1)\%$ of the initial sample mass was determined, which significantly exceeds the 6.5 wt% expected for the release of 1 mol H_2 per mol BH_3NH_3. This additional mass loss is explained by the production of additional volatile products [9, 59, 61, 62]. Gas-phase investigations by FTIR and mass spectrometry demonstrated the simultaneous release of monomeric amino borane BH_2NH_2 (approximately 0.2 mol (mol AB)$^{-1}$), borazine $(BHNH)_3$ (0.06 mol (mol AB)$^{-1}$), and diborane (0.06 mol (mol AB)$^{-1}$) [62]. The emission of volatile products is strongly dependent on the heating rate. Rates lower than $\leq 1\ K\ min^{-1}$ result in almost no detectable by-products. Rates in the range of 3–10 K min^{-1} leave the amount of hydrogen released almost unaltered while displaying for the first decomposition step still fairly low by-product emissions but increased ones for the second step. At high rates, mass losses at the end of the experiments at approximately 180 °C of up to 25 wt.% of the initial AB material were observed [60, 62]. Since the by-product emission of boron- and nitrogen-containing species is problematic for a hydrogen source material because of the hydrogen feed stream contamination and the sensitivity of catalytic devices such as fuel cells, understanding these processes is very important.

A possible explanation for the rate dependence of the volatile by-product formation is given by Scheme 8.2 [47]:

In addition to solid polymeric amino borane, gaseous monomeric amino borane is formed. At low heating rates, there is enough time for the unstable BH_2NH_2 to oligomerize before it can escape the sample. At fast decomposition rates, a larger amount of BH_2NH_2 escapes the sample and splits off another hydrogen molecule at

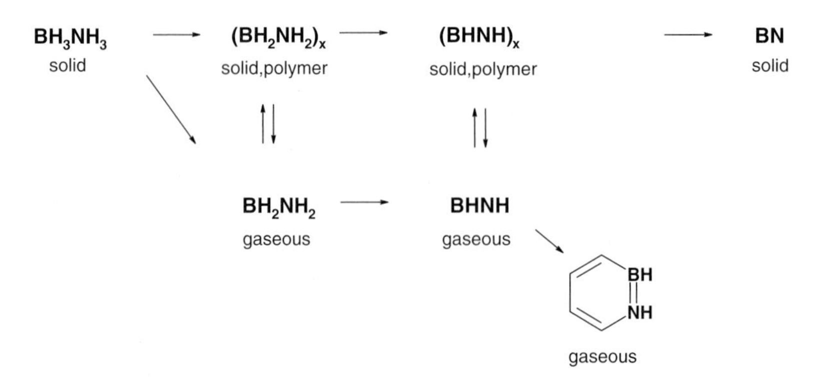

Scheme 8.2 Pyrolytic decomposition pathways of ammonia borane.

elevated temperatures by forming the even more reactive monomeric BHNH, which subsequently trimerizes to borazine $(BHNH)_3$. This view is based on the assumption that the initial hydrogen release occurs intramolecularly, that is, the released hydrogen comes from the neighboring boron and nitrogen atom of the same ammonia borane molecule. There is, however, strong experimental evidence that it actually occurs intermolecularly. Theoretical and experimental studies have identified in ammonia borane the existence of dihydrogen bonds, as a consequence of the presence of protic and hydridic hydrogen atoms in the ammonia borane molecule (see Section 8.2). Because of this particular hydrogen–hydrogen interaction an intermolecular hydrogen release mechanism seems likely [73]. Since a detailed understanding of this mechanism may be beneficial for the development of dehydrogenation catalysts, isotope exchange experiments have been conducted to clarify the question as to whether intra- or inter-molecular hydrogen release dominates. Smith *et al.* reported thermal decomposition experiments of BH_3NH_3/BD_3ND_3 mixtures and detected, via mass spectrometry, all three species H_2, D_2, HD in the gas phase [74]. A fact that counts as proof of an intermolecular mechanism if one assumes that no parallel exchange reactions occur. Similar experiments were conducted with substituted AB material, such as BH_3NHMe_2/BD_3NDMe_2 and BD_3NHMe_2/BH_3NDMe_2 [75]. Again all three combinations H_2, D_2, HD were found, supporting the intermolecular mechanism.

Similarly to the hydrogen gas contamination, the presence of an induction period for the hydrogen release is not acceptable for a practical hydrogen source material. During the induction period, which can last for several hours in the low temperature region (70–80 °C), no significant hydrogen release is detectable [47]. The kinetics of the hydrogen release follows a sigmoidal course, which is typical for nucleation and growth mechanisms of solid particles. Detailed knowledge of the underlying mechanism may eventually help to overcome this deficiency. Stowe *et al.* carried out detailed *in situ* solid state ^{11}B MAS-NMR investigations of the thermal decomposition of BH_3NH_3 and BH_3ND_3 [76]. The material was decomposed under isothermal conditions at 88 °C. The protocol of these experiments is as follows. After 8 min signals of the solid AB compound shrank and signals of a new species containing a BH_3 group appeared. From these results it was concluded that a new BH_3NH_3 phase, in which the hydrogen–hydrogen bonds are broken, was formed, resulting in much more mobile BH_3NH_3 units. The induction period, during which no hydrogen emission was observed, correlates to the time needed for the transformation of solid AB material into a new phase, indicated by an unidentified NMR signal. After 14 min, the formation of $[BH_2(NH_3)_2]^+$ $[BH_4]^-$ (DADB) and the release of hydrogen, sets in. After 18 min the NMR signal of the initial AB material has disappeared, the H_2 release is strongly enhanced, and polymeric BNH products are formed, possessing not well characterized compositions and structures. Stowe's group also suggests a more detailed mechanism summarized by Scheme 8.3:

DADB is created by an AB molecule abstracting a hydride ion from another AB unit. The process is essentially a base exchange, that is, the exchange of H^- for NH_3. The created DADB molecule reacts with another AB unit via H_2 formation. In the scheme above, DADB appears to be more reactive than the original AB material,

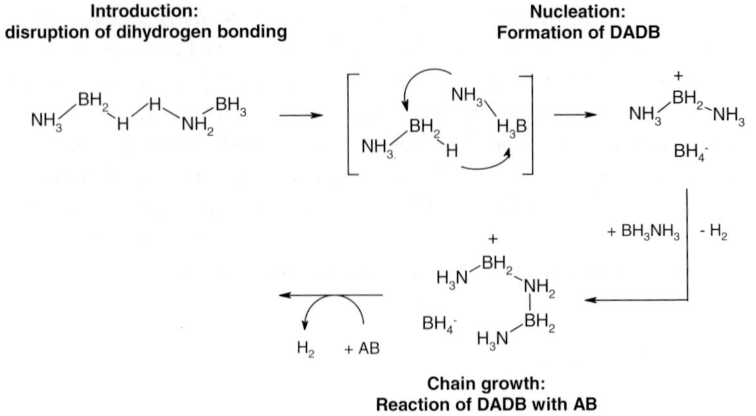

Introduction:
disruption of dihydrogen bonding

Nucleation:
Formation of DADB

+ BH_3NH_3 | - H_2

H_2 + AB

Chain growth:
Reaction of DADB with AB

Scheme 8.3 Proposed mechanism for the pyrolytic decomposition of solid ammonia borane.

which implies a lower activation barrier for the reaction of DADB + AB to a new intermediate. This new intermediate reacts in turn with another AB molecule via hydrogen release to give the final polymeric product.

The proposed mechanism does not explain the release of the gaseous components observed during the various AB decomposition experiments, such as monomeric amino borane and borazine. The mechanism was, however, deduced from the online monitoring of the dehydrogenation process under isothermal conditions at 88 °C under which very little monomeric BH_2NH_2 release was detected by mass spectrometry [62]. As pointed out above, fast temperature ramping was a prerequisite for emitting large amounts of monomeric BH_2NH_2 and borazine. The mechanism for this case must therefore be a somewhat altered version of that displayed in Scheme 8.3.

Monomeric amino borane BH_2NH_2, which is the BN analog of ethylene, is a highly reactive, at ambient conditions, gaseous compound. Besides being detected as product during the pyrolysis of solid AB, it was also found during the low-pressure pyrolysis of gaseous AB [46, 77], during the gas-phase reaction between diborane and ammonia [78], and during the electric discharge in a borazine-containing gaseous atmosphere [79]. The monomer is only stable at low temperature (−196 °C). With increasing temperatures the monomer oligomerizes and forms solid nonvolatile products [78, 79]. The cyclic compound borazine, $(BHNH)_3$, which is the BN analog of benzene, is a liquid at ambient conditions with the boiling point at 55 °C. During the pyrolysis of solid AB only small amounts of this compound are released [9, 62]. Larger amounts are formed if AB is thermally decomposed in organic solvents, such as in glymes (see Section 8.4.2). The structure of borazine was solved with electron diffraction [80] and XRSD [81]. Liquid borazine decomposes at 70–110 °C with release of hydrogen and the formation of polyborazylenes, which consist of cross-linked borazine units [82–84]. Besides the polymeric products, small amounts of compounds with lower molecular weights, for example, BN analogs of biphenyl and naphthalene, were also found [82].

From the presence of monomeric amino borane in the gas phase, one can postulate that it may be the precursor for the formation of polymeric amino borane as well as for borazine, since it is known to be unstable at room temperature [79]. Another process not fully understood is the formation of cyclic products. A strong influence of the reaction environment is observed. The decomposition of AB material in the solid form almost exclusively results in solid polymeric BNH_x-products. Cyclic products such as borazine or cycloaminoborane ($(BH_2NH_2)_3$) are then only formed in minute amounts, whereas in solution cyclic products are preferred. The exact parameters that control this selection process have not yet been determined.

8.4.2 Decomposition in Organic Solvents

The decomposition of AB material in solution could be attractive for many reasons. From a chemical engineering point of view, the use of a liquid fuel is generally preferred because of the simplified transport compared to solid materials. Liquid AB hydrogen sources can be used to improve the controllability of the AB decomposition by separating the decomposition zone (reactor) from the storage tank. This measure would be an important improvement with respect to safety, since the dehydrogenation reaction potentially bears the risk of a thermal run-away due to its exothermic nature.

From a chemical point of view, the AB decomposition in solution may allow one to adjust the type of decomposition product obtained after dehydrogenation and thus may influence the recycling strategy to be applied. Furthermore, benefits from improved kinetics can also be expected. Of course, the fundamental prerequisite is a high solubility and the stability of ammonia borane in the solvent in question. Corresponding data are given in Section 8.2.

Early work addressing the AB decomposition in solution was carried out by Wang and Geanangel, who investigated the decomposition of dilute (0.15 M) solutions of AB in several aprotic solvents [85]. They found three types of reactions (i) AB decomposition with hydrogen loss, (ii) base displacement by the solvent, and (iii) the hydroboration of the solvent. In pyridine a complete exchange between ammonia and the base pyridine was observed, leading to a $C_5H_5NBH_3$ adduct. The decomposition in acetonitrile under reflux resulted, after 12 h, in the formation of a mixture of $(BH_2NH_2)_3$, $(BHNH)_3$, μ-aminodiborane $NH_2B_2H_5$ as well as *N*-ethyl ammonia borane C_2H_5-NH_2BH_3. The observation of the latter in the product mixture indicates that the dehydrogenation of AB is accompanied by either a hydroboration or hydrogenation of the solvent. The decomposition of AB in ethers (monoglyme, diglyme, tetrahydrofuran (THF), and 2-methyltetrahydrofuran) leads to a mixture of dehydrogenated BHN-compounds, which were detected by ^{11}B NMR. Major products are the cyclic compounds $(BH_2NH_2)_3$ and $(BHNH)_3$, the formation of which corresponds to the release of 1 and 2 equivalents of H_2, respectively. Contrary to the pyrolysis of solid AB, only negligible amounts of the polymeric materials $(BH_2NH_2)_x$ and $(BHNH)_x$ were obtained. The kinetics of the AB decomposition in ether solvents considered in [85] is comparable to the kinetics of the pyrolysis of solid

AB. For example, in diglyme and monoglyme AB is not entirely decomposed at 85 °C even after 9 and 25 h, respectively. Dixon and coworkers [86] reported that AB in monoglyme (0.14 M) releases only 0.05 equiv H_2 at 60 °C after 24 h. As expected, the H_2 release can be fast at higher temperatures, the complete conversion of AB into borazine $(BHNH)_3$ requires less than 3 h at 130–140 °C [85]. Wideman and Sneddon [87] used the thermal decomposition of AB in solution for the laboratory scale preparation of borazine. For this purpose, solid AB was added slowly, over the course of 3 h, to tetraglyme at 140–160 °C. Borazine was removed from the reaction mixture, trapped at −78 °C, and isolated, resulting in a 67% yield.

The decomposition of AB in solution offers new routes to influence the decomposition kinetics catalytically and the selectivity of the dehydrogenation pathway. From a practical point of view heterogeneous catalysts would be preferred because they can effectively be used for the control of the reaction. Manners and coworkers and the Berke group reported on the catalytically supported release of hydrogen from substituted ammonia boranes [88–92], most extensively from BH_3NHMe_2. The dehydrogenation of BH_3NHMe_2 in toluene is accompanied by the release of 1 equiv H_2 and the formation of the cyclic dimer $(BH_2NMe_2)_2$. Although the material itself would not be interesting as a hydrogen source material due to a low storage density, the work is interesting with respect to mechanistic questions. Many catalysts were tested, among them various Rh, Ir, Ru, Pd complexes [89, 90], $TiCp_2$ [92] and Pd/C as a heterogeneous catalyst [89]. The most active catalyst, $Rh(cod)(\mu\text{-}Cl)]_2$ (cod = 1,5-cyclooctadiene), leads to the quantitative formation of $(Me_2NBH_2)_2$ after 8 h at 25 °C. The nature of the active catalytic species is still under discussion. When the precatalyst $Rh(cod)(\mu\text{-}Cl)]_2$ is added to a solution of a substituted ammonia borane, a black suspension of metallic colloidal particles appears after an induction period. The generation of hydrogen gas sets in immediately after the occurrence of colloids, indicating that metallic Rh is the catalytic active species. Transmission electron microscopy (TEM) studies revealed the formation of Rh(0) particles approximately 2 nm in size [89, 93]. Autrey and coworkers demonstrated via X-ray absorption fine structure (XAFS) data that rhodium particles formed under similar conditions exist as toluene-soluble Rh_6-clusters surrounded by BNH_x ligands [94, 95]. The loss of hydrogen from the ligands towards the end of the reaction decreases the solubility of the clusters and leads to their precipitation. The precipitation process can be reversed if fresh BH_3NHMe_2 is added, presumably due to ligand exchange.

Unfortunately, the activity of the $[Rh(cod)(\mu\text{-}Cl)]_2$ precatalyst for the dehydrogenation of the parent AB is significantly lower [88–90]. Addition of $[Rh(cod)(\mu\text{-}Cl)]_2$ (0.6 mol%) to a solution of AB in diglyme or tetraglyme (26 wt.%) leads, via the intermediates $(BH_2NH_2)_3$ and $\mu\text{-}NH_2B_2H_5$, to a product mixture containing borazine $(BHNH)_3$ as well as several oligomeric and/or polymeric compounds. The reaction proceeds slowly. AB is entirely decomposed after 48–84 h at 45 °C, in contrast to the dehydrogenation of BH_3NHMe_2, which is already complete after 2 h at 45 °C [89]. These differences in the catalytic activity can perhaps be explained by the fact that, in contrast to the BH_3NHMe_2 dehydrogenation, the rhodium clusters formed in the presence of AB precipitate already at the beginning of the reaction [95].

Two factors concerning the solubility of the Rh-clusters need to be taken into account:

1) The influence of the solvent
   It cannot be excluded that the lower activity of the $[Rh(cod)(\mu\text{-}Cl)]_2$ precatalyst for the dehydrogenation of AB is caused by an unsuitable solvent choice (diglyme, tetraglyme). Mertens' group observed that, in the case of BH_3NHMe_2, the use of diglyme as solvent also leads to a strong reduction in the Rh-catalyzed decomposition rate. In diglyme, at 25 °C no decomposition was observed during the 12 h observation period. In toluene, however, the complete conversion of BH_3NHMe_2 into $(BH_2NMe_2)_2$ took only 8 h at 25 °C [89]. Generally, the selection of an appropriate solvent is not an easy task because of the low solubility of AB in most organic solvents (see Table 8.2).
2) The influence of the substituent:
   Manners and coworkers reported that under similar conditions (glymes as solvent, nearly equal amounts of $[Rh(cod)(\mu\text{-}Cl)]_2$ precatalyst) the Rh-catalyzed dehydrogenation of BH_3NH_2Ph is much faster (the formation of the cyclic $(BHNPh_3)_3$ is finished after 16 h at 25 °C) than the dehydrogenation of BH_3NH_3 and BH_3NH_2Me, which both required >48 h at 45 °C for complete decomposition [89]. The reason for such a strong influence of the substituent is still unknown. A possible explanation could be a different solubility (and activity) of the corresponding Rh colloids.

Goldberg and coworkers reported that the Ir "pincer" complex $Ir(H)_2POCOP$ ($POCOP = \eta^3\text{-}1,3\text{-}(OPt\text{-}Bu_2)_2C_6H_3$), known as an effective homogeneous catalyst for the dehydrogenation of alkanes, also shows an unprecedented high activity for the AB dehydrogenation [96]. The addition of $Ir(H)_2POCOP$ (0.5 mol%) to a solution of AB in THF (0.5 M) results in the release of 1 equiv H_2 after just 14 min at room temperature. In contrast to the Rh-catalysts, the Ir-catalyzed dehydrogenation leads to the formation of a white precipitate, identified by X-ray powder diffraction and infrared spectroscopy as the cyclic pentamer $(BH_2NH_2)_5$. So far the benefit of fast kinetics must be traded against incomplete dehydrogenation and the production of solid dehydrogenation products. Although the mechanism is not yet understood, it is suggested by theoretical investigations [97] that a simultaneous transfer of a hydride from the boron to the iridium center and of a proton from the nitrogen atom to a hydride bound to the iridium center occurs, leading to the formation of the $Ir(H)_4POCOP$ intermediate.

Baker and coworkers [98] investigated catalysts based on complexes of Ni, Ru, and Rh with strong electron-donating ligands, such as *N*-heterocyclic carbenes (NHCs). Most active is an *in situ* generated Ni complex with the Enders' carbene, 1,3,4-triphenyl-4,5-dihydro-1H-1,2,4-triazol-5-ylidene. The catalyst performance reads as follows: 2.5 H_2 equiv in 4 h at 60 °C for a dilute solution of AB in diglyme (0.14 M, catalyst amount 10 mol%); 2.5 H_2 equiv in 2.5 h at 60 °C for a highly concentrated solution of AB in diglyme (25 wt.%, catalyst amount 1 mol%). The obtained product mixture consists mainly of soluble polyborazylene. Besides the high yield of more than 2 equiv H_2, this catalytic system is also beneficial in respect to a very low borazine

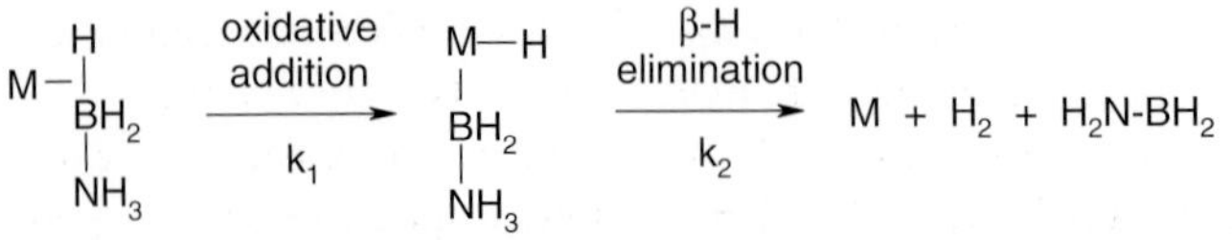

$$\mathrm{M{-}H{\cdot}BH_2{-}NH_3} \xrightarrow[k_1]{\text{oxidative addition}} \mathrm{H{-}M{-}BH_2{-}NH_3} \xrightarrow[k_2]{\beta\text{-H elimination}} \mathrm{M + H_2 + H_2N{-}BH_2}$$

Scheme 8.4 Proposed mechanism for the metal catalyzed dehydrogenation of ammonia borane.

release. Other $M(NHC)_2$ catalysts with M = Rh, Ru as well as imidazolium-based NHCs showed less activity. The catalytic mechanism is still unknown, but Baker and coworkers proposed the following sequence [98]: (i) formation of a σ-complex, (ii) B–H activation at the metal center, (iii) β-H elimination from the nitrogen atom (Scheme 8.4).

Investigations of the kinetic isotope effect with NH_3BD_3, ND_3BH_3, ND_3BD_3 indicate that the rate-determining step involves either the simultaneous cleavage of the B–H and the N–H bonds or that both steps occur subsequently but with similar rate constants. The last case would be compatible with the mechanism outlined in Scheme 8.4.

Dixon and coworkers recently demonstrated that strong Brønsted or Lewis acids catalyze the AB dehydrogenation in glymes [82]. The results of the investigations performed at 60 °C are summarized in Table 8.4.

A particular emphasis was put on the investigation of $B(C_6F_5)_3$ as a strong Lewis acid. $B(C_6F_5)_3$ is known as an excellent activator component for olefin polymerization reactions in combination with homogeneous dialkylmetallocene Ziegler–Natta catalysts [99, 100]. In these systems, it is employed to abstract an alkyl carbanion from the respective catalyst precursor. Aspects of $B(C_6F_5)_3$ as catalyst or cocatalyst can be found in [101, 102]. Recently, it was demonstrated that this strong Lewis acid is even capable of heterolytically splitting molecular hydrogen in conjunction with nitrogen bases [103] and phosphines [104].

Table 8.4 Acid-catalyzed dehydrogenation of AB in glymes at 60 °C (data from [86]).

Acid	Acid amount [mol%][a]	AB concentration [M]	Reaction time [h]	H_2 equivalents[b]
none	0	0.14[c]	24	0.05
$B(C_6F_5)_3$	25	0.14[c]	24	0.6
$B(C_6F_5)_3$	0.5	2.6[d]	20	1.1
$HOSO_2CF_3$	25	0.13[c]	18	0.8
$HOSO_2CF_3$	0.5	6.2[e]	18	1.3
HCl	0.5	2.9[e]	20	1.2

a) Related to AB.
b) mol H_2 per mol AB.
c) In monoglyme.
d) In tetraglyme.
e) In diglyme.

Previously, Denis *et al.* demonstrated that $B(C_6F_5)_3$ also catalyzes the dehydrogenation of the borane–phosphine adduct, BH_3PH_2Ph [105]. They proposed the formation of a reactive intermediate, $B(C_6F_5)_3PH_2Ph$, by an exchange reaction. Since $B(C_6F_5)_3$ is more Lewis acidic than BH_3, this exchange increases the protic character of the P–H proton and thus enhances its reactivity with the hydridic B–H hydrogen.

In contrast to the dehydrogenation of borane–phosphine adducts, Dixon and coworkers [86] clearly identified a different mechanism for the $B(C_6F_5)_3$-catalyzed dehydrogenation of BH_3NH_3. The activation proceeds via hydride abstraction, according to $BH_3NH_3 + B(C_6F_5)_3 \rightarrow [BH_2NH_3]^+[HB(C_6F_5)_3]^-$. The $[HB(C_6F_5)_3]^-$ anion is known and was previously characterized by NMR spectroscopy [106], where it was obtained by the reaction between $PhMe_2SiH$ and $B(C_6F_5)_3$. The boronium cation $[BH_2(NH_3)L]^+$ (L = solvent) formed is similar to the $[BH_2(NH_3)_2]^+$ cation in DADB. Dixon and coworkers postulated that the formed boronium cation $[BH_2(NH_3)L]^+$ reacts with an additional AB molecule with the release of 1 equiv H_2 (Scheme 8.5, step a) [86]. This step is related to the reaction between DADB and

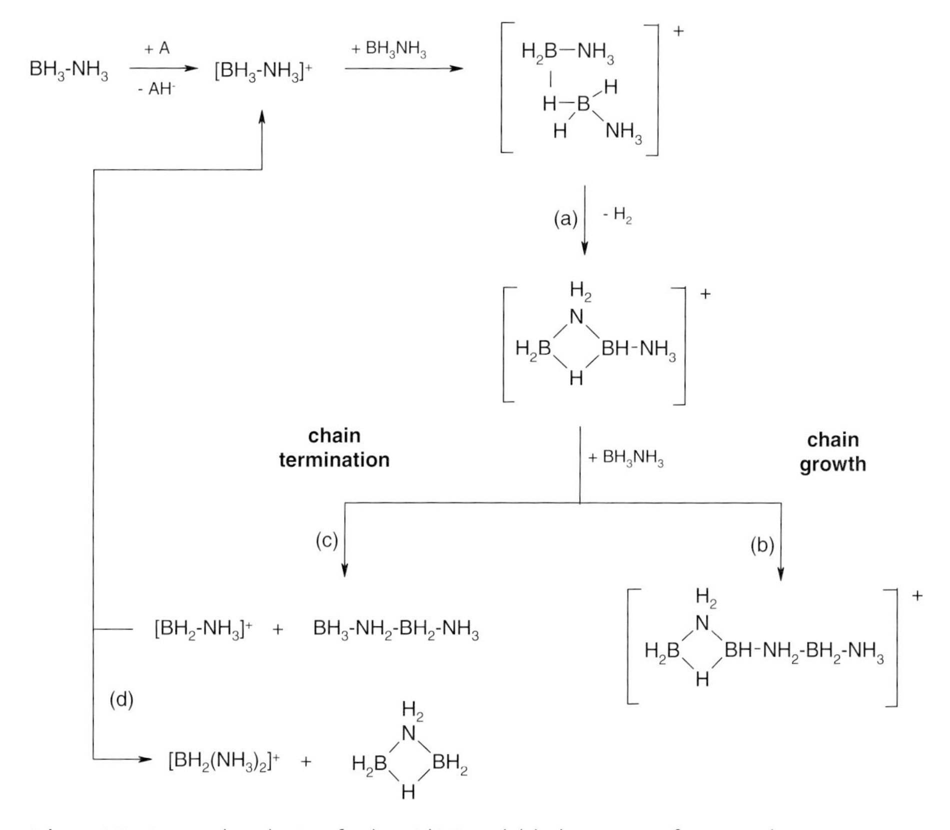

Scheme 8.5 Proposed mechanism for the acid-initiated dehydrogenation of ammonia borane in solution.

AB in the thermal dehydrogenation of solid AB (Section 8.4.1). Subsequent additions of AB to the cationic growing chain lead to the formation of linear and branched μ-NH_2-bridged species that maintain fourfold-coordinated boron, as was concluded from a DFT study. (Scheme 8.5, step b).

Chain termination was assumed to proceed via a hydride abstraction from AB, resulting in a neutral aminoborane oligomer and regenerating the cationic boronium initiator (Scheme 8.5, step c). This pathway explains the formation of linear and branched acyclic BNH_x products. In glymes these oligomers undergo a subsequent dehydrocyclization reaction. Alternatively, an ammonia transfer from a neutral oligomer to the boronium cation can occur, (Scheme 8.5, step d). The contributions of this side reaction, resulting in μ-amino diborane, become more important if the acid concentration is 10 mol% or more.

8.4.3
Decomposition of Ammonia Borane in Heterogeneous Systems

Similarly to the decomposition in solution, it is conceivable that the decomposition of AB material in heterogeneous, nanostructured systems, like in a solid micro- or meso-porous solid matrix, may effect the dehydrogenation properties in several beneficial ways: (i) a significant improvement of the kinetics of the dehydrogenation of solid AB, (ii) the reduction of the emission of borazine to avoid the H_2 gas stream contamination, and (iii) a change in the overall thermodynamics to enhance the chance of finding an energy efficient rehydrogenation scheme. For all applications of AB in composite materials as hydrogen source materials, it will be crucial that the AB content is still high enough to obtain an acceptable gravimetric hydrogen density. Autrey and coworkers have noticed a positive effect on the dehydrogenation kinetics for AB in nanophase scaffolds [107]. In one case they loaded mesoporous high surface silica with 50% AB. The half-time for the dehydrogenation at 50 °C in this system was only 85 min, compared to 290 min at 80 °C for neat AB. This significantly increased reaction rate indicates that the decomposition pathway differs mechanistically. A strongly altered product spectrum compared to the neat AB case confirms this view. Particularly beneficial in this respect is the reduction of the borazine emissions. The decomposition of AB in a silica scaffold is also less exothermic than that of pure solid AB. The group also investigated a carbon-cryogel – AB nanocomposite with 50% AB load [108, 109] and found a further acceleration of the decomposition reaction and an even lower production of borazine compared to the silica experiment, but since the process was much more exothermic a reaction with the scaffold material may have occurred (formation of surface O–BX_2 bonds) [109].

Sneddon's group investigated AB dehydrogenation in the ionic liquid 1-butyl-3-methylimidazolium chloride (bmimCl) with an AB content of 50% [110]. Because of the low solubility of AB in bmimCl, one can assume that this system was essentially heterogeneous in nature. Similarly to the solid nanocomposites the change in the reaction environment induces several positive effects on the AB decomposition, such as an increase in the reaction kinetics, the absence of an induction period, a strong reduction in the borazine emissions, and with 1.6 equiv of H_2 at 98 °C in contrast to

0.9 equiv with solid AB, a strong increase in the H_2 yield. It was assumed by Sneddon that the activating effect by the ionic liquid may be related to its ability to induce the formation of ionic species, like DADB, $[BH_2(NH_3)_2]^+BH_4^-$, the presence of which was confirmed by ^{11}B NMR during H_2 release from AB in bmimCl [110].

8.5 Hydrolysis of AB

Instead of using volatile organic solvents, it would be beneficial, for compatibility with fuel cells and the environment, to use just water as the solvent. The solubility of AB in water is high, 25 wt.% [14]. Aqueous AB solutions are also relatively stable if kept under an inert gas atmosphere [10, 111]. In air the solution possesses a much lower stability which is believed to result from an increase in acidity in the solution due to the dissolution of CO_2 [111]. The stability of the aqueous AB solution is obviously kinetically controlled since the decomposition is thermodynamically favored due to the formation of strong B–O bonds according to:

$$BH_3NH_3 + 2H_2O \rightarrow NH_4BO_2 + 3H_2 \tag{8.1}$$

The formula suggests a maximum hydrogen release of 9 wt.% with respect to the AB and to the consumed water weight. This value is obtained neglecting that significantly more water will be needed in order to keep the reactants and products in solution and to keep the viscosity of the solution low. Thus, the practical hydrogen storage density will be much lower and there is a general interest in minimizing the amount of water needed. A comparison of the pH value and the ^{11}B-NMR resonance of the obtained aqueous solution with corresponding NH_4OH/H_3BO_3 solutions at different concentrations indicates that there is an equilibrium between the H_3BO_3, BO_2^- and other borate species, which convert into one another according to $BO_2^- + H^+ + H_2O \rightleftharpoons H_3BO_3$ too rapidly on the NMR timescale for individual detection [112].

It is well known that AB hydrolysis, and thus H_2 release, proceeds quickly in acidic conditions, even at room temperature [3, 5]. Mechanistic investigations suggest that the rate-determining step is the splitting of the BN bond by an electrophilic H^+ attack with the formation of an NH_4^+ ion. In a subsequent step, the remaining BH_3 is hydrolyzed to boric acid [113–115]. Complete hydrolysis requires, therefore, a stoichiometric amount of acid and a less than stoiciometric H^+ supply leads to reduced hydrogen release [47]. Consequently, the acid does act as a catalyst but as a reactant. Investigations of acid-induced hydrogen liberation encompassed solid acids, cation exchange resins (Dowex, Amberlyst, Nafion) in protonated form, and H-type zeolites [111]. Dowex and Amberlyst exhibit the highest activity, the reaction being complete in about 4 min at room temperature using a 0.33 wt.% aqueous AB solution and producing $3H_2$ per mol AB. Besides the fast kinetics, a special advantage of the acid-induced hydrogen release results from the excellent control possibilities. In the case of solid acids, however, a disadvantage will be that the solid acid material needs to be reprotonated or exchanged because it is not simply removed together with the spent fuel.

Alternatively, the hydrogen release rate can be accelerated by the use of metal catalysts. Various noble transition metals (Pt, Rh, Ru, Pd, and Au) on different supports (Al_2O_3, C, and SiO_2) have been tested for this purpose [112, 116, 117]. The highest catalytic activity was obtained with a 2 wt.% Pt/Al_2O_3 catalyst (mol Pt/mol AB = 0.018), resulting in the release of 3 mol H_2/mol AB from a 0.33 wt.% aqueous AB solution at room temperature within 45 s [116]. By increasing the AB concentration from 0.33 to 25 wt.%, the relative reaction rate, expressed as mol H_2/mol AB per minute, decreases, but the absolute hydrogen release rate, in molH_2/min, does not change significantly, indicating a high activity of the Pt/Al_2O_3 catalyst also at high AB concentrations [116]. The mol H_2/mol AB ratio reaches the same value of 3.0 at different AB concentrations, independent of the water excess (see Figure 8.2). For an aqueous AB solution of 25%, this ratio corresponds to an effective hydrogen storage density of 4.9 wt.%.

Efforts have been made to replace the noble transition metal catalyst with a cheaper on-noble transition metal catalyst such as Co, Ni, Cu, Fe [118].Generally the reaction rates are slower than with Pt. The most active variants of these catalysts (10 wt.% Co/Al_2O_3, 10 wt.% Co/C, and 10 wt.% Ni/Al_2O_3) liberate nearly 3 mol H_2/mol AB within 50–70 min at ambient temperature (1 wt.% aqueous AB solution, mol metal/mol AB = 0.018).

Xu and coworkers have demonstrated that even non-noble metals can exhibit very high catalytic activity for the AB hydrolysis reaction [119]. They found that with *in situ* synthesized amorphous Fe nanoparticles, the hydrolysis of AB (0.16 M aqueous AB solution, mol Fe/mol AB = 0.12) is complete within only 8 min. The as-prepared Fe-catalyst was reused up to 20 times with no obvious loss of activity. It was concluded

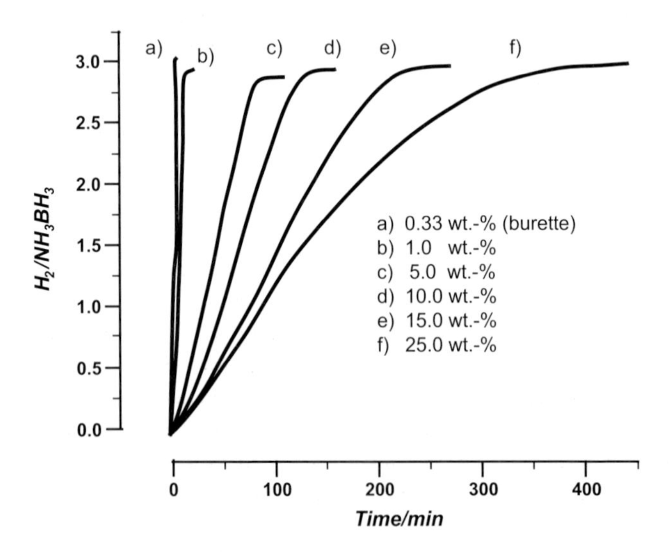

Figure 8.2 Concentration dependence of the H_2 release (from [116]) of Pt/Al_2O_3 catalyzed ammonia borane in aqueous solution.

that the amorphous character is essential to the high activity of Fe nanoparticles. A strong influence of crystallinity, particle size, and thus of the preparation conditions on its activity towards the hydrolysis of AB was also found by others [92, 118, 120]. To a certain degree the presented catalytic results can be expected to be applicable to AB-related compounds. Yoon and Sneddon have demonstrated that the hydrolysis of ammonia triborane $NH_3B_3H_7$ can be performed with concentrations up to 22.7 wt.% [121] using a 5 wt.% Rh catalyst at 21 °C, achieving a hydrogen release of 6.1 wt.% solution based.

Knowing the exact nature of the hydrolysis products will be an important requirement for the identification of the minimum amount of water still needed for a full conversion of the AB material without sacrificing too much of the usable hydrogen density. In contrast to most organic solvents, water is not only the AB carrier but also a reactant. To what degree depends on the products formed. Matthews and coworkers demonstrated that the hydrolysis of $NaBH_4$ with water vapor at 100 °C can be carried out with 100% conversion only for $H_2O/NaBH_4$ molar ratios larger than 4, since the hydrolysis product $NaBO_2{\cdot}2H_2O$ is formed instead of $NaBO_2$ [122–124]. At a ratio of $H_2O/NaBH_4 < 4$ a mixture of the unreacted $NaBH_4$ and $NaBO_2{\cdot}2H_2O$ is obtained and less hydrogen is released. Corresponding investigations for AB addressing the minimization of the required amount of water have not yet been performed. Depending on whether or not the product NH_4BO_2 exists only as a hydrate ($NH_4BO_2{\cdot}xH_2O$) at 100 °C, it may be necessary to use additional water compared to the amount suggested by Eq. (8.1) to liberate the 3 H_2 indicated there. This, in turn, determines whether or not hydrolysis appears beneficial over the well introduced system $NaBH_4/H_2O$. In most other concepts of hydrogen release from AB, the hydrogen carried by the nitrogen part is released together with one from the boron part, and, thus, contributes to the overall very high usable hydrogen density of AB. It therefore seems to be disadvantageous to apply a concept where the hydrogen of the nitrogen moiety remains unused. In this respect it may help to view the ammonia part just as a means to moderate the otherwise rather violent hydrolysis reaction of boranes.

8.6 Substituted Ammonia Boranes

AB-materials have gained attention as potential hydrogen storage materials primarily because of their high hydrogen content and the low release temperature. Other properties, however, are not very favorable, such as being a solid, being not rechargeable, and having a tendency to release diborane and ammonia – the latter being harmful to the catalytic materials of many fuel cells types. It may, therefore, be beneficial to trade, if possible, some of the hydrogen content for better performance in the other categories. For example, a weakening of the dihydrogen bond should be possible by substituting one of the hydrogen atoms located at the nitrogen atom by an electron-donating group or atom.

The number of substituted AB-compounds with one or more hydrogen atoms replaced is legion. A large number are well described in the Gmelin series, where

Table 8.5 Hydrogen release from substituted AB materials.

Substituted AB material	T_{Decomp} [°C]	H_2 equivalents[a)]	H_2 density[b)] [wt.%]	$\Delta_R H$[c)] [kJ mol^{-1}]
BH_3NH_2-Me[d)]	90–122	1.3	5.8	−75
BH_3NH_2-CH_2CH_2-NH_2BH_3[d)]	110–180	4	9.1	−50
BH_3NH_2-Li[e)]	91	2	11.0	−4
BH_3NH_2-Na[e)]	91	2	7.5	−4
BH_3NH_2-Ca-NH_2BH_3[f)]	170	3.6	7.2	3

a) mol H_2 per mol substituted AB material.
b) Hydrogen density of the material based on the H_2 equivalents released.
c) per mol substituted AB materials.
d) Data determined by the authors via coupled DSC/pressure measurements at a heating rate of 5 K min^{-1}.
e) Reported in [125].
f) Reported in [129].

synthesis protocols and even thermal decomposition data can be found. However, the number of potential candidates as hydrogen storage materials is greatly reduced because of storage density considerations. Obviously, materials with large and heavy substituents are excluded, but also those with more than one hydrogen atom substituted. Among possible candidates are methyl-substituted ammonia boranes. By applying the standard synthesis for laboratory quantities in which the ammonium salt is replaced by methyl ammonium chloride, BH_3NH_2Me is easily obtained. It releases hydrogen at temperatures lower than 120 °C (see Table 8.5). A particularly interesting feature is its low melting point, 50 °C, which may allow its transport in liquid form in a hydrogen generator. Our stability tests with respect to melting and recrystallization cycles (up to 10) did not reveal any material changes. At 90 °C the release of molecular hydrogen sets in. With respect to the construction of a practical hydrogen generator, the liquid state also introduces the problem of volatile species, the material itself, but also the dimer $(BH_2NHMe)_2$, the trimer $((BH_2NHMe)_3$, and trimethyl borazine $(BHNMe)_3$. In principle, the problem of the reduction of the hydrogen weight percentage by a substituent can be partly ameliorated by the creation of bridge substituted materials, such as $(BH_3NH_3)_2Mg$ and $(BH_3NH_3)_2Ca$, since the substituents are then shared by two BN units. So far, not many systems of this class suitable as potential hydrogen source materials have been investigated (Table 8.5).

Electron-donating substituents replacing the hydrogen atoms at the nitrogen center will lower the protic character of the remaining hydrogens at the nitrogen atom and thus reduce the dihydrogen bond strength. Similarly, electron-withdrawing groups (such as halides) at the boron atom, should have the same effect on the dihydrogen bond by reducing the hydridic character of the remaining hydrogens at the boron center. Gas phase DFT calculations for *N*-lithium-substituted AB material at the B3LYP/6-311++g(d,p) level resulted in a change in the sign of the Gibbs energy ($\Delta_R G = +6.0$ kJ mol^{-1}, $\Delta_R H = -6.0$ kJ mol^{-1} at 298 K) for the hydrogen release with dimer formation ($2BH_3NLiH_2 \rightarrow B_2N_2Li_2H_8 + H_2$) in comparison to the original AB case ($2BH_3NH_3 \rightarrow B_2N_2H_{10} + H_2$) ($\Delta_R G = -56.4$ kJ mol^{-1},

$\Delta_R H = -67.3\,\text{kJ mol}^{-1}$ at 298 K). The obtained value, which is close to zero, can be seen as a sign that the thermodynamic situation for the lithinated AB materials is improved with respect to a possible rehydrogenation of spent material, but of course for the polymeric solid, with the presence of dihydrogen bonds, the situation will be very different. So far, two independent reports on the preparation of *N*-lithium-substituted AB material by ball-milling of stoichiometric solid AB/LiH mixtures have been published [125, 126]. The created material displays improved kinetics and thermochemical properties for the dehydrogenation reaction compared to the parent AB material. A hydrogen release of 1.8 mol H_2/mol BH_3NH_2Li (10.5 wt.%) after 5 h at 100 °C [126] and of 2 mol H_2/mol BH_3NH_2Li (11 wt.%) after 19 h at 91 °C [125] was observed. In contrast, only 1 equiv H_2 can be extracted from neat AB at temperatures below 110 °C (see Section 8.4). As a positive side effect, it was noticed that no gaseous by-products, such as diborane or borazine, were emitted into the hydrogen stream. Another advantage is the reduced reaction enthalpy for the hydrogen release. Depending on the author, the corresponding $\Delta_R H$ values for 2 mol H_2/mol BH_3NH_2Li were estimated as –8 kJ/mol BH_3NH_2Li by [126] or as −3 to −5 kJ/mol BH_3NH_2Li by [125]. In comparison, the enthalpy for the dehydrogenation of AB to $(BHNH)_x$ is much more exothermic (<-36 kJ/mol AB) (see Section 8.4.1).

Since the material obtained by ball-milling is X-ray amorphous, the identification as BH_3NH_2Li is not straightforward. It was eventually carried out by by Xiong *et al.* by applying high resolution powder X-ray diffraction analysis [125].

Additional support for the successful BH_3NH_2Li formation is given by solid state ^{11}B-NMR spectra of the assumed product BH_3NH_2Li, which differs from those of the parent compound AB. The interpretation of these data, however, can also be difficult. For example, the AB-related compound $BH_3NH(CH_3)_2$ and its lithinated form $BH_3N(CH_3)_2Li$, the structure of the latter was unambiguously obtained by XRSD investigations [128], display identical ^{11}B-NMR chemical shifts and only differ in their coupling constants [129]. The same effect in the ^{11}B-NMR appearance, identical chemical shift but different coupling constants, is displayed by BH_3NH_3 and the calcium-substituted AB $(BH_3NH_2)_2Ca$ [127]. The authors of this article found an indication that these phenomena may also occur in BH_3NH_2Li. They synthesized BH_3NH_2Li according to the method of Myers [130] by reacting AB with butyllithium C_4H_9Li in Et_2O at 0 °C and realized that indeed the chemical shift stayed the same but the coupling constants shifted from 98 Hz for the parent AB to 84 Hz for the obtained product. It is remarkable that the BH_3NH_2Li product seems to be stable only in solution. All attempts to obtain it in pure form lead to the precipitation of an amorphous solid which is not soluble in Et_2O and thus cannot be BH_3NH_2Li. Whether this precipitate is similar to the products obtained by the AB/LiH ball-milling procedure is not yet known.

The sodium-substituted BH_3NH_2Na can be prepared by the reaction between AB and Na in liquid ammonia [4] and, in analogy to BH_3NH_2Li, by ball-milling of a solid AB/NaH mixture [125]. The dehydrogenation properties of BH_3NH_2Na are quite similar to those reported for BH_3NH_2Li [125] (see Table 8.5). The calcium-substituted AB, $(BH_3NH_2)_2Ca$, was synthesized by the reaction between AB and CaH_2 in THF [127]. The structure of the $(BH_3NH_2)_2Ca$ THF adduct was determined by XRSD. The dehydrogenation was estimated to be even endothermic (see Table 8.5). The authors of [127] noted that the amount of released hydrogen depends strongly on temperature, presumably indicating the occurrence of an equilibrium state. However, there is no clear evidence yet for the reversibility of the dehydrogenation process.

8.7 Recycling Strategies

The application of AB materials as hydrogen source materials requires a cost effective production or at least an economical regeneration procedure for spent material. As pointed out, the type of spent material depends very much on the dehydrogenation procedure. Decomposition of AB in the solid state and in inert solvents leads, depending on the conditions, to products of various B:N:H rations, divers degree of polymerization, and of very different reactivity.

Unless procedures are developed that generate a singular type of waste product, a generic recycling procedure that can process all, or at least most, of different types of spent materials is desirable.

Most of the tested BNH systems are considerably exothermic in respect to hydrogen release (see Table 8.6) and, therefore, cannot be recycled simply by exposure to high-pressure hydrogen gas at practically obtainable pressures. Some

Table 8.6 Experimental reaction enthalpy values and data calculated for some conceivable dehydrogenation processes[a)].

Reaction	$\Delta_R H$ [kJ mol^{-1}]
BH_3NH_3 (s) → $1/x$ $(BH_2NH_2)_x$ (s) + H_2 (g)	−22[b)]
$1/x$ $(BH_2NH_2)_x$ (s) → $1/x$ $(BHNH)_x$ (s) + H_2 (g)	−15 to −24[c)]
BH_3NH_3 (aq) + 3 H_2O (l) → $(NH_4)_3BO_3$ (aq) + 3 H_2 (g)	−179[d)]
BH_3NH_2Li (s) → $1/x$ $(BHNLi)_x$ (s) + 2 H_2 (g)	−4[e)]
$(BH_3NH_2)_2Ca$ (s) → $1/x$ $(B_2H_2N_2Ca)_x$ (s) + 4 H_2 (g)	+3.5[f)]
BH_3NH_3 (s) → 1/3 $(BH_2NH_2)_3$ (s) + H_2 (g)	−15[g)], −24[h)]
1/3 $(BH_2NH_2)_3$ (s) → 1/3 $(BHNH)_3$ (l) + H_2 (g)	−3[g)], −34[h)]

a) Reaction enthalpy values, which were measured using differential scanning calorimetry (DSC), correspond to the reaction equation in the left column, although the detected amount of the released hydrogen may deviate slightly from the stochiometric value.
b) Measured in [61] for thermal decomposition in the range 70–90 °C.
c) Measured in [63] for thermal decomposition in the range 120–170 °C, enthalpy value depends on the heating rate.
d) Measured in [47] for the Pt-catalyzed hydrolysis of aqueous AB at 25 °C.
e) Measured in [125].
f) Measured in [127].
g) Calculated using DFT [131].
h) Calculated using DFT [132].

modified AB-materials have been shown to be at or close to thermal neutrality, that is, a reaction enthalpy close to zero, which also should lower the free energy barrier for the direct regeneration via molecular hydrogen at high pressure (Table 8.6).

Besides the ability to cope with the very inhomogenous nature of the spent materials, an energy efficient procedure for the regeneration of BH_3NH_3 must be developed.

Ideally, typical production considerations, such as cost of the chemicals and the technological issues (temperature level, reaction rates etc.) should also be taken into account. Especially, any use of costly auxiliary agents that cannot be recycled or production of by-products that require their disposal needs to be avoided. In this respect, the most favorable recycling procedure would be self-contained, that is, the regeneration is only carried out with molecular hydrogen and the BNH waste. To date, none of the proposed recycling schemes fulfill all of these requirements.

8.7.1
Recycling from B-O-Containing Materials

Although hydrogen liberation via hydrolysis or alcoholysis from AB materials, which leads to the formation of B–O bonds, is not very favorable because of discarding hydrogen in the nitrogen moiety, there is, nevertheless, a strong interest in regenerating B–H bonds from B–O bond-containing materials because of the decomposition of metal borohydrides via hydrolysis [133, 134].

Obviously, in terms of the energy expense, no recycling procedure can be better than the primary hydrogen release reaction. In this respect the generation of B–H bonds from B–O-containing materials will be very energy costly, but the processes for these conversions are, nevertheless, the subject of investigation, especially in a semi-industrial setting [135–137]

The most common processes are:

1) Schlesinger process [138, 139]:

$$H_3BO_3 + 3CH_3OH \rightleftharpoons B(OCH_3)_3 + 3H_2O$$

$$4Na + B(OCH_3) + H_2 \rightleftharpoons NaBH_4 + 3NaOCH_3$$

   This process is the standard method for the production of sodium borohydride, which is a key component for the AB synthesis, from boric acid.

2) via elemental carbon as the reducing agent

   Millennium Cell, inc., a company that intends to market a sodium borohydride hydrolysis-based hydrogen delivery system for large scale automotive propulsion applications, proposed a process designed under energy efficiency aspects based on the utilization of carbon as reducing agent and the intermediate generation of halogenated boron species from B_2O_3 followed by a direct hydrogenation [135].

   Other patented procedures utilizing B_2O_3 or $B(OR)_3$ as intermediates as well as other metals than sodium are:

3) via B_2O_3 [140–142]

$$H_3BO_3 \xrightleftharpoons{260\,^\circ C} B_2O_3 + H_2O$$

$$B_2O_3 + 3H_2 + 2Al \xrightleftharpoons{1000\,^\circ C} B_2H_6 + Al_2O_3$$

Other reactants such as Fe, Mg, Ca, Na, B,C and so on, are also possible

4) via $B(OR)_3$ [143–149]

$$H_3BO_3 + 3ROH \xrightleftharpoons{260\,^\circ C} B(OR)_3 + 3H_2O$$

$$B(OR)_3 + 3H_2 + Al \xrightleftharpoons{120-300\,^\circ C} 1/2\,B_2H_6 + Al(OR)_3$$

or

$$B(OR)_3 + + AlR'_3 \rightleftharpoons BR'_3 + Al(OR)_3$$

$$BR'_3 + 3H_2 \xrightleftharpoons{200\,^\circ C, Pd} 1/2\,B_2H_6 + 3R'H$$

All these processes use either metals or carbon as reducing agents and eventually form strongly oxidized species. It is obvious that all of the metal-based recycling procedures are very energy costly, since in a self-contained recycling scheme these metals need to be regenerated from their oxides or alkoxides. It is therefore clear that the formation of boric acid (or other B–O bonds)-containing compounds during the hydrogen generation from borohydrides including AB-materials can only be accepted if energy aspects are not very important.

8.7.2
Recycling of BNHx-Waste Products

Because of the variety of possible dehydrogenation products a homogenization procedure to generate a well-defined starting material is desired. Several groups have proposed different homogenization (digestion) procedures as the starting point for possible AB recycling schemes. Their generic form can be described in four steps.

1) Digestion of spent material (B–X and ammonium formation)
2) Reduction of B–X bonds to form B–H bonds
3) Recovery of ammonia
4) Formation of AB materials by the ammoniation of $BH_3$with X being a halogen atom.

An early proposal by Sneddon applied CF_3COOH as digestion agent and used NMe3-AlH3 as hydride for the hydrogenation step. The development of this procedure was soon abandoned to avoid the formation of the strong B–O bonds.

Later the group worked on a digestion method utilizing $HBr/AlBr_3$ and forming B–Br bond materials that were subsequently treated with $HSiEt_3$ or $HSnBu_3$[2]. The latter concept exploits the fact that HBr, which by itself is only a weak acid in organic solvents, forms, in the presence of $AlBr_3$, a superacid [150]. The digestion method by Tumas utilizes dithiols leading to B–S bonds followed by hydrogenation via $HSi(OMe)_3)$ or HSnBu3.

The hydrogenation steps in all of these proposals are carried out by using a specific hydride donor, which exchanges the hydride for a halide ion during the B–X to B–H conversion. To arrive at a closed recycling scheme, it is eventually necessary to convert a spent hydride donor, that is, its halogenated form, back to the corresponding hydrides simply by treatment with molecular hydrogen. It would be a great achievement if this indirect hydrogenation could be saved and a direct hydrogenation route for converting the BX_3 species to BH_3 could be found, especially since it can be assumed that the more recycling steps there are the larger the energy expense will be.

These considerations suggest for the development of an energetically effective procedure, the realization of the direct hydrogenation of the boron species created in the digestion process with molecular hydrogen. Up to now very little work concerning the direct molecular hydrogenation of BX_3 species has been carried out.

The direct dehydrochlorination of BCl_3 was reported by Murib *et al.* at 600–700 °C using silver meshes as catalyst [151]:

$$BCl_3 + H_2 \rightleftharpoons BHCl_2 + HCl \tag{8.2}$$

A continuously operating pilot system was built in the1960s producing 1 kg $B_2H_6\,h^{-1}$ at 99% purity. The procedure includes the subsequent rapid quenching of the reaction mixture to a temperature of −196 °C, followed by the decomposition of $BHCl_2$ at −80 °C according to

$$6BHCl_2 \rightleftharpoons B_2H_6 + 4BCl_3 \tag{8.3}$$

The diborane gas is easily separated from the liquid BCl_3 and the latter is fed back to the primary process (8.2). The temperature of 750 °C is an approximate upper bound for reaction (8.2) because at higher temperatures the decomposition to elemental boron sets in. The process is, nevertheless, carried out at these high temperatures because of the thermodynamic limitations (see Table 8.7).

In order to carry out the hydrogenation under milder conditions, the thermodynamic situation needs to be altered by the use of an auxiliary agent that can be easily retrieved. Because amines easily form HCl adducts, they are a good choice for shifting the dehydrochlorination equilibrium. An experiment using NMe_3 was carried out by Taylor and Dewing, who reported the hydrodechlorination of BCl_3 at 200 °C at a pressure of 2000 bar with a 25% yield of BH_3NMe_3 [152]. Even under these harsh conditions, the conversion to BH_3NMe_3 seems remarkable, since no metal catalyst was used. It can be speculated, however, that the autoclave material

2) Results were obtained in the framework of the DOE Hydrogen Innovation Program. It can be expected that they will be published in the program's annual reports.

Table 8.7 $\Delta_R H^\circ$ and $\Delta_R G^\circ$ values for the three individual steps of BCl_3 hydrodechlorination calculated by DFT at the B3LYP/g-311 + + (d,p) level.

Reaction	$\Delta_R H^\circ$/kJ mol^{-1}	$\Delta_R G^\circ$/kJ mol^{-1}
$BCl_3 + H_2 \rightleftharpoons BHCl_2 + HCl$	+52.56	+43.15
$BHCl_2 + H_2 \rightleftharpoons BH_2Cl + HCl$	+72.19	+67.14
$BH_2Cl + H_2 \rightleftharpoons BH_3 + HCl$	+84.47	+81.38

may have provided some catalytic influence. The heterolytic splitting of molecular hydrogen without a metal center by main group elements has just recently become an active field of research [103, 104]. It can be expected that by choosing an appropriate catalyst and an optimized base the hydrodechlorination process can be carried out under much milder conditions. In this case, an idealized self-contained recycling scheme can be formulated (Scheme 8.6) [153].

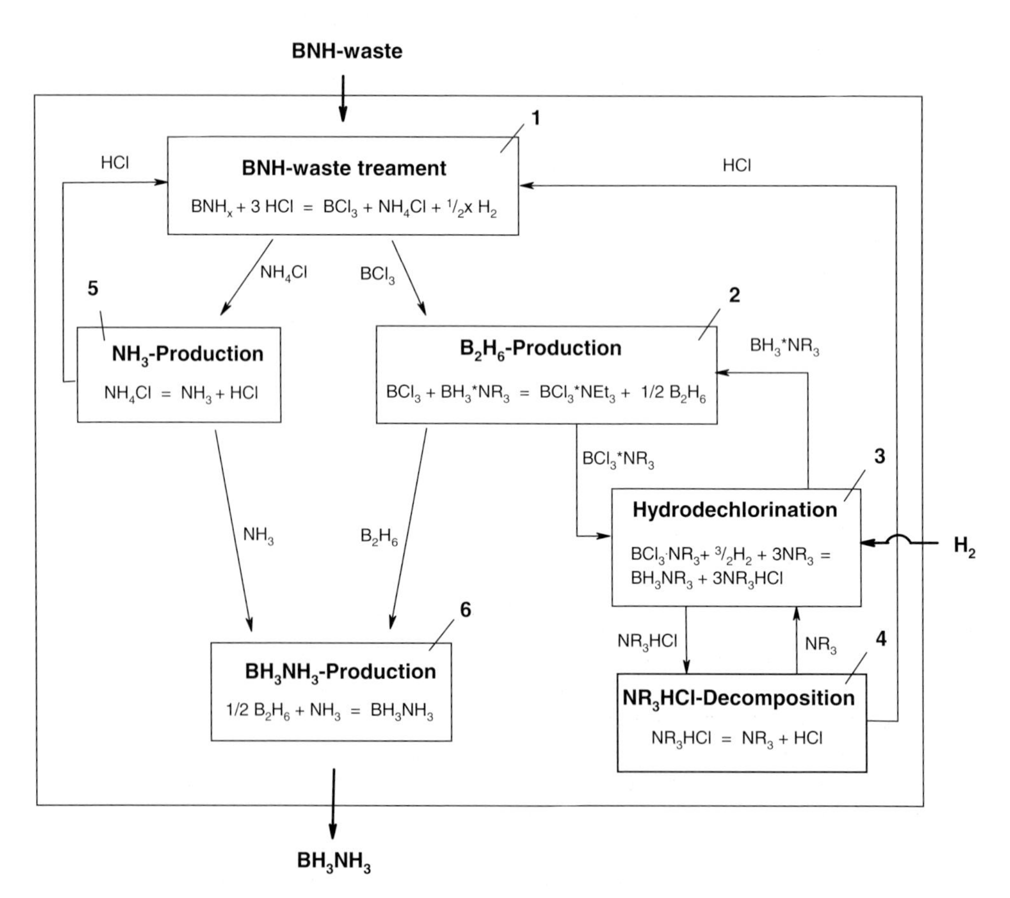

Scheme 8.6 Proposed elements of a self-contained recycling scheme.

The outlined recycling scheme, however, can only be seen as a cartoon indicating the route to a possible solution, since many challenges still need to be met in order to arrive at a practical recycling procedure. The fundamental problem of finding an adequate digestion method is still unsolved because, so far, only conversions of BNH waste to BX_3 species somewhat higher than 60% have been achieved. The inhomogeneous nature of the pyrolytically dehydrogenated material may even remain a block to this regeneration strategy. If so, steering the dehydrogenation reaction to a uniform product may be an alternative way around the problem. As outlined before, decomposition in organic solvents or molecular scaffolds may provide a possible means for doing so. Steering the dehydrogenation reaction to an almost stoichiometric conversion to borazine, for example, should initiate research activities for the development of a direct hydrogenation procedure to cyclo trimeric borazane (CTB), the BN analog of cyclohexane, and of digestion procedures with almost complete conversion, in conjunction with the corresponding recycling strategies, in order to arrive again at hydrogen-rich species.

With respect to the dehydrohalogenation of BX_3 species as a likely digestion product, much work remains to explore the wide range of options. Amines used as thermodynamic drivers by binding the dehydrohalogenation product HX can influence the reaction in different ways, depending not only on their base strength but also on their steric influences.

For example, if the BN bond were to break during the dehydrohalogenation process amine bases could be chosen that selectively capture a proton while the halogenated boron species remains uncoordinated due to steric hindrance. This concept is related to the recently published frustrated Lewis pair concept used for hydrogen activation by sterically demanding borane and amine species [104].

8.8 Summary

Ammonia borane as a very hydrogen-rich material with benign release properties has attracted wide attention as a possible hydrogen source material. It has been intensively investigated but remains, as BNH compounds in general, an active field of research. The hydrogen release can be carried out in different ways and the conditions chosen have a strong influence on the resulting product distribution. For example, the thermal dehydrogenation of solid ammonia borane preferentially leads to polymeric products while the non-hydrolytic decomposition in solution results primarily in cyclic compounds. Mechanistic studies concerning the corresponding decomposition pathways have been carried out and basic mechanistic elements are identified but, due to the large variety of possible systems, there remain many open questions, such as the relationship between the decomposition in diluted systems and that in the solid phase. Ultimately, it could be assumed that a deeper understanding of the decomposition mechanisms could be used to steer the dehydrogenation reactions to selected products and to arrive at a more uniform product distribution supporting different recycling strategies. Although, for the use of ammonia borane as hydrogen source material in large scale applications, the cost and the unsolved recycling problem may be prohibitive, small scale applications should, nevertheless, be further explored since the exceptionally high hydrogen content and its release at low temperatures make ammonia borane a unique material for this purpose.

References

1 Ferreira-Aparicio, P., Benito, M., and Sanz, J. (2005) *Catal. Rev. – Sci. Eng.*, **47** (4), 491–588.

2 Welch, G.C., San Juan, R.R., Masuda, J.D., and Stephan, D.W. (2006) *Science*, **314** (5802), 1124–1126.

3 Shore, S.G. and Parry, R.W. (1955) *J. Am. Chem. Soc.*, **77**, 6084–6085.

4 Shore, S.G. and Parry, R.W. (1958) *J. Am. Chem. Soc.*, **80**, 8–12.

5 Hu, M.G., van Paasschen, J.M., and Geanangel, R.A. (1977) *J. Inorg. Nucl. Chem.*, **39**, 2147–2150.

6 Mayer, E. (1973) *Inorg. Chem.*, **12**, 1954–1955.

7 Stafford, F.E., Pressley, G.A., and Baylis, A.B. (1968) *Adv. Chem. Ser.*, **72**, 137.

8 Mayer, E. (1972) *Inorg. Chem.*, **11**, 866–869.

9 Hu, M.G., Geanangel, R.A., and Wendlandt, W.W. (1978) *Thermochim. Acta*, **23**, 249–255.

10 Sorokin, V.P., Vesnina, B.V., and Klimova, N.S. (1963) *Zh. Neorg. Khim.*, **8**, 66–68.

11 Shaulov, Y.K., Shmyreva, G.O., and Tubyankaya, V.S. (1966) *Zh. Fiz. Khim.*, **40**, 122–124.

12 Briggs, T.S. and Jolly, W.L. (1975) *Inorg. Chem.*, **14**, 2267.

13 Shore, S.G. and Boeddeker, K.W. (1964) *Inorg. Chem.*, **3**, 914–915.

14 Storozhenko, P.A., Svitsyn, R.A., Ketsko, V.A., Buryak, A.K., and Ulyanov, A.V. (2005) *Zh. Neorg. Khim.*, **50**, 1066–1071.

15 Beres, J., Dodds, A., Morabito, A.J., and Adams, R.A. (1971) *Inorg. Chem.*, **10**, 2072–2074.

16 Lippert, L. and Lipscomb, W.N. (1956) *J. Am. Chem. Soc.*, **78**, 503–504.

17 Hoon, C.F. and Reynhardt, E.C. (1983) *J. Phys. C*, **16**, 6129–6136.

18 Buehl, M., Steinke, T., Schleyer, P.v.R., and Boese, R. (1991) *Angew. Chem. Int. Ed.*, **30**, 1160–1161.

19 Klooster, W.T., Koetzle, T.F., Siegbahn, P.E.M., Richardson, T.B., and Crabtree, R.H. (1999) *J. Am. Chem. Soc.*, **121**, 6337–6343.

20 Suenram, R.D. and Thorne, L.R. (1981) *Chem. Phys. Lett.*, **78**, 157–160.

21 Thorne, L.R., Suenram, R.D., and Lovas, F.J. (1983) *J. Chem. Phys.*, **78**, 167–171.

22 Dillen, J. and Verhoeven, P. (2003) *J. Phys. Chem. A*, **107**, 2570–2577.

23 Merino, G., Bakhmutov, V.I., and Vela, A. (2002) *J. Phys. Chem. A*, **106**, 8491–8494.

24 Hoon, C.F. and Reynhardt, E.C. (1983) *J. Phys. C*, **16**, 6129–6136.

25 Yang, J.B., Lamsal, J., Cai, Q., James, W.J., and Yelon, W.B. (2008) *Appl. Phys. Lett.*, **92**, 091916/1–3.

26 Reynhardt, E.C. and Hoon, C.F. (1983) *J. Phys. C*, **16**, 6137–6152.

27 Gunaydin-Sen, O., Achey, R., Dalal, N.S., Stowe, A., and Autrey, T. (2007) *J. Phys. Chem. B*, **111**, 677–681.

28 Penner, G.H., Chang, Y.C.P., and Hutzal, J. (1999) *Inorg. Chem.*, **38**, 2868–2873.

29 Cho, H., Shaw, W.J., Parvanov, V., Schenter, G.K., Karkamkar, A., Hess, N.J., Mundy, C., Kathmann, S., Sears, J., Lipton, A.S., Ellis, P.D., and Autrey, S.T. (2008) *J. Phys. Chem A*, **112**, 4277–4283.

30 Hess, N.J., Bowden, M.E., Parvanov, V.M., Mundy, C., Kathmann, S.M., Schenter, G.K., and Autrey, T. (2008) *J. Chem. Phys.*, **128**, 034508.

31 Wolf, G., van Miltenburg, J.C., and Wolf, U. (1998) *Thermochim. Acta*, **317**, 111–116.

32 Richardson, T.B., Gala, S.d., Crabtree, R.H., and Siegbahn, P.E.M. (1995) *J. Am. Chem. Soc.*, **117**, 12875–12876.

33 Custelcean, R. and Dreger, Z.A. (2003) *J. Phys. Chem. B*, **107**, 9231–9235.

34 Trudel, S. and Gilson, D.F.R. (2003) *Inorg. Chem.*, **42**, 2814–2816.
35 Merino, G., Bakhmutov, V.I., and Vela, A. (2002) *J. Phys. Chem. A*, **106**, 8491.
36 Epstein, L.M., Shubina, E.S., Bakhmutova, E.V., Saitkulova, L.N., Bakhmutov, V.I., Chistyakov, A.L., and Stankevich, I.V. (1998) *Inorg. Chem.*, **37**, 3013–3017.
37 Popelier, P.L.A. (1998) *J. Phys. Chem. A*, **102**, 1873–1878.
38 Li, J., Zhao, F., and Jing, F. (2002) *J. Chem. Phys.*, **116**, 25–32.
39 Kulkarni, S.A. (1999) *J. Phys. Chem. A*, **103**, 9330–9335.
40 Cramer, C.J. and Gladfelter, W.L. (1997) *Inorg. Chem.*, **36**, 5358–5362.
41 Morrison, C.A. and Siddick, M.M. (2004) *Angew. Chem. Int. Ed.*, **43**, 4780–4782.
42 Crabtree, R.H., Siegbahn, P.E.M., Eisenstein, O., Rheingold, A.L., and Koetzle, T.F. (1996) *Acc. Chem. Res.*, **29**, 348–354.
43 Alton, E.R., Brown, R.D., Carter, J.C., and Taylor, R.C. (1959) *J. Am. Chem. Soc.*, **81**, 3550–3551.
44 Weaver, J.R., Shore, S.G., and Parry, R.W. (1958) *J. Chem. Phys.*, **29**, 1–2.
45 Weaver, J.R. and Parry, R.W. (1966) *Inorg. Chem.*, **5**, 713–718.
46 Kusnesof, P.M., Shriver, D.V., and Stafford, F.E. (1968) *J. Am. Chem. Soc.*, **90**, 2557–2560.
47 Baumann, J. (2003) PhD thesis, University of Freiberg.
48 Parry, R.W. and Shore, S.G. (1958) *J. Am. Chem. Soc.*, **80**, 15–20.
49 Burg, A.B. (1947) *J. Am. Chem. Soc.*, **69**, 747–750.
50 Schlesinger, H.I. and Burg, A.B. (1938) *J. Am. Chem. Soc.*, **60**, 290–299.
51 Shore, S.G. and Parry, R.W. (1958) *J. Am. Chem. Soc.*, **80**, 12–15.
52 Parry, R.W., Schultz, D.R., and Girardot, P.R. (1958) *J. Am. Chem. Soc.*, **80**, 1–3.
53 Schultz, D.R. and Parry, R.W. (1958) *J. Am. Chem. Soc.*, **80**, 4–8.
54 Shore, S.G., Girardot, P.R., and Parry, R.W. (1958) *J. Am. Chem. Soc.*, **80**, 20–24.
55 Onak, T.P. and Shapiro, I. (1960) *J. Chem. Phys.*, **32**, 952.
56 Taylor, R.C., Schultz, D.R., and Emery, A.R. (1958) *J. Am. Chem. Soc.*, **80**, 27–30.
57 Parry, R.W., Kodama, G., and Schultz, D.R. (1958) *J. Am. Chem. Soc.*, **80**, 24–27.
58 Ramachandran, P.V. and Gagare, P.D. (2007) *Inorg. Chem.*, **46**, 7810–7817.
59 Sit, V., Geanangel, R.A., and Wendlandt, W.W. (1987) *Thermochim. Acta*, **113**, 379–382.
60 Geanangel, R.A. and Wendlandt, W.W. (1985) *Thermochim. Acta*, **86**, 375–378.
61 Wolf, G., Baumann, J., Baitalow, F., and Hoffmann, F.P. (2000) *Thermochim. Acta*, **343**, 19–25.
62 Baitalow, F., Baumann, J., Wolf, G., Jaenicke-Rößler, K., and Leitner, G. (2002) *Thermochim. Acta*, **391**, 159–168.
63 Baumann, J., Baitalow, F., and Wolf, G. (2005) *Thermochim. Acta*, **430**, 9–14.
64 Komm, R., Geanangel, R.A., and Liepins, R. (1983) *Inorg. Chem.*, **22**, 1684–1686.
65 Boeddeker, K.W., Shore, S.G., and Bunting, R.K. (1966) *J. Am. Chem. Soc.*, **88**, 4396–4401.
66 Pusatcioglu, S.Y., McGee, H.A. Jr., Fricke, A.L., and Hassler, J.C. (1977) *J. Appl. Polym. Sci.*, **21**, 1561–1567.
67 Kim, D.-P., Moon, K.-T., Kho, J.-G., Economy, J., Gervais, C., and Babonneau, F. (1999) *Polym. Adv. Technol.*, **10**, 702–712.
68 Lory, E.R. and Porter, R.F. (1973) *J. Am. Chem. Soc.*, **95**, 1766–1770.
69 Thompson, C.A. and Andrews, L. (1995) *J. Am. Chem. Soc.*, **117**, 10125–10126.
70 Baitalow, F., Wolf, G., Grolier, J.-P.E., Dan, F., and Randzio, S.L. (2006) *Thermochim. Acta*, **445**, 121–125.
71 Khvostov, V.V., Konyashin, I.Yu., Shouleshov, E.N., Babaev, V.G., and Guseva, M.B. (2000) *Appl. Surf. Sci.*, **157**, 178–184.
72 Konyashin, I.Yu., Aldinger, F., Babaev, V.G., Khvostov, V., Guseva, M.B., Bregadze, A., Baumgärtner, K.-M., and Räuchle, E. (1999) *Thin Solid Films*, **355–356**, 96–104.
73 Patwari, G.N. (2005) *J. Phys. Chem. A*, **109**, 2035–2038.
74 Smith, R.S., Kay, B.D., Schmid, B., Li, L., Hess, N., Gutowski, M., and Autrey, T. (2005) *Prepr. Symp. – Am. Chem. Soc., Div. Fuel Chem.*, **50**, 112–113.
75 Ryschkewitsch, G.E. and Wiggins, J.W. (1970) *Inorg. Chem.*, **9**, 314–317.

76 Stowe, A.C., Shaw, W.J., Linehan, J.C., Schmid, B., and Autrey, T. (2007) *Phys. Chem. Chem. Phys.*, **9**, 1831–1836.

77 Gerry, M.C.L., Lewis-Bevan, W., Merer, A.J., and Westwood, N.P.C. (1985) *J. Mol. Spectrosc.*, **110**, 153–163.

78 Sugie, M., Takeo, H., and Matsumura, C. (1979) *Chem. Phys. Lett.*, **64**, 573–575.

79 Kwon, C.T. and McGee, H.A. (1970) *Inorg. Chem.*, **9**, 2458–2461.

80 Harshbarger, W., Lee, G.H. II, Porter, R.F., and Bauer, S.H. (1969) *Inorg. Chem.*, **8**, 1683–1689.

81 Boese, R., Maulitz, A.H., and Stellberg, P. (1994) *Chem. Ber.*, **127**, 1887–1889.

82 Fazen, P.J., Remsen, E.E., Beck, J.S., Carroll, P.J., McGhie, A.R., and Sneddon, L.G. (1995) *Chem. Mater.*, **7**, 1942–1956.

83 Fazen, P.J., Beck, J.S., Lynch, A.T., Remsen, E.E., and Sneddon, L.G. (1990) *Chem. Mater.*, **2**, 96–97.

84 Wideman, T., Fazen, P.J., Lynch, A.T., Su, K., Remsen, E.E., Sneddon, L.G., Chen, T., and Paine, R.T. (1998) *Inorg. Synth.*, **32**, 232–242.

85 Wang, J.S. and Geanangel, R.A. (1988) *Inorg. Chim. Acta*, **148**, 185–190.

86 Stephens, F.H., Baker, R.T., Matus, M.H., Grant, D.J., and Dixon, D.A. (2007) *Angew. Chem. Int. Ed.*, **46**, 746–749.

87 Wideman, T. and Sneddon, L.G. (1995) *Inorg. Chem.*, **34**, 1002–1003.

88 Jaska, C.A., Temple, K., Lough, A.J., and Manners, I. (2001) *Chem. Comm.*, 962–963.

89 Jaska, C.A., Temple, K., Lough, A.J., and Manners, I. (2003) *J. Am. Chem. Soc.*, **125**, 9424–9434.

90 Jaska, C.A., Bartole-Scott, A., and Manners, I. (2003) *Dalton Trans.*, 4015–4021.

91 Jiang, Y. and Berke, H. (2007) *Chem. Comm.*, 3571–3573.

92 Clark, T.J., Russel, C.A., and Manners, I. (2006) *J. Am. Chem. Soc.*, **128**, 9582–9583.

93 Jaska, C.A., Temple, K., Lough, A.J., and Manners, I. (2004) *Phosphorus, Sulfur Silicon Relat. Elem.*, **179**, 733–736.

94 Chen, Y.S., Fulton, J.L., Linehan, J.C., and Autrey, T. (2005) *J. Am. Chem. Soc.*, **127**, 3254–3255.

95 Fulton, J.L., Linehan, J.C., Autrey, T., Balasubramanian, M., Chen, Y.S., and Szymczak, N.K. (2007) *J. Am. Chem. Soc.*, **129**, 11936–11949.

96 Denney, M.C., Pons, V., Hebden, T.J., Heinekey, D.M., and Goldberg, K.I. (2006) *J. Am. Chem Soc.*, **128**, 12048–12049.

97 Paul, A. and Musgrave, C.B. (2007) *Angew. Chem. Int. Ed.*, **46**, 8153–8156.

98 Keaton, R.J., Blacquire, J.M., and Baker, R.T. (2007) *J. Am. Chem. Soc.*, **129**, 1844–1845.

99 Yang, X., Stern, C.L., and Marks, T.J. (1991) *J. Am. Chem. Soc.*, **113**, 3623–3625.

100 Yang, X., Stern, C.L., and Marks, T.J. (1994) *J. Am. Chem. Soc.*, **116**, 10015–10031.

101 Piers, W.E. and Chievers, T. (1997) *Chem. Soc. Rev.*, **26**, 345–354.

102 Erker, G. (2005) *Dalton Trans.*, 1883–1890.

103 Sumerin, V., Schulz, F., Nieger, M., Leskelä, M., Repo, T., and Rieger, B. (2008) *Angew. Chem. Int. Ed.*, **47**, 6001–6003.

104 Welch, G.C. and Stephan, D.W. (2007) *J. Am. Chem. Soc.*, **129**, 1880–1881.

105 Denis, J.-M., Forintos, H., Szelke, H., Toupet, L., Pham, T.-N., Madec, P.-J., and Gaumont, A.-C. (2003) *Chem. Commun.*, 54–55.

106 Blackwell, J.M., Sonmor, E.R., Scoccitti, T., and Piers, W.E. (2000) *Org. Lett.*, **2**, 3921–3923.

107 Gutowska, A., Li, L., Shin, Y., Wang, C.M., Li, X.S., Linehan, J.C., Smith, R.S., Kay, B.D., Schmid, B., Shaw, W.J., Gutowski, M., and Autrey, T. (2005) *Angew. Chem. Int. Ed.*, **44**, 3578–3582.

108 Feaver, A., Sepehri, S., Shamberger, P., Stowe, A., Autrey, T., and Cao, G. (2007) *J. Phys. Chem. B*, **111**, 7469–7472.

109 Sepehri, S., Feaver, A., Shaw, W.J., Howard, C.J., Zhang, Q., Autrey, T., and Cao, G. (2007) *J. Phys. Chem. B*, **111**, 14285–14289.

110 Bluhm, M.E., Bradley, M.G., Butterick, R. III, Kusari, U., and Sneddon, L.G. (2006) *J. Am. Chem. Soc.*, **128**, 7748–7749.

111 Chandra, M. and Xu, Q. (2006) *J. Power Sources*, **159**, 855–860.

112 Chandra, M. and Xu, Q. (2006) *J. Power Sources*, **156**, 190–194.

113 Kelly, H.C. and Marriott, B. (1979) *Inorg. Chem.*, **18**, 2875–2878.
114 Kelly, H.C. and Underwood, J.A. (1969) *Inorg. Chem.*, **8**, 1202–1204.
115 Ryschkewitsch, G.E. (1960) *J. Am. Chem. Soc.*, **82**, 3290–3294.
116 Chandra, M. and Xu, Q. (2007) *J. Power Sources*, **168**, 135–142.
117 Xu, Q. and Chandra, M. (2007) *J. Alloys Compd.*, **446–447**, 729–732.
118 Chandra, M. and Xu, Q. (2006) *J. Power Sources*, **163**, 364–370.
119 Yan, J.M., Zhang, X.B., Han, S., Shioyama, H., and Xu, Q. (2008) *Angew. Chem., Int. Ed.*, **47**, 2287–2289.
120 Cheng, F.Y., Ma, H., Li, Y.M., and Chen, J. (2007) *Inorg. Chem.*, **46**, 788–794.
121 Yoon, C.W. and Sneddon, L.G. (2006) *J. Am. Chem. Soc.*, **128**, 13992–13993.
122 Aiello, R., Sharp, J.H., and Matthews, M.A. (1999) *Int. J. Hydrogen Energy*, **24**, 1123–1130.
123 Marrero-Alfonso, E.Y., Gray, J.R., Davis, T.A., and Matthews, M.A. (2007) *Int. J. Hydrogen Energy*, **32**, 4717–4722.
124 Marrero-Alfonso, E.Y., Gray, J.R., Davis, T.A., and Matthews, M.A. (2007) *Int. J. Hydrogen Energy*, **32**, 4723–4730.
125 Xiong, Z.T., Yong, C.K., Wu, G.T., Chen, P., Shaw, W., Karkamkar, A., Autrey, T., Jones, M.O., Johnson, S.R., Edwards, P.P., and David, W.I.F. (2008) *Nat. Mater.*, **7**, 138–141.
126 Kang, X.D., Fang, Z.Z., Kong, L.Y., Cheng, H.M., Yao, X.D., Lu, G.Q., and Wang, P. (2008) *Adv. Mater.*, **20**, 2756–2759.
127 Diyabalanage, H.V.K., Shrestha, R.P., Semelsberger, T.A., Scott, B.L., Bowden, M.E., Davis, B.L., and Burrell, A.K. (2007) *Angew. Chem., Int. Ed.*, **46**, 8995–8997.
128 Nöth, H., Thomas, S., and Schmidt, M. (1996) *Chem. Ber.*, **129**, 451–458.
129 Fischer, G.B., Fuller, J.C., Harrison, J., Alvarez, S.G., Burkhardt, E.R., Goralski, C.T., and Singaram, B. (1994) *J. Org. Chem.*, **59**, 6378–6385.
130 Myers, A.G., Yang, B.H., and Kopecky, D.J. (1996) *Tetrahedron Lett.*, **37**, 3623–3626.
131 Matus, M.H., Anderson, K.D., Camaioni, D.M., Autrey, S.T., and Dixon, D.A. (2007) *J. Phys. Chem. A*, **111**, 4411–4421.
132 Miranda, C.R. and Ceder, G. (2007) *J. Chem. Phys.*, **126**, 184703/1–11.
133 Zhang, Q., Smith, G.M., and Wu, Y. (2007) *Int. J. Hydrogen Energy*, **32**, 4731–4735.
134 Amendola, S.C., Sharp-Goldman, S.L., Janjua, M.S., Spencer, N.C., Kelly, M.T., Petillo, P.J., and Binder, M. (2000) *Int. J. Hydrogen Energy*, **25**, 969–975.
135 Amendola, S.C., Kelly, .M.T., and Wu, Y. (2002) PCT Int. Appl. WO 2002083551.
136 Wu, Y., Brady, J.C., Kelly, M.T., Ortega, J.V., and Snover, J.L. (2003) *Prepr.Symp. – Am. Chem. Soc., Div. Fuel Chem.*, **48**, 938–939.
137 Snover, J. and Wu, Y. (2004) US Patent 6706909
138 Banus, M.D. and Bragdon, R.W. (1955) US Patent 2720444.
139 Fedor, W.S., Banus, M.D., and Ingalls, D.P. (1957) *Ind. Eng. Chem.*, **49**, 1664–1672.
140 Muettertis, E.L. (1962) US Patent 3019086
141 Clark, C.C., Kanda, F.A., and King, A.J. (1962) US Patent 3022138.
142 Ford, T.A., Kalb, G.H., McClelland, A.L., and Muettertis, E.L. (1964) *Inorg. Chem.*, **3**, 1032–1035.
143 Ashby, E.C. and Podall, H.E. (1964) US Patent 3161469
144 Hunt, M.W., Carter, L.M., and Tillman, R.M. (1965) US Patent 3219412
145 Klein, R., Nadeau, H.G., Schoen, L.J., and Bliss, A.D. (1960) US Patent 2946664.
146 Klein, R., Bliss, A., Schoen, L., and Nadeau, H.G. (1961) *J. Am. Chem. Soc.*, **83**, 4131–4134.
147 Köster, R. (1960) GB Patent 854919.
148 Köster, R. (1957) *Angew. Chem.*, **69**, 94.
149 Köster, R., Bruno, G., and Binger, P. (1961) *Liebigs Ann. Chem.*, **644**, 1–22.
150 Farcasiu, D., Fisk, S.L., Melchior, M.T., and Rose, K.D. (1982) *J. Org. Chem.*, **47**, 453–457.
151 Murib, J.H., Horvitz, D., and Bonecutter, C.A. (1965) *Ind. Eng. Chem.*, **4**, 273–280.
152 Taylor, F.M. and Dewing, J. (1963) U.S. Patent, US 3 103 417.
153 Hausdorf, S., Baitalow, F., Wolf, G., and Mertens, F.O.R.L. (2008) *Int. J. Hydrogen Energy*, **33**, 608–614.

9
Aluminum Hydride (Alane)

Ragaiy Zidan

9.1
Introduction

Aluminum hydride is a covalently bonded hydride that exists in a metastable state as a solid at room temperature. It is a trihydride material (AlH_3) with three hydrogen atoms to every aluminum atom, giving rise to a large gravimetric and volumetric hydrogen capacity (10.1 wt% and 149 kg m^{-3}, respectively), which makes it an attractive material to be used in hydrogen storage. Discovering efficient and economic means for storing hydrogen is critical to realizing the hydrogen economy. Extensive research work has been conducted, worldwide, to develop new materials or revisit known materials in an effort to achieve the material properties needed for hydrogen storage. Unfortunately, the majority of these compounds fail to fulfill the thermodynamic and kinetic requirements for an onboard storage system. However, aluminum hydride not only has the required gravimetric and volumetric density but also has thermodynamic and kinetic conditions of operation compatible with the use of waste heat from PEM fuel cells or hydrogen burning internal combustion engines.

The main drawback to using aluminum hydride in hydrogen storage applications is its unfavorable hydriding thermodynamics. The direct hydrogenation of aluminum to aluminum hydride requires over 10^4 MPa of hydrogen pressure at room temperature. The impracticality of using high hydriding pressure has precluded aluminum hydride from being considered as a reversible hydrogen storage material. Alternate methods of making aluminum hydride are using chemical synthesis that typically produces stable metal halide byproducts that make it practically unfeasible to regenerate the aluminum hydride. However, new economical methods and novel techniques have been perused by a number of researchers to reversibly form aluminum hydride and form aluminum hydride adducts. These methods will be discussed in this chapter. Aluminum adducts are also of great interest because they can be used as hydrogen storage compounds, for the formation of derivatives of other metal complexes and as precursors for chemical vapor deposition of aluminum meta1, especially in the electronic and optical industries. It is important to note that in

Handbook of Hydrogen Storage. Edited by Michael Hirscher

ISBN: 978-3-527-32273-2

this chapter the terms aluminum hydride, alane, AlH_3 and aluminum trihydride are used depending on the discussed source publication and reference, but all three expressions are synonyms.

9.2 Hydrogen Solubility and Diffusivity in Aluminum

Understanding the solubility of hydrogen in metals is important because the formation of a metal hydride from the elements starts with hydrogen solution in the metal. In the aluminum industry, however, hydrogen solubility can be a problem for aluminum and aluminum alloys casting. The growth of the aluminum industry has been based on aluminum castings since its beginning in the late nineteenth century. Commercial aluminum products, from cooking utensils to airplanes, are modern daily life products, exploiting the novelty of the aluminum metal. Casting processes were developed to mass produce aluminum and aluminum alloy products. The technology of molten metal processing, solidification, and properties control has advanced to a wide range of applications. Hydrogen is the only gas that is considerably soluble in aluminum and its alloys, for this reason the solubility of hydrogen in aluminum has been studied. Hydrogen solubility is directly a function of temperature and the square root of pressure. When molten aluminum starts to cool and solidify, dissolved hydrogen in excess of the extremely low solid solubility will precipitate in molecular form, resulting in the formation of primary and/or secondary voids. There are several sources of hydrogen from which hydrogen can dissolve in aluminum during casting. Moisture in the atmosphere dissociates at the molten metal surface, resulting in concentration of atomic hydrogen that can diffuse into the melt. Although the barrier oxide of aluminum resists hydrogen solution by this mechanism, the disturbances of the melt surface during casting breaks the oxide barrier, allowing rapid hydrogen dissolution. Alloying elements such as magnesium can affect hydrogen solution by forming oxidation reaction products that offer reduced resistance to the diffusion of hydrogen into the melt. The disposition of hydrogen in a solidified structure depends on hydrogen concentration and the solidification conditions. Hydrogen porosity occurs as a result of diffusion-controlled nucleation and growth. Therefore, by decreasing the hydrogen concentration and increasing the rate of solidification, void formation and growth can be suppressed. That explains why, when castings are made in expendable mold processes, they are more susceptible to hydrogen-related defects than when products are produced using permanent mold or pressure die casting.

Hydrogen solubility in solid aluminum has also been studied from the perspective of hydride formation. The solubility of hydrogen in a metal is essential for nucleation and growth to occur, followed by hydride formation. The solubility of hydrogen in metals cited by Smith [1] can be expressed as

$$\ln S = A-(B/T) + \frac{1}{2}\ln P_{H_2} \qquad (9.1)$$

Smith's expression is a combination of solubility's laws; at isothermic conditions based on Sievert's law, $S_T = kP^{1/2}{}_{H2}$, and under the solubility isobaric condition by Borelius law, $S_p = A\exp(-b/T)$ [2]. The solubility of hydrogen in aluminum has been studied for years, starting with the early work conducted by the Ransley and Neufeld group [3], the Grant and Opie group [4], Sharov [5], and Eichenauer *et al.* [6]. These groups found the solubility of hydrogen in aluminum to be very small and hydrogen diffusion very slow. They reported that the solubility of hydrogen at room temperature is negligible and it increases at higher temperatures becoming relatively significant when aluminum melts.

The results obtained by Ransley and Neufeld were expressed in terms of Eq. (9.1) as:

$$\log S\,(\mathrm{cm}^3/100\,\mathrm{g\,Al}) = -0.652 - 2080/T + 1/2\log P_{\mathrm{mm}}(\text{for solid Al})$$

$$\log S(\mathrm{cm}^3/100\,\mathrm{g\,Al}) = 1.356 - 2760/T + 1/2\log P_{\mathrm{mm}}(\text{for liquid Al})$$

The results obtained by Eichenauer *et al.* were expressed in terms of Eq. (9.1) as:

$$\log S(\mathrm{cm}^3/100\,\mathrm{g\,Al}) = 0.521 - 3042/T + 1/2\log P_{\mathrm{mm}}(\text{for solid Al})$$

$$\log S(\mathrm{cm}^3/100\,\mathrm{g\,Al}) = 1.356 - 3086/T + 1/2\log P_{\mathrm{mm}}(\text{for liquid Al})$$

The reported results on the solubility of hydrogen, in terms of cm^3 of hydrogen per 100 g, reported by the above groups were similar. They found that the solubility can change by an order of magnitude as the temperature increases. For example going from 10^{-8} at 300 °C to 10^{-7} at 600 °C and becoming 10^{-5} when the aluminum melted.

These results were obtained, using hydrogen absorption into pure aluminum between 300 and 1050 °C and were calculated by temporal process degassing. Eichenauer *et al.* also reported that the square root Sievert's law is satisfied within the limit of error in the solid and in the liquid state. Utilizing the fact that diffusion in the metal is responsible for the hydrogen degassing rate of the solid aluminum they deduced the diffusion coefficient to be $D = 0.11\exp(-9780/RT)\ \mathrm{cm}^2\,\mathrm{s}^{-1}$. The heat of solution was calculated from these measurements by Birnbaum *et al.* [7] and Ichimura *et al.* [8] and found to be in the range of +0.6 to +0.7 eV.

Later, Hashimoto and Kino [9] measured hydrogen diffusion in aluminum by the permeation method, quadrupole mass spectrometry (temperature range 300–400 °C), and electrochemical methods at room temperature. An interesting finding from this work was that the room temperature diffusivity was considerably higher than extrapolated from high temperature measurements. This discrepancy was attributed to a change in vacancy concentration as a function of temperature. Dissolved hydrogen migrates through interstitial sites at room temperature where the hydrogen concentration is much higher than that of the vacancies. However, at high temperatures the number of vacancies becomes large enough to couple with almost all of the dissolved hydrogen and then hydrogen diffuses together with a vacancy due to a large binding energy between hydrogen and the vacancy. The combination of hydrogen–vacancy would have a larger diffusivity than a vacancy alone as result of lattice vibration enhancement around the vacancy due to the existence of hydrogen.

The diffusivity of hydrogen in aluminum was also studied, using first-principle calculations, by Wolverton *et al.* [10]. They calculated the heat of solution of H in preferred tetrahedral sites to be a large positive value (0.71 eV), consistent with experimental solubility data. Their calculations of diffusion took into account the influence of vacancies on diffusion and the binding energy between hydrogen and vacancy. They performed the calculation for the H/vacancy in several different positions (i) tetrahedral H, nearest neighbor to vacancy and (ii) octahedral H, nearest neighbor to vacancy, (iii) octahedral H, next nearest neighbor to vacancy and (iv) substation H (on the vacancy site). The results indicated that, even in the presence of vacancies, the interstitial tetrahedral positions are favored over the substitutional positions. However, they found that atomic relaxation and anharmonic vibrational effects play a significant role in the preferred sites. In the case of unrelaxed static calculation hydrogen prefers the larger octahedral sites. However, atomic relaxation is larger in tetrahedral sites and when invoked it overwhelms the static site preference and the tetrahedral sites become favored. The vibrational frequency in a tetrahedral site is larger than in an octahedral site. When all these effects were added up in this calculation a preference for hydrogen in tetrahedral sites was predicted and was found to be consistent with experimental data.

9.3 Formation and Thermodynamics of Different Phases of Alane

Aluminum hydride (alane) has proved to be a very attractive compound for many applications, such as in organic chemistry, and has been used as a rocket fuel for its high hydrogen capacity. Several groups reported that alane can be formed as different polymorphs: α-alane, α′-alane, β-alane, δ-alane, ε-alane, θ-alane, and γ-alane. α-Alane is found to be the most thermally stable polymorph. β-Alane and γ-alane are the two other common types of alane. β-Alane and γ-alane are transformed into α-alane upon heating. δ, ε, and θ-Alane have been reported to be produced in different crystallization condition. δ, ε, and θ-Alane were described to be less thermally stable and they do not convert into α-alane upon heating. However, it is important to note that only α-alane, α'-alane, β-alane and γ-alane have been synthetically reproduced. Until its first synthesis, published in 1947 by Finholt, Bond and Schlesinger [11], aluminum hydride (or alane) was known as an impure solvated solid complexed with amine or ether [12]. The reaction reported by Finholt *et al.* involved ethereal reaction of lithium aluminum hydride ($LiAlH_4$) and $AlCl_3$. The reaction yields an ether solution-aluminum hydride, AlH_3, Et_2O, after precipitation and filtration of lithium chloride.

$$3\,LiAlH_4 + AlCl_3 \xrightarrow{\text{ether}} 4\,AlH_3 \cdot Et_2O + 3\,LiCl \tag{9.2}$$

$$3\,LiAlH_4 + AlCl_3 \xrightarrow{\text{ether}} 4\,AlH_3 \cdot n[(C_2H_5)_2O] + 3\,LiCl$$

The ether solution of aluminum hydride, $AlH_3{\cdot}Et_2O$, requires immediate use or processing because aluminum hydride etherate solutions are known to degrade over a few days. Caution should be taken when attempting to remove the ether from the complex since the product can decompose into ether, aluminum and hydrogen. The nonsolvated aluminum hydride may be prepared by precipitation of the solid hydride etherate, $AlH_3{\cdot}Et_2O$, from the solution, by adding pentane or ligroin, followed by a vacuum treatment of the precipitate to remove the diethyl ether [13, 14]. Aluminum hydride produced by this procedure was usually obtained as partially desolvated product. It is important to note that the solid isolated at first is not alane but aluminum hydride etherate, and needs to be converted into alane. The crucial issue is separating nonsolvated alane from the $AlH_3{\cdot}Et_2O$, without the formation of decomposition products ether, aluminum, and hydrogen.

The preparation of crystalline phases of nonsolvated aluminum hydride by desolvation of $A1H_3$ etherate was reported by Brower *et al.* α-AlH_3 phase, was prepared both from the solid metastable phases and by crystallizing directly from a refluxing diethyl ether–benzene solution in what is known as the DOW method, published in 1975 as a contribution from The Dow Chemical Company [15]. Brower *et al.* explained that diethyl etherate can be removed, without decomposition, to a nonsolvated phase only in the presence of excess $LiAlH_4$. They prepared the nonsolvated metastable β- or γ-phases by heating solid AlH_3 etherate mixed with excess $LiBH_4$ or $LiAlH_4$ *in vaccuo*. These metastable phases, upon further heating, are converted to the more stable α-phase. The α- AlH_3 phase can be crystallized from a diethyl ether and benzene solution if conditions are carefully controlled. Brower *et al.* also reported that crystalline δ- and ε-AlH_3 can form in the presence of traces of water. They claimed the nonsolvated phases, α', δ, and ε did not convert to α-AlH_3 and are much less thermally stable than the α- phase. Ethereal solutions of AlH_3 containing excess $LiAlH_4$ when heated between 70 and 80 °C yield a nonsolvated AlH_3 phase they designated as the α'-AlH_3, structure [16, 17]. In their work they reported the formation of an ξ-AlH_3 phase that is either nonsolvated or slightly solvated and was prepared from hot di-*n*-propyl ether. This phase was reported not to convert to α -AlH_3 and to be less thermally stable. Again, only α-alane, α'-alane, β-alane and γ-alane phases have been confirmed/reproduced by others.

More recently, the preparation and thermodynamics of different aluminum hydride phases (α -AlH_3, β-AlH_3, and γ-AlH_3), prepared by means of the DOW method, were reported by Graetz *et al.* Hydrogen capacities approaching 10 wt% at desorption temperatures less than 100 °C have been demonstrated with freshly prepared AlH_3 [18]. Alane was synthesized using ethereal reaction of $AlCl_3$ with $LiAlH_4$ to produce an etherated compound of aluminum hydride, $AlH_3{\cdot}0.3$-$[(C_2H_5)_2O]$. The removal of the attached ether from the complex was achieved by heating the solvated AlH_3 in the presence of a complex metal hydride (e.g., $LiAlH_4$ and $LiBH_4$ depending on the desired phase) under vacuum. The synthesis is described to be extremely sensitive to the desolvation conditions (e.g., temperature and time). A small alteration can lead to the precipitation of a different AlH_3 polymorph phase. γ -AlH_3 is formed in the presence of excess $LiAlH_4$, while β-AlH_3 forms in the presence of excess $LiAlH_4$ and $LiBH_4$. Importantly noted is that freshly prepared nonpassivated AlH_3 is pyrophoric, reacts violently in water and must be treated with caution.

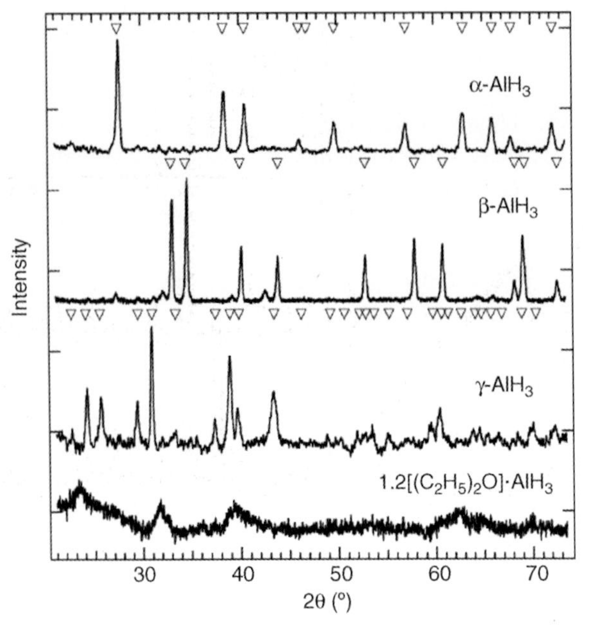

Figure 9.1 Powder XRD of AlH_3 etherate ([$(C_2H_5)_2O$]. AlH_3), α-AlH_3, β-AlH_3, and γ-AlH_3 prepared by organometallic synthesis. Markers indicate expected peak positions for each phase, Ref. [18].

The identification of these phases is depicted by their structures, as shown in Figure 9.1. The positions of the peaks (markers on graph) are compared to structures reported by Brower *et al.* [15].

Graetz *et al.*'s results show that at around 100 °C the decomposition of the β- and γ-polymorphs occurs by an initial phase transition to the α-polymorph followed by decomposition of the α-phase. Using a scanning calorimeter, the total heat evolved during the transition from β- to α-phase is found to be 1.5 ± 0.4 kJ/mol.AlH_3 and 2.8 ± 0.4 kJ/mol·AlH_3 for the γ- to α-phase transition [19]. The results of *ex situ* X-ray diffraction of β- and γ-AlH_3, during thermal treatment, have shown that the decomposition occurs via a transition to the more stable α-phase at around 100 °C followed by decomposition of the α-phase to Al and H_2. The enthalpies of phase transformations of the polymorph are depicted in Figures 9.2 and 9.3. The decomposition enthalpy for α-AlH_3 was determined to be approximately −10 kJ/mol.AlH_3, which is consistent with other experimental and calculated results [19].

Kato *et al.* also described the formation of AlH_3-etherate and its desolvation reaction into AlH_3 of different phases and the thermal relationship between these phases [20]. AlH_3-etherate was prepared based on the methods in [15]. Kato *et al.* identified the XRD profile of the powder to be AlH_3-etherate, as shown in Figure 9.4. This work was focused on identifying different phases of alane, phase transformations and the stabilities of these phases. Their characterization effort started with the AlH_3-etherate, as shown in Figure 9.4 in order to identify the progression of alane at different stages.

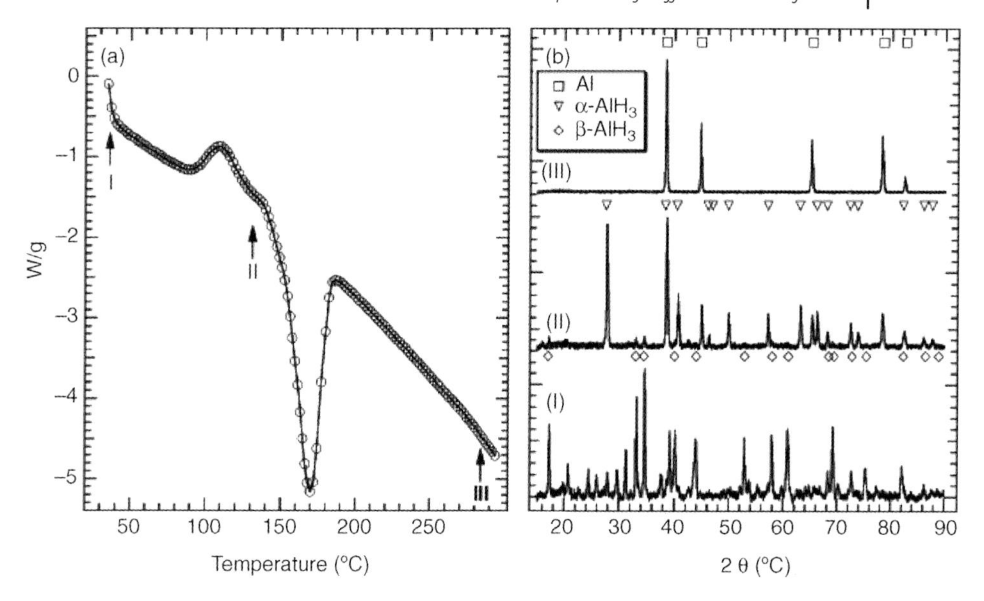

Figure 9.2 (a) Differential scanning calorimetry plot of β-alane in the temperature range 35–300 °C ramped at a rate of 10 °C min^{-1}; (b) *ex situ* diffraction patterns acquired at room temperature. (I) Before thermal treatment; (II) after a temperature ramp to 130 °C; (III) after a temperature ramp to 300 °C, Ref. [19].

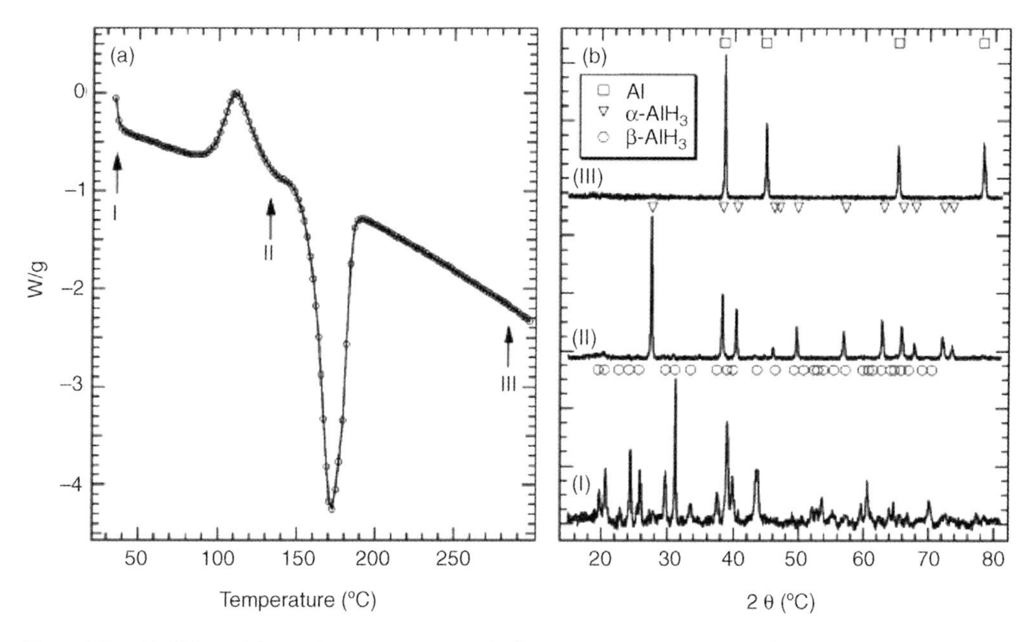

Figure 9.3 (a) Differential scanning calorimetry plot of γ-AlH_3 in the temperature range 35–300 °C ramped at a rate of 10 °C min^{-1}; (b) *Ex situ* diffraction patterns acquired at room temperature. (I) Before thermal treatment; (II) after a temperature ramp to 130 °C; (III) after a temperature ramp to 300 °C, Ref. [19].

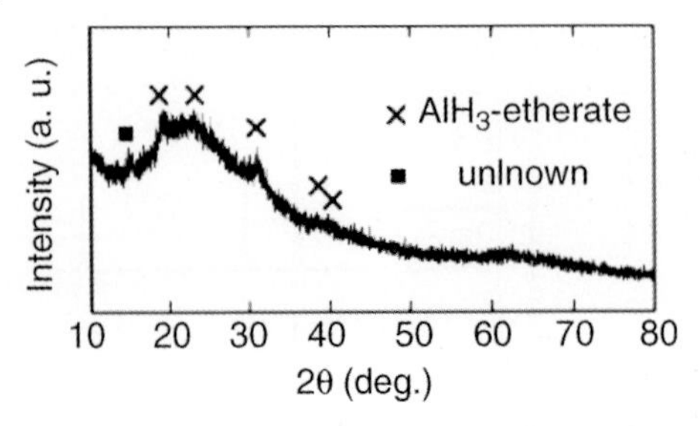

Figure 9.4 Powder X-ray diffraction identified as AlH_3-etherate Ref. [20].

They reported the presence of γ-AlH_3, $(C_2H_5)_2O$ and $LiAlH_4$ based on Raman modes, as shown in Figure 9.5. The data were compared and found to be in good agreement with previous work [21, 22].

Phases and products resulting from heat treatment of AlH_3-etherate in *ex situ* measurements of XRD and Raman spectrum at temperatures 77, 100, 132, and 172 °C were identified in this work. The XRD profiles in Figure 9.6 show the presence of γ-AlH_3 and metallic Al at 132 °C (Temp. C) and an increase in the amount of aluminum at 172 °C (Temp. D), respectively. The Raman spectra in Figure 9.7 correspond to the XRD profiles (excluding the metallic Al). At 100 °C (Temp. B) only the CH modes of ether (around 3000 cm^{-1}) were weakened, indicating that AlH_3-etherate was partially desolvated into γ-AlH_3. At 132 °C (Temp. C) two modes were observed in the range 500–800 cm^{-1} and they were attributed to the presence of the α-AlH_3 phase.

A detailed decomposition process was observed by *in situ* XRD measurements at 76 and 87 °C which depicts the conditions of phase transformation and decomposition. The *in situ* X-ray diffraction measurements showed that AlH_3-etherate formed γ-AlH_3 upon desolvation with an endothermic reaction starting at 77 °C [20]. The γ-AlH_3 was immediately transformed into α-AlH_3 through an exothermic reaction above 100 °C. The α-AlH_3 was dehydrided, resulting in aluminum and hydrogen through an endothermic reaction at 132–167 °C. These data provide information on

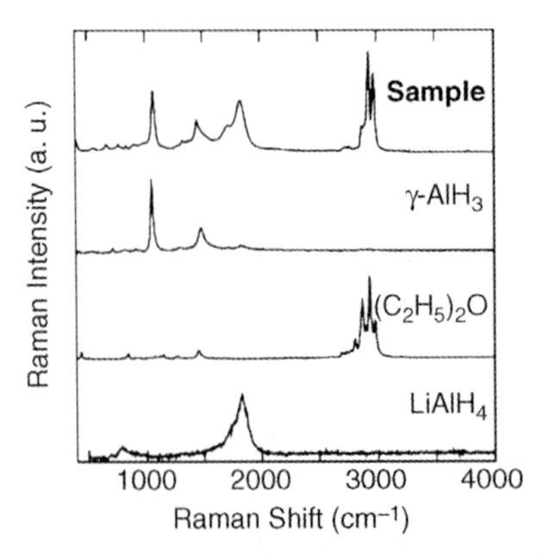

Figure 9.5 Raman spectra of samples compared to Raman spectra of γ-AlH_3, $(C_2H_5)_2O$ and $LiAlH_4$ for references. In this work the Raman spectrum of γ-AlH_3 was observed for the first time, Ref. [20].

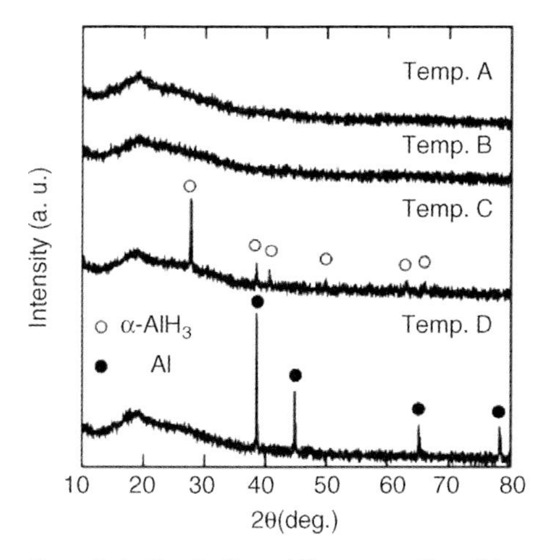

Figure 9.6 Powder X-ray diffraction profiles of the samples that were heated to 77, 100, 132, and 172 °C (*ex situ* measurements), Ref. [20].

the conditions needed to obtain γ- and α-AlH_3 by restraining the occurrence of the dehydriding reaction. The results of the *in situ* XRD measurements at 76 and 87 °C are shown in Figure 9.8(a) and (b), respectively.

The results reported by Kato *et al.* can be summarized as follows; AlH_3-etherate is desolvated to form γ-AlH_3 (probably due to the similarity in the Al–H bonding character between them), through an endothermic reaction starting at 77 °C and then γ-AlH_3 is immediately transformed into α-AlH_3 in an exothermic reaction above 100 °C. The dehydriding of α-AlH_3 proceeds together with an endothermic reaction starting at 132 °C, producing aluminum metal.

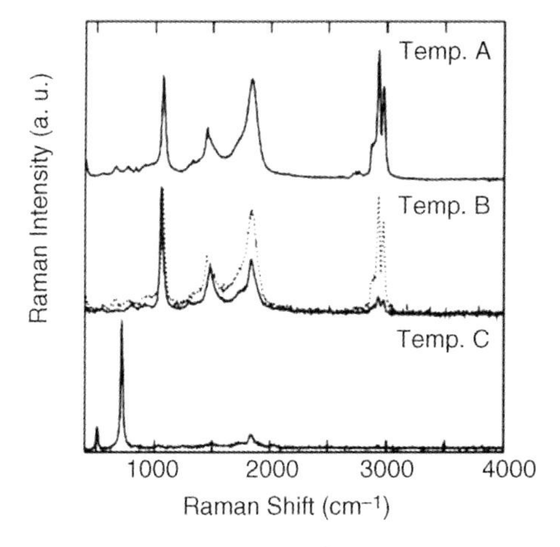

Figure 9.7 Raman spectra of samples that were heat treated corresponding to A–C (*ex situ* measurements). Raman spectrum at 132 °C is in good agreement with that of α-AlH_3 reported in the literature, Ref. [20].

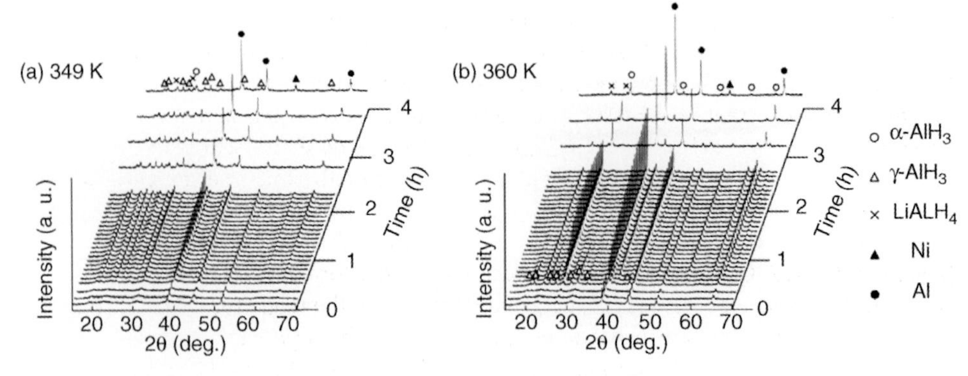

Figure 9.8 Powder X-ray diffraction profiles (*in situ* measurements) at (a) 76 and (b) 87 °C. These measurements were carried out under vacuum, Ref. [20].

The transformation of different phases of alane and the effect of ball-milling on the thermal stabilities of these phases were investigated by Orimo *et al.* [23]. Similarly to pervious studies, three phases of AlH_3 (α-, β- and γ-) were prepared by the reaction described before [15]. Orimo *et al.* investigated the thermal stabilities in relation to the phase transformation of the various phases of AlH_3. Figure 9.9 depicts the results of the thermal gravimetric analysis (TGA) and differential thermal analysis (DTA) under a helium flow. The TGA profile of the α-phase shows that the dehydriding

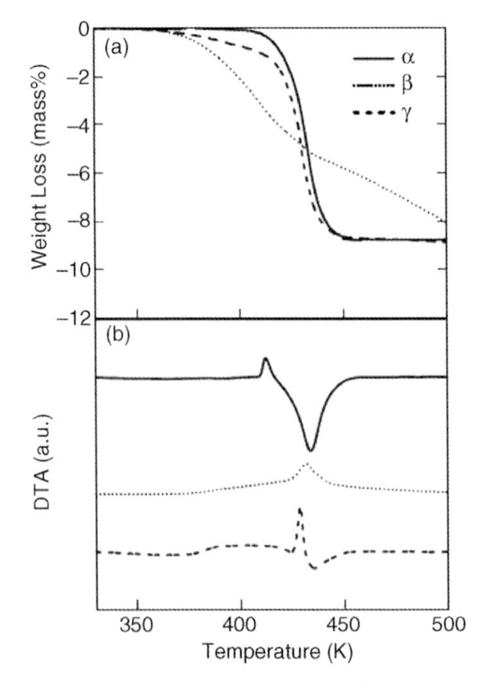

Figure 9.9 Thermogravimetric and differential thermal analysis (TGA-DTA) of the three phases of AlH_3; α-phase (solid line), β-phase (dotted line) and γ-phase (dashed line) Ref. [23].

reaction started around 127 °C and the amount of hydrogen released was nearly 9 wt %, close to 10.1 wt%, the theoretical amount of hydrogen in AlH_3. The TGA profile of the γ-phase was found to be similar to that of the α-phase. The β-phase, however, was found to have a different TGA profile where a continuous dehydriding started at a lower temperature, around 97 °C. The DTA profile of the α-phase shows a single and large endothermic peak at 147–177 °C, following a small exothermic peak which could be attributed to the transformation of the γ- into the α-phase, consistent with early work by Claudy *et al.* [24]. A small exothermic reaction peak was reported for the β-phase at 107–167 °C, which can be attributed to a phase transformation from β to α. An exothermic reaction peak was also observed for γ-alane in the range 107–167 °C and was attributed to the transformation of the γ- phase to the α-phase.

Phase transformation and hydrogen release were detected by DSC, as shown in Figure 9.10. Mg_2NiH_4 was used as a reference.

The enthalpy values of dehydriding AlH_3 have been determined. The α-phase dehydriding reaction was found to be endothermic with $6.0 \pm 1.5\,kJ\,(molH_2)^{-1}$ consistent with values reported elsewhere of $7.6\,kJ\,(molH_2)^{-1}$ [19]. Similar to early reports, the γ-phase showed exothermic transformation to the α-phase followed by dehydriding to aluminum with a net endothermic heat of reaction of $1.0 \pm 0.5\,kJ\,(molH_2)^{-1}$. The β-phase, however, was found to dehydride directly to aluminum, which is a different result from results reported elsewhere [15, 16, 19].

Particle size, surface conditions, and defects in the structure of a hydride are known to have great effect on its stability. Particle size and grain boundaries of hydrides have been shown to affect the kinetics and diffusion of hydrogen in hydrides [25–27]. Obtaining a smaller particle size, modifying the surface, and creating defects in alane crystals can all be achieved by ball milling. Orimo *et al.* showed that the thermal stabilities of AlH_3 and phase transformation are affected by mechanical milling. Mechanical milling methods were employed with each of the three phases (α-, β- and γ-) of AlH_3. The effects of ball milling on the phase

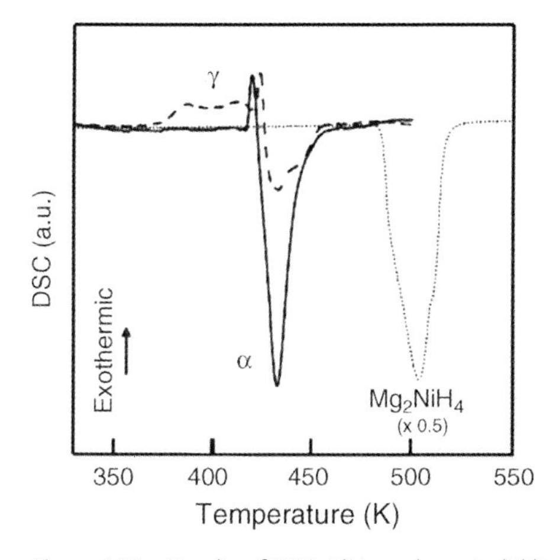

Figure 9.10 Results of DSC; AlH_3 α-phase (solid line) and AlH_3 γ-phase (dashed line). Mg_2NiH_4 (dotted line), used as a reference, Ref. [22].

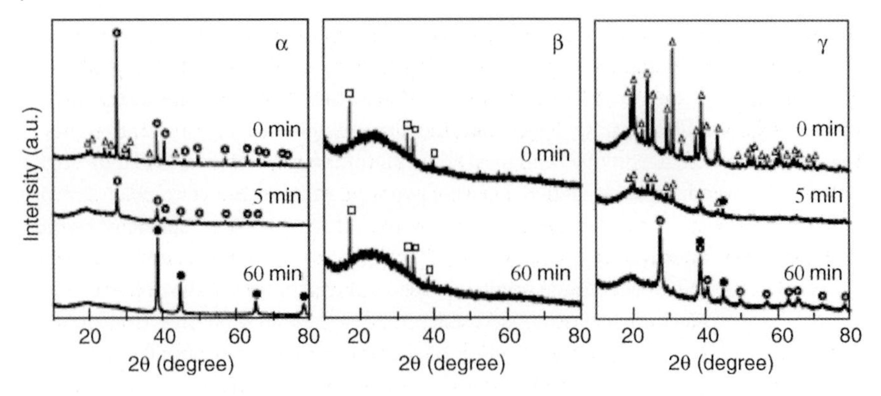

Figure 9.11 Powder X-ray diffraction profiles of the three phases of AlH_3 before and after mechanical milling. ○ = α –phase, • = Aluminum, □ = β-phase and △ = γ-phase, Ref. [23].

transformation and stability of AlH_3 are shown in Figure 9.11 of the XRD profiles. The α-phase was found to dehydride during mechanical milling, and only the metallic Al remained after milling for 60 min. The dehydriding reaction was also observed for the γ-phase. However, the α -phase transformed from the γ -phase remained in a hydride state after 60 min of milling. There was almost no change in the β-phase, even after 60 min of milling.

It is important to note that the thermodynamic stability of alane is not the only factor responsible for the stability of alane. Kinetics has the main role in producing stable alane at room temperature. The value of the enthalpy measured by researchers is of the order of 8 kJ $(molH_2)^{-1}$. This enthalpy value is not large enough to keep alane from decomposing at temperatures below room temperature since the entropy is the typical value, approximately 130 J $(mol\ H_2)^{-1}\ K^{-1}$, of the difference between hydrogen in the solid and gaseous states. However, one of the main rate-limiting steps in the hydrogen desorption process is the association of hydrogen atoms forming hydrogen molecules at the surface. This step is dependent mainly on the type of surface and its conditions. It seems that the surface of alane is very inactive, either due to the presence of an oxide layer and/or an intact layer of aluminum hydride that hinders the process of hydrogen atoms associating into molecules. This inactive surface creates a high energy barrier, preventing alane from decomposing at low temperature. This explains why ball milling causes alane to decompose at room temperature, by breaking the protective surface layer and also by creating defects that act as spots for localized electrons, assisting in associating hydrogen atoms into molecules. Adding catalysts would also be helpful to lower the association barrier and allow the alane to decompose at a lower temperature.

9.4 Stability and Formation of Adduct Organo-Aluminum Hydride Compounds

Alane adducts (AlH_3-adducts) are of interest because they can store hydrogen, can be utilized for the formation of derivatives of other metal complexes and can be used as precursors for chemical vapor deposition of aluminum meta1. These areas of aluminum hydride adduct applications require a detailed understanding of the structure and properties of such adducts. When aluminum hydride is prepared by the reaction shown in Eq. (9.2), and after the precipitation of lithium chloride, the product is $AlH_3{\cdot}Et_2O$, namely an ether solution-aluminum hydride adduct. In ether solution aluminum hydride was reported to complex with the solvent and was determined to be monomeric in ether before polymerization [28]. However, a stronger base such as tetrahydrofuran (THF) forms a more stable complex with AlH_3, hence polymerization of AlH_3 is slowed in the THF solvent [29]. Other stable aluminum hydride adducts include 1: 1 and a 2: 1 complexes with trimethylamine [30]. Several other 1:1 complexes with tertiary alkyl amines have also been made [31]. Amine adducts of alane are usually prepared by a reaction forming aluminum hydride in direct combination with the appropriate amine.

Ruff *et al.* in a series of publications described the synthesis of amine complexes of aluminum hydride [32, 33]. Their study investigated the reaction of these materials with typical Lewis bases in order to define the conditions for the stability of aluminum hydride derivatives in which the aluminum atom exhibits a coordination number of five. They first described methods for making tertiary alkyl amine complexes of aluminum hydride utilizing lithium aluminum hydride and an amine hydrochloride. A finely ground lithium aluminum hydride was placed together with trimethylammonium chloride (ratio 1: 2). They prepared other trialkylamine alanes and the *N*-dialkylaminoalanes, in a similar fashion. These adducts of alane were found to sublime readily at temperatures up to 40 °C except for the tri-n-propylamine alane, which sublimed very slowly and could also be recrystallized from hexane at −80 °C.

Tertiary amine adducts of alane show a diverse range of structures based on four and five coordinate species, hydride-bridging dimeric and polymeric species, and ionic species.

Ruff *et al.* [32] described the formation of bis-trialkylamine alanes in solution and discussed the stability of these complexes. The reactions of amines with trialkylamine alanes or dimethylamino alane could be categorized in two classes. The first is secondary amines reacting predominantly as compounds containing an acidic hydrogen. The second is tertiary amines reacting as Lewis bases, resulting in either displacement or addition. The bond hybridization in the bis-amine alanes was determined most likely to be sp^3d with the nitrogen atoms bonded to the apices of the bipyramid since this configuration was determined to be the one of least steric hindrance. The support for their conclusion was based on earlier reported measurement of the dipole moment of bis-trimethylamine alane, which gave a very low value, as would be expected for the symmetrical structure [34]. Atwood *et al.* studied the structure of tertiary amine adducts of alane [35], using X-ray diffraction. They showed dimeric species with two bridging hydrides are a common solid state structural unit among several alane adducts with unidentate tertiary amines, including the well-known compound H_3AlNMe_3. The structure of H_3AlNMe_3 was found to be severely disordered, when compared to the well-behaved dimeric structures of the adduct $[\{H_3Al(NMe_2CH_2Ph)\}_2]$ and $[\{H_3Al(NMeCH_2CH_2CH{=}CHCH_2)\}_2]$. These two previous adducts were reported to start decomposing at 130 °C forming aluminum mirrors on the side of the vial. This decomposition temperature is higher than those of other monodentate tertiary amine derivatives of alane, including H_3AlNMe_3 (starts decomposing at 100). Atwood *et al.* also pointed out that polydentate tertiary amines such as tetramethylethylenediamine (TMEDA) have a higher thermal stability (TMEDA decomposes around 200 °C). Trimethylaluminum is reported to be monomeric in the vapor phase and dimeric in the solid, which was also reported in other studies [36, 37].

Metalorganic chemical vapor deposition (MOCVD) is a very important technique for a variety of applications, including the fabrication of electronic devices. This technique has been used in the fabrication of microelectronic devices to deposit semiconductors, insulators, and metal layers. Frigo *et al.* examined the properties of alane dimethylethylamine (DMEAl), with the chemical formula $AlH_3.NMe_2Et$, specifically to be used as a precursor for the MOCVD of aluminum-containing layers [38]. They reported that MOCVD of aluminum layers for metallization was also achieved, using A1 trialkyls (R_3Al) as precursor; where R = i-Bu [39] and also R = Me [40]. They reported a synthesis route of DMEAl not involving ether solvents. They also described DMEAl to be a colorless volatile liquid at ambient temperatures but to have limited thermal stability, slowly decomposing to generate hydrogen. They explained that the use of ether in the preparation of DMEAl would very likely result in contaminating the end product with the ether, which could lead to incorporation of oxygen into the Al-containing layers. The main advantage of making such compounds for use in MOCVD is that they do not contain carbon directly bonded to the metal. The tertiary amine ligands are thermally stable under the conditions of MOCVD and will not be sources of carbon contamination [41]. Their synthesis method used only alkane solvents, which are inert in the deposition processes, and are volatile enough to affect easy separation from the product. They described the DMEAl adduct at ambient temperatures as a crystal-clear colorless liquid with melting point of 5 °C and density of $0.78\,g\,cm^{-3}$ at 25 °C, as compared with $0.68\,g\,cm^{-3}$ at 20 °C for the free amine. The DMEAl adduct was found to be only moderately soluble in pentane or hexane at room temperature but readily dissolved upon mild heating. Similar to solid alane adducts with trimethylamine (TMAA1) with chemical formula AlH_3-NMe_3, DMEAl decomposed very rapidly upon exposure to the atmosphere. However, the exposure of small quantities of DMEAl to the atmosphere did not result in spontaneous ignition, in contrast to most of the trialkyls. Ignition was observed after about 1 min when $6\,cm^3$ of DMEA1 was dropped onto vermiculite with a depth of 0.5 cm.

In 1964, Ashby reported the first successful direct synthesis of an amine alane, triethylenediamine (AlH_3 -TEDA) by direct reaction of aluminum and hydrogen in the presence of triethylenediamine (TEDA, $C_6H_{12}N_2$) using activated aluminum powder and TEDA in THF under a hydrogen pressure of 35 MPa [42].

$$Al + H_3 \text{ TEDA} \longrightarrow AlH_3 \leftarrow \text{TEDA}$$

More recently, and in an attempt to obtain a compound useful for hydrogen storage, direct and reversible synthesis of AlH_3-triethylenediamine (AlH_3-TEDA) from Al and H_2 was reported by Graetz *et al.* This work demonstrated that activated aluminum doped with titanium $Al_{(Ti)}$ could be used to form AlH_3-TEDA under low hydrogen pressure of 3.6 MPa. They started with $Al_{(Ti)}$ and TEDA in THF (or undecane) slurry and the slurry reacted reversibly with H_2 under mild conditions of temperature and pressure to form AlH_3-TEDA, which is much more stable than crystalline AlH_3 [43]. The activated aluminum was prepared in an argon-purged drybox, adapting a doping procedure used to introduce Ti dopant to $NaAlH_4$ [44]. They, first, prepared AlH_3 using the DOW synthesis route mentioned above, which involves reacting $LiAlH_4$ with $AlCl_3$ in ether [15]. In order to obtain Ti-doped aluminum they modified the DOW synthesis to produce a final product of $Al_{0.98}Ti_{0.02}$ ratio. They obtained pressure–composition isotherms in the temperature range 70–90 °C and they described a possible reaction mechanism. They demonstrated hydrogenation and dehydrogenation reactions of a slurry of solid $Al_{(Ti)}$ and TEDA dissolved in THF. A reference experiment was performed, using undoped Al at 3.6 MPa to show the effect and the importance of the Ti dopant on the hydrogen uptake and release by the $Al_{(Ti)}$-TEDA/THF system. Hydrogen uptake and release did not occur in the absence of Ti dopants. They also found that if there was no agitation the system approached equilibrium very slowly. This was attributed to the fact that the dense solid phases sink to the bottom of the liquid phase and diffusion of the H_2 molecules in the still liquid is very slow. Since equilibrium was attained very slowly in the non-agitated slurry system, H_2 could be introduced without any significant dissolution in the liquid phase and the pressure could be recorded before and after the reaction took place. When the appropriate rate of stirring was resumed, equilibrium between the liquid phase and the gas phase was achieved very quickly. That made it possible to estimate the amount of dissolved hydrogen in the liquid phase. After hydrogenation and a relatively slow formation of amine-alane the solid amine-alane was recovered in a drybox by filtration. Figure 9.12 shows the uptake of hydrogen by a slurry of Al plus TEDA and THF, the plot demonstrates that AlH_3-TEDA can be readily prepared at room temperature and under hydrogen pressure less than 3.6 MPa, using Ti-doped Al, which is far below the 35 MPa cited by Ashby [42].

When the reaction reached completion the grayish white product was recovered by filtration followed by drying. The product was examined using XRD and the obtained pattern showed only the presence of AlH_3-TEDA, as indicated in Figure 9.13. The hydrogen content of the product of the experiment was determined using thermal analysis. Graetz *et al.* obtained equilibrium pressure–composition isotherms (PCI) for Al-TEDA/THF and hydrogen. A series of equilibrium absorption isotherms was

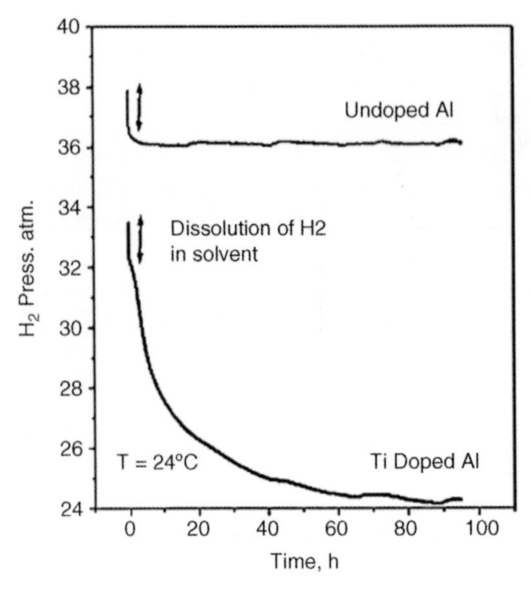

Figure 9.12 (b) Uptake of hydrogen in a TEDA/THF solution + Ti-doped Al powder in suspension. (a) Illustrates the absence of any reactivity using undoped Al powder. Note dissolution of H_2 in the liquid phase in both plots upon H_2 addition, Ref. [43].

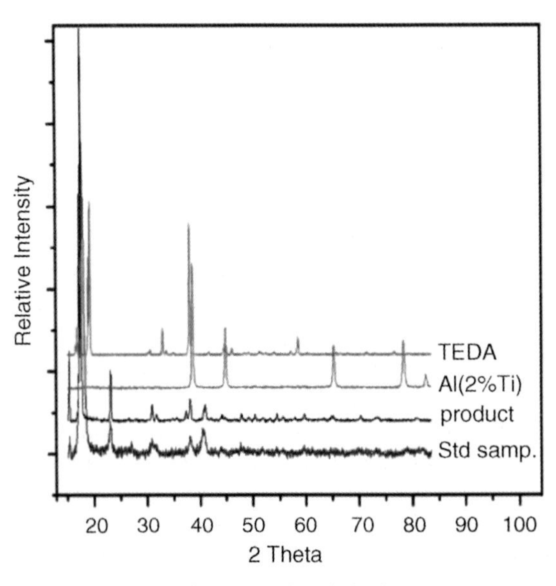

Figure 9.13 XRD of TEDA, Ti-doped Al, AlH_3-TEDA reaction product compared to a standard sample of AlH_3-TEDA, Ref. [43].

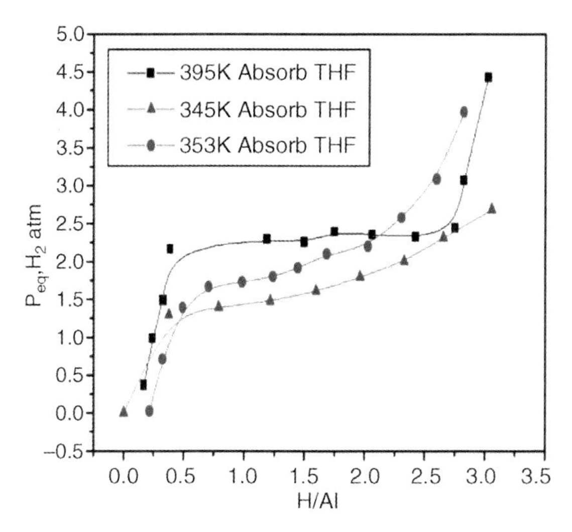

Figure 9.14 Equilibrium PC isotherms corrected for the THF vapor pressure and H_2 solubility in the solvent, Ref. [43].

generated at 72, 80, and 86 °C, as shown in Figure 9.14. The pressure axis refers to the partial pressure of hydrogen after correcting for the vapor pressure of the liquid phase (THF + dissolved TEDA). The vapor pressure of the liquid changes slightly as the TEDA reacts with aluminum and hydrogen but this did not affect the determination of the H_2 partial pressure because a TEDA concentration well exceeding the stoichiometric amount required was used.

Although TEDA is not completely soluble in n-undecane, nevertheless, they observed the reactions proceeded without difficulty but at a slower rate than in THF. The equilibrium isotherms were generated at 88 °C at which point the vapor pressure of the solvent could be neglected. They used the results obtained from the above equilibrium isotherms to calculate the heat (ΔH) and entropy (ΔS) of formation. ΔH was determined to be -39.5 ± 4.1 kJ (mol $H_2)^{-1}$ and ΔS to be -117 ± 11 J (mol $H_2)^{-1}$ K^{-1} resulting in ΔG_{298K}(adduct) = −4.63 kJ (mol $H_2)^{-1}$. Therefore, the stability of AlH_3-TEDA is higher than the stability of α-AlH_3 ($\Delta G_{298K}(AlH_3)$ = 32.3 kJ (mol $H_2)^{-1}$) making it impossible to separate AlH_3 from the TEDA, instead AlH_3-TEDA decomposes into aluminum hydrogen and TEDA. A complete cycle of an equilibrium absorption–desorption isotherm was obtained using n-undecane ($C_{11}H_{24}$) as the solvent and is shown in Figure 9.15. n-Undecane was used to take advantage of its low solvent vapor pressure.

The titanium catalyst could be assisting the formation of Al_xH_y intermediates that can be stabilized by binding with TEDA to form AlH_3-TEDA, which was calculated to be the most stable configuration. It was speculated that the product is not monomeric, because of the difunctional nature of the amine [42]. The easy formation of this amine alane (TEDA-AlH_3) from the elements, compared to the difficulty of preparing monofunctional amine alanes, was attributed to the heat of polymerization (or crystallization) of the product as well as to the basicity of the amine [42].

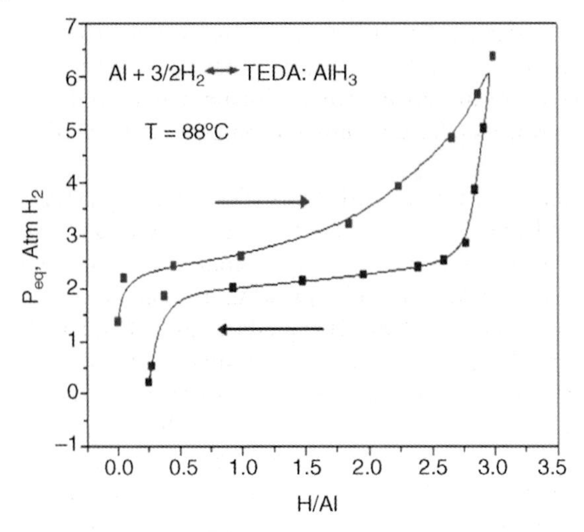

Figure 9.15 Equilibrium absorption–desorption isotherms generated in $C_{11}H_{24}$ slurry, corrected for H_2 solubility in solvent, Ref. [43].

9.5 Phases and Structures of Aluminum Hydride

As previously discussed, non-solvated AlH_3 was prepared by Brower *et al.* [15], who reported the existence of at least 7 nonsolvated phases, namely α, α′, β, γ, δ, ε, and ξ. However only α-, α′-, β-, and γ-phases have been synthesized reproducibly, in contrast to the other structures that may have been identified as a result of impurities.

Ke *et al.* conducted a study on the structural properties of AlH_3 using density functional theory (DFT) methods [45]. They reported that the α′-(orthorhombic) and β-(cubic) modifications are more stable than α-(hexagonal) at 0 K (no phonons). The finding is, however, in contrast to experimental results, which have shown that the α-crystalline phase is the most stable phase at ambient conditions.

Early work by Turley *et al.* to study the structure of α-alane used powder X-ray and powder neutron diffraction [46]. The A1–H (bridging) distance was reported to be 1.715 Å, shorter than the 2.1 Å value for $Al(BH_4)_3$ [47]. Turley *et al.* described the structure of α-AlH_3 as having aluminum atoms surrounded by six hydrogen atoms that bridge to six other aluminum atoms. The structure was identified with space group *R-3c*, encompassing alternating planes of aluminum and hydrogen atoms stacked perpendicular to the *z*-axis and spaced a distance $z/12 = 0.984$ Å apart, center to center. Each aluminum is surrounded by six hydrogen atoms in an octahedral structure, three in the plane above and three in the plane below, participating in six three-center A ⋯ H ⋯ Al bonds. Such bonds are a consequence of the electron deficiency of the elements of Group 3. This can be self-stabilized by the formation of 3-center 2-electron bridge bonds as a covalent compound, in which the aluminum is also surrounded by hydrogen atoms in an octahedral assembly. However, the value 1.715 Å of the A1–H (bridging) distance is larger than the short-lived species Al–H (1.648 Å) and longer than the distance found in lithium aluminum hydride (1.547 Å) in which each aluminum is assembled in a tetrahedral fashion [46].

More recent work to determine the accurate structures of α and α', by powder neutron diffraction was reported by Brinks *et al.* [48]. AlD_3 was synthesized by ball milling $3LiAlD_4 + AlCl_3$. When ball milling was done at room temperature the product was found to be a mixture of AlD_3 (α and α') and Al, in addition to LiCl. However, when the ball milling was done at –196 °C the product was AlD_3 and LiCl. The AlD_3 product was a mixture of about 2/3 α and 1/3 α'. Using powder neutron diffraction the structure of α' was determined to take the β-AlF_3 structure with space group *Cmcm* and $a = 6.470(3)$, $b = 11.117(5)$, and $c = 6.562(2)$ Å. The structure is arranged with corner-sharing AlD_6 octahedra in an open structure with hexagonal holes of radius 3.9 Å. They reported that α' was found to slowly decompose when heated to 40 °C. They described α-AlD_3 to also be a corner-sharing AlD_6 network but in a more dense ReO_3-type arrangement, see Figure 9.16. Both AlD_3 (α and α') modifications have slightly shorter Al–D distances than Na_3AlD_6, Na_2LiAlD_6, and K_2NaAlH_6.

The β-AlD_3 detailed structure has also been determined by Brinks *et al.* using combined powder neutron diffraction (PND) and synchrotron X-ray diffraction (SRPXD) [49]. The β alane, β-AlD_3, was synthesized by the method described by Brower *et al.* [15]. The DFT optimized model reported by Ke *et al.* [45] was used as a starting point for the Rietveld refinements of β-AlD_3. The β-AlD_3 structure was found to take the pyrochlore-type structure (space group $Fd\bar{3}m$), resembling FeF_3.

The β-AlD_3 structure contains corner-sharing AlD_6 octahedra, similar to the α- and the α'-phases. It should be noted that a high purity β-AlD_3 sample is difficult to prepare [15]. However, the β-AlD_3 sample studied by Brinks *et al.* was considered to

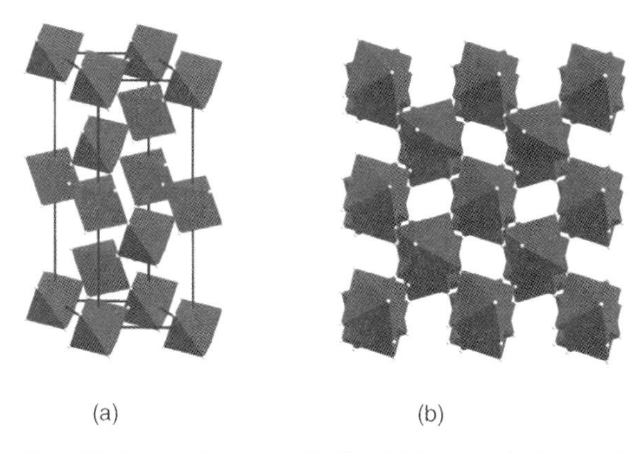

Figure 9.16 Crystal structure of α-AlD_3 (a) The unit cell of α-AlD_3, (b) illustration of the connectivity of the octahedra. Each octahedron shares one corner with one other octahedron, building a distorted primitive Al sublattice, Ref. [48].

Figure 9.17 The framework of AlD_6 octahedra for β-AlD_3, depicting the channels through the structure, Ref. [49].

be of high purity and good crystallinity. The crystallite size, based on broadening of SR-PXD reflections, was estimated to be 175 nm and had an isotropic strain of 0.04%. Based on Rietveld refinements, the sample was found to contain 89.5 mol% β-AlD_3, 9.4 mol% γ-AlD_3, 0.9 mol% α-AlD_3 and 0.1 mol% Al. The structure consisted of corner-sharing AlD_6 octahedra where all octahedra are connected to two of the neighboring octahedra. The surrounding octahedra form two groups of three octahedra that are interconnected. They indicated that α'-AlD_3 can be regarded as an intermediate between α -AlD_3 and β -AlD_3. They explained that the connectivity of the octahedra results in an open framework, leading to the formation of channels in several directions. The diameter of the channels is about 3.9 Å (see Figure 9.17). The volume per formula unit of β-AlD_3 is larger (45.6 $Å^3$) than α-AlD_3 (33.5 $Å^3$) and α'-AlD_3 (39.3 $Å^3$).

The structure of γ-alane was studied by two different groups; Yartys *et al.* using SR-PXD and Brinks *et al.* using combined PND and SR-PXD [50, 51]. In both cases, samples were prepared following the methods described by Brower *et al.* [15]. The space group of γ-alane was determined to be *Pnnm*. A structure with corner-sharing AlD_6 in γ-AlD_3, similar to the α-, α-$'$, and β-alane phases, and containing edge-sharing octahedra was reported. Two types of Al in the structure were described as one sharing corners only and one sharing both edges and corners. The crystal structure may be described as chains formed by pairs of edge-sharing octahedra connected via corner sharing in the chain and with octahedra with corner-sharing only connecting the different chains. The average Al–D distance was found to be 1.71 Å. This value is similar to the Al–D distance in the other structure modifications of alane [51]. Another feature of the crystal structure of γ-alane is the formation of large cavities between the AlH_6 octahedra. Strands of Al_2H_6 are always separated by differently oriented Al_1H_6 octahedra, resulting in a packing structure of γ-AlH_3 that has a smaller density than α-AlH_3 (by 11%) [50]. Tabulated structure data are shown in Table 9.1 (Figure 9.18).

Table 9.1 Different aluminium hydride structures and their space groups and unit cell parameters.

Structure	Space group	*a* (Å)	*b* (Å)	*c* (Å)	Ref.
α-AlD_3	*R-3c*	4.4364(3)		11.7963(10)	[48]
α'-AlD_3	*Cmcm*	6.470(3)	7.3360(3)	6.562(2)	[48]
β-AlD_3	*Fd-3m*	9.0037(1)			[49]
γ-AlD_3	*Pnnm*	7.3360(3)	5.3672(2)	5.7562(1)	[51]

Figure 9.18 Structures of different phases compared. α'-AlH_3 [101], Ref. [48], β-AlH_3 [101], Ref. [49], γ-AlH_3 [010], Ref. [51].

9.6
Novel Attempts and Methods for Forming Alane Reversibly

The direct hydrogenation of aluminum to AlH_3 (alane) requires over 10^5 bar of hydrogen pressure at room temperature. The impracticality of using high hydriding pressure has precluded alane from being considered as a reversible hydrogen storage material. For these reasons, development of a novel cycle to cost effectively regenerate AlH_3 under practical conditions is needed.

The typical formation route for alane is through the chemical reaction of lithium alanate with aluminum chloride in diethyl ether:

$$3\,LiAlH_4 + AlCl_3 \xrightarrow{\text{ether}} 4\,AlH_3 \cdot Et_2O + 3\,LiCl$$

This reaction yields dissolved alane etherate, $AlH_3{\cdot}Et_2O$, and precipitates lithium chloride. Alane can then be separated from the ether by heating *in vacuo*. The synthesis of AlH_3 by this method results in the formation of alkali halide salts such as LiCl. The formation of alkali halide salts is, however, a thermodynamic sink because of their stability. For a cyclic process, lithium metal must be recovered from lithium chloride by electrolysis of a LiCl/KCl melt at 600 °C, requiring at least $-429\ \mathrm{kJ\,mol^{-1}}$ of energy equivalent to satisfy the LiCl heat of formation and heat of fusion. The large amount of energy required to regenerate AlH_3 from spent aluminum and the alkali halide makes this chemical synthesis route economically impractical for a reversible AlH_3 storage system.

Graetz *et al.* suggested that since direct formation of amine-alane such as alane triethyldiamine (AlH_3-TEDA), works well but forms a strong Al–N bond [43], it may be possible to form a less-stable amine alane such as triethylamine (AlH_3-TEA) [52] and eventually separate AlH_3 from the adduct.

They proposed a route of regeneration of alane that involves the use of $LiAlH_4$

$$3\,LiAlH_4 + 2\,NR_3 \rightarrow 2\,NR_3 \cdot AlH_3 + Li_3AlH_6 \tag{9.4}$$

Where NR_3 represents an amine that forms a weak Al–N bond, such as triethylamine (NEt_3)

The goal is to form amine alanes with a weak Al–N bond because that can be easily separated to recover pure AlH_3. A complete regeneration scheme for AlH_3 can be envisioned as:

$$LiH + Al + \frac{3}{2}H_2 \rightarrow LiAlH_4 \cdot 4\,THF + 2\,NR_3 \rightarrow 2\,NR_3 \cdot AlH_3 + Li_3AlH_6\downarrow \rightarrow AlH_3 + 2NR_3\uparrow \tag{9.5}$$

In the proposed scheme depicted by reaction (9.5) Li_3AlH_6 will precipitate in solution to be filtered out and the alane will be recovered from the alane amine adduct by heating under vacuum.

The second possible route that was suggested by Graetz *et al.* [53] makes use of two reported reactions using trimethylamine (TMA) [54].

$$Al + \frac{3}{2}H_2 + N(CH_3)_3 \rightarrow (AlH_3) \cdot N(CH_3)_3 \tag{9.6}$$

$$(AlH_3) \cdot N(CH_3)_3 + NR_3 \rightarrow (AlH_3) \cdot NR_3 + N(CH_3)_3 \tag{9.7}$$

The final step is expected to be

$$(AlH_3) \cdot NR_3 \rightarrow AlH_3 + 2\,NR_3\uparrow \tag{9.8}$$

Zidan *et al.* developed and demonstrated a cycle that uses electrolysis and catalytic hydrogenation of products to avoid both the high hydriding pressure needed for aluminum to form aluminum hydride and the formation of stable by-products such as LiCl [55]. The cycle, shown in Figure 9.19, utilizes electrochemical potential to drive the formation of alane and alkali hydride from an ionic alanate salt. The starting alanate is regenerated by reacting spent aluminum with the by-product alkali hydride in the presence of titanium catalyst under moderate hydrogen pressure.

The electrolysis reaction is performed in an electrochemically stable, aprotic, polar solvent such as THF or ether. $NaAlH_4$ is dissolved in this solvent, forming the ionic solution ($Na^+/AlH_4^-/THF$) which is used as an electrolyte. Though not directed at the regeneration of alane, elaborate research and extensive studies on the electrochemical properties of this type of electrolyte have been reported [56, 57]. Although attempts in the past were made to synthesize alane electrochemically [58–60] none have shown isolated material or a characterized alane product.

The thermodynamic properties of the $NaAlH_4$/THF electrolyte, together with the cyclic voltammetry of the electrochemical cell were the basis for performing this electrochemical process [55]. Thermodynamic calculations were made to determine

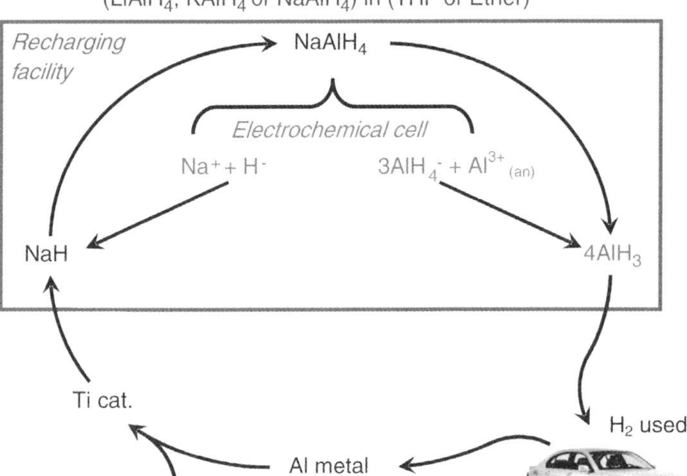

Figure 9.19 Reversible cycle for alane generation. All components of the electrochemical process can be recycled to continually afford a viable solid state storage material, Ref. [55].

the reduction potentials of the possible electrochemical reactions of $NaAlH_4$ in an aprotic solution (THF) with an aluminum electrode. From the half-reaction potentials, the cell voltage for alane formation was calculated and a theoretical cyclic voltammagram was constructed (see Figure 9.20)

In the electrochemical cell two separate reaction mechanisms can produce alane at the aluminum electrode, as shown in Figure 9.20. One possible mechanism is the oxidation of the alanate ion to produce alane, an electron, and hydrogen, as shown in Eq. (9.9):

$$AlH_4^- \rightarrow AlH_3 \cdot nTHF + \frac{1}{2}H_2\uparrow + e^- \quad (9.9)$$

Another possible mechanism is the reaction of AlH_4^- with the aluminum anode to form alane. In this reaction route, the evolution of hydrogen is suppressed and the reaction is expected to consume the Al electrode, as in Eq. (9.10):

$$3\,AlH_4^- + Al \rightarrow 4\,AlH_3 \cdot nTHF + 3e^- \quad (9.10)$$

Zidan *et al.* have reported dissolving of the aluminum electrode when the conditions for reaction (9.10) were applied [55]. In addition to determining the electrochemical processes for producing AlH_3, recovering AlH_3 from the THF solution was a major step of this cycle. Although the separation of alane from $AlH_3{\cdot}Et_2O$ is well established and affords pure AlH_3 [11, 15, 20] the separation of alane from the $AlH_3{\cdot}THF$ adduct is more complicated because it can easily decompose, when heated under vacuum, to THF, aluminum and hydrogen. Therefore, in

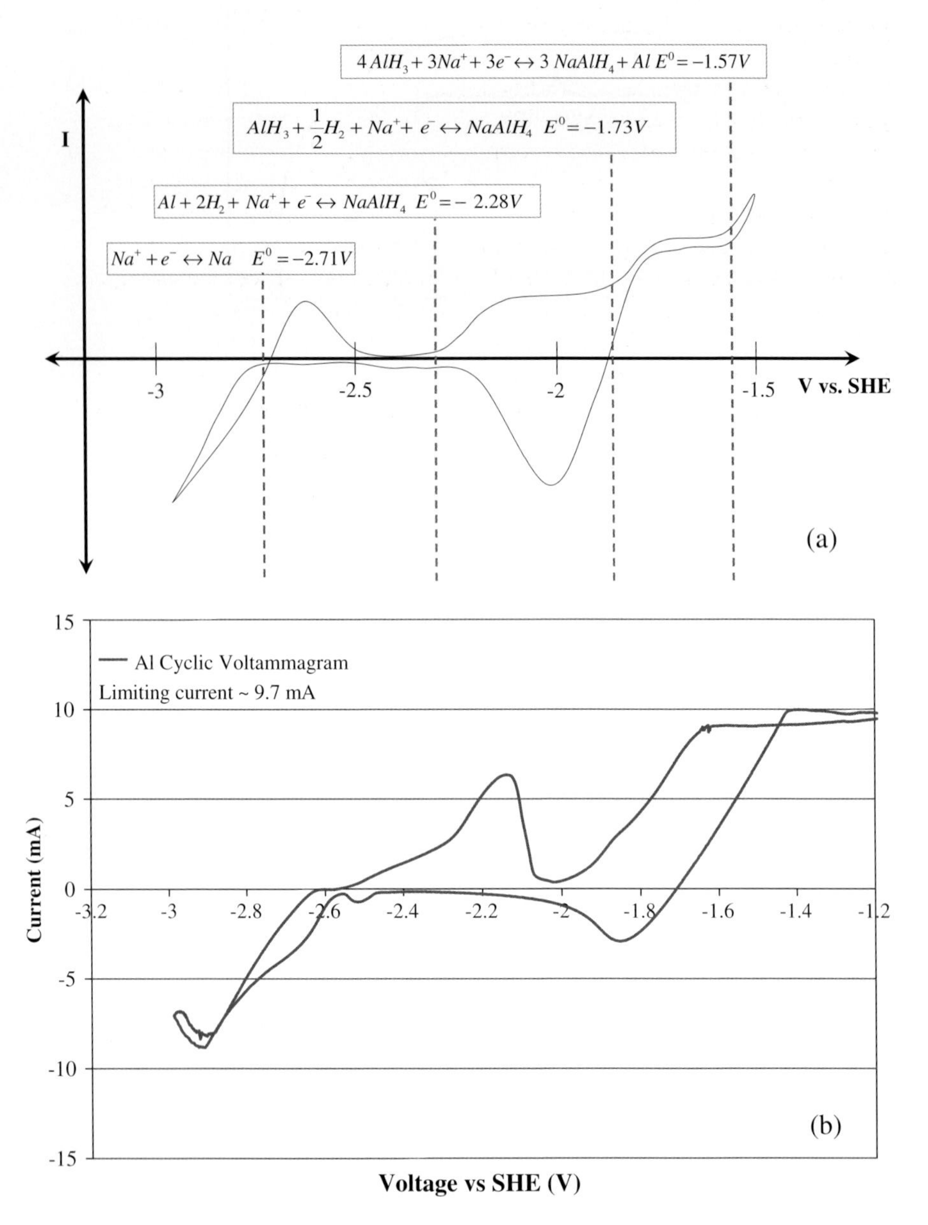

Figure 9.20 Theoretical and experimental cyclic voltammagrams for the electrochemical formation of alane. (a) A hypothetical cyclic voltammagram was formulated from the equilibrium potential data for possible reactions and the anticipated state of each species generated. (b) Bulk electrolysis experiment at an aluminum wire electrode for a cell containing a 1.0 M solution of $NaAlH_4$ in THF at 25 °C, Ref. [55].

this electrochemical cycle adducts such as TEA were added to the reaction product to stabilize the alane during purification. Adduct-free alane is recovered by heating the neat liquid AlH_3·TEA *in vacuo*.

Alane recovered from the electrochemical cells was characterized by powder X-ray diffraction, Raman spectroscopy, and TGA. Powder X-ray diffraction patterns for two different separation methods are shown in Figure 9.21. When AlH_3·THF product was heated without adding TEA a noticeable amount of aluminum was present with the α-alane (Figure 9.21a). Separation using the TEA yields only α-alane (Figure 9.21b). The unrefined unit cell parameters from indexing of this pattern were $a = 4.446$ Å and $c = 11.809$ Å. Based on the systematic absences, the space group was assigned as *R-3c* and is consistent with α-alane [46].

Raman spectra and TGA data collected for AlH_3 isolated from the electrochemical cell are shown in Figures 9.22 and 9.23 respectively. Raman modes present at 510, 715, 1050, and 1515 cm^{-1} are consistent with the literature values for α-alane [61].

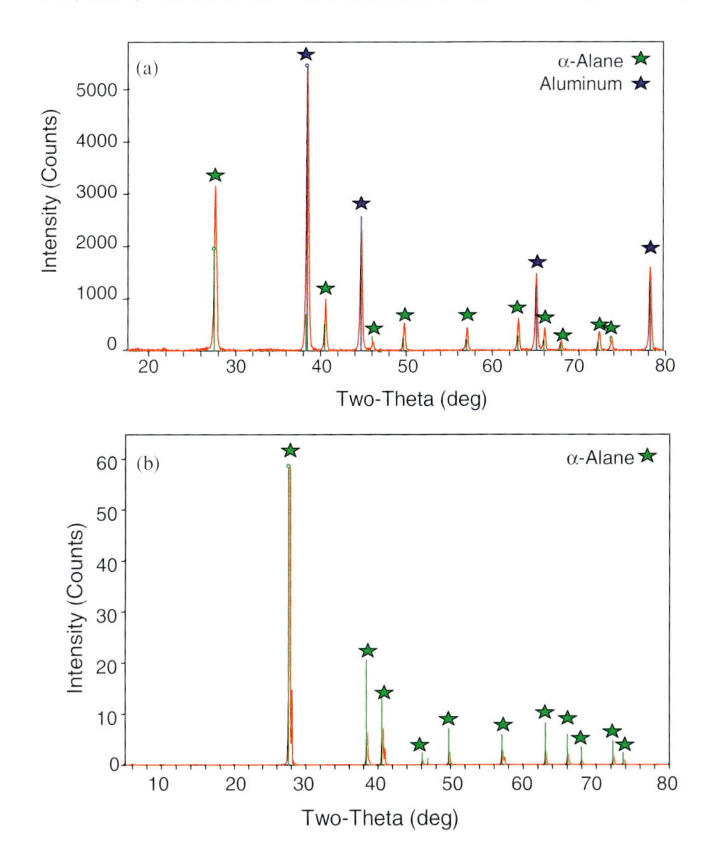

Figure 9.21 XRD patterns for products recovered from an electrochemical cell. (a) Alane separated from reaction mixture as the THF adducts. When heated under vacuum to remove THF, the solid partially decomposes, losing hydrogen and affording aluminum. (b) Alane is separated using triethylamine to stabilize the adduct, prevent polymerization, and increase the yield, Ref. [55].

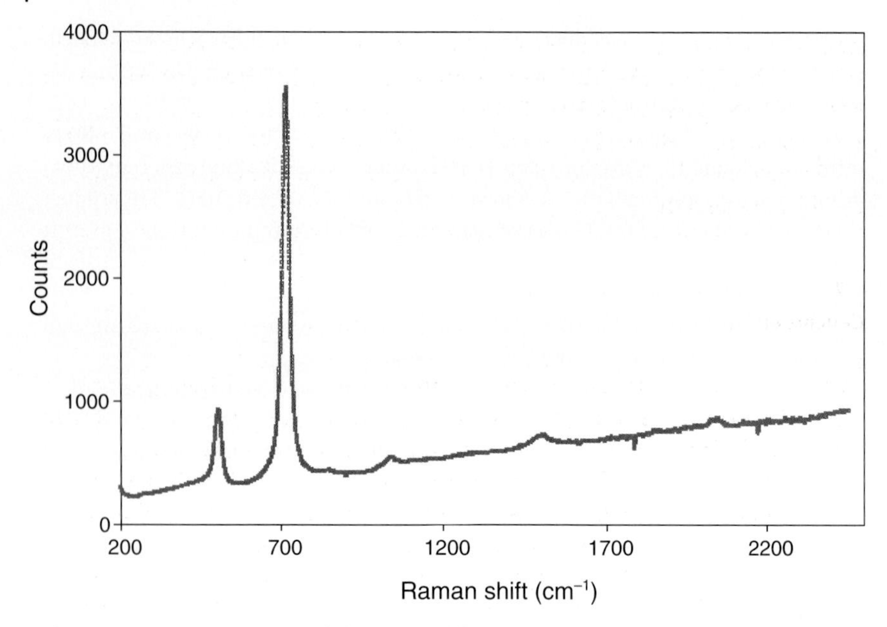

Figure 9.22 Raman spectrum of alane isolated from an electrochemical cell, Ref [55].

As part of this work [55] it was also possible to produce AlH_3-TEDA, described previously, using the electrochemical method without the use of catalysts. Producing AlH_3-TEDA was originally used as a mean to confirm the formation of AlH_3 molecules in the electrochemical cell. Although AlH_3 cannot be separated from

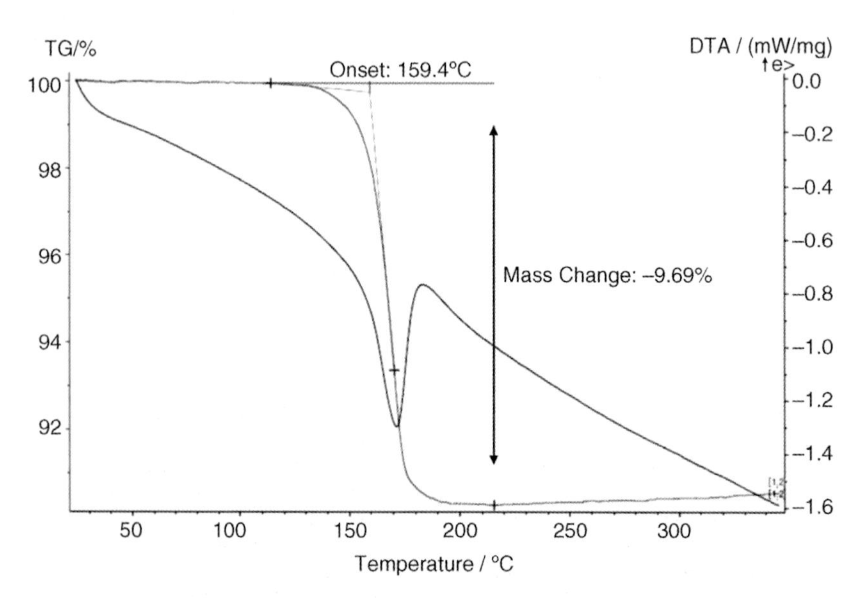

Figure 9.23 Thermal gravimetric analyzer (TGA) decomposition of electrochemically generated alane, releases almost full H_2 capacity expected in AlH_3, Ref [55].

AlH_3-TEDA it was selected because it precipitates as a white solid and its formation can be observed visually. The visual observation of the high rate of formation of relatively large quantities of AlH_3-TEDA was a good qualitative indication of the high rate of producing AlH_3 in the electrochemical cell. The formation of the AlH_3-TEDA adduct indicates that a whole range of other adducts could be formed, using the electrochemical route.

9.7 Conclusion

In this chapter the potential of using aluminum hydride as a hydrogen storage material has mainly been emphasized. Other applications of aluminum hydride and aluminum hydride adducts were mentioned. The main thrust behind the use of aluminum hydride for hydrogen storage is it capacity of 10.1 wt% hydrogen and the mild conditions of hydrogen release. The thermodynamic properties of different phases of alane and the transformations and stabilities of these phases were discussed, which can be useful when matching a specific application. It was emphasized that the stability of alane at room temperature and higher is due to hindered kinetics, possibly the inhibited hydrogen association at the surface. The structure and phases of alane are not only of great importance for understanding the thermodynamics and kinetics of alane but also for understanding a whole class of aluminum-based compounds such as alanates. The existing methods to make alane – either by direct hydrogenation or by chemical means – are not practical for hydrogen storage applications. If alane is to be used as a hydrogen storage medium new and novel methods of regenerating alane must be found. In this chapter a proposed method by Graetz *et al.* was shown. Our method (Zidan *et al.*) of generating alane using electrochemistry in conjunction with direct hydrogenation was reported. The characterization methods and results of alane product from the electrochemical cell were also described. However, more work in the areas of chemical regeneration and electrochemical regeneration is ongoing and is planned to continue to achieve higher yields and better efficiency. The alane research effort is predicted to grow and attract the interest of many researchers and more innovative ideas for forming alane are expected to emerge.

References

1 Smith, D.P. (1948) *Hydrogen in Metals*, The University of Chicago Press, Chicago.
2 Borelius, G. (1927) *Ann. Phys.-Berlin*, **83**, 0121–0136.
3 Ransley, C.E. and Neufeld, H. (1948) *J. Inst. Met.*, **74**, 599–620.
4 Opie, W.R. and Grant, N.J. (1950) *Trans. Am. Inst. Min. Metall. Eng.*, **188**, 1237–1241.
5 Sharov, M.V. and Legkie, S. (1958) *Akad. Nauk SSSR Inst. Met.*, **1955**, 365.
6 Eichenauer, W., Hattenbach, K., and Pebler, Z. (1961) *Z. Metallk.*, **52**, 682–684.
7 Birnbaum, H.K., Buckley, C., Zeides, F., Sirois, E., Rozenak, P., Spooner, S., and Lin, J.S. (1997) *J. Alloys Compd.*, **253**, 260–264.

8 Ichimura, M., Katsuta, H., Sasajima, Y., and Imabayashi, M. (1988) *J. Phys. Chem. Solids*, **49**, 1259–1267.
9 Hashimoto, E. and Kino, T. (1983) *J. Phys. F Met. Phys.*, **13**, 1157–1165.
10 Wolverton, C., Ozolins, V., and Asta, M. (2004) *Phys. Rev. B*, **69**, 144109-1–144109-16.
11 Finholt, A.E., Bond, A.C., and Schlesinger, H.I. (1947) *J. Am. Chem. Soc.*, **69**, 1199–1203.
12 Stecher, O. and Wiberg, E. (1942) *Ber. Dtsch. Chem. Ges.*, **75**, 2003–2012.
13 Chizinsky, G., Evans, G.G., Gibb, T.R.P., and Rice, M.J. (1955) *J. Am. Chem. Soc.*, **77**, 3164–3165.
14 Rice, M.J. and Chizinsky, G. (1956) Non-solvated aluminum hydride, U.O.o.N. Research Contract ONR-494(04) ASTIA No. 106967.
15 Brower, F.M., Matzek, N.E., Reigler, P.F., Rinn, H.W., Roberts, C.B., Schmidt, D.L., Snover, J.A., and Terada, K. (1976) *J. Am. Chem. Soc.*, **98**, 2450–2453.
16 Matzek, N.E. and Musinski, D.F. (1975) United States Pat. 3 883 644.
17 Matzek, N.E. and Musinski, D.F. (1975) *Chem. Abstr.*, **83**, 45418.
18 Graetz, J. and Reilly, J.J. (2005) *J. Phys. Chem. B*, **109**, 22181–22185.
19 Graetz, J. and Reilly, J.J. (2006) *J. Alloys Compd.*, **424**, 262–265.
20 Kato, T., Nakamori, Y., Orimo, S., Brown, C., and Jensen, C.M. (2007) *J. Alloys Compd.*, **446**, 276–279.
21 Wieser, H., Laidlaw, W.G., Krueger, P.J., and Fuhrer, H. (1968) *Spectrochim. Acta, Part A*, **A24**, 1055–1089.
22 Chellappa, R.S., Chandra, D., Gramsch, S.A., Hemley, R.J., Lin, J.F., and Song, Y. (2006) *J. Phys. Chem. B*, **110**, 11088–11097.
23 Orimo, S., Nakamori, Y., Kato, T., Brown, C., and Jensen, C.M. (2006) *Appl. Phys. A-Mater.*, **83**, 5–8.
24 Claudy, P., Bonnetot, B., and Letoffe, J.M. (1979) *J. Therm. Anal.*, **15**, 129–139.
25 Zaluski, L., Zaluska, A., and Stromolsen, J.O. (1995) *J. Alloys Compd.*, **217**, 245–249.
26 Liang, G., Huot, J., Boily, S., Van Neste, A., and Schulz, R. (1999) *J. Alloys Compd.*, **291**, 295–299.
27 Liang, G., Huot, J., Boily, S., and Schulz, R. (2000) *J. Alloys Compd.*, **305**, 239–245.
28 Wiberg, E. and Uson, I.R. (1955) *Rev. Acad. Cienc. Exact. Fis. Quim. Nat. Zaragoza*, **10**, 35–41.
29 Wiberg, E. and Gosele, W. (1956) *Z. Naturforsch. B*, **11**, 485–491.
30 Wiberg, E., Graf, H., and Uson, R. (1953) *Z. Anorg. Allg. Chem.*, **272**, 221–232.
31 Wiberg, E. and Noth, H. (1955) *Z. Naturforsch. B*, **10**, 237–238.
32 Ruff, J.K. and Hawthorne, M.F. (1960) *J. Am. Chem. Soc.*, **82**, 2141–2144.
33 Ruff, J.K. and Hawthorne, M.F. (1961) *J. Am. Chem. Soc.*, **83**, 535–538.
34 Schomburg, G. and Hoffmann, E.G. (1957) *Z. Elektrochem.*, **61**, 1110–1117.
35 Atwood, J.L., Bennett, F.R., Elms, F.M., Jones, C., Raston, C.L., and Robinson, K.D. (1991) *J. Am. Chem. Soc.*, **113**, 8183–8185.
36 Lewis, P.H. and Rundle, R.E. (1953) *J. Chem. Phys.*, **21**, 986–992.
37 Almennin, A., Halvorse, S., and Haaland, A. (1969) *J. Chem. Soc. D-Chem. Commun.*, 644–648.
38 Frigo, D.M., Vaneijden, G.J.M., Reuvers, P.J., and Smit, C.J. (1994) *Chem. Mater.*, **6**, 190–195.
39 Bent, B.E., Nuzzo, R.G., and Dubois, L.H. (1989) *J. Am. Chem. Soc.*, **111**, 1634–1644.
40 Masu, K., Tsubouchi, K., Shigeeda, N., Matano, T., and Mikoshiba, N. (1990) *Appl. Phys. Lett.*, **56**, 1543–1545.
41 Moss, R.H. (1984) *J. Cryst. Growth*, **68**, 78–87.
42 Ashby, E.C. (1964) *J. Am. Chem. Soc.*, **86**, 1882.
43 Graetz, J., Chaudhuri, S., Wegrzyn, J., Celebi, Y., Johnson, J.R., Zhou, W., and Reilly, J.J. (2007) *J. Phys. Chem. C*, **111**, 19148–19152.
44 Bogdanovic, B. and Schwickardi, M. (1997) *J. Alloys Compd.*, **253**, 1–9.
45 Ke, X.Z., Kuwabara, A., and Tanaka, I. (2005) *Phys. Rev. B*, **71**, 184107-1–184107-7.
46 Turley, J.W. and Rinn, H.W. (1969) *Inorg. Chem.*, **8**, 18.
47 Bauer, S.H. (1950) *J. Am. Chem. Soc.*, **72**, 622–623.
48 Brinks, H.W., Istad-Lem, A., and Hauback, B.C. (2006) *J. Phys. Chem. B*, **110**, 25833–25837.

49 Brinks, H.W., Langley, W., Jensen, C.M., Graetz, J., Reilly, J.J., and Hauback, B.C. (2007) *J. Alloys Compd.*, **433**, 180–183.

50 Yartys, V.A., Denys, R.V., Maehlen, J.P., Frommen, C., Fichtner, M., Bulychev, B.M., and Emerich, H. (2007) *Inorg. Chem.*, **46**, 1051–1055.

51 Brinks, H.W., Brown, C., Jensen, C.M., Graetz, J., Reilly, J.J., and Hauback, B.C. (2007) *J. Alloys Compd.*, **441**, 364–367.

52 Graetz, J., Wegrzyn, J., and Reilly, J.J. (2008) *J. Am. Chem. Soc.*, **130**, 17790–17794.

53 Graetz, J., Wegrzyn, J., Reilly, J., Johnson, J., Celebi, Y., and Zhou, W.M. (2008) Annual Review Meeting EERE-DOE.

54 Marlett, E., Frey, F., Johnston, S., and Kaesz, H. (1972) United States Pat. 3642853.

55 Zidan, R., Garcia-Diaz, B., Fewox, C., Stowe, A., Gray, J., and Harter, A. (2009) *Chem. Comm.* doi:

56 Senoh, H., Kiyobayashi, T., Kuriyama, N., Tatsumi, K., and Yasuda, K. (2007) *J. Power Sources*, **164**, 94–99.

57 Senoh, H., Kiyobayashi, T., and Kuriyama, N. (2008) *Int. J. Hydrogen Energy*, **33**, 3178–3181.

58 Adhikari, S., Lee, J.J., and Hebert, K.R. (2008) *J. Electrochem. Soc.*, **155**, C16–C21.

59 Alpatova, N.M., Dymova, T.N., Kessler, Y.M., and Osipov, O.R. (1968) *Russ. Chem. Rev.*, **37**, 99–114.

60 Clasen, H. (1962) Germany Pat. 1141 623.

61 Wong, C.P. and Miller, P.J. (2005) *J. Energetic Mater.*, **23**, 169–181.

10
Nanoparticles and 3D Supported Nanomaterials

Petra E. de Jongh and Philipp Adelhelm

10.1
Introduction

Many light metal hydride systems are discussed in this book. However, none of them is currently able to meet all the demands for practical on-board hydrogen storage: high volumetric and gravimetric density, reasonably high hydrogen equilibrium pressure at room- or fuel cell operating-temperature, fast kinetics for both loading and unloading and ample reversibility. A variety of strategies is being pursued to meet these goals: ball-milling to improve the kinetics and add catalysts, searching for new yet unknown material compositions, and mixing several different compounds ("reactive hydride composites" or "destabilized hydrides") [1–6]. Some approaches are remarkably successful, such as ball-milling in general to improve the kinetics [2, 3], and the addition of a small amount of a Ti-based catalyst to improve the kinetics of both hydrogenation and dehydrogenation of $NaAlH_4$ [1]. However although steady progress is reported, we are still far from meeting simultaneously all criteria for on-board storage.

In this chapter we discuss an alternative approach: altering the properties of a given material by nanosizing and/or supporting the material. Although this approach is relatively new for hydrogen storage applications, it has been known for a long time in other fields such as heterogeneous catalysis, where a high surface/volume ratio is essential. Interesting material classes are unsupported clusters, nanoparticles and nanostructures, and 3D supported (or scaffolded) nanomaterials. In general the crystallite size of the materials discussed is below 10 nm. This is a clear distinction from materials prepared by ball-milling, presently the most common processing technique, by which crystallite sizes of 10–30 nm or above (depending on the material) are achieved. Furthermore, in general (though not always) the alternative preparation techniques used to obtain these nanosized materials (gas-phase deposition, melt infiltration, or solution-based synthesis techniques) allow a much better

Handbook of Hydrogen Storage. Edited by Michael Hirscher

ISBN: 978-3-527-32273-2

control over morphology and particle size than ball-milling. However, most important is that entering this size regime, one can expect important changes in hydrogen sorption properties, such as improved kinetics and reversibility, and, possibly, a change in thermodynamics. The study of unsupported clusters, nanoparticles and nanostructures is mostly aimed at advancing the fundamental knowledge and understanding of how these effects may be used to the benefit, involving studies on relatively simple binary ionic or interstitial hydrides. However, supporting or confining the materials might be especially relevant for the recently developed more complex systems. These bring new challenges such as slow kinetics and lack of reversibility due to phase segregation for multiple component systems, and the release of unwanted gasses such as NH_3 and B_2H_6 for novel compositions that have a high hydrogen content, but also contain nitrogen or boron.

We will start this chapter by discussing the potential impact of size on the hydrogen sorption thermodynamics and kinetics for metal (hydride) nanoparticles. We will then turn to the experimental results, and, first, treat the literature that deals with the production of unsupported metal (hydride) clusters, nanoparticles and nanostructures and their hydrogen sorption properties. Small particles of light metals are especially difficult to prepare and stabilize, as the sensitivity to oxidation is enhanced by the large volume to surface ratio. We will briefly report on size effects for clusters of transition metals. Then we discuss in more detail Pd(H) nanoparticles. Although strictly speaking not a light metal hydride, extended research has been performed on Pd(H), and it is, hence, an interesting model system to illustrate size effects in metallic (or interstitial) hydrides. As the last type of materials we will treat the ionic hydrides, formed from alkali and alkaline earth metals. In detail we will show results on preparation strategies and first hydrogen sorption results for magnesium-based compounds, being the most investigated example of an ionic binary hydride. Until now, as far as we are aware, it has not been possible to prepare clusters or unsupported nanoparticles (<10 nm) of more complex materials such as alanates and boronates.

Recently much interest has been aroused by materials that we will denote as "3D supported nanomaterials". An alternative term that is used, and that denotes roughly the same class of materials is "scaffolded materials". The term "nanocomposites" is also found in the literature, but it is confusing as it is used for several quite different classes of materials. The field is relatively new, and not many experimental results are available yet. We will first discuss which effects might be expected due to addition of a support. This comprises both fundamental effects (such as influence on the particle shape and electronic interaction) as well as more practical implications (such as potentially improved thermal and mechanical properties of the hydrogen storage system). Of course a disadvantage is the weight penalty due to the presence of the support, which can be minimized by using light and highly porous materials. The characteristics of the two most commonly used supports (nanoporous silica and carbon) and the deposition of an active material onto these supports are discussed in detail. Finally, we will turn to experimental results for different supported nanomaterials: ammonia borane, sodium alanate, magnesium-based materials and lithium boronate, and end this chapter with a short outlook.

10.2
Particle Size Effects

10.2.1
Thermodynamics

It is well known that the fundamental physical and chemical properties of materials can change drastically when entering the nanosize regime (usually regarded as below 100 nm). A famous illustration is gold, which upon nanosizing shows different colors (due to plasmon resonance light absorption), a strong reduction in melting temperature (657 °C for 2.5 nm particles compared to 1063 °C for bulk gold) and increased chemical reactivity (with gold nanoparticles being an efficient oxidation catalyst, for instance for the epoxidation of propene) [7]. For metal nanoparticles these changes in properties upon decreasing the feature size are relatively well understood. However, for metal hydride nanoparticles, both theoretical and experimental data are scarce. It is not possible to sketch the full picture, as we are interested in reversible hydrogen sorption, and, hence, in the difference in thermodynamics of the metal hydride with respect to the metal upon nanosizing. Nevertheless, it is useful to briefly discuss general trends and concepts. The changes that can be observed when progressing from bulk to nanosized materials can be schematically divided into two types (or two size regimes). For small clusters (roughly below 10^4–10^5 atoms, so, depending on the type of material, up to ~5 nm) the number of atoms and valence electrons per cluster is countable. This leads to profound effects of the exact cluster size on the electronic and structural properties, and hence stability. At larger sizes these effects become less significant, and an impact on stability is observed which scales smoothly with the particle size, and is related to the surface to volume ratio of the particles. These effects are significant up to roughly 10–30 nm, depending on the material.

Starting with the smallest sizes, a first important parameter determining the stability is the total number of valence atoms in a cluster or nanoparticle. In a series of seminal experiments in the 1980s, Knight *et al.* [8] measured clearly periodic abundance patterns in the mass spectrum when generating gas-phase alkali metal clusters (see Figure 10.1). Increased stability was observed for metal clusters containing 2, 8, (18), 20, 34, 40, 58, and so on, valence electrons. The occurrence of these so-called magic numbers is an electronic effect, which can adequately be described by a simple physical model (the Jellium model) which assumes a uniform background of positive charges and a delocalized cloud of valence electrons. The magic numbers are due to the electronic shell filling of the orbitals with increasing number of cluster valence electrons, similar to the stability effects observed for the elements when progressing through the periodic table. These effects are particularly strong and well understood for the alkali metals. Alkaline earth metals, such as magnesium, in which also van der Waals bonding between the atoms contributes significantly, are less well understood, but still show relatively strong effects experimentally. For the transition metals (including the noble metals) which possess strongly oriented and partly filled d-type orbitals with relatively small differences in energy levels, electronic effects are much less pronounced and quantitative under-

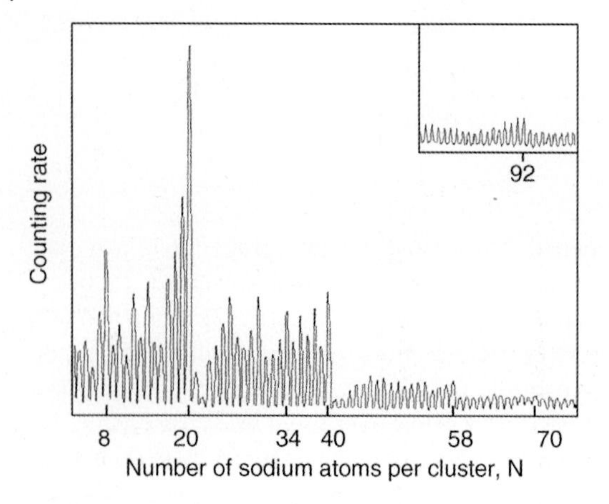

Figure 10.1 Illustration of electronic magic numbers: abundance of sodium clusters as a function of size measured by mass spectrometry in a molecular beam of sodium seeded in Ar. Relatively high stability is found for cluster sizes corresponding to filled electron shells such as those containing 8 (1p), 20 (2s), 34 (1f), and 40 (2p) valence electrons. Reprinted with permission from [8]. Copyright (1984) by the American Physical Society.

standing is still under development, although clearly increased stability for clusters with an even number of electrons is observed for Cu, Ag, Au, V and Cr [9].

A second important effect for small clusters is a geometrical effect. A large fraction of the atoms resides at the surface, and has a lower coordination number than bulk atoms. As a result, the average bonding energy per atom, and hence the thermodynamic stability of clusters, is strongly influenced by the surface atoms. The most stable cluster sizes are those corresponding to geometrically closed shells (shell number given by n), hence minimizing the number of surface atoms. For instance, for particles with an icosahedral shape, the most stable configurations are given for atoms numbers N satisfying the formula

$$N = 1 + \sum_{1}^{n} 10n^2 + 2 \tag{10.1}$$

This effect on the stability of clusters is also experimentally observed for alkali (earth) metals (see Figure 10.2) and, in general, extends up to larger cluster sizes than the electronic magic numbers.

For metal hydride clusters, much less is known about electronic and geometrical size effects, except for very small clusters that can be relatively easily traced with computational methods. Results on other ionic compounds, such as ZnS clusters, indicate less well defined size effects, and only for small cluster sizes [11]. As the size dependence will be clearly different than for metal clusters, potentially there is a large influence of size on the stability *differences* between small hydride and the corresponding metal clusters. However, the practical impact of these effects is probably limited, as in the bulk experimental preparation of light metal (hydride) clusters, generally, polydisperse samples are obtained, and hence pronounced effects of

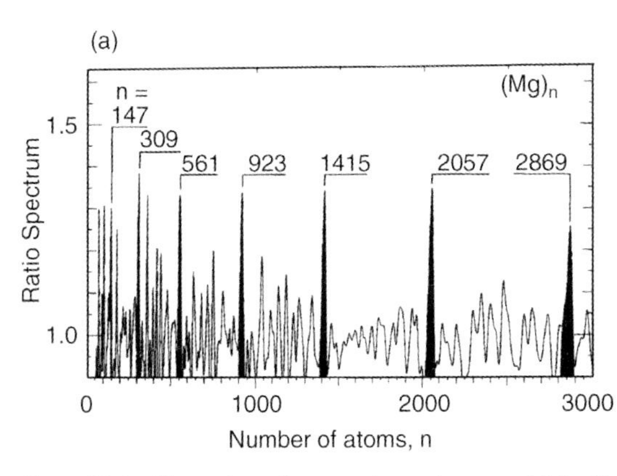

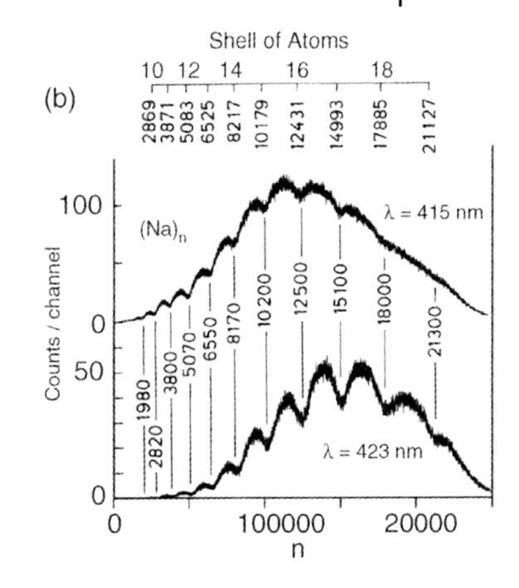

Figure 10.2 Illustration of geometric magic numbers: (a) mass spectra indicating abundance of Mg clusters in the gas phase, maxima indicate the most stable clusters. Reproduced with permission from [10], published by Elsevier, 1991; (b) mass spectra of photo-ionized Na clusters, minima indicate the most stable structures, shell closings listed at the top. Reproduced with permission from [11], published by Elsevier, 1996.

an exact given number of atoms or electrons will not be significantly observed. Furthermore, the geometrical magic numbers are valid for clusters with the thermodynamically most stable shape. However, in practice, particle shapes vary substantially, as they are also determined by the presence and local nature of supports or stabilizing agents, or related to an intrinsic nonequilibrium nature of the preparation method (and usually the kinetic barriers to change the particle shape are large).

As a result of surface minimalization, small nanoparticles can have different crystal shapes than bulk crystals. A commonly observed shape is the icosahedron, but cuboctahedral, truncated decahedra, and other more complicated geometries are also observed. However, some of these crystal shapes, like the icosahedron with its fivefold symmetry, are not space-filling symmetries, causing additional internal strain in the particle. When the particles grow larger the stability loss due to internal strain becomes larger than the energy gain due to surface minimalization, and a transition to the bulk crystal lattice (for metals mostly fcc or bcc) is observed. The size at which the transition takes place depends on the material; for instance 1 nm gold particles are polytetrahedral, while 1 nm Pt particles have an fcc structure, and the transition is at around 100 atoms for Al, but at tens of thousands of atoms for Na [11]. One of the few materials for which the impact of particle shape on the hydrogen sorption properties has been investigated experimentally is palladium (see Section 10.3.2). It should be noted that, in practice, the surface energy (and hence nanoparticle shape) is also strongly influenced by surface stabilization effects due to surfactants or ligands in solution, passivation of the surface, or the presence of a substrate.

Although the exact number of atoms, and hence shell filling effects, becomes less important for larger particles, Figure 10.3(a) illustrates that the fraction of atoms at the surface (the so-called dispersion) remains an important factor up to relatively

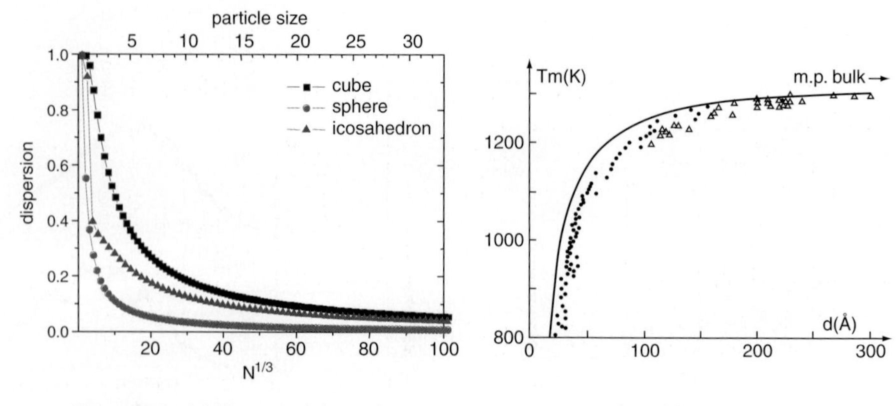

Figure 10.3 (a) Theoretical dispersion as a function of the number of atoms for different geometries, top gives a rough indication of particle sizes (based on 3 Å atoms). (b) Melting temperatures for gold nanoparticles as a function of size (points = experimental values). Reproduced with permission from [12], published by Elsevier, 1981.

large sizes. Figure 10.3(b) shows the experimental melting point depression of gold nanoparticles scaling inversely with the nanoparticle size. The overall contribution of the surface energy remains significant up to sizes of 10–30 nm, and in the simplest approximation (assuming an incompressible spherical particle) the decreasing thermodynamic stability can be described as [12]:

$$\mu = \mu_\infty + \frac{2\gamma}{\varrho r} \tag{10.2}$$

where r is the radius of a spherical particle, μ the chemical potential, γ the surface free energy and ϱ the specific mass.

To evaluate whether the hydrides become relatively less stable with decreasing particle size, the important factor is the difference in surface energy between the metal and the hydride phase (while taking changes in molar volume into account). If the surface energy of the hydride is larger than that of the corresponding metal, this will generally result in higher hydrogen equilibrium pressures for smaller particles, and vice versa if the situation is reversed.

The energies of several bulk metal surfaces have been experimentally determined, either in the liquid phase near the melting temperature, or for the solid phase. In general the values are 20–30% higher for the solid than for the liquid [13]. The surface energies for metals directly correlate with the heat of evaporation (respectively, sublimation for solid metals). This can be readily understood: if you assume you could perfectly cleave a metal without any energy change (neglecting surface reconstruction), the resulting energy of the two surfaces should be the same as the work needed to break the bonds. Very schematically, for the alkali metals small surface energies are found (ranging from 0.5 J m^{-2} for Li to 0.1 J m^{-2} for Cs), for the alkaline earth metals somewhat higher values (0.8 J m^{-2} for Mg to 0.4 J m^{-2} for Ba), for the transition metals typically 1.5–2.5 J m^{-2}, with the exception of the noble metals which have somewhat lower surface energies (1.0–2.0 J m^{-2}) [13–15].

Unfortunately, for the hydrides virtually no experimental data for the surface energies are available. Assuming again a direct correlation with the cohesive energies, the surface energies for the ionic hydrides would be expected to be much larger than for the corresponding metals (as these hydrides are much more stable than the corresponding metals), while for the interstitial hydrides, such as palladium hydride, smaller differences would be expected. However, the simple approximation to directly link the surface energy to the cohesive energy is expected to break down for the hydrides, as it is also very relevant how the surface is terminated, and surface reconstruction plays an important role. For ionic compounds very different effective surface energies can be found for compounds with similar cohesive energies, as illustrated by the example of anatase TiO_2 (0.7 J m^{-2}) and rutile TiO_2 (2.2 J m^{-2}), for which the cohesive energy difference is less than 10%. The surface energies also depend on the exact exposed crystallographic plane. Although, as far as we are aware, no reliable experimental data for the surface energies of hydrides are available, there are some data on other ionic compounds, such as metal oxides, which have widely different surface energies in the range of 0.5–3 J m^{-2}. Some data for the hydrides have recently become available from density functional calculations, such as 0.1–0.3 J m^{-2} for $LiBH_4$ [16] and 1.8 J m^{-2} and 0.5 J m^{-2} for the (101) and (110) surfaces of MgH_2 [17].

Generally stating, it would be expected that for metals that more readily form surface hydrides than bulk hydrides (such as palladium) the hydride would be expected to become relatively more stable when decreasing the particle size, while for metals which more readily form bulk hydrides than surface hydrides (such as magnesium) a destabilization of the hydride with decreasing particle size would be expected. As a rough indication: if the surface energy of the hydride is assumed to be 1.5 J m^{-2} higher than that of the corresponding metal, and considering a hydride that has an enthalpy of formation of 75 kJ mol^{-1}, and shows 20–30% volume expansion on hydriding, the enthalpy of formation of the hydride could be 20% lower for particles with radii smaller than 5 nm compared to the bulk, if we assume that no surface reconstruction takes place [18]. It does not seem very useful to attempt similar calculations for complex hydrides, as in this case the situation is complicated by the presence of multiple phases with unknown crystallite sizes upon dehydriding.

An important drawback of the macroscopic surface energy approach to calculate the relative stability of small particles is that the *effective* surface energies of both metal and hydride will depend on particle size, at least for particles smaller than 10 nm. For instance for Pd, Au and Pt nanoparticles, surface energies of about 20–80% higher are reported for the metal nanoparticles than for the bulk [19, 20]. This is in line with expectation. An important assumption in this macroscopic approach is that the particles are spherical, but this spherical drop model seriously underestimates the real surface area for small particles (see Figure 10.3(a)). Furthermore, for small particles edge, kink and step atoms are also a major fraction of the surface atoms, and having a lower coordination number than an atom in a perfect surface they will have a much higher "effective surface energy". Another major problem is that surface reconstruction upon decreasing the particle size is not taken into account.

A different approach to the stability of hydride nanoparticles is bottom-up, using computational techniques to construct equilibrium (lowest energy) configurations for the clusters, nanoparticles or thin films of the metal and corresponding hydride, and evaluate the difference in stability between the two as a function of particle size. Although the possibilities are rapidly increasing, these calculations are still limited to small cluster sizes due to computational restraints. As a relevant example we will only briefly discuss various types of calculations for the case of the ionic hydrides MgH_2 and NaH.

Magnesium hydride is cheap and contains 7.7 wt% hydrogen, making it one of the most attractive hydrogen storage materials. However, one of the fundamental problems is that thermodynamics dictate that hydrogen desorption from bulk magnesium hydride only takes place at or above 300 °C for 1 bar H_2, which is a major impediment for practical application. Hence, an interesting question is whether, for nanoparticles or clusters, the hydride might thermodynamically be destabilized with respect to the metal compared to the bulk compounds. Figure 10.4, showing the thermodynamic stability of the hydride as a function of cluster size, illustrates the results of calculations on small clusters [21, 22]. Although different computational approaches were used to address the issue (Goddard *et al.* developed a reactive force field for magnesium and magnesium hydride from quantum mechanical data on magnesium clusters, while Wagemans *et al.* used density functional calculations) in both cases a clear relationship was found between the cluster size of MgH_2 and the heat of formation. The stability of the hydride decreased rapidly for particle sizes below 1.0 nm, and a convergence to bulk stability values for particle sizes above 2 nm was observed. It might be relevant to state that these results for the stability as a function of particle size cannot be fitted by just assuming a fixed difference in surface energy between magnesium and magnesium hydride spherical

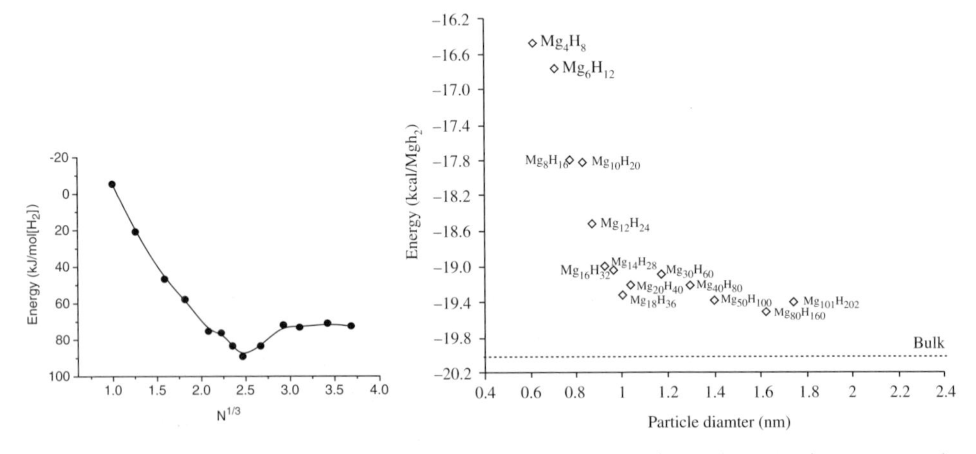

Figure 10.4 Stability of MgH_2 clusters (a) from DFT calculations as a function of the number of Mg atoms in the cluster N. Adapted with permission from [21]. Copyright (2005) American Chemical Society; (b) versus particle diameter. Reprinted with permission from [22]. Copyright (2005) American Chemical Society.

particles, indicating that the simple surface energy approach falls short in this size region. Other groups also did calculations for $Mg(H_2)$ thin films and nanowires with DFT [23] and the project augmented plane wave method [17] and obtained similar results. Recently, calculations were also performed for NaH cluster sizes up to 400 atoms. In this case also, a clearly lower thermodynamic stability was found for smaller clusters than for the bulk hydride [24].

Giving only the average value for the dehydriding enthalpy is not accurate in the case of clusters, as it will depend on the amount of hydrogen left in the sample. In general, the first hydrogen molecules are easy to extract from a fully hydrided cluster, but the last hydrides are significantly more difficult to extract. It is remarkable that both for MgH_2 and NaH, calculations indicate that specific hydride clusters intermediate between very small clusters and bulk seem more stable than the bulk hydride [21, 24]. Although the details of the stability of even small hydride clusters are not yet well understood, computational methods at least predict that, up to the size of 1–2 nm, the thermodynamics of hydrogen sorption will be different from that of bulk hydrides, which might be very relevant for the application of these ionic hydrides as hydrogen storage materials.

10.2.2 Kinetics

Not only thermodynamic factors, but also the slow kinetics of hydrogen sorption are a barrier to practical application for many materials. In this section we will briefly discuss the expected impact of the particle size on kinetics. First, it is important to realize which step in the hydrogen sorption process is rate limiting, which depends not only on the type of material, but also on the specific experimental conditions. Taking the absorption of hydrogen (which is usually slower than the desorption at a given temperature) as an example, the following steps can be discerned:

1) hydrogen physisorption at the surface
2) dissociation of the hydrogen molecule/chemisorption
3) surface penetration of the hydrogen into the material
4) diffusion through the hydride layer to the interface with the metallic phase
5) conversion of metal into metal hydride

In some cases the situation can be further complicated, for instance by the presence of subsurface sites with distinctly different absorption energies, such as for PdH_x (see Figure 10.9). For a mathematical analysis of the different steps we refer to the literature [25, 26]. Hydrogen physisorption (step 1) is a fast process, and hence generally not rate limiting. Hence, an equilibrium concentration at the surface (depending on the heat of adsorption and the hydrogen pressure) can be generally assumed. Also steps 3 and 5 are rarely rate limiting. The rate-limiting step for absorption of hydrogen is usually either the dissociation of hydrogen at the surface (which depends on the surface concentration and, hence, indirectly on the physisorption parameters), or the diffusion of hydrogen through the hydride to the metal/hydride interface, or a combination of both. Which step dominates also depends on the stage in the hydriding process: upon increasing hydride layer thickness (assuming hydriding from the outside to the core of the particles), rate limitation due to diffusion through the hydride layer will gradually become more important [27]. For hydrogen desorption analogous steps can be discerned.

When considering the potential impact of decreasing particle size on these properties, two major factors are important: a decrease in solid state diffusion distances, and enhanced hydrogen dissociation and association rates (due to an increase in specific surface area, and to the fact that step, kink and corner atoms are present in higher concentrations in nanoparticles, and are more reactive than atoms in a perfect, flat, surface). In specific cases other factors might play a role. For instance it is possible that mechanical strain in the bulk material poses an activation barrier for hydriding. This strain might be absent in small, freely expanding nanoparticles, but it might also be enhanced due to boundary conditions. It is also possible that phase conversion depends on nucleation of the hydride phase in the metal, which is determined by defects. Nucleation rates might be enhanced in small particles due to nucleation at the surface, but, on the other hand, diminished due to a lower concentration of defects. These specific cases will not be discussed in detail in this section.

So far, a major step forwards in the processing of hydrogen storage materials was the application of ball milling in the 1990s [2, 3]. It can also serve as an illustration of the impact of particle (or more accurately grain) size on the kinetics of hydrogen sorption. For magnesium (hydride) the absorption is significantly slower than the desorption of hydrogen. Figure 10.5 shows the impact of ball-milling, which can reduce the crystallite size down to 30 nm, on hydrogen sorption. For μm-sized particles the absorption is so slow that it is negligible even after 2 h, while for 50 and 30 nm-sized crystallites 80% of the hydrogen content is attained in 70 and 30 min, respectively.

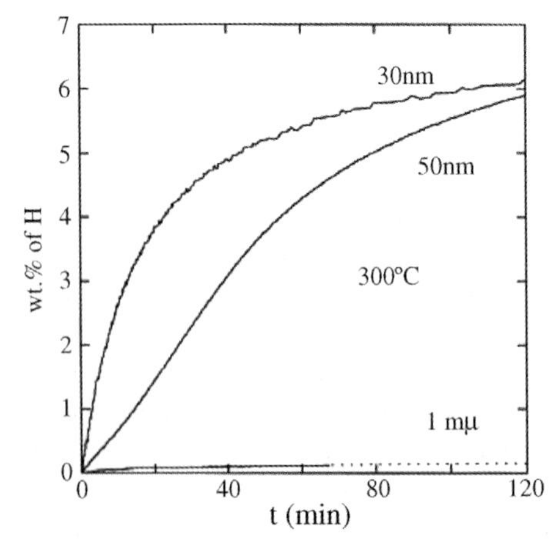

Figure 10.5 Effect of crystallite size reduction by ball-milling on the hydrogen desorption rates in vacuum at 300 °C. Reproduced with permission from [2], published by Elsevier, 1999.

Let us consider movement of hydrogen in the solid state. Following Fick's diffusion laws an estimate of the diffusion time t for hydrogen to diffuse over a distance x is:

$$t = \frac{x^2}{2D} \qquad (10.3)$$

where D is the effective diffusion coefficient for a given temperature and hydrogen concentration. The real situation is more complicated than this 1D situation due the particle shape and other factors. However, the diffusion time is expected to scale roughly with the square of the distance (which is related to the particle radius), which is not in disagreement with the results shown in Figure 10.5. Unfortunately, it is not straightforward to use ball-milled samples to investigate particle size effects. A problem is that not only the average crystallite size is reduced, but also the effective diffusion coefficient might be altered by, for instance, the introduction of metastable phases (such as γ-MgH_2 instead of β-MgH_2) or high defect densities due to the amount of mechanical energy added. Furthermore, for materials that are easily passivated in air, like magnesium, breaking up of the passivating oxide layer can also play an important role, especially for the first hydrogen absorption run.

In general, the diffusion of hydrogen in metals, and in interstitial hydrides such as Pd, is relatively fast, as the lattice is essentially not changing, and the neutral hydrogen atom can move from one interstitial position to another [28]. However, for the ionic hydrides the situation is quite different, as the mobility of the H^- anion is much lower, especially in the hydride phase, and the metal phase typically has very limited hydrogen solubility. It is difficult to reliably measure diffusion coefficients, as they depend strongly on morphology (grain boundaries). Varying values are reported, but, for instance, for the β-phase MgH_2 diffusion coefficients are of the order of 10^{-14}–$10^{-13}\,m^2\,s^{-1}$ at 300 °C, while being 3–4 orders of magnitude faster in the metallic α-phase [29–31]. This means that diffusion into particles of 1 μm or larger will take more than a few minutes. Hence, decreasing the effective diffusion distances, and hence crystallite size, is essential for ionic hydrides to achieve acceptable hydrogen release and uptake rates.

Another rate-limiting step can be the hydrogen dissociation/absorption at the surface [32, 33]. This is generally not expected to be a problem for compounds involving transition metals (including noble metals such as Pd and Pt) that are classical (de)hydrogenation catalysts. The oriented d-electron orbitals have a strong interaction with hydrogen molecules and atoms, resulting in ready adsorption, and destabilization of the bond between the two hydrogen atoms in the molecule, and hence acceptable hydrogen association and dissociation rates. However, the alkali and alkaline earth (or the s-type) metals generally do not chemisorb hydrogen. Due to the limited interaction, for well-defined flat surfaces the association and dissociation of hydrogen is very slow. In practice, the rate is often determined by defects, kink and step sites, or contamination in the samples. Bringing the particle size into the low nanometers regime is expected to result in a strong increase in surface reaction rates, not only due to an increase in effective surface area, but probably even more importantly due to an increase in the average reactivity of the surface atoms. However, it should be realized that, in several cases, it is more practical to use catalysts at the surface to increase these reaction rates, although effective and compatible catalysts have not been identified for all metal hydrides.

For alanates, boronates and systems combining several compounds, the situation is much more complex. The determining step for the kinetics and reversibility in these systems is usually poorly understood. Probably solid-state transport in heterogeneous phases and phase segregation play an important role. Some complex systems start with a mixture of compounds, and, generally, on dehydrogenation several different solid phases are formed (for instance NaH and Al in the case of $NaAlH_4$). Classical hydrogenation catalysts usually do not work, pointing towards the fact that the association or dissociation of hydrogen is not the rate-limiting factor in these cases. This is also illustrated by the fact that Ti-based catalysts are the best known today for $NaAlH_4$, greatly improving both the hydrogenation and dehydrogenation rates, while the addition of Pd or Ni has no large effect. However, in most cases the slow reaction kinetics seem related to diffusion or transport in the solid phases, and for these complex systems it has already been experimentally shown that bringing down the effective crystallite size can have a large impact (see Section 10.6).

10.3
Non-Supported Clusters, Particles and Nanostructures

Metal or metal oxide nanoparticles and colloids are widely used for catalytic, electronic, optical and magnetic applications, and hence an extended range of preparation techniques is available [34–36]. Direct preparation of metal hydride nanoparticles or colloids is rare. The preparation techniques can roughly be divided into three categories: gas-phase preparation, liquid-based syntheses and solid-state reactions. For nanoparticles the gas-phase and solution-based techniques are the common ones, as solid-state synthesis (except ball-milling) usually involves high temperatures and hence larger crystallite sizes. In general, gas-phase techniques only produce small quantities of metal nanoparticles, which are mostly used for fundamental studies and specialty applications. It is relatively easy in the gas phase to grow small clusters, and have adequate control over the number of atoms in a cluster. Also, it is possible to have controlled reaction in the gas phase, for instance by adding small amounts of hydrogen or oxygen. The growth of nanoparticles in solution is also widely used. Especially, noble metal nanoparticles can been grown in solution with good control over size and shape. However, in general, the growth of light metal hydrides such as magnesium or lithium in solution is cumbersome, as most of them react very readily with common solvents.

In this part of the chapter we will discuss the preparation and available information on the hydrogen sorption properties of both nanoparticles and nanostructures, classified according to the type of material. We will first brieflyly discuss very small clusters (generally not more than a few 100 atoms) produced in the gas phase. This comprises both metallic hydrides, and metals that only form surface hydrides. Nevertheless, also for the latter, the hydrogen sorption capacity of nanoparticles can be appreciable due to the large surface to volume ratio. The second part will deal with interstitial hydrides, discussing palladium (hydride) nanoparticles as the most prominent example. The last part discusses ionic hydrides, focussing on magnesium hydride, which so far is the most widely studied nanosized alkaline (earth) metal that is relevant for hydrogen storage. For magnesium, both growth in solution and using gas- and liquid-based techniques have been experimentally demonstrated. However, understanding the influence of the particle size on hydrogen storage properties is still in its infancy, as only very recently were the first results on the hydrogen sorption properties reported.

10.3.1 Transition Metal Clusters

Gas phase clusters of almost any metal can be conveniently formed using so-called cluster sources. Metal atoms in the gas phase can be created, for instance, by laser ablation, evaporation or ion- or magnetron-sputtering of a target. Clusters are then formed either by spontaneous nucleation and growth, or the supersaturation is increased by an adiabatic expansion. Typically, a molecular beam is created in which the metal particles are entrained in an inert gas flow. Depending on the source, neutral and/or charged clusters are produced. In addition, it is possible to manipulate the charge state of clusters after production and to mass select them. The (mass-selected) clusters can then be characterized or deposited using soft-landing techniques. A wide range of characterization techniques is available, such as mass spectroscopy, photoionization spectroscopy, and optical spectroscopy (IR, Raman or UV), and, after deposition, diffraction experiments, scanning tunneling microscopy and electron microscopy.

It is expected that the applicability of this type of clusters for practical hydrogen storage applications is limited. It is difficult to envision how the gas-phase method of cluster generation could be upscaled to produce clusters in industrial quantities and become cost-effective. Furthermore, gas phase clusters are present in relatively low concentrations, and increasingly tend to sinter when the concentration is increased. However, stabilization might be achieved by condensation in a cooled liquid, or by capturing the clusters in a support material. The unique control over the size and the ample characterization possibilities make small metal (hydride) clusters very important for fundamental research on size effects. For the smallest sizes direct comparison with DFT-based calculations is possible. However, such calculations become prohibitively expensive for systems containing more than 10–20 transition metal atoms. This might be resolved by using semi-empirical potentials and/or molecular dynamics to assess the somewhat larger sized clusters.

The reactivity of the smallest clusters towards hydrogen molecules is directly related to the stability fluctuations discussed in Section 10.2.1. For metal clusters with open d-shells (most of the transition metals) such as Ni, Co and Fe, large variations in cluster reactivity are observed, which are related to the valence electronic structure of the specific cluster sizes. Although the kinetics of reaction are strongly dependent on the exact cluster size, for clusters larger than about 10 metal atoms the binding energies show only a weak size dependence [9, 37]. For closed shell systems, for instance, for Cu, Ag, and Au, a clear correlation between the stability predicted by the electronic and geometrical magic numbers, and the experimental reactivity towards small molecules is observed [38].

Even for metals that do not form bulk hydrides, very appreciable hydrogen storage capacities might be achieved due to their high surface/volume ratio. In several studies of transition metal clusters loadings much higher than one hydrogen (or deuterium) atom per metal have been reported, as illustrated in Figure 10.6 [39, 40].

It is clear that Ni, Pt and Rh clusters containing less than 50 atoms can be significantly enriched in hydrogen compared to flat surfaces, containing up to 8, 5 and 3 deuterium atoms per Rh, Pt and Ni metal atom, respectively. This can probably be ascribed to a combination of enriched stoichiometry of the corner and edge metal ions, a preference of hydrogen to bind to bridge and face sites instead of higher coordination sites, and, in some cases, a contribution of subsurface hydrogen sorption. Furthermore, certain small metal clusters can bind hydrogen molecularly with appreciable binding energies. The high hydrogen coordination number raises the question whether very highly dispersed (either very small clusters or even monoatomically dispersed) transition metals on carbon might be suitable as hydrogen storage materials, as proposed by Yilderim *et al.* [41].

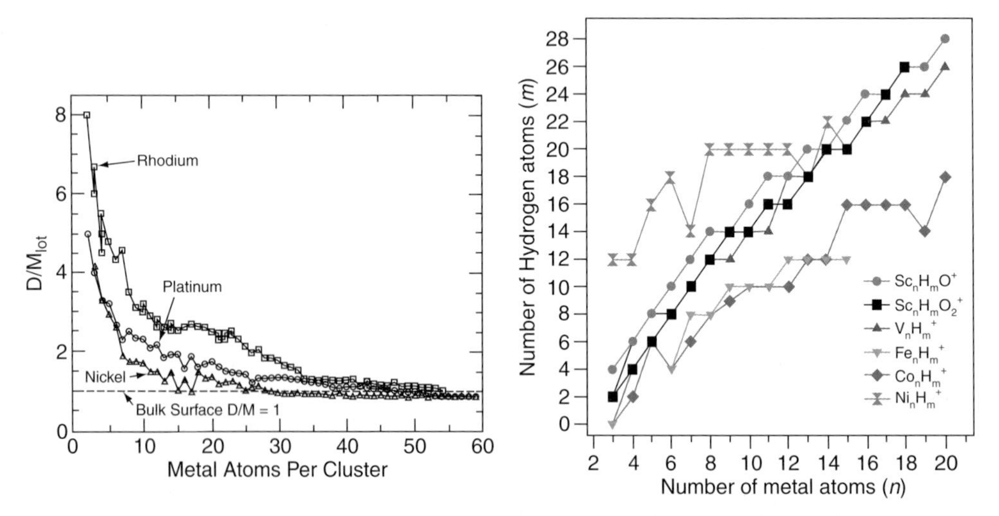

Figure 10.6 (a) Deuterium uptake D/M versus the number of metal atoms for clusters of Pt_n^+, Ni_n^+, and Rh_n^+. With permission from Springer Science and Business Media [39]; (b) largest number of hydrogen atoms present in $Sc_nH_mO^+$, $Sc_nH_mO_2^+$, $V_nH_m^+$, $Fe_nH_m^+$, $Co_nH_m^+$, and $Ni_nH_m^+$ complexes as a function of cluster size ($n = 3$–20). Reprinted with permission from [40]. Copyright (2008) American Chemical Society.

10.3.2
Interstitial Hydrides, Focussing on Palladium Hydride

Several transition metals such as V, Nb, Ta, and Pd can form stable bulk hydrides, so-called interstitial hydrides; the bonding in the hydride phase is not ionic but mostly metallic in character, and the hydrogen to metal ratio is not necessarily stoichiometric. Especially, nanoparticles of noble metals such as Pd are relatively easy to prepare by various methods, such as vapor phase deposition on substrates, reductions of salts in solution (electrochemically or electroless), and the inverse micelle templated growth. They are not easily oxidized, and, in recent years, several methods have been developed to precisely control the size of the particles or clusters. Furthermore, growth in solution in the presence of surfactants and stabilizers allows control over the shape of the final particles [35, 36, 42].

Palladium nanoparticles with a size of a few nanometers supported on carbon are widely used as catalysts, for instance in three-way automotive exhaust catalysts and fuel cells, and can easily be prepared by impregnation of a porous support body with a precursor solution, followed by drying, decomposition of the precursor and, if necessary, reduction. It is well-known that the activity and selectivity of these catalysts for hydrogenation reactions depend on the palladium dispersion for particles sizes in the range 1–10 nm. It is, hence, not surprising that the interaction of Pd with hydrogen, and the influence of nanosizing, have been widely studied.

Palladium is a special case in the Group 10 metals: in contrast to Ni and Pt it forms bulk hydrides under mild conditions. The schematic Pd-H phase diagram is shown in Figure 10.7. At low hydrogen concentrations the atoms occupy a small fraction of the interstitial octahedral positions in the palladium fcc lattice (the diluted PdH or α-phase), up to about 0.015 H per Pd atom at room temperature. With increasing hydrogen gas pressure (and below the critical temperature T_c, which is 295 °C for

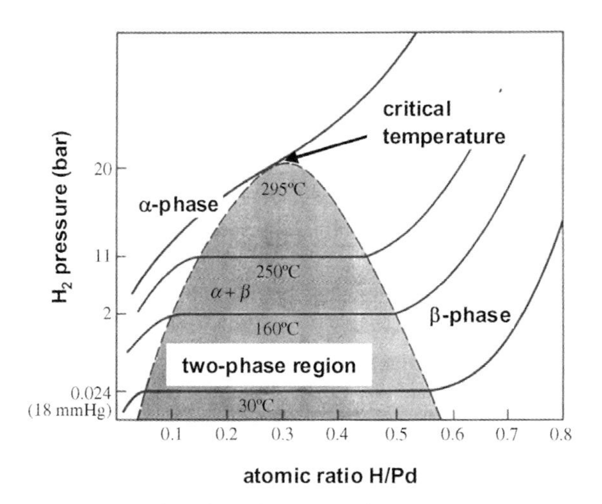

Figure 10.7 Pd–H phase diagram, isotherms are indicated (after [43]).

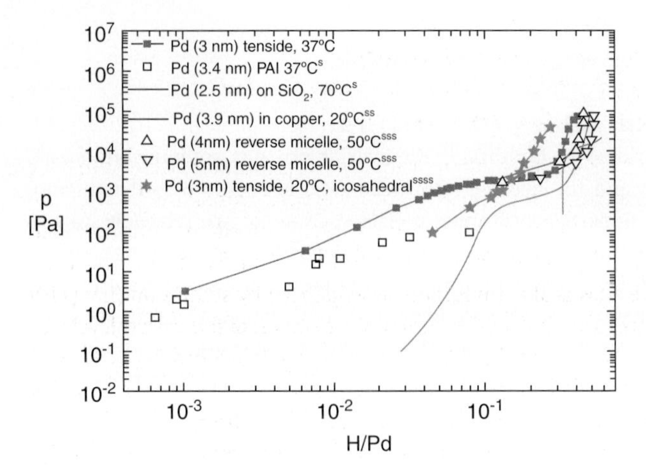

Figure 10.8 Overview of hydrogen sorption equilibrium isotherms measured for Pd nanoparticles of different sizes, shapes, and stabilizers [44].

bulk Pd), the concentration of the hydrogen in the palladium increases, and nucleation of a second phase occurs (the concentrated or β-phase) which contains hydrogen in an H/Pd ratio up to 0.6–0.7. This non-stoichiometric amount is not determined by the lattice symmetry (only some of the octahedral sites are filled), but by the electronic interaction; the energy gain is due to sharing of the electron density of the hydrogen atom with the s and d-band electron density of the palladium.

The α-phase to β-phase transition is accompanied by a lattice expansion of approximately 3.5%. A large two-phase region exists, in which the α- and β-phases are in thermodynamic equilibrium. If, at relatively high pressures, all PdH_x is converted into the β-phase, the incorporation of further small amounts of extra hydrogen is associated with rapidly increasing hydrogen pressures and a slight further expansion of the Pd fcc lattice. The coexistence of two phases (a hydrogen diluted and concentrated phase) is a common feature of most metal-hydride systems. Diffusion of hydrogen in palladium is relatively fast at room temperature, and the heat of absorption of hydrogen in palladium is slightly less than 40 kJ $(\text{mol } H_2)^{-1}$, both very favorable for on-board hydrogen storage applications. However, palladium is a relatively heavy metal and the bulk hydride contains less than 1 wt% hydrogen. As it is also very expensive, it is not of practical relevance as a bulk hydrogen storage material. Nevertheless, it is one of the most important model systems for hydrogen–metal interaction. An extensive review of the effect of the crystallite size on the interaction between palladium and hydride has been recently published by Pundt *et al.* [44].

Figure 10.8 gives an overview of some hydrogen sorption measurements for nanoparticles of different sizes and shapes. Two striking observations are that there are important differences between nanoparticulate palladium and bulk palladium, and that the scatter in the data for the palladium nanoparticles is very large. All nanoparticulate samples have a few characteristics in common compared to the bulk:

- higher hydrogen uptake at low pressures
- a lowering of the hydrogen uptake in the beta phase
- a decrease in the width of the two-phase coexistence region
- a slope in the isotherm in the coexistence region

To explain the different behavior for nanoparticles, it is important to realize that for palladium not only absorption in the bulk of the material is important, but also the sorption of hydrogen near the surface should be considered. It is known that palladium forms surface hydrides, with about 1 H per Pd surface atom. However, sorption in subsurface sites is also important, as evidenced by studies on (flat (110) surfaces [45]. Figure 10.9 shows a schematical representation of the energy levels for the hydrogen atom at the surface, in subsurface sites, and absorbed in the bulk.

Many of the experimental observations can also be explained by considering that hydrogen sorption near to the surface is more favorable than absorption in the bulk. The significant surface chemisorption and subsurface absorption explain the apparent increased solubility of hydrogen in the α-phase (the increase in hydrogen sorption at low pressures) as this is due to the relatively large surface/volume ratio. Similarly, the decreased absorption in the bulk β-phase can be explained, as there are simply less bulk absorption sites available per mole palladium in nanoparticles compared to the bulk. A combination of these two factors results in a decrease in the width of the two-phase coexistence region. The sloping isotherm in the two-phase region is still under discussion. It might be related to a distribution in particle sizes or geometries, strain, or an intrinsic property of nanoparticles [42, 44].

A very interesting question is what the influence is of particle size on the relative stability of the hydride. As indicated before, it is not really useful to talk about "the" heat of sorption, as there is a clear distribution in energies. Nevertheless, very generally speaking, one would expect the stability of the hydride to increase with decreasing particle size, due to the higher stability of the (sub)surface hydride. However, as can be seen there is a very large scatter in the data obtained experimentally. Pd nanoparticles as such are not stable at elevated temperatures as they tend

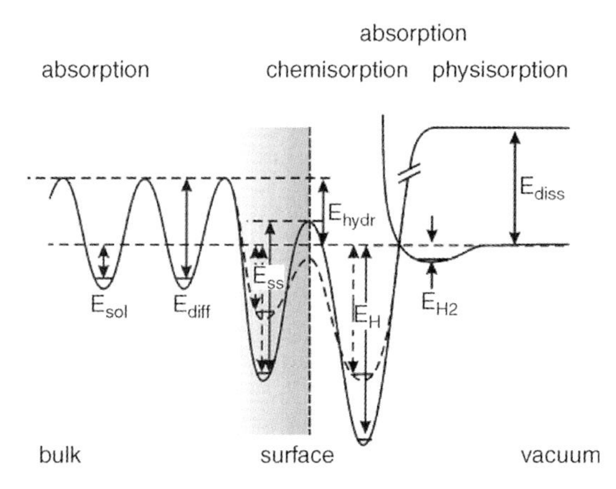

Figure 10.9 Energy levels for hydrogen at the surface of and in palladium ([44] redrawn from [45]).

to sinter, especially when hydrogen is present. A support material can prevent sintering by stabilizing the particles and inhibiting coalescence. For particles in solution, they have to be stabilized with a soft or harder matrix, for instance by employing surfactants or polymers. The scatter in the results is most likely due to the effect of the different supports, matrices and stabilizers present. It is theoretically well-known that internal stress due to clamping or confinement, or interfacial tension between particles and stabilizing polymers or surfactants has a large influence on the thermodynamics of hydrogen sorption. Furthermore, depending on the presence of stabilizers, the particle shape can be different from that of bulk crystals up to relatively large particle sizes.

10.3.3 Ionic Hydrides, Focussing on Magnesium Hydride

The binary metal hydrides that are most interesting for practical application in hydrogen storage are those of the light metals, more specifically the alkali and alkaline earth metals such as lithium, magnesium, and sodium. Unfortunately, these binary hydrides are all very stable, resulting in low hydrogen equilibrium pressures at room temperature. Magnesium hydride is the most relevant one as it is cheap and abundantly available and the least stable hydride of the series. It is also the only one for which the hydrogen sorption properties of nanoparticles or structures have been studied in somewhat more detail.

Unfortunately all alkali (earth) metals have low reduction potentials, which means that they react readily not only with oxygen and water, but also with many other compounds containing oxygen or nitrogen atoms. Due to their large surface to volume ratio the nanosized materials are especially reactive, hence the synthesis and handling is a major challenge. Furthermore bulk metals can often be "passivated" by the formation of a protecting oxide layer (typically one to a few nanometers thick), which, for materials with dimensions of only a few nanometers, would mean that most of the material is lost.

Knowledge of the synthesis, handling, and properties of alkaline (earth) nanoparticles and compounds in solution traditionally comes from two areas: organometallic chemistry and batteries. Li-ion batteries are an important class of rechargeable and lightweight batteries [46], and magnesium-based batteries are under development [47]. For these applications the low reduction potential and reactivity of nanoscale lithium and magnesium are an important challenge, which led to the development of specialized solvents and electrolytes, often based on carbonates or ethers [48, 49]. Lithium and magnesium compounds are also commonly used reactants for organometallic synthesis (the so-called Grignard reagents) while the hydrides of sodium, magnesium, potassium, and also aluminum hydrides such as $NaAlH_4$ and $LiAlH_4$ and boron hydrides such as $NaBH_4$ and $LiBH_4$ are in common use as reductants.

Already since the 1960–1970s it was known that Li, Na and Mg could be hydrided in solution under very mild conditions (typically 1 bar H_2 and room temperature or 0 °C in the presence of a catalyst). Using this type of reaction, finely divided and very

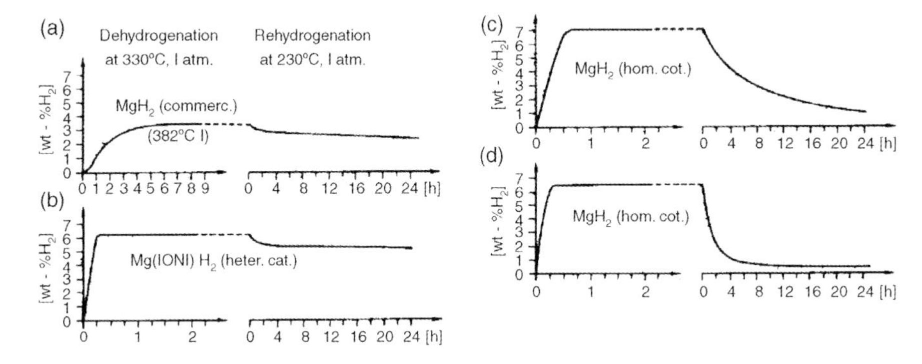

Figure 10.10 Hydrogen desorption at 330 °C, 1 bar, and rehydrogenation at 230 °C, 1 bar for (a) and (b) commercial MgH_2 powder, and (c) and (d) chemically prepared MgH_2., either without, (a) and (c), or in the presence of a Ni catalyst, (b) and (d). With permission from the IAHE [52].

reactive LiH, NaH and MgH_2 could be prepared in solution [50, 51]. One of the first to realize the potential of these chemically-prepared light metal hydrides for hydrogen storage was Bogdanović [52, 53]. He prepared MgH_2 by combining magnesium with anthracene in THF solution (forming a magnesium-anthracene complex), and then hydriding this complex in the presence of halides of chromium, titanium at room temperature at 1 bar H_2 pressure, or in a faster process at 60–65 °C under 1–80 bar H_2 pressure. The finely divided MgH_2 was isolated from the solution as a powder. Testing showed that it decomposed at 330 °C and rehydrided at 230 °C under atmospheric hydrogen pressure, much milder conditions than previously reported for hydrogen (see Figure 10.10). Furthermore, high pressure cycling experiments were performed, showing no significant loss of capacity over 30–50 cycles (using 99.9% pure H_2). The material showed a fractal-like morphology, with agglomerates of 500 nm and 50 nm crystals, and a specific surface area of 70–130 $m^2 g^{-1}$.

The hydrogen sorption properties were further investigated by Zidan *et al.* [54], who confirmed the rapid uptake and release of the hydrogen (>80% in 10 min uptake at 330 °C and 8 bar H_2 pressure) with an improvement by a factor of two in the presence of a Ni catalyst. The rapid kinetics were mostly attributed to shorter diffusion distances of the hydrogen in the solid (hydride) material, although a contribution from enhanced hydrogen dissociation due to the large specific surface area was also likely. Equilibrium sorption measurements showed that the thermodynamics were similar to those of bulk magnesium hydride, with an enthalpy of formation of the hydride of 75 kJ mol^{-1} derived from the absorption isotherms at four different temperatures. This is in agreement with what would be expected for particles with a minimum size of 50 nm.

From research into battery applications its is known that magnesium salts are unstable in most polar aprotic solvents, such as esters, acetonitrile, and the standard alkyl carbonate lithium-battery electrolyte solvents. In the literature it is reported that it is possible to have magnesium in liquid ammonia solutions, either prepared electrochemically [55] or by trapping gas-phase nanoparticles by co-condensation [56]. It is known that alkali metals can be dissolved in liquid ammonia, resulting in cations associated with solvated electrons (leading to blue solutions). However, a problem for magnesium solutions is that they are intrinsically unstable, due to the formation of the amide:

$$[2e^-, Mg^{2+}] + 2NH_3 \rightarrow Mg(NH_2)_2 + H_2$$

which upon further heating, or evaporation of the ammonia, might be expected to react toward the more stable nitride:

$$3Mg(NH_2)_2 \rightarrow Mg_3N_2 + 4NH_3$$

Imamura *et al.* claimed that this reaction could be avoided by evacuating the solution at low temperatures ($< -45\,^{\circ}C$) and adding THF to the solution. They obtained relatively large (17–23 nm) magnesium nanoparticles after removal of the solvent. However, if this was done in the presence of activated carbon much smaller crystallites (below the XRD diffraction limit) were obtained. The first hydrogen desorption measurement was complicated by the potential reaction of the magnesium with the strongly bound ammonia upon heating (also yielding hydrogen), and sample changes upon heating were not clear. However, the particles could be rehydrided quickly under mild conditions, especially in the presence of 1% Ni on carbon (>80% within 5 h at 0.5 bar and 200 °C) [56, 57].

Apart from liquid ammonia, only ethers have been proposed as suitable solvents for magnesium salts [48]. Electrochemical growth of magnesium from these electrolyte solutions leads to relatively large scale, dendritic nanostructures, and is not suitable for the production of significant amounts of nanoparticles. However, recently, Gedanken *et al.* [58] reported the combination of electrochemical deposition with ultrasonic methods to facilitate the formation of small particles. Precursors were the Grignard reagents EtMgCl and BuMgCl, which were dissolved in tetrahydrofuran (THF) or dibutyldiglyme. To increase the conductivity of the electrolyte $AlCl_3$ could be added, leading to the formation of electrochemically active $MgCl^+$ species. Electrical current pulses (forming a high density of fine metal nuclei) were followed by ultrasonic pulses that removed these metal particles from the electrode. XRD (via applying the Debye Scherrer equation) yielded average crystallite sizes of 20–30 nm, while transmission electron microscopy showed that 4 nm particles were present in the sample (see Figure 10.11). Hydrogen sorption properties were not reported [58].

Very recently Aguey-Zinsou *et al.* [59] prepared 5 nm magnesium particles in solution using tetrabutylammonium as a capping agent and stabilizer. The magnesium was generated electrochemically using a magnesium anode and cathode. The nanoparticles were separated from the solution by centrifugation, washed with THF and dried under vacuum at 50 °C. The particles were hydrided in 5 h using 20 bar H_2 at 60 °C. The thermal desorption and DTA spectra were complicated by a phase transition, melting and decomposition of the TBA upon heating (also releasing hydrogen). However, comparing the hydrided colloids to the non-hydrided ones, an extra hydrogen release starting at 100 °C was observed, and a more endothermic peak in the DTA spectrum around 165 °C. The extra amount of hydrogen released agreed with that expected for full hydrogenation of the Mg colloids. Furthermore, the

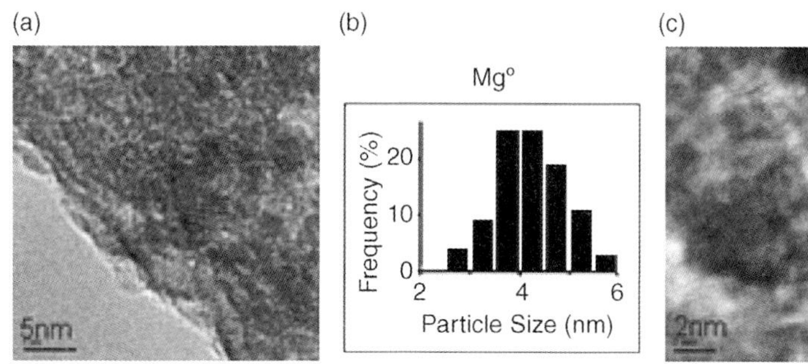

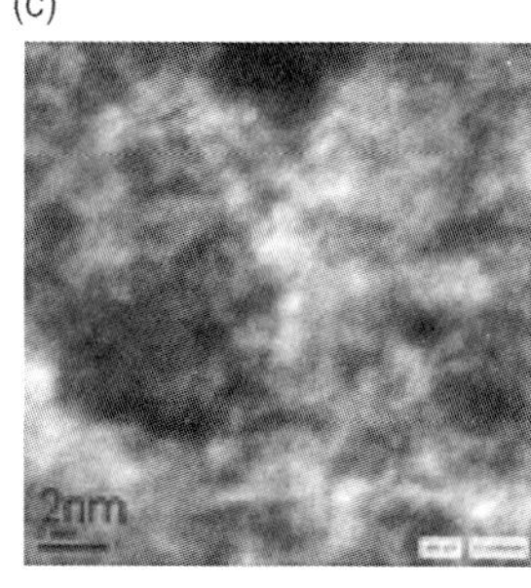

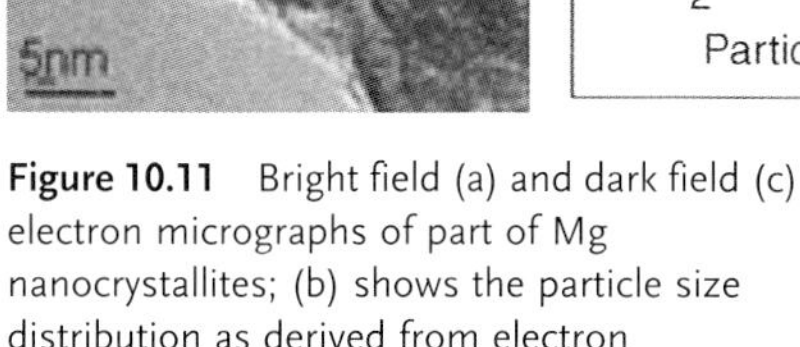

Figure 10.11 Bright field (a) and dark field (c) electron micrographs of part of Mg nanocrystallites; (b) shows the particle size distribution as derived from electron micrographs. The crystallite size derived from XRD was 20–30 nm [58]. Reproduced with permission of The Royal Society of Chemistry.

thermodynamic stability seemed different than that of bulk MgH_2, as dehydriding was possible at 85 °C and 3 mbar H_2 pressure (while for bulk MgH_2 this would need pressures below 0.1 mbar assuming 74 kJ (mol H_2)$^{-1}$ enthalpy of formation). The decrease in stability could be due to the high surface/volume ratio, but might also be induced by the stabilizer [59]. In an alternative approach Aymard *et al.* first electrochemically deposited Li on (ball-milled) MgH_2/C electrodes, which led to the formation of 10–50 nm Mg particles embedded in a LiH matrix. Subsequent removal of the LiH (leaching with a 1 M triethylborane in THF solution) led to nanosized Mg which could be fully hydrided at 100 °C under 10 bar of H_2 in 1 h [60].

Several groups reported the gas-phase growth of magnesium nanoparticles in a vacuum or inert atmosphere [61–64]. Particle sizes ranged from 1.4 to 100 nm. Usually a hexagonal platelet shape was found, and if no special measures were taken, the particles were passivated by the base pressure of oxygen, growing MgO shells of about 3 nm [63, 64]. The shape was clearly influenced by the passivating oxygen layer. Kimura *et al.* [62] protected the metal particles by trapping them in a matrix of hexane at cryogenic temperatures. Synthesis was performed by subliming a Mg tip in an atmosphere of 6N He gas. Mg particles with different sizes, ranging from 1.4 to 40 nm were formed. The particles were then transferred to a nitrogen atmosphere, and magnetic susceptibility measurements were performed in a glove box in a quartz cell. Excellent control over the particle sizes for the smaller particles was obtained (1.4 ± 0.3 nm, 1.7 ± 0.3 nm), 2.5 ± 0.8 nm, 5.2 ± 2.9 nm, and 40 ± 25 nm), with the 1.4 nm particles containing about 60 atoms. The authors state that the measurements were not hampered by significant oxidation of the nanoparticles. Clear quantum size effects of the small magnesium particles were observed (enhancement of the paramagnetism), which could be ascribed to the importance of the electronic state of the corner atoms of the particles. Unfortunately, hydrogen sorption properties were not reported.

Recently, several authors have reported first results on the hydrogen sorption properties of magnesium (hydride) nanostructures with different shapes, such as wires, platelets and spheres grown from the vapor phase [65–67]. In general, the hydrogen sorption kinetics were very fast compared to bulk, non-catalyzed magnesium (hydride), and no activation cycles were needed. A note of caution is that it might be difficult to exclude metal contamination during growth, as the investigators typically used stainless steel or Inconell sample containers and substrates, which contain high amounts of different metals, most notably nickel. The XRD characterization generally shows only magnesium (hydride) as a crystalline phase. Often the wires disintegrated due to mechanical strain on cycling more than 5–15 times, and hence might be used as a source to generate nanoparticles [65, 67].

Materials with a relatively well-defined morphology were obtained by Chen *et al.* [65] Commercial Mg was evaporated in a stainless steel tube, and deposited on a cooler (300 °C) stainless steel substrate. The morphology of the deposits was tuned by varying the Ar flow (400–800 ml min^{-1}), temperature (750–950 °C), and heating time. At 950 °C Mg nanowires were formed predominantly, and their aspect ratio could be tuned by varying the flow rate. The hydrogen sorption properties of nanowires with diameters of 30–50, 80–100, and 150–170 nm were measured (see Figure 10.12) The activation energies for hydriding/-dehydriding were dependent on the nanowire diameter, and were for the smallest diameter only 33.5/38.8 kJ mol^{-1}

Figure 10.12 Scanning electron micrographs (a) and hydrogen sorption properties (b) for MgH_2 nanowires with diameters of 30–50, 80–100, and 150–170 nm (decreasing size indicated with arrow). Hydrogen pressure 4–20 bar for absorption, and 0.2–6 bar for desorption. Reprinted with permission from [65]. Copyright (2007) American Chemical Society.

(compared with that of catalyzed and ball-milled MgH_2/Nb_2O_5 of 63–65 kJ mol^{-1}). Furthermore, the formation enthalpy for the hydride depended on the nanowire diameters, and was 63.3, 65.9 and 67.2 kJ (mol H_2)$^{-1}$ for the 30–50, 80–100 and 150–170 nm diameter wires, respectively. This suggests a lower enthalpy of formation than that of bulk MgH_2 (74 kJ mol^{-1}) [65].

In conclusion, we can say that both with solution-based and vapor-based methods it is possible to prepare magnesium (hydride) nanoparticles and structures. In solution the synthesis is complicated by the low reduction potential of the magnesium, while in the gas phase preventing the oxidation of the nanoparticles is a major challenge. Reports on the relation between size and structure and hydrogen sorption properties are still scarce, but fast kinetics and low activation barriers are clearly proven, and there are indications of a change in thermodynamic stability. We are still far from a fundamental understanding of these effects, as different factors such as not only size, but also shape, oxidation and mechanical stress in the nanostructures might play a role. Given the rapid progress in recent years, it will be very exciting to see how the field develops in the future.

10.4 Support Effects

The idea of using porous materials as a support for nanosized active species is widely used in the field of catalysis. Depending on the application, a support material can fulfill a number of different purposes, of which Figure 10.13 gives a schematic overview. It is very difficult to separate all these effects and in a practical experiment one will always end up with a combination of factors, depending on the system (choice of metal hydride, support and synthesis conditions).

We will now briefly discuss all the different effects, first those that affect the intrinsic hydrogen sorption properties of the active phase deposited on the support

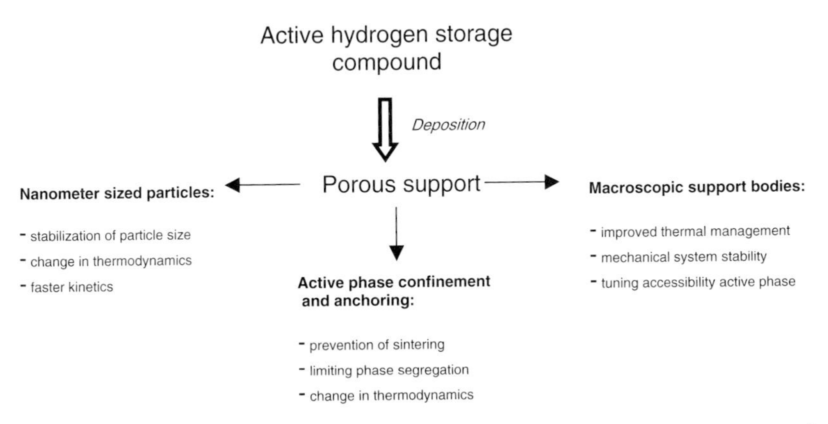

Figure 10.13 Schematic illustration of the potential impact of using nanoporous supports for hydrogen storage materials.

(like the influence on particle size or physical confinement) and then those related to impact on the overall system characteristics (such as overall thermal conductivity or mechanical stability of the material packed into the hydrogen storage tank).

10.4.1
Stabilization of Small Particle Sizes

In heterogeneous catalysis, porous support bodies are widely used to obtain a good dispersion of the nanosized active metal. Only the surface of the metal is active and as often expensive metals of limited worldwide resources are involved (Pd, Pt, Rh, Co), it is essential to maximize the surface to volume ratio. Metal surface atoms gain significant mobility at temperatures of roughly half the melting temperature (the "Tamman temperature"). Due to the large surface energy contribution, small metal particles have a strong tendency to sinter at around this temperature or above. To prevent sintering, the nanoparticles are anchored on a support, usually either an oxide such as SiO_2, Al_2O_3 ZrO_2, or MgO, or porous carbon materials. The support plays multiple roles: it facilitates the preparation of very small particles, prevents sintering during catalytic activity that is often taking place at relatively high temperatures and under chemically reactive atmospheres, and also improves the overall mechanical properties and stability of the system, allowing packing in large reactors. Furthermore, there are several examples in which the support–metal interaction also influences the catalytic activity of the metal nanoparticles.

Figure 10.14 shows two examples of supported nanoparticles from heterogeneous catalysis. A 20 wt% loading of 4 nm NiO particles has been deposited inside the mesopores of SBA-15, an ordered mesoporous silica. Reduction results in highly

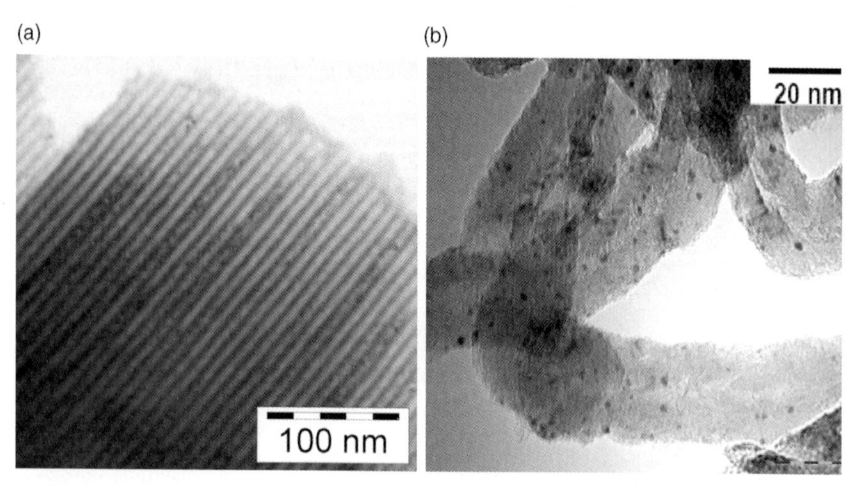

Figure 10.14 Two examples of <10 nm particles deposited on supports: (a) 4 nm NiO nanoparticles inside the mesopores of SBA-15 SiO_2. Reproduced with permission from [68]. Copyright Wiley-VCH Verlag; (b) 2 nm Pt particles on carbon nanofibers, after heating for 2 h to 700 °C in flowing N_2. Reproduced with permission from [69], published by Elsevier, 2004.

dispersed Ni nanoparticles of the same size, which are an active catalyst in hydrogenation reactions. The NiO is formed by impregnation of the porous silica with an aqueous precursor solution containing nickel nitrate, followed by drying and a special heat treatment [68]. The presence of a nanoporous support facilitates the preparation of small particles: if the solution is dried and the precursor is decomposed in the absence of a support, significantly larger NiO crystallites result.

An example showing the influence on size stability is the 5 wt% loading of 2 nm Pt particles deposited on carbon nanofibers (Figure 10.14 (b)). The platinum was deposited on the fibers by precipitation from an aqueous $Pt(NH_3)_4(NO_3)_2$ solution, where the deposition was controlled by a gradual pH change due to the decomposition of urea. After heat treatment for 2 h at 700 °C in N_2 flow, only a very modest increase in average particle size was observed (from 1.4 to 2.3 nm) which illustrates the stabilizing influence of the support on the particle size. [69]. Apart from the stabilization of particle sizes by anchoring, stabilization by confinement or trapping can also play an important role. One of the most logical examples of this phenomenon is that of metal clusters in zeolite cages [70].

In general, the active metal particle size is influenced by the experimental preparation procedure, the pore size and chemical nature of the support, and the loading of the active phase. Impregnation of a porous support with a precursor (solution) of the active phase is the most widely used method in catalysis to prepare supported nanoparticles. However, other methods, such as deposition precipitation, ion-exchange and melt-infiltration have also been developed. In the past few years these types of methods have also begun to be explored for the preparation of nanosized metals for hydrogen storage. An additional complication is that the relevant light metal hydrides are not easily reduced from the oxidic state while for catalysts first forming the oxide, and then reducing it to metal nanoparticles is a common approach. Furthermore, (the precursors of) most of the light metals are not compatible with the aqueous solutions widely used in the preparation of catalytic materials. Nevertheless the experience and insight developed in the preparation of supported nanoparticles in heterogeneous catalysis can serve as a useful basis for the preparation of light metal (hydride) nanoparticles for hydrogen storage.

10.4.2
Limiting Phase Segregation in Complex Systems

A second type of effect is that restricting mobility by anchoring onto a support, or confining in the pores of a support, can limit macroscopic phase segregation. This effect is related to the particle size stabilization but is especially relevant in more complex systems involving multiple phases. In the quest for the optimum thermodynamic properties for a hydrogen storage system, increasingly complex systems are studied, basically moving from binary hydrides such as PdH and MgH_2, via complex hydrides such as $NaAlH_4$ or $LiBH_4$, to even mixed systems (so-called "reactive hydrides" or "destabilized hydrides") such as $LiBH_4 + MgH_2$ [4, 5] and $MgH_2 + LiNH_2$ [71, 72]. For these systems it is increasingly difficult to achieve

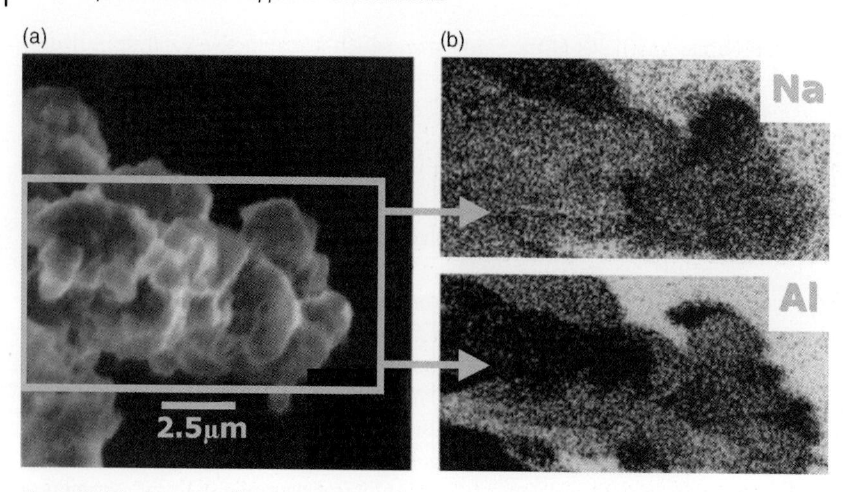

Figure 10.15 Scanning electron microscopy (a) combined with elemental analysis, (b) of Ti-doped $NaAlH_4$ after dehydrogenation, showing the about 3 μm-sized Na- and Al-rich regions (black areas). Reproduced with permission from [73], published by Elsevier, 2000.

ready reversibility of the hydrogen sorption, even if the thermodynamics are favorable. Macroscopic phase segregation probably plays an important role. For instance when dehydriding $NaAlH_4$, in the first step two different solid phases are formed:

$$NaAlH_4 \rightarrow NaH(s) + Al(s) + \frac{3}{2}H_2(g)$$

Electron microscopy elemental analysis shows that upon dehydriding phase segregation indeed occurs in μm-sized domains containing preferentially Na or Al, as illustrated in Figure 10.15.

This means that during the reverse reaction both components have to diffuse back to the interfacial region, and interdiffuse to again form one single alanate phase. In general, the diffusion of solids over μm-distances is very slow. Confinement of the phases in small pores, or anchoring of the phases to restrict their mobility might be an important factor in improving the reversibility.

Another effect is that confinement in pores might influence the phases formed for complex systems. Although not completely understood this is found experimentally, for instance, by Autrey *et al.* for the confinement of ammonia borane in mesoporous SiO_2 and carbon (see Section 10.6.1) [74]. For complex materials often not only thermodynamic considerations but also kinetics have an import influence on which phases are formed upon hydriding or dehydriding. For instance, on decomposition of $LiBH_4$ the thermodynamically most favorable reaction is:

$$LiBH_4 \rightarrow LiH + B + 2H_2$$

Nevertheless, in the decomposition of boronates, the formation of B_2H_6 and other boron hydride species is also often observed.

$$2LiBH_4 \rightarrow 2LiH + B_2H_6$$

This probably proceeds via the formation of BH_3 for kinetic reasons. Another well-known example is the decomposition of $LiNH_2$, which usually results in the formation of significant amounts of ammonia:

$$2LiNH_2 \rightarrow Li_2NH + NH_3$$

This reaction path changes completely, and the formation of ammonia is suppressed if the $LiNH_2$ is in close contact with an equimolar amount of LiH [71]:

$$2LiNH_2 + LiH \rightarrow LiH + Li_2NH + H_2$$

In these examples the formation of unwanted gaseous products can only be prevented if the kinetics for the desired reaction path are favorable, and intimate contact between the relevant reactants is ensured. Confining the material in nanopores might be very useful for increasing the effective contact area and time.

10.4.3
Metal–Substrate Interaction

From catalysis it is well-known that the metal–substrate interaction influences the reactivity of supported nanoparticles. For instance, for noble metal particles on oxidic supports, the hydrogenation and hydrogenolysis activity is much greater if the support has a higher acidity (high concentration of acidic −OH groups at the surface) than for neutral or alkaline oxidic supports. The influence of the presence of a support on the catalytic activity of metal nanoparticles has been ascribed to [70, 75–79]:

- variations in metal nanoparticle shape and structure due to the support
- electronic support effects
- specific sites at the metal–support boundary and spill-over effects

The impact of the presence of a support on the stability and reactivity of the nanoparticles is greater if the bonding between nanoparticle and substrate is strong, and if the nanoparticle is small. The nature of the bonding depends on the type of metal. Free-electron metals, like the alkali and alkaline earth metals and also aluminum, bind to the surface of an oxide through the Coulomb interaction between the surface ions and their screening charge density in the metal. Transition metals on the other hand can form quite covalent bonds across the interface, by intermixing of the metal 3d with the O 2p orbitals of the oxidic substrate. For noble metals with filled d-shells the binding is of an intermediate nature. The interaction with carbon supports is usually weaker than that for oxidic supports due to the absence of polarity [79–82]. Complete wetting of a support, or total flattening of the nanoparticles seldom occurs. Nevertheless, a distortion of the isolated 3D particle shape is usually observed [82, 83]. Equilibrium particle shapes are hard to predict computationally, as it is intrinsically difficult to couple cluster calculations for the small metal (hydride) cluster to periodical calculations for the support. The shape is most strongly

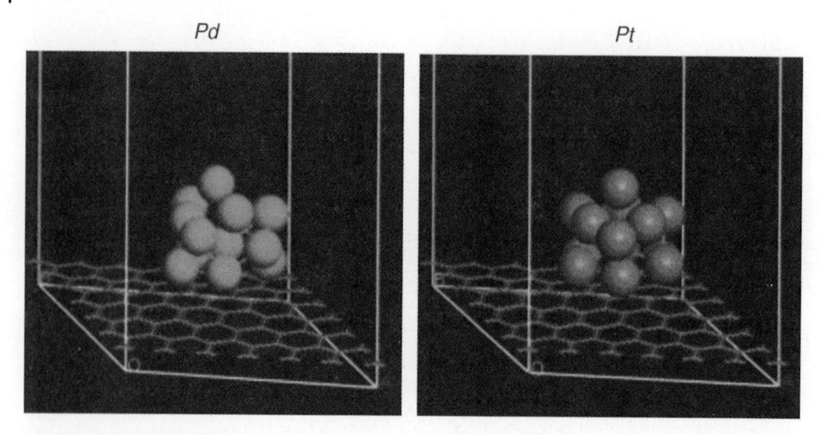

Figure 10.16 Calculated structures for 13 atom Pd and Pt clusters on an extended graphene sheet, binding energies 1.2 eV for Pd and 0.7 eV for the Pt cluster. Reproduced with permission from [84], published by Elsevier, 2004.

influenced for covalently binding metals on polar supports. The interaction and, hence, influence on the particle shape is expected to be smallest for filled-shell metals on carbon supports. As an example Figure 10.16 shows the calculated structures for 13 atom Pd and Pt clusters on an extended graphene sheet [84]. Binding energies of 1.2 eV for the Pd and 0.7 eV for the Pt cluster are found. The difference between the two noble metals can be ascribed to weaker metal–metal bonds within the Pd cluster, which result in a more open structure and more adaptation to the support structure. In this case only a slight deformation of the clusters, and a slight elongation of the bond lengths were predicted.

Calculations give information about the thermodynamically most favored morphology. However, these morphologies can only be expected to be observed for pure, uniform and well-defined support and cluster combinations, and after heating to temperatures high enough to allow the atoms of the nanoparticles to gain enough mobility to attain the equilibrium position. In practice, particle shapes are typically not determined purely by thermodynamics, but are strongly influenced by the preparation or deposition procedure [78]. Furthermore, defects and contamination often play an important role in particle anchoring, and temperature and gas atmosphere are also factors determining the particle shape. Techniques like STM and high-resolution TEM give experimental information about the morphologies on flat model substrates, while for 3D nanoporous supports the particle shapes can be derived from TEM or indirectly from EXAFS. A variety of particle shapes is reported for given nanoparticle–support systems, sometimes even within one sample [85]. In general, rather flat particle shapes are observed, for instance for Pt/Al_2O_3 and Pd/Al_2O_3 [79]. Even for systems for which theoretically almost spherical particle shapes are predicted, like Pt or Pd on carbon, experimentally clearly flattened particles are often observed, such as for the Pt/C system [86].

The changes in metal nanoparticle reactivity due to a support are often explained in terms of electronic effects. Experimentally a shift in electronic levels, especially around the Fermi level, is observed by XPS, for instance for Pd/SiO_2 [78]. However, the exact nature of this electronic effect is a matter of debate. Several models propose the transfer of electron density between metal and oxidic support. For instance, for acidic oxide supports it is proposed that the support withdraws electron density from the metal nanoparticle, leaving it electron deficient [87, 88]. Other models do not involve electron transfer between metal and support but assume that the primary interaction is a Coulomb attraction between metal particle and support oxygen ions [89]. This then leads to a redistribution of the electron density within the metal nanoparticle, leaving the surface furthest away from the support electron deficient, and, hence, more reactive than for an isolated metal nanoparticle. In general carbon support materials are considered to be neutral carriers. As carbon has such as low surface polarity, the presence of defects is often crucial for anchoring nanoparticles. These defects might be unintentional: structural defects or caused by contamination. Also, intentional modification of the surface (for instance introducing carboxylic groups by oxidation with aqueous HNO_3 solution) will significantly increase the possibility to anchor active materials to the support [90, 91].

A third factor to be considered is whether there is the possibility of reaction between the active hydrogen storage phase and the support. Especially for oxidic supports, compounds (silicates, titanates, zirconates, etc.) can often be formed. On the one hand this reduces the amount of active material present in the system. On the other hand, if the reaction is limited to an interfacial layer, it might present an effective strategy for strong bonding of the active phases. Carbon is generally not reactive, although carbides or intercalation compounds can be formed for certain metals.

The presence of hydrogen and the formation of hydrides will influence the surface energies, and hence the equilibrium shape of the metal particles [83]. The mobility of the nanoparticle atoms is also changed by the presence of hydrogen. In general, for the alkali and alkaline earth metals, a lower mobility of the metal hydride compared to the metal would be expected, while, for instance, for the noble metals the hydrides are usually significantly more mobile. Catalytically active metals such as Ni on carbon at higher temperatures are a special case where entire crystallites have been observed to "float" over the carbon substrate. This is due to reaction of hydrogen with the carbon, with the methane formation being catalyzed by the Ni nanoparticles. Effectively, the carbon is etched, lifting entire crystallites [92]. However, this phenomenon is generally expected to occur at temperatures $>300\,^{\circ}C$, higher than that relevant for reversible on-board hydrogen storage.

10.4.4 Physical Confinement and Clamping

The (de)hydrogenation process involves a change in volume, and, hence, the boundary conditions at the interface of the nanoparticles are expected to have an influence on the relative stability of the hydride with respect to the metal. For thin films it is known that 2D clamping at the thin film/substrate interface can lead to very drastic changes in, for instance, ferro-electric properties [93]. For strongly adhering Nb films on Al_2O_3 substrate, it has been shown that hydrogenation can lead to a highly anisotropic stress build up of several GPa [94]. As a result the hydrogenation properties of thin films can differ strongly from those of bulk compounds, generally showing smaller absolute values for the desorption enthalpy, a sloping plateau in the isotherm, and a large hysteresis between absorption and desorption. If, due to the mechanical strain, the film is delaminated from the substrate, the stress is released and hydrogen sorption properties close to those of the bulk materials are observed [95–98]. Also for 2D clamped nanoparticles stresses might have an influence on the hydrogen sorption properties. However, the stresses are expected to be less severe, and it is unlikely that they will exceed the limit above which the nanoparticle will detach from the substrate.

More relevant for nanoparticles might be the possibility of confinement in three dimensions. 3D boundary conditions can have a dramatic impact on hydrogen sorption properties. For instance Liang *et al.* used density functional theory to calculate the impact of physical confinement in a (10,10) single-walled carbon nanotube on the thermodynamical stability of magnesium hydride (see Figure 10.17). They predicted that confinement could lead to a large reduction in decomposition enthalpy of up to 50% [99]. However, it should be remarked that in these calculations it was assumed that the carbon nanotube was completely rigid. In practice, the effect achievable will also depend on the mechanical strength of the porous support of matrix.

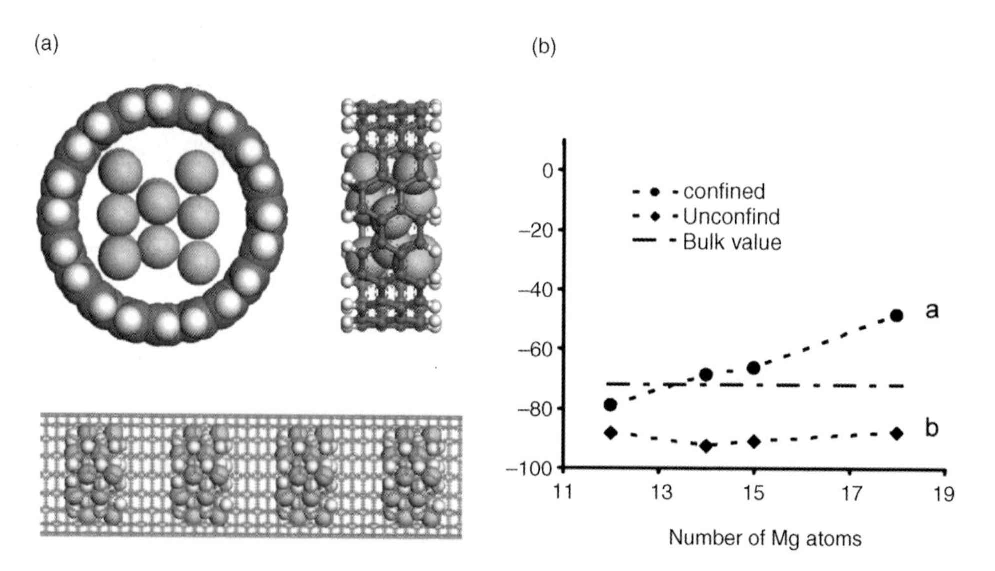

Figure 10.17 (a) schematic representation of Mg/MgH_2 particles confined in a carbon nanotube and (b) calculated hydrogen sorption energy at 0 K for a confined (curve a) and non-confined system (curve b) as a function of increasing confinement (here represented as the increasing number of Mg atoms in the confined phase). Reprinted with permission from [99]. Copyright (2005) American Chemical Society.

10.4.5 Thermal Properties of the System

Hydrogenation/dehydrogenation reactions involve the release/uptake of a large amount of heat and the transport of heat can be a limiting step during reaction. Heat transfer issues are less relevant for small "laboratory size" sample quantities, but they are of great importance when up-scaling a system to an actual tank size. In particular, during hydrogenation, the heat transport must be fast and efficient since "refilling of the tank" should be possible within minutes. For example, around 4 kg of hydrogen needs to be stored in a fuel-cell powered car with an average cruising range of 400 km [100]. Assuming a magnesium hydride tank, this would amount to 52.5 kg MgH_2. With an enthalpy of $-75\,kJ\,mol^{-1}$, the total amount of heat released during hydrogenation is around 150.000 kJ, corresponding to a power of ~500 kW (refilling in 5 min). In the case of insufficient heat transport, the resulting temperature increase could cause serious safety issues or inhibition of the hydrogenation reaction, that is, the refilling process would be prolonged.

For metal hydride powders, effective thermal conductivities of only $\sim 0.1\,W\,m^{-1}\,K^{-1}$ are found [101]. It is obvious that there is need for an additional conducting agent in a metal hydride tank. However, there is always a weight penalty that has to be paid for such systems. Possible solutions are for example the use of metallic foams [102, 103] or metal wires [104] with high thermal conductivity. Also, a support material could contribute considerably to a better heat exchange. In particular, carbon-based support materials show high thermal conductivities (the thermal conductivity of graphite is found to be around $12–175\,W\,m^{-1}\,K^{-1}$ [105]. Though the thermal conductivity depends on many parameters (particle size, porosity, tank design etc.), it was shown that the addition of even small amounts of expanded graphite significantly improved the effective thermal conductivity in metal hydride compacts. Kim *et al.* found a thermal conductivity of $>3\,W\,m^{-1}\,K^{-1}$ for compacts of $LaNi_5$ containing 2% graphite) [106]. Sanchez *et al.* found the thermal conductivity of metal hydride/expanded graphite compacts to be higher than when aluminum foams were used as conducting agent [103].

10.4.6 Mechanical Stability and Pressure Drop

The use of powders in a metal hydride tank is limited by the fact that the development of mechanical stresses during cycling (due to the different densities of the metal and the corresponding hydride) can affect the stability of the tank system. The mechanical stability of macroscopic support bodies can prevent mechanical stress and significant overall changes in shape and size, as is well known from catalytic reactors.

Another practical advantage of using a support material is that the pressure drop over a column can be significantly reduced. This is especially relevant during the hydrogenation reaction, which is usually conducted at high hydrogen pressures (>10 bar) to promote hydride formation. The pressure drop over a column can be

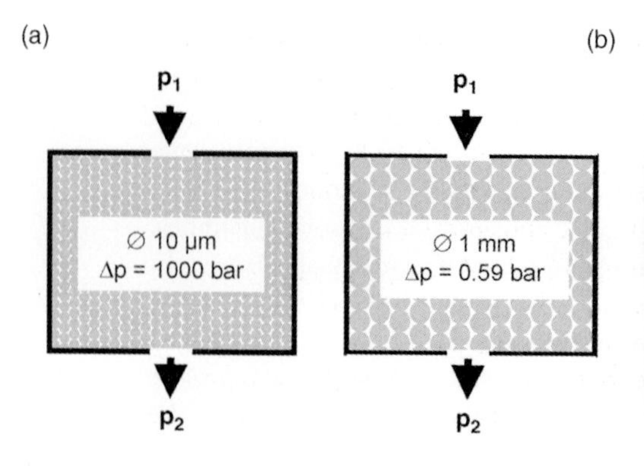

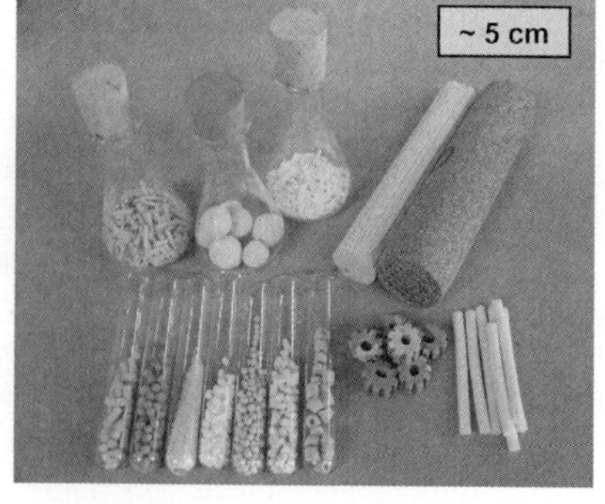

Figure 10.18 (a) Pressure drop Δp over a column for different particle sizes using the Ergun equation (GHSV = 2000 h^{-1}, reactor diameter = 2 m, reactor height = 0.5 m, spherical particle shape, 25 °C, air as mobile phase, static conditions) (b) Examples for shaped carrier bodies used as catalyst support.

calculated using the Ergun equation [107] and increases significantly with decreasing particle size (Figure 10.18a). As an example from catalysis (assuming a gas hourly space velocity of 2000 h^{-1}, a reactor of 2 m in diameter and 0.5 m in height, spherical particle shape, 25 °C, air as mobile phase and 1 bar pressure at the bottom of the column), the pressure drop over the column for particles of 1 mm in diameter is 0.59 bar. If 10 μm-sized particles are used instead, the pressure drop increases to 1000 bar, that is, a pressure of 1001 bar has to be applied to reach 1 bar at the bottom of the reactor. Even though the conditions in this example are different from refilling a metal hydride hydrogen storage tank (non-static conditions), it is obvious that one has to take potential pressure drops into account when going from laboratory size experiments to actual metal hydride tank size (especially when working with nanoparticles). In catalysis, the problem of pressure drop is solved by using highly porous support materials ("carrier bodies"). Mechanically stable, macroscopic carrier bodies can be produced in almost any form and shape for example, by extrusion or granulation. In this way, the active phase remains finely dispersed inside the support material, while the macroscopic shape of the support keeps the pressure drop at low levels. Figure 10.18(b) shows a collection of oxidic carrier bodies of different shapes. For hydrogen storage purpose, carrier bodies could be tailored to combine good thermal conductivity and minimal pressure drop at the same time. In this way, the refilling time of the tank could be effectively reduced.

The support could also prevent large scale phase segregation that might occur when the hydrogen storage compound decomposes into phases of different densities. For example, accumulation of Al ($\varrho = 2.7\,g\,cm^3$) and NaH ($\varrho = 1.39\,g\,cm^3$) at the bottom of the tank during decomposition of $NaAlH_4$ ($\varrho = 1.24\,g\,cm^3$). Such processes are especially problematic when liquid phases occur during the hydrogenation/dehydrogenation process, since they strongly enhance the mass transport. A porous support could prevent such phase segregation due to anchoring, and in the case of liquid phases additionally by capillary forces.

Counterbalancing these advantages, the support adds weight and volume to the system and hence decreases the overall system capacity (hydrogen content per weight and volume). The following example illustrates the weight penalty that has to be paid when using a support. Assuming a material with a total porosity of 1.5 $cm^3 g^{-1}$ is used as a support for MgH_2 ($\varrho_{MgH_2} = 1.45\ gcm^{-3}$, 7.7 wt% H) and the pore filling is 80%. Then, the total weight per gram stored MgH_2 would be 1.57 g and the capacity would drop from 7.7 to 4.9 wt%. Thus, a high weight loading of the active phase on the support is crucial in order to meet the requirements for on-board hydrogen storage. However, the beneficial effects of the support material on the overall system performance (as described above) could more than compensate for the additional weight and volume.

10.5 Preparation of Three-Dimensional Supported Nanomaterials

Three-dimensional supported nanomaterials basically consist of an active, nanosized species that is deposited on a three-dimensional support material exhibiting a large surface area. The final properties of such supported nanomaterials are determined by the nanoparticle species itself, the support material and the interactions between them. Depending on the support, these interactions can be physical and/or chemical in nature. In the first part of this section, the most commonly used support materials in the field of hydrogen storage, namely carbon and silica, are introduced in terms of their relevant properties. In the second part, the different synthesis strategies for the preparation of 3D supported nanomaterials are discussed.

10.5.1 Support Materials

For a support material that is suitable for use in hydrogen storage systems, the following key requirements have to be considered:

- light weight (keeping the weight penalty low)
- chemical inertness (avoiding unwanted side reactions with the active phase)
- porosity (to maintain small particle size and allow a high loading of the active phase)
- mechanical stability (compensating stress/strain during cycling)
- low price

The economic importance of catalysis led to the development of a large variety of support materials with different properties. Most frequently, alumina, silica, zeolites and carbon-based materials are used for this purpose [91, 108]. Taking advantage of this large pool of possible supports and by adopting the synthesis strategies from catalyst preparation, a simple and cost effective way of nanosizing materials for hydrogen storage could be feasible.

Table 10.1 A selection of support materials and their key properties.

	Surface area ($m^2 g^{-1}$)	Predominant porosity	Purity	Price
Fumed silica	50–400	micro, meso	high	high
Silica gel based on alkali silicates	100–800	micro, meso	medium	low
Silica gel based on metal-organic precursors	100–800	micro, meso	high	high
Templated silica (e.g., SBA-15)	200–1000	meso, (micro)	high	very high
Aerogels	250–2000	micro, meso, macro	high	high
Activated carbon	500–1200	micro, (meso)	poor	low
Carbon Black	10–400	micro, meso	high	low
Multiwall carbon nanotubes, carbon nanofibers	100–400	micro, meso	high-medium	very high
Templated carbons (e.g., CMK-1)	200–1500	meso, micro	High	very high
Zeolites	100–300	micro	high	low
γ-Alumina	15–300	micro, meso	medium	low
Metal Organic Frameworks (MOF)	500–4500	micro	high/medium	medium

IUPAC defines pores according to their diameter d as

- Macropores ($d > 50$ nm)
- Mesopores (2 nm $< d < 50$ nm)
- Micropores ($d < 2$ nm)

Obviously, the presence of micro- and meso-pores generates large surface areas whereas macropores strongly increase the total porosity of the material. Pores in the mesopore range are often desired since they combine large surface areas and good accessibility. Here, we focus on silica and carbon as support materials, since they are comparatively light and are produced on a large scale. A summary of possible supports and their key properties is given in Table 10.1.

10.5.1.1 **Silica**

Silica (SiO_2) is the most abundant inorganic compound within the earth's crust and is used (as a constituent or in its pure form) in many applications. Amorphous silica materials of commercial relevance are fumed silica and silica gel and they are widely used as filler material, thickening agent, desiccant, catalyst support or as a stationary phase in chromatography. By using synthetic SiO_2 precursors, both types can be produced in high purity. The porosity can also be adjusted over a wide range by changing the synthesis parameters. Fumed silica (e.g., Aerosil®, Cabosil®) is synthesized by vapor-phase hydrolysis of silicon tetrachloride in a flame (flame hydrolysis)

$$SiCl_4 + O_2 + 2H_2 \rightarrow SiO_2 + 4HCl$$

Commercial products typically exhibit surface areas between 50 and 400 $m^2 g^{-1}$. Fumed silica is comparatively expensive but exhibits a high purity [109]. Silica gel is synthesized by sol–gel methods using aqueous silicate solutions [110]

$$Na_2SiO_3 + 2H^+ \xrightarrow{H_2O} SiO_2 + H_2O + 2Na^+$$

After drying, the materials usually exhibit large surface areas, typically between 100 and 900 $m^2 g^{-1}$ due to a high micropore content and large total porosities. Compared to fumed silica, the obtained material is cheaper but less pure. Silica gel of higher purity can be obtained by hydrolysis and condensation using metal-organic precursors as the silica precursor [111]

$$Si(OR)_4 + 4H_2O \rightarrow Si(OH)_4 + 4(ROH)$$

and

$$Si(OH)_4 \rightarrow SiO_2 + 2\,H_2O$$

An advantage of the sol–gel method is that it allows casting of the sol into molds to obtain monolithic materials directly. After casting the sol and its gelation, the final properties of the material depend strongly on the drying procedure. Drying at atmospheric pressures leads to a large volume shrinkage and so-called xerogels are obtained [111]. Removal of the solvent by freeze drying gives rise to cryogels. Aerogels are obtained by drying under supercritical conditions inhibiting the volume shrinkage, thus exhibiting large total porosities (up to 98%) [112].

Silica materials with high mesopore content can be obtained by sol–gel techniques using surfactants and polymers as additional templates. Surfactant self-assembly during condensation of the silica precursor allows the synthesis of ordered mesoporous materials (OMS) with its most prominent members MCM-41 (*M*obil *C*omposition of *M*atter) [113] and SBA-15 (*S*anta *B*arbara *A*morphous) [114]. Both materials exhibit one-dimensional mesopores of defined size which are aligned in a hexagonal structure, although different pore symmetries can be obtained. The pore size of these materials can be tuned between 2 and 30 nm, making them ideal materials to study nano-confinement effects. However, it should be mentioned that these materials are model supports and so far not applied on a larger scale. Hierarchical macro-/meso-porous silica materials with defined pore sizes can also be obtained via the Nakanishi process [115]. This process takes advantage of spinodal decomposition between polymer and silica precursor. The resulting material can be synthesized in monolithic form and exhibits a disordered but interconnected pore structure with excellent pore accessibility. A commercial version of such a material (Chromolith®) exhibits pore sizes in the range of 13 nm to 2 μm and a surface area around 300 $m^2 g^{-1}$. SEM and TEM images of some silica materials can be seen in Figure 10.19.

Silica materials are thermally stable and exhibit ample chemical stability. They dissolve in strong basic solutions and hydrofluoric acid. The surface chemistry of silica

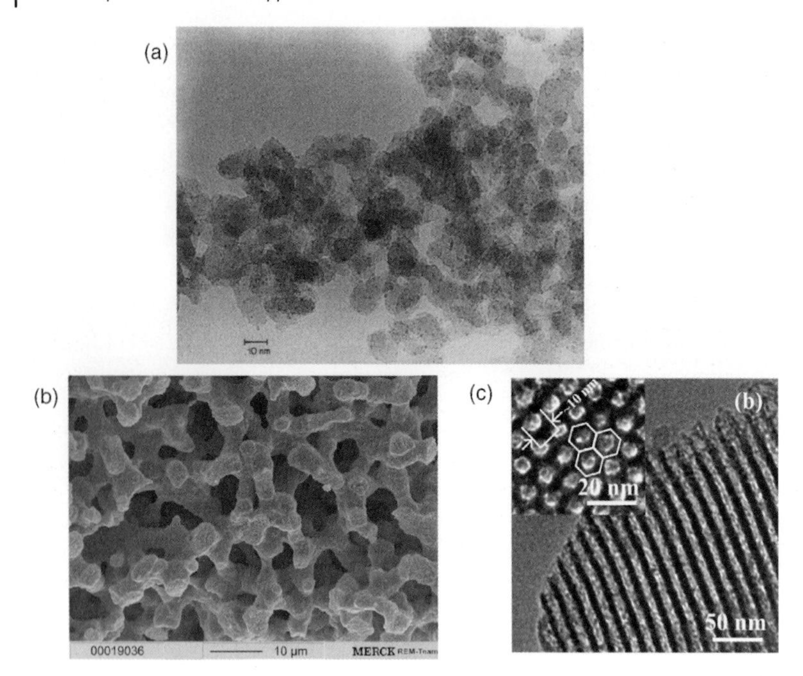

Figure 10.19 Examples of porous silica materials. (a) TEM image of rhodium nanoparticles deposited on fumed silica. Reproduced with permission from [116], published by Elsevier, 1990; (b) SEM image of a porous silica network (Chromolith) [117]. Copyright Wiley-VCH Verlag. Reproduced with permission; (c) TEM image of ordered mesoporous silica. Reprinted with permission from [118]. Copyright (2005) American Chemical Society.

materials mainly depends on the presence of silanol groups ($-Si(OH)_x$). The higher their concentration, the more hydrophilic the surface. The hydrophilic character can be reduced by removal of the hydroxy groups by thermal treatment. The hydrophilic/hydrophobic character of the support surface is an important parameter since it determines for example, the wetting behavior and the interaction with the active phase. In this respect, carbon materials complement silica materials because of their hydrophobicity. As possible reactions between the active phase and the silica support, formation of oxides, silicates or silicides have to be considered. For example, it is known from catalyst preparation that nickel and cobalt form unwanted silicates when deposited on silica supports. In principle, all metals with high reduction potential (e.g., alkali, alkaline earth) can reduce Si^{4+} with formation of Si and the corresponding metal oxide. Thus, the number of possible reactions between silica and the active phase might favor carbon materials as support.

10.5.1.2 **Carbon**

Carbon materials are widely used in industry in form of for example, porous powders, fibers, fabrics, pellets, extrudates or composites for very different purposes. This is because carbon materials can, depending on their structure, exhibit very different properties.

For example, diamond is used as an abrasive due to its hardness and is electronically insulating. In contrast, graphite is very soft and can be used as dry lubricant or as conducting agent. However, most of the carbon materials used are noncrystalline and are so-called nongraphitic (or turbostratic) carbons. Compared to the crystalline forms of carbon, their microstructure is quite complex. Briefly, nongraphitic carbons consist of graphene sheets that pile up to form stacks of a few nanometers in size. It is important to note that the graphene sheets are only more or less parallel packed upon each other. Only in graphite, a regular hexagonal stacking and larger stack sizes are found. Small stacks of high polydispersity give rise to a very disordered carbon material (larger porosity), whereas large stacks of small polydispersity increase the degree in order (less porosity). Figure 10.20 gives an overview of the structure of different carbon materials.

Activated carbons are the most prominent example of disordered carbon materials. Usually, they are synthesized by heat treatment of organic matter in an inert atmosphere. The use of natural materials such as coal, wood or coconut shell as the carbon source allows a low cost production and is the reason for their widespread use in purification applications. The high porosity is obtained by activation processes (e.g., steam activation) giving rise to surface areas typically between 500 and 1200 m^2 g^{-1} (the properties vary widely depending on the carbon precursor and treatment). Due to their disorder, activated carbons have a large number of inherent slit-shaped micropores.

Synthetic carbon precursors are used if a high purity is required. Carbon black is a colloidal carbon consisting of spheres and their fused aggregates synthesized by thermal decomposition of hydrocarbons [119]. Carbon blacks can be synthesized

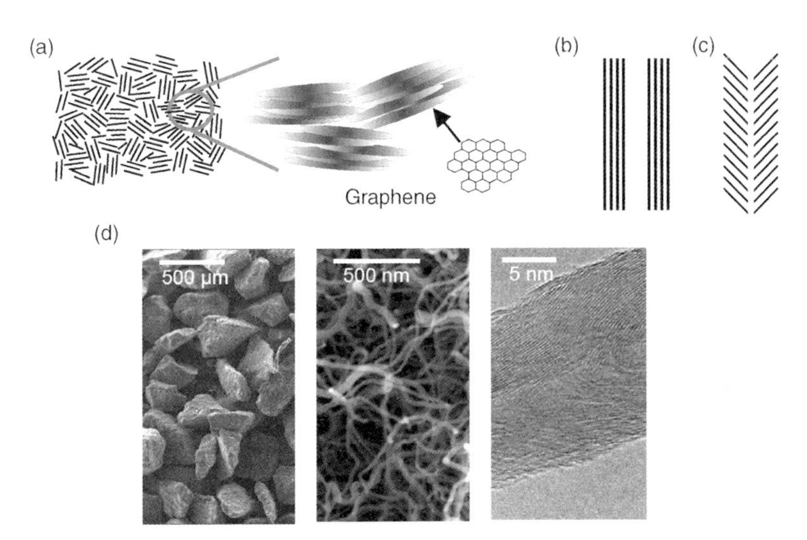

Figure 10.20 Structure model of different carbon materials. (a) nongraphitic carbon exhibiting inherent slit-shaped micropores, (b) multiwall carbon nanotube (MW-CNT), (c) fishbone-type carbon nanofibers (CNF), (d) SEM and TEM images of carbon nanofibers (courtesy Harry Bitter/Krijn de Jong).

with surface areas up to 1500 $m^2 g^{-1}$ and are used industrially as pigments, for example, in toners, or as conducting agent.

Besides the large amount of industrially produced carbon materials, there are several more that are produced on a small scale. For example, more recent developments are carbon nanofibers (CNF) and carbon nanotubes (CNT). Both can be synthesized by the catalytic decomposition of carbon-containing gases (e.g., CH_4, CO, C_2H_4) on small metal particles. Using different synthesis conditions, the structure can be parallel (multiwall-CNT) or fishbone type with typical surface areas between 10 and 400 $m^2 g^{-1}$ and diameters between 10 and 200 nm [110, 120]. As can be seen from the illustration in Figure 10.20(b–d), their surfaces exhibit either basal planes or edge planes. As their surface energy differs, the interaction with other phases (e.g., wetting behavior) will be different from each other. For graphite, for example, a surface energy of 0.077 $J m^{-2}$ was found for the basal planes (at 970 °C) whereas 4 $J m^{-2}$ were found for the edge planes [121]. Ordered mesoporous carbon materials can be synthesized by using ordered mesoporous silica materials such as SBA-15 as template. The porous silica template is first impregnated with a carbon precursor. After carbonization of the precursor and dissolution of the silica template, a mesoporous carbon replica is obtained [122]. Depending on the template, carbon replicas with different pore size can be synthesized. Recently, (ordered) mesoporous and hierarchical porous carbon materials could also be synthesized in less elaborate processes using polymers as porogen [123, 124].

Another class of highly porous carbon materials is the carbon aerogels. They are synthesized by sol–gel polymerization of resorcinol/formaldehyde or phenol/furfural solutions, followed by gel drying under supercritical conditions and subsequent pyrolysis. Similar to silica, different types of materials (e.g., carbon cryogel) can be obtained by variation in the drying procedure. Average pore sizes smaller than 50 nm, high total porosities (>80%) and surface areas between 400 and 1200 $m^2 g^{-1}$ are typically found for these materials [125]. The large porosity of carbon aerogels gives rise to very low total densities, usually between 0.05 and 0.8 $g cm^{-3}$ (2.06 $g cm^{-3}$ for the carbon skeleton) [126].

Carbon materials are often used as support materials due their low price, chemical inertness, high thermal stability in inert atmosphere and low weight. Additionally, their ability to conduct heat effectively is an important advantage over other inorganic support materials. Carbon is useful because most of the light metals will not react with the support. The few possible reactions between active phase and the carbon matrix involve the formation of carbides, intercalation compounds and reactions with residual oxygen (e.g., reactions with terminating carboxylic groups on the carbon surface). Also, methane formation might occur under the presence of hydrogen gas and catalytic species, but it is generally not expected at temperatures relevant for hydrogen storage. The inertness is somewhat ambivalent since anchoring of the active phase on the support is only possible when sufficient interaction is present, so surface modification might be necessary.

10.5.1.3 Other Support Materials

Other industrially applied materials that might be used as a support in hydrogen storage are zeolites and alumina. Zeolites are crystalline aluminosilicates with the general composition $M_{y/z}[(SiO_2)_x(AlO_2)_y]nH_2O$ where M is an exchangeable uni- or bivalent cation (valence z) or H^+. The crystal structure of zeolites is based on SiO_4^{4-} and AlO_4^{5-} tetrahedra that are connected by oxygen bridges. Due to their crystalline nature, zeolites exhibit a defined pore structure consisting of interconnected pores in the micropore range giving rise to large surface areas (typically $>200\,m^2\,g^{-1}$ with pore volumes between 0.1 and $0.5\,cm^3\,g^{-1}$). Since the pores are molecular in dimension, zeolites are also referred to as molecular sieves. Even though their defined pore size makes zeolites an interesting support material to study size effects, their use as a support for hydrogen storage compounds has not yet been reported. This is very likely due to their very high reactivity, making deposition difficult and unwanted side reactions probable. Because the walls have uniform thickness (<1 nm), zeolites have been used as a template for the synthesis of ordered microporous carbon materials [127–130] and they are tested for hydrogen physisorption [131].

Alumina (Al_2O_3) is also a widely used catalyst support and is produced at low cost with different porosities. Mostly, γ-Al_2O_3 with typical surface areas between 15 and $300\,m^2\,g^{-1}$ is used for this purpose. The structure of γ-Al_2O_3 can be described as a defective spinel type structure, with oxygen forming an fcc structure and aluminum occupying partly the tetrahedral and octahedral positions. Compared to silica, γ-Al_2O_3 shows a higher reactivity for unwanted side reactions (formation of aluminates) between the active phase and the support. γ-Al_2O_3 can be transformed by heat treatment to the less reactive α-Al_2O_3 but at the cost of much lower porosity [109].

Metal organic frameworks (MOFs) exhibit the highest surface areas among all classes of porous materials [132]. MOFs are crystalline materials consisting of metal complexes (nodes) linked together by rigid organic molecules (linkers), forming a mechanically robust, highly porous framework ($\sim$0.8–2.5 $cm^3\,g^{-1}$ [133]). By varying the chemical composition of nodes and linkers, nearly 5000 different MOF structures with a large variety in pore size and structure have been synthesized. The most prominent MOF is "MOF-5", consisting of Zn_4O^{6+} clusters connected with benzene-1,4-dicarboxylate as linker, forming a three-dimensional cubic network with interconnected pores with a diameter of 1.2 nm and a surface area of $2900\,m^2\,g^{-1}$[132]. The highest pore volume and surface areas so far were found for MOF-177 exceeding $0.6\,cm^3\,g^{-1}$ and $4000\,m^2\,g^{-1}$, respectively. MOFs are investigated for hydrogen storage using physisorption [134–136] (see Chapter 2 by Michael Hirscher), but can also be used as a support for metal nanoparticles [137, 138].

10.5.2 Preparation Strategies

Due to the widespread application of heterogeneous catalysis in industry, many strategies for preparing nanosized particles on porous support materials have been developed. The final result (e.g., particle size, particle distribution, activity, stability, loading, etc.) depends on many parameters of which only the most important ones will be addressed here. Especially, when catalytic materials are synthesized on an industrial scale, the exact synthesis procedures are based on experience and optimization rather than on fundamental understanding. In heterogeneous catalysis, the main preparation techniques used are impregnation, ion exchange, homogeneous deposition precipitation and coprecipitation. Gas-phase deposition techniques (CVD) can also be used, as demonstrated, for example, with Pd on MOFs [137]. Most relevant for hydrogen storage materials are *solution impregnation* and *melt infiltration*.

10.5.2.1 Solution Impregnation

In solution impregnation, a porous support is infiltrated with a solution of the active phase or its precursor. After careful removal of the solvent, a high dispersion of the active phase over the whole support is obtained (if a precursor of the active phase is used, a conversion step after drying is necessary). The basic principle of the process is shown in Figure 10.21.

There are two possible ways of impregnating a porous support with a solution. One is to add a volume of the solution that equals (or is less than) the pore volume of the support. This technique is called *incipient wetness* (also known as dry impregnation or pore volume impregnation). In practice, the solution is gradually added to the support until the powder gets somewhat sticky, that is, the whole pore volume is filled with the solution. Due to the large capillary forces present, this process is rather fast, usually in the range of seconds. If an excess of solution is used and the support is soaked in the solution, the term *wet impregnation* is used instead. The final result will depend on many parameters, so a careful choice of appropriate solvent, support, infiltration procedure and drying procedure is necessary. In the following, the main aspects of solution impregnation are discussed, as this method was already successfully used to synthesize 3D supported nanomaterials for hydrogen storage. Figure 10.22 illustrates the required and unwanted interactions between support, solvent and active phase.

First, it is crucial that the active phase shows reasonable solubility in the solvent and sufficient interaction with the support. If the interaction with the support is too strong, there is a risk that most of the active phase will be deposited on the outer part of the support particle. If the interaction is too weak, there is a risk of redistribution of

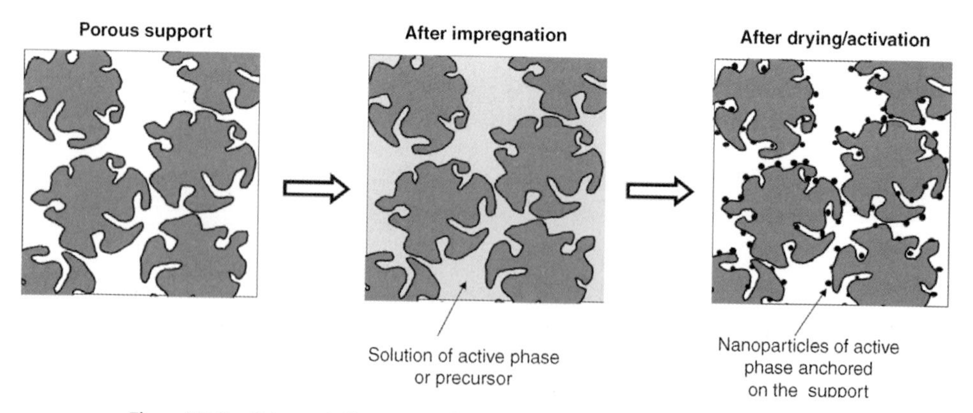

Figure 10.21 Schematic illustration of solution impregnation used to synthesize highly dispersed nanoparticles on a porous support.

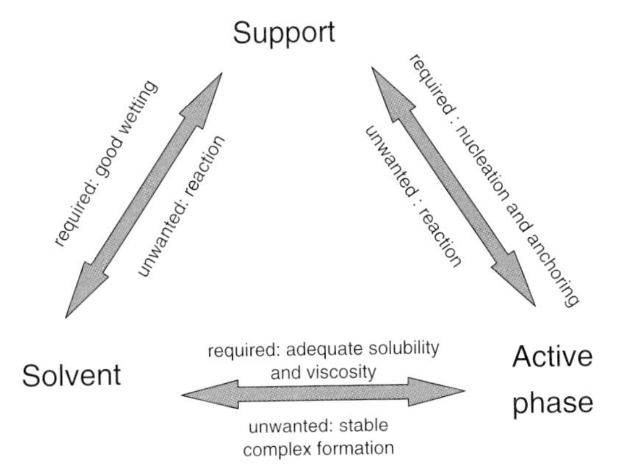

Figure 10.22 Interactions during solution impregnation between support, solvent and active phase.

the active phase during drying (as described later) or further processing (e.g., during heat treatments). If the solubility of the active phase is low, only a small quantity of the active phase can be deposited at once, making repeated infiltration and drying procedures necessary if higher loadings are needed. On the other hand, highly concentrated solutions can lead to an unwanted increase in viscosity, making infiltration more difficult. Also, the wetting behavior of the support by the solvent is very important. One can imagine that, due to their hydrophobicity, carbon supports are easily infiltrated by nonpolar solvents. In contrast, polar solvents wet silica-based supports more easily. If the wetting behavior is not sufficient, surface modification of the support is an option to improve the impregnation process. For example, carbon supports can be slightly oxidized (by acid treatment) to increase the number of surface carboxyl groups, thus reducing their hydrophobic character and providing more anchoring places for the active phase. On the other hand, silica supports can be heat treated in order to reduce their number of hydroxy groups.

The wetting behavior of a solid by a liquid can be described using the Young equation

$$\cos\theta = \frac{\gamma_{SV} - \gamma_{SL}}{\gamma_{LV}} \tag{10.4}$$

where γ_{SV}, γ_{SL}, γ_{LV}, are the interfacial tensions between solid (S), liquid (L) and vapor (V) and θ is the contact angle, or wetting angle (see Figure 10.23a). Depending on the magnitude of the interfacial tensions, $\cos\theta$ can be either positive and wetting occurs (i.e., $\theta < 90°$), or negative and wetting does not occur (i.e., $\theta > 90°$).

Another important factor is capillary forces. The pressure increase Δp across the liquid–vapor meniscus in a cylindrical pore of radius r is described by the Young–Laplace equation.

$$\Delta p = \frac{2\gamma_{LV} \cdot \cos\theta}{r} = \frac{2}{r}(\gamma_{SV} - \gamma_{SL}) \tag{10.5}$$

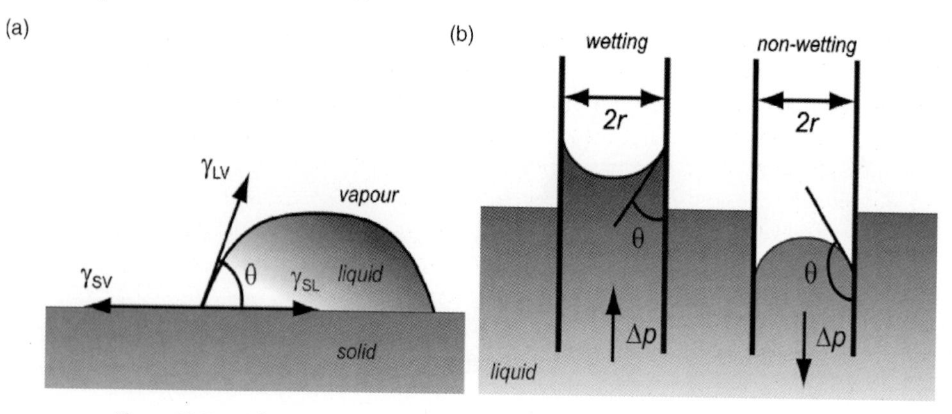

Figure 10.23 Schematic illustration of the wetting behavior of a solid by a liquid (a) liquid on a planar surface and (b) liquid in a cylindrical pore (with contact angle θ and γ_{SV}, γ_{LV},γ_{SL} being the interfacial tensions between solid, liquid and vapor (courtesy Rudy Wagemans).

Depending on the contact angle, the capillary pressure can either promote ($\theta < 90°$) or suppress ($\theta > 90°$) the pore filling, as illustrated in Figure 10.23b. Especially for small nanopores, the capillary pressure can easily reach hundreds of bars. However, the contact angle of a liquid in a pore depends on the curvature, making quantitative use of Eq. (10.5) difficult.

Solution impregnation can be used for both powdered supports and industrial carrier bodies (e.g., extrudates). The use of a powder usually leads to a homogeneous distribution of the active phase, whereas the use of larger carrier bodies can, depending on the interaction between active phase and support, lead to a more inhomogeneous distribution. As mentioned before, the solution has to be carefully removed after impregnation, since the drying process can lead to a redistribution of the active phase. Looking at a support particle, one can assume that the evaporation of the solvent starts at its outer part. Imagine that the outer part exhibits small pores which are connected to larger pores towards the inner part of the particle. Then, the capillary force leads to transport of the solution towards the outside of the particle. If in this case, the interaction between active phase and support is too weak, the active phase is transported towards the particle surface and thus a so-called egg-shell distribution is obtained. To overcome this problem, one can use high drying rates, or solvents with higher viscosity [139]. Also, if the interaction between solvent and support or solvent and active phase is too strong (for example complex formation between $NaAlH_4$ or $LiBH_4$ and THF), one might not be able to remove all the solvent during the drying procedure. The residual solvent can then undergo unwanted reactions with the active phase during dehydrogenation, complicating accurate hydrogen sorption measurements.

10.5.2.2 **Melt Infiltration**

In catalysis expensive noble metals are usually used as the active phase. Thus, solution impregnation methods are optimized on maximum utilization, that is, small particle size and low weight content. However, for hydrogen storage a high weight loading of the active phase is necessary in order to keep the weight penalty of the support small. A method that allows easy preparation of samples with higher weight loading is infiltration of the support by a melt of the active phase (or its precursor). Once impregnated, the supported nanomaterial is simply obtained by cooling/solidification (or by conversion, if a precursor is used). The feasibility of melt infiltration for hydrogen storage materials was recently demonstrated by synthesizing nanosized Mg nanoparticles using carbon as support [140]. By selecting carbon materials with different porosity, magnesium nanoparticles of different size could be prepared. For example, the use of a microporous carbon allowed the synthesis of particles smaller than 2 nm, a regime where significant changes in kinetics, and eventually thermodynamics, can be expected.

It depends on the nature of the active phase, whether the hydrogenated form (i.e., metal hydride) or the dehydrogenated form (i.e., metal) is more suitable for the synthesis. In the first case, it has to be considered that hydrides can decompose before or during the melting process (e.g., $NaAlH_4$), so it might be necessary to conduct the infiltration process under hydrogen pressure. For a successful melt infiltration, the following key requirements should be met:

- no reaction between support and active phase
- thermal stability of the support under the conditions used
- good wetting behavior (i.e., contact angle $<90^\circ$)
- low viscosity of the molten phase
- preferably low melting point and low vapor pressure of the active phase.

The wetting behavior can be described with Eqs. (10.4) and (10.5). However there is only a little information about the contact angles between molten metals/metal hydrides and support materials. Compared to liquids used in solution impregnation (e.g., THF, H_2O), the surface tension of liquid metals is high. For example, surface tensions between 0.855 and 0.873 J m^{-2}, and 0.525 and 0.583 J m^{-2} are found for aluminum and magnesium at their melting point [141]. Generally speaking, large values for γ_{LV} will not favor wetting of a surface (as is illustrated for mercury, $\gamma_{Hg,20^\circ C} = 0.425$ J m^{-2}). Dujardin *et al.* found that carbon nanotubes can only be wetted/filled by elements exhibiting surface tensions below ~0.1–0.2 J m^{-2}, such as Rb and Cs [142]. However, the surface tension, and hence the wetting properties, depends also on the gas atmosphere. Under wetting conditions, the use of porous supports with small pore sizes is advantageous to promote pore filling by capillary force. For the magnesium–carbon system, the few reported contact angles are just below 90°, that is, the wetting behavior is poor but the capillary force would still promote wetting. An example for molten magnesium (assuming cylindrical pores, $\theta = 85^\circ$, $\gamma_{LV} = 0.581$ N m^{-1} and $T_m = 667\,^\circ$C), the capillary pressure reaches up to ~1000 bar for a pore radius of 1 nm. Obviously, capillary forces of such magnitude can strongly enhance the infiltration process. Since pore filling is only promoted in the case of $\theta < 90^\circ$, specific surface modification might be essential for a successful melt infiltration process.

10.6 Experimental Results on 3D-Supported Nanomaterials

This chapter gives an overview of the progress that has been made in recent years on the synthesis of 3D-supported nanomaterials for hydrogen storage. Even though the research field is relatively new, the examples clearly prove that the concept of using supported nanomaterials is a powerful tool to improve certain key properties of hydrogen storage compounds. A summary of the compounds discussed in this chapter can be found in Table 10.2.

Table 10.2 Key properties of compounds discussed in this chapter and the main issues for practical application.

Compound	Hydrogen content (wt%)	Enthalpy of formation ΔH (kJ (mol hydride)$^{-1}$)	Main issues
NH_3BH_3	12 (19.6)	−115.06	decomposition exothermic irreversible formation of borazine
$NaAlH_4$	5.6 (7.4)	−112.9	low hydrogen content medium reversibility
$1/3\ Na_3AlH_6 + 2/3\ Al + H_2 \rightarrow NaAlH_4$		−36.7[a)]	
$3\ NaH + Al + 3/2\ H_2 \rightarrow Na_3AlH_6$		−69.6[b)]	
$1/2\ H_2 + Na \rightarrow NaH$		−56.4	
MgH_2	7.7	−75.7	high thermodynamic stability relatively slow absorption
Mg_2NiH_4	3.6	−128[c)]	low hydrogen content
$LiBH_4$	13.9 (18.5)	−190.46	high stability poor reversibility formation of diborane
$LiH + B + 3/2\ H_2 \rightarrow LiBH_4$		−52–76[d)]	
$1/2\ H_2 + Li \rightarrow LiH$		−90.5	
PdH_x	<1	−40[e)]	low hydrogen content high price

a) Enthalpy of formation from the elements taken from HSC chemistry database. For compounds decomposing into their elements (MgH_2), the absolute value corresponds to the enthalpy of decomposition. Complex hydrides decompose in several steps, thus the enthalpy of decomposition depends on the reaction step. More detailed information about the thermodynamic behavior of complex hydrides can be found for example, in [143].
b) Calculated values [144].
c) [145].
d) [143].
e) per mole H_2.

10.6.1 Ammonia Borane (NH_3BH_3)

Ammonia borane is a classical chemical hydride, thus hydrogen is released via an exothermic chemical reaction and off-board regeneration using energy intensive chemical processing is necessary. This is in contrast to metal hydrides, where the dehydrogenation is an endothermic process and reversible dehydrogenation/hydrogenation is achieved by changes in temperature and hydrogen pressure. Nevertheless, ammonia borane is discussed here as it was the first example of a supported nanomaterial reported in the open literature.

The N–B-unit in ammonia borane is isoelectronic with C–C and thus ammonia borane can be also seen as an inorganic analog of ethane. It exhibits a theoretical hydrogen content of 19.6 wt% but the decomposition is complex and full decomposition is not obtained until high temperatures. For example, heating to 200 °C releases ~14.3 wt% of hydrogen [146]. Ammonia borane decomposes in several steps involving polymerization to polyaminoborane $(NH_2BH_2)_n$ and polyiminoborane $(NHBH)_n$, cross-linking and finally decomposition to boron nitride:

$$
\begin{array}{ll}
n\mathrm{NH_3BH_3} \rightarrow (\mathrm{NH_2BH_2})_n + (n-1)\mathrm{H_2} & < 120\ ^\circ\mathrm{C} \\
(\mathrm{NH_2BH_2})_n \rightarrow (\mathrm{NHBH})_n + \mathrm{H_2} & \sim 150\ ^\circ\mathrm{C} \\
2(\mathrm{NHBH})_n \rightarrow (\mathrm{NHB{-}NHB})_x + \mathrm{H_2} & \sim 150\ ^\circ\mathrm{C} \\
(\mathrm{NHB{-}NHB})_n \rightarrow \mathrm{BN} + \mathrm{H_2} & > 500\ ^\circ\mathrm{C}
\end{array}
$$

Before polymerization, ammonia borane melts at around 112–114 °C. The reactions are exothermic, thus the release of hydrogen is accompanied by the evolution of heat. For example, a reaction enthalpy of $\Delta H = -21.7 \pm 1.2$ kJ (mol $(NH_3BH_3))^{-1}$ was found for the first decomposition step [147]. A drawback of using ammonia borane as a hydrogen storage material is the possible formation of volatile products during dehydrogenation, such as borazine ($B_3N_3H_6$), that can damage the PEM fuel cell.

In 2005, Autrey and coworkers reported significant improvements in the hydrogen sorption properties when ammonia borane was deposited in ordered mesoporous silica (SBA-15, pore diameter 7.5 nm, surface area ~900 $m^2 g^{-1}$, total pore volume 1.2 $cm^3 g^{-1}$) [74]. Samples containing 50 wt% ammonia borane were prepared by solution impregnation using methanol as solvent. Figure 10.24 a shows the desorption profile of hydrogen and borazine for pure and supported NH_3BH_3. Pure NH_3BH_3 releases hydrogen in two steps, corresponding to the formation of the two different polymeric compounds. The decomposition of ammonia borane is accompanied by not only the release of hydrogen, but also the release of large amounts of borazine, in particular during the second polymerization step. The supported NH_3BH_3 shows two significant differences compared to the reference. First, the hydrogen is released at lower temperatures. Secondly, the evolution of unwanted borazine is suppressed, indicating that the decomposition reaction is altered. A possible change in the reaction pathway was further supported by DSC and NMR measurements. It was found that hat the reaction enthalpy is significantly reduced from 21 ± 1 kJ (mol $(NH_3BH_3))^{-1}$mol^{-1} (pure NH_3BH_3) to -1 ± 1 kJ (mol $(NH_3BH_3))^{-1}$ (supported NH_3BH_3). Also, NMR measurements showed different

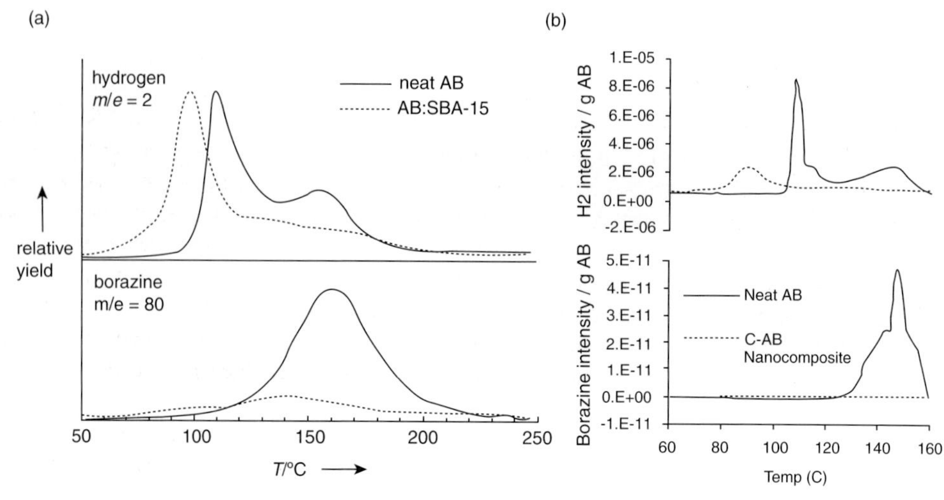

Figure 10.24 (a) Combined TPD/MS measurements of pure NH_3BH_3 (denoted as neat AB) and silica-supported NH_3BH_3 (denoted as AB: SBA-15, 50 wt% loading), (1 K min^{-1}, Ar atmosphere) [74]. Copyright Wiley-VCH Verlag. Reproduced with permission. (b) Combined DTA/MS measurements of pure NH_3BH_3 and carbon-supported NH_3BH_3 (denoted as C-AB, 24 wt% loading) (1 K min^{-1}, atmosphere unknown). Reprinted with permission from [148]. Copyright (2007) American Chemical Society.

decomposition products (heat treatment at 85 °C led to the formation of polyaminoborane for the supported NH_3BH_3 whereas three additional boron compounds were detected for the pure NH_3BH_3).

Isothermal kinetic measurements showed that the hydrogen release from supported NH_3BH_3 was much faster than from pure NH_3BH_3 (e.g., 50% reaction conversion at 80 °C in less than 10 min for supported NH_3BH_3, compared to 290 min for pure NH_3BH_3). Also, a much lower activation energy was found for the hydrogen release from supported material ($E_a = 67 \pm 5$ kJ mol^{-1} compared to 184 ± 5 kJ mol^{-1} for the reference), however, the total amount of hydrogen released from the samples was not specifically mentioned. Although there was strong indication of pore filling, it could not be elucidated whether all ammonia borane was deposited within the pores or if a fraction remained on the outside.

In a following study, mesoporous carbon cryogel was used as support (pore diameter 2–20 nm, surface area 300 m^2 g^{-1}, pore volume 0.7 cm^3 g^{-1}) [148]. Samples with a NH_3BH_3 loading of 24 wt% were obtained using wet impregnation and THF as solvent. Nitrogen physisorption revealed a pore volume loss of ~30%. For different weight loadings, the loss in pore volume corresponded well to the amount of NH_3BH_3 added. Interestingly, the pore size distribution was found to be independent of the loading. Figure 10.24b shows the DTA/MS results for the pure and carbon-supported (C-AB) NH_3BH_3. Carbon as a support has an even larger influence than SBA-15. The desorption maximum shifts to even lower temperatures and borazine formation can be completely prevented. However, the amount of hydrogen released from supported NH_3BH_3 was significantly less than from the NH_3BH_3 reference (9 wt% compared to 13 wt%). Surprisingly, the reaction enthalpy was measured to be −120 kJ

$(mol\ (NH_3BH_3))^{-1}$ (compared to $-21\ kJ\ mol^{-1}$ for pure NH_3BH_3 and $-1 \pm 1\ kJ\ mol^{-1}$ for SBA-15-supported NH_3BH_3). NH_3BH_3 underwent the expected dehydrogenation reactions when confined in the carbon cryogel, thus which process caused the change in reaction enthalpy remains so far unexplained. The authors suggest that their findings (for both SiO_2 and carbon support) are due to the nano-confinement (change in surface energy of NH_3BH_3, higher number of defects) and/or catalytic effects (induced by the terminating hydroxy groups in silica and carboxylic groups in carbon supports), however it remains difficult to discriminate between the effects. If using carbon cryogel as support, a smaller pore size correlated with a shift towards lower dehydrogenation temperatures. Further research on pore size effect and structural characterization is currently being pursued in order to understand the findings. The experimental results are a strong example of the fact that the nature of the support can influence the dehydrogenation behavior of a hydrogen storage compound.

10.6.2 Sodium Alanate, ($NaAlH_4$)

Sodium alanate ($NaAlH_4$) remains so far the most studied complex hydride. It is found that $NaAlH_4$ decomposes in several steps:

$$3NaAlH_4 \rightarrow Na_3AlH_6 + 2Al + 3H_2$$

$$Na_3AlH_6 \rightarrow Al + 3NaH + \frac{3}{2}H_2$$

$$NaH \rightarrow Na + \frac{1}{2}H_2$$

Also, Na_3AlH_6 undergoes a phase transformation at 252 °C [149, 150] (monoclinic α-Na_3AlH_6 to fcc β-Na_3AlH_6). Since decomposition of NaH occurs at temperatures above 300 °C (under Ar atmosphere), the last step is not considered and thus the hydrogen content of $NaAlH_4$ is 5.6 wt% for practical application instead of 7.4 wt%. Even though $NaAlH_4$ should decompose in equilibrium conditions already at low temperatures (e.g., 33 °C for $NaAlH_4$ and 110 °C for Na_3AlH_6 at 1 atm H_2, see Figure 10.25), its decomposition is kinetically hindered and substantial dehydrogenation only starts (even under Ar atmosphere) around its melting point ($T_m = 180\ °C$, $\Delta_m H = 23\ kJ\ mol^{-1}$).

An important improvement to the $NaAlH_4$ system was the discovery in 1997 that Ti species are effective catalysts [1]. Compared to pure $NaAlH_4$ the decomposition of Ti-catalyzed $NaAlH_4$ shifted by ~80–85 °C towards lower temperatures (heating rate of $4\ K\ min^{-1}$) and reversibility of the system was achieved.

The advantages of combining $NaAlH_4$ with a porous support were first mentioned in a patent dating from 2004 [151]. Several samples of Ti-doped $NaAlH_4$ on different carbon aerogels (pore sizes of 15 and 23 nm) were synthesized using solution and melt infiltration. Compared with non-infiltrated Ti-doped $NaAlH_4$, the patent states an improvement in the kinetic behavior, better safety, improved cycling behavior and even alteration in the thermodynamics. The described samples contained up to ~50 wt% $NaAlH_4$. However, the exact measurement conditions and particle dimensions are not mentioned.

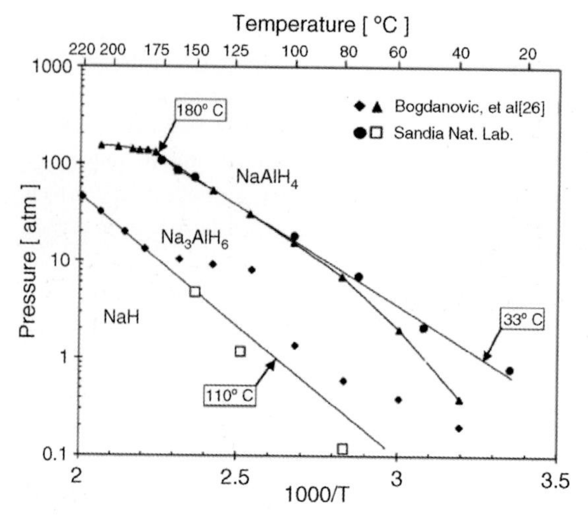

Figure 10.25 Phase diagram of $NaAlH_4$ and its decomposition products. Reproduced with permission from [154], published by Elsevier, 2002.

In 2006, the hydrogen desorption characteristics were reported for nanosized $NaAlH_4$ deposited on fishbone-type carbon nanofibers (CNF) using solution impregnation and THF as solvent [152]. The carbon nanofibers used were 20–50 nm in diameter and exhibited a surface area of $130\,m^2\,g^{-1}$ and a total pore volume of $\sim 1\,cm^3\,g^{-1}$. The surface of the carbon nanofibers was oxidized by HNO_3 treatment in order to improve the anchoring and wetting (CNF_{ox}). The weight loading of $NaAlH_4$ on the carbon nanofibers was around 9 wt%.

Figure 10.26 a shows the first dehydrogenation curves (room temperature to 160 °C) of the supported $NaAlH_4$ in comparison to a physical mixture ($NaAlH_4$ mixed with carbon nanofibers). As can be seen, dehydrogenation of the supported $NaAlH_4$ (denoted as $NaAlH_4/CNF_{ox}$) starts already at around 40 °C compared to 150 °C for the physical mixture. 2.9 wt% H_2 is desorbed from the supported $NaAlH_4$ compared to only 0.15 wt% for the physical mixture. In order to study the effect of surface modification, $NaAlH_4$ was also deposited on carbon nanofibers that were not surface oxidized (CNF_{as}). The results shown in Figure 10.26a indicate that the surface oxidation of the carbon support was not beneficial since the shapes of the dehydrogenation curves are very similar. However, it can be seen that the amount of hydrogen released from the non-oxidized sample is higher (3.7 wt%), probably due to fewer side-reactions between $NaAlH_4$ and the support. Also, the supported $NaAlH_4$ showed partial rehydrogenation under relatively mild conditions (48 h, 115 °C, 90 bar H_2). Using the same deposition technique, a relation between particle size and the activation energy of the $NaAlH_4$ decomposition was recently reported [153]. $NaAlH_4$ particles of different sizes were deposited on CNF by varying the total $NaAlH_4$ loading and drying conditions. For example, using a $NaAlH_4$ loading of 2 wt% and drying at low temperatures (−40 to −15 °C), particles between 2 and 10 nm were obtained. Figure 10.26b shows the first dehydrogenation curve for CNF-supported $NaAlH_4$ of different particle sizes. Clearly, the temperature of maximum desorption correlates inversely with the particle size. Using Kissinger analysis, an activation energy of

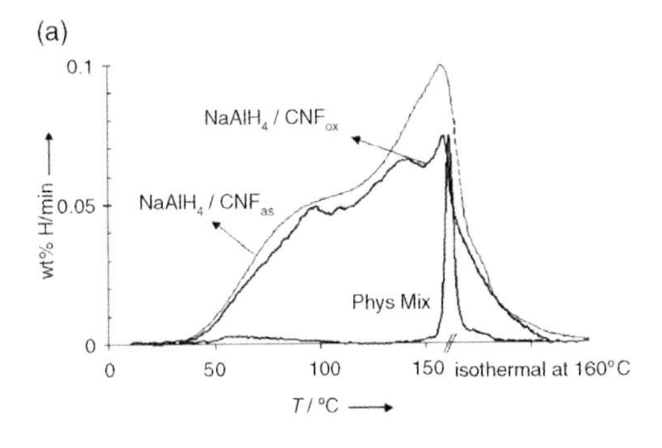

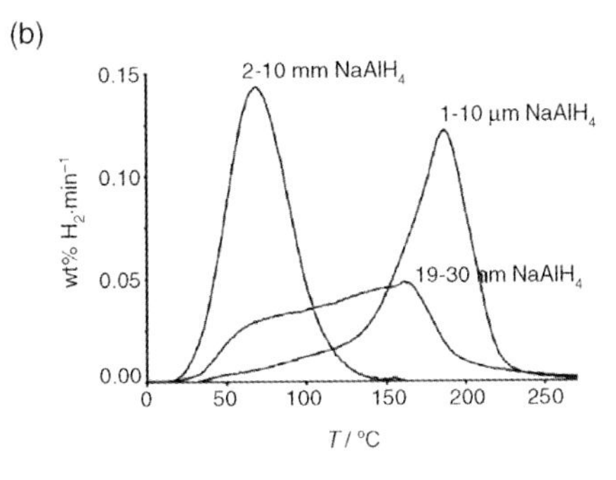

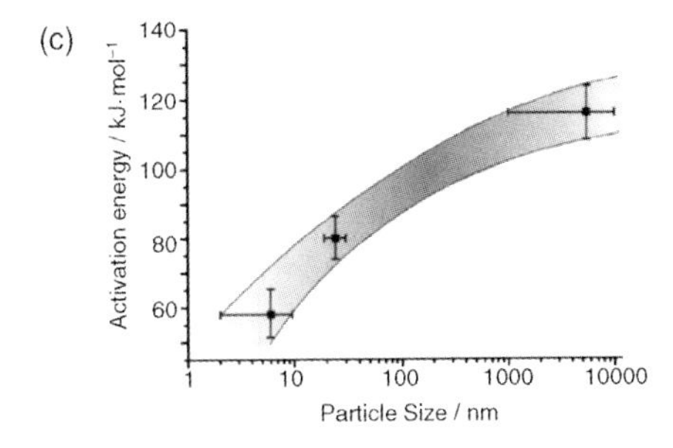

Figure 10.26 (a) Temperature programmed desorption of carbon nanofiber-supported $NaAlH_4$ (denoted as $NaAlH_4/CNF_{as}$, $NaAlH_4/CNF_{ox}$) and physical mixture. (Ar atmosphere, $2\,K\,min^{-1}$) [152]. Copyright Wiley-VCH Verlag. Reproduced with permission; (b) Particle size effect on temperature programmed desorption of $NaAlH_4/CNF_{as}$ (Ar atmosphere, $5\,K\,min^{-1}$ (c) Relation between particle size and activation energy for the decomposition of supported $NaAlH_4$ (using Kissinger analysis). Reprinted with permission from [153]. Copyright (2008) American Chemical Society.

$58\,kJ\,mol^{-1}$ was found for particles in the range 2–10 nm, whereas $\sim 120\,kJ\,mol^{-1}$ was found for bulk $NaAlH_4$, thus a smaller particle size correlates with a decrease in activation energy (Figure 10.26c). It is worth noting that, for the smallest particles, the activation energy was even lower than that of ball-milled titanium doped $NaAlH_4$ ($\sim 80\,kJ\,mol^{-1}$ [154]). The authors explain their findings with possible changes in the $NaAlH_4$ structure when the particle size is reduced to the nanosize [155], or shifts in the rate-determining steps due to the shorter diffusion lengths. Also, the larger surface/volume ratio of the $NaAlH_4$ particles could cause the shift towards lower desorption temperatures.

Using this synthesis, only small weight loadings have been achieved so far. Also, the desorption measurements were performed under argon atmosphere and no thermodynamic data were reported. Despite the finding, that the first rehydrogenation already starts at comparably low pressures (20 bar H_2), the synthesized samples showed poor reversibility, probably caused by the phase separation of NaH and Al on the surface of the carbon nanofibers (elemental mapping of Na and Al after dehydrogenation indicated that Al segregated into nanometer-sized domains whereas Na was found to be well dispersed). It should be noted that in this work, $NaAlH_4$ was deposited on the outside of the carbon nanofibers, that is, on a convex surface.

Confinement in a nanopore might be more reasonable, that is, deposition on a concave surface.

The idea of using nano-confinement for stabilization was recently adopteded by Zheng *et al.* [156]. They synthesized nanosized $NaAlH_4$ by solution impregnation using THF as solvent and ordered mesoporous silica (SBA-15) as support. The reaction between the silica matrix and $NaAlH_4$ is thermodynamically favored, leading to, for example, formation of Al_2SiO_5. Prior to impregnation, the silica surface was modified with methyl groups in order to reduce unwanted side-reactions between $NaAlH_4$ and residual silanol groups. This approach had already been used successfully for the synthesis of TiN nanoparticles in SBA-15 by solution impregnation using $Ti(NMe_2)_4$ as precursor and toluene as solvent [157]. However, it was not further reported to what extent this treatment suppressed unwanted side-reactions between $NaAlH_4$ and the support. A total weight loading of 20 wt% $NaAlH_4$ was obtained by multiple impregnations. The maximum $NaAlH_4$ particle size was limited by the pore diameter of the support to approximately 10 nm. Using TEM, it was shown that the pores of the support were partially filled with $NaAlH_4$ and the ordered mesoporous structure was maintained after synthesis and dehydrogenation, indicating sufficient stability of the support. Figure 10.27 a shows the dehydrogenation curve of SBA-15-supported $NaAlH_4$ and pure $NaAlH_4$ for comparison, monitored by DSC and TG. The DSC traces for both samples show basically three peaks. The first peak can be assigned to the melting of $NaAlH_4$, followed by decomposition over a wide temperature range to Na_3AlH_6 (second peaks marked by arrows). The last peak can be attributed to the decomposition of Na_3AlH_6 to NaH. Obviously, a shift towards lower temperatures and peak broadening is observed for supported $NaAlH_4$. The total hydrogen loss is comparable for both samples (~4.8 wt%) but is less than the theoretical value (5.6 wt%).

Figure 10.27b compares the dehydrogenation behavior of $NaAlH_4$ and supported $NaAlH_4$ at constant temperature. Clearly, pure $NaAlH_4$ decomposes very slowly until reaching the melting temperature, whereas the supported $NaAlH_4$ sample shows

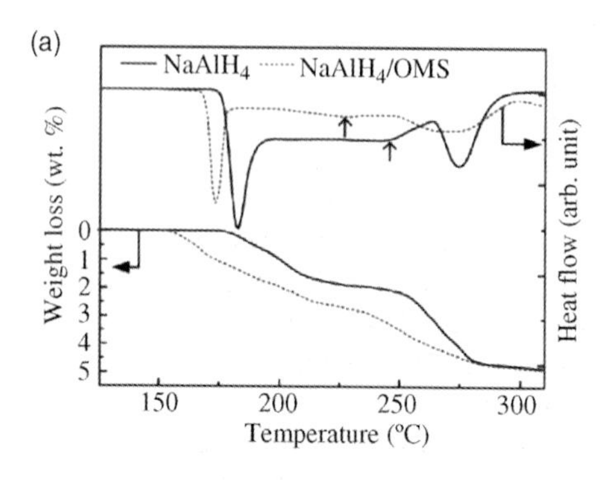

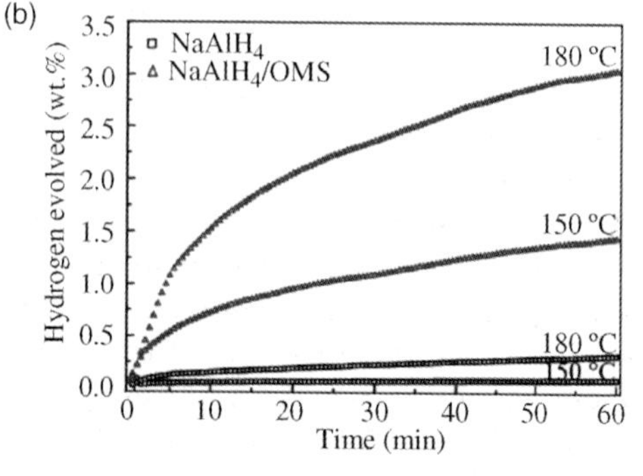

Figure 10.27 (a) DSC and TG curves for supported $NaAlH_4$ ($NaAlH_4$/OMS) and pure $NaAlH_4$ (5 K min^{-1}, Ar atmosphere). (b) Dehydrogenation curves for $NaAlH_4$ and $NaAlH_4$/OMS at 150 °C and 180 °C (10 Pa back pressure). Reprinted with permission from [156]. Copyright (2008) American Chemical Society.

much higher hydrogen release rates. No large particles of the decomposition products (Al, NaH) were detected by XRD and TEM after dehydrogenation, supporting the concept that large-scale phase segregation was prevented by the nano-confinement. Also, the rehydrogenation process was found to be strongly facilitated. The amount of hydrogen absorbed under the given conditions (3 h at 180 °C and 55 bar H_2) was around nine times higher than the non-supported sample, however full rehydrogenation was not obtained.

Comparing both supports, carbon seems to affect the dehydrogenation behavior much more strongly than silica. Even for larger particle sizes of $NaAlH_4$ on carbon (e.g., 19–30 nm), dehydrogenation starts at much lower temperatures compared to SBA-15-supported $NaAlH_4$ (measurement conditions are comparable). A synergetic or catalytic effect of the carbon cannot be excluded. However, the loading of $NaAlH_4$ on carbon nanofibers was much smaller. The theoretical hydrogen content (5.6 wt%) was not reached for either support, but it is interesting to note that the use of silica yielded a higher hydrogen content and also showed better reversibility (samples were one time rehydrogenated), probably due to the higher loading. Oxidation and phase segregation of carbon nanofiber-supported $NaAlH_4$ might be facilitated due to the smaller particle size and the lack of confinement but also the surface modification of SBA-15 might have reduced the number of unwanted side-reactions.

10.6.3
Magnesium Hydride (MgH_2)

Magnesium hydride is one of the most promising hydrogen storage materials and has already been extensively studied due to its low price and high hydrogen content (7.7 wt%, 110 $g\,l^{-1}$). However, the Mg/MgH_2 system suffers from slow kinetics and unfavorable thermodynamics that hamper its use in practical application for automotive purposes. For example, the enthalpy of formation of the hydride ($\Delta H = -75$ $kJ\,mol^{-1}$) corresponds to an equilibrium temperature of $\sim$300 °C (at 1 bar H_2 pressure). In practice, dehydrogenation temperatures well above 400 °C are found. The kinetic constraints can be improved by ball-milling or use of suitable catalysts [2], however the dehydrogenation temperature still remains too high.

In an earlier work of Imamura *et al.*, solutions of magnesium in ammonia were used for impregnation of activated carbon (1040 $m^2\,g^{-1}$) [56]. In this synthesis, magnesium is first evaporated under vacuum and condensed in ammonia matrices at −196 °C. Melting of the ammonia resulted in a blue solution of dissolved magnesium (proposed dissolution reaction: $Mg + nNH_3 \rightarrow Mg^{2+} + 2e(NH_3)_n^-$), and impregnation of the carbon was conducted at −75 °C, followed by drying. Studies on the hydrogenation behavior of the prepared material (at 200 °C) showed little hydrogen uptake under 0.53 bar hydrogen pressure (below 20% of the theoretical capacity after 5 h). Deposition of nickel as catalyst on the carbon prior to impregnation (1 wt% Ni) resulted in a much faster hydrogen uptake (more than 80% of the theoretical capacity within 5 h) compared to the unsupported references with much higher nickel fraction. This suggests that nickel as catalyst in combination with the carbon support was effective regarding the hydrogenation kinetics. However, no

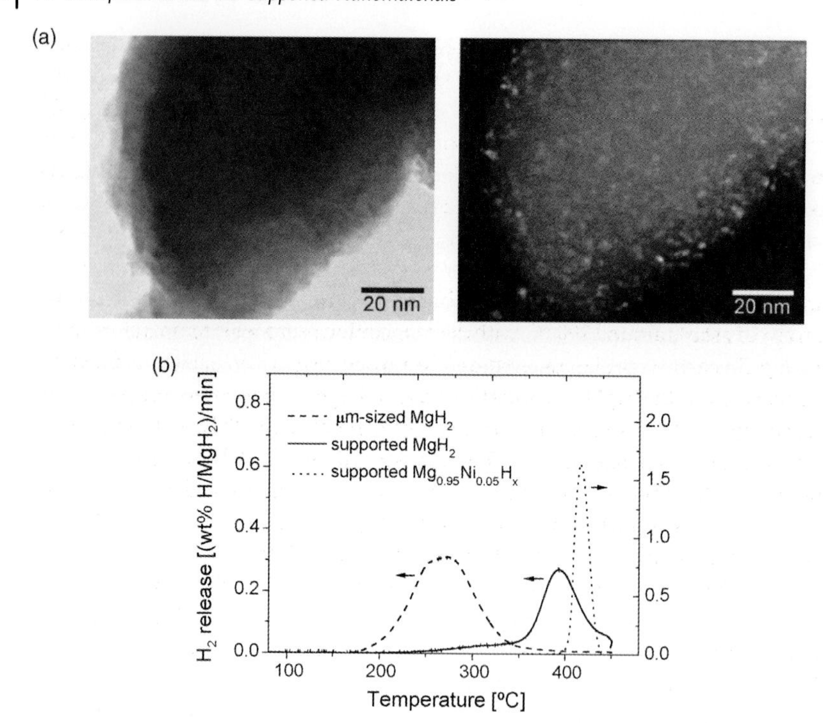

Figure 10.28 (a) TEM images of magnesium nanoparticles on carbon support, left bright field and right dark field [140]. (b) Thermal programmed desorption (TPD) heating with 5 °C min^{-1} in argon flow comparing a 10 wt% MgH_2 with carbon physical mixture, 10 wt% MgH_2 melt-infiltrated into carbon, and 50 wt% MgH_2 on carbon containing 5 atom% Ni [158].

absolute values of the hydrogen uptake, comparison with bulk Mg or the size of the supported magnesium particles were reported.

Using melt infiltration, 2–5 nm Mg/MgH_2-nanoparticles were recently synthesized using micro-/mesoporous carbons as support [140]. Starting point was a physical mixture of μm-sized MgH_2 particles and carbon. In a first step, full dehydrogenation of the MgH_2 was ensured by heating up to 600 °C. In this way, Mg of high purity was obtained. In a second step, the melt infiltration was conducted at 667 °C, slightly above the melting point of Mg (T_m = 650 °C). Figure 10.28 a shows TEM images of the nanoparticles that are distributed over the carbon support. Because of their similar atomic weight, the nanocrystallites are hardly visible in bright field TEM. However, due to the diffraction contrast they can be seen in the dark field micrograph as bright spots. The crystallites are well separated from each other, which could help to avoid unwanted phase segregation during cycling of the material. The size of the nanoparticles was influenced by the pore size of the different carbons. Using microporous carbon, nanoparticles smaller than 2 nm could be synthesized. The carbon matrix was not damaged by the melt infiltration, as evidenced by N_2-physisorption after leaching out the active material. Above a certain loading (10–15 wt % for a specific porous carbon material) not only nanocrystalline, but also larger crystallites were formed outside the pores, as shown by XRD and physisorption [140].

Hydrogenation of the supported Mg nanoparticles was performed in an autoclave (2 h at 325 °C, 85 bar). Figure 10.28b shows the dehydrogenation behavior of a hydrided 10 wt% Mg/carbon sample, a 50 wt% Mg/carbon sample doped with 5 atom% Ni, and a reference physical mixture. The 5 atom% Ni (with respect to the amount of magnesium) was added by first depositing nickel nanoparticles on the carbon support by impregnation, drying and reduction, followed by melt infiltration with Mg. The reference (a physical mixture consisting of μm-sized MgH_2 particles (~50 μm) and carbon), starts desorbing above 400 °C. Nanosizing and supporting clearly influences the desorption since, for the 10 wt% MgH_2 on carbon, decomposition already starts around 200 °C, with decomposition over a wide temperature range following. Since the pore size and shape of the carbon support are not well defined, the decomposition of the MgH_2 nanoparticles over a larger temperature range could be explained by variations in particle size and confinement. Up to 80% of the theoretical hydrogen capacity was released. For the nickel-doped sample all hydrogen is released below 350 °C. Even though the stoichiometric ratio between Mg and Ni is 17: 1, this release behavior is similar to that of pure Mg_2NiH_4. It was already reported before, by Zaluska and coworkers for ball-milled samples, that Ni-doped MgH_2 can release hydrogen at temperatures similar to that for Mg_2NiH_4, even if the nickel content is significantly lower than that corresponding to stoichiometric Mg_2NiH_4 [159]. No thermodynamic sorption measurements have been reported so far for these systems.

Recently, intriguing results on the hydrogen sorption properties of Mg/C thin films were reported. XRD-amorphous magnesium/carbon thin films (50: 50 atom%) with 25 nm thickness were deposited by co-sputtering [160]. The films were covered with a 5 nm layer of Pd to prevent oxidation and enhance the absorption of hydrogen. Nevertheless, the films exhibited an oxygen content of ~10 atom%. Hydrogen absorption/desorption experiments were conducted by measuring the electrical resistance of the film. The kinetics of hydrogen uptake were much faster than for pure Mg films, which might be explained by faster diffusion along the Mg/C/Mg interface, smaller domain size and/or clamping effects. The authors report the measurement of reversible hydrogen switching at 70 °C, at which temperature the hydride is formed at pressures between 0.2 and 7.5 mbar. This would suggest that the stability of the magnesium hydride phase formed is significantly less than that of bulk MgH_2. However, the authors also report that it is difficult to measure sorption properties reproducibly, and that it is not possible to measure full isotherms and obtain thermodynamic information from van t' Hoff plots.

10.6.4 Lithium Borohydride ($LiBH_4$)

Lithium borohydride recently gained much attention due to its high hydrogen content. Theoretically, $LiBH_4$ contains 18.5 wt% of hydrogen but forms, similar to $NaAlH_4$, a stable hydride during decomposition. Thus, its decomposition yields a hydrogen content of 13.9 wt% according to the following reaction: $LiBH_4 \rightarrow LiH + B + \frac{3}{2}H_2$.

The decomposition reaction also includes several intermediate steps, however, the exact details are not known [161]. When heated, the crystal structure of $LiBH_4$

changes from orthorhombic to tetragonal (at 108 °C), before reaching its melting point at 278 °C. Even though $LiBH_4$ starts to decompose around its melting point, most of the hydrogen is released above 400 °C [161]. Rehydrogenation is difficult and high pressures and temperatures are needed, that is, at 600 °C under 155 bar hydrogen pressure [162]. Another problem for practical application is the formation of volatile and poisonous diborane (B_2H_6) during dehydrogenation. Züttel *et al.* found that the dehydrogenation temperature of $LiBH_4$ is greatly reduced by the addition of SiO_2 [163].

Lohstroh and coworkers investigated the thermal decomposition of $LiBH_4$/silica-gel mixtures. Samples containing 50 wt% $LiBH_4$ were prepared by simply mixing $LiBH_4$ with silica gel (3 nm pores) in a mortar [161], followed by TG-MS experiments. The authors report a very efficient wetting of the support by molten $LiBH_4$ (or its reaction products). Thus it can be assumed that the porous support was effectively infiltrated during heating, even though no further information about pore filling was given. Compared to pure $LiBH_4$, the use of silica-gel as support led to a strong decrease in the dehydrogenation temperature, starting even below the melting point. For example, using a heating rate of 5 K min^{-1} and He atmosphere, a 5% weight loss was obtained at 300 °C. However, unwanted diborane was released during the whole dehydrogenation process. The authors found that by using $LaCl_3$ and $TiCl_3$ as additive, diborane formation was significantly reduced above 350 °C. Thus, diborane formation might be suppressed by the right additive.

Vajo *et al.* demonstrated the possibility of infiltrating melted $LiBH_4$ into carbon aerogels [164]. In their work, carbon aerogels of different pore size (exhibiting a pore sizes of 13 and 25 nm and surface areas of 770 and 810 $m^2\,g^{-1}$, respectively) and activated carbon (microporous, 1200 $m^2\,g^{-1}$) were infiltrated by heating at 280–300 °C under an Ar atmosphere. For the aerogels, weight loadings up to 50% were obtained, corresponding to a maximum pore filling of 90%. After melt

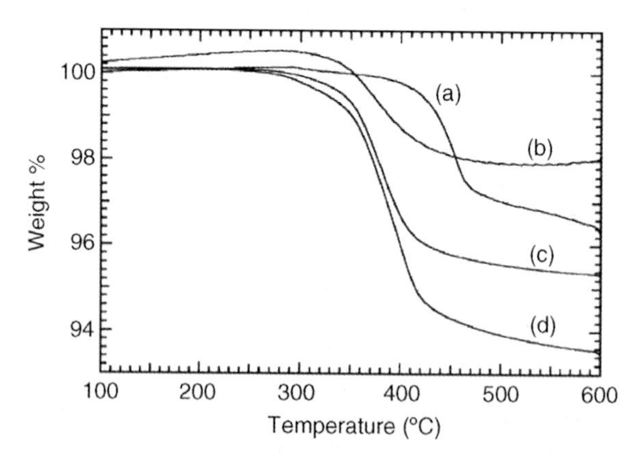

Figure 10.29 TGA data for $LiBH_4$ supported by different carbon supports. (a) $LiBH_4$ mixed with nonporous graphite. (b) $LiBH_4$ incorporated into activated carbon. Curve (c) and (d) $LiBH_4$ incorporated into carbon aerogel with average pore size of 13 and 25 nm. The total amount of weight losses scale with the $LiBH_4$ loading. (heating rate 10 K min^{-1}, Ar atmosphere). Reprinted with permission from [164]. Copyright (2008) American Chemical Society.

infiltration, no additional phases were detected, indicating that $LiBH_4$ does not react with the carbon matrix. Figure 10.29 shows the thermal decomposition of $LiBH_4$ incorporated into the different carbons. The use of a nonporous carbon reference shows decomposition similar to pure $LiBH_4$, whereas the use of porous carbon as support significantly lowers the dehydrogenation temperature. The activation energy was also found to be smaller for the infiltrated samples. For example, an activation energy of $E_a = 146 \pm 3\,kJ\,mol^{-1}$ was found using non-porous carbon as support whereas $103 \pm 4\,kJ\,mol^{-1}$ was found for the aerogel with 13 nm pores. Also, dehydrogenation rates up to 50 times faster than those for bulk $LiBH_4$ were measured. Preliminary tests on the evolution of unwanted side-products showed no diborane formation during dehydrogenation, although small amounts of methane were detected during hydrogenation. Improved cycling behavior (hydrogenation at 400 °C, 100 bar H_2) was reported.

A direct comparison between porous carbon and porous silica supports is difficult since the measurement conditions are different. In the case of silica the hydrogen release temperature is lower but a possible chemical reaction between the silica and the $LiBH_4$ has to be considered. It is interesting that the porous carbon support seems to suppress the formation of diborane.

10.6.5
Palladium

The use of Pd as a hydrogen sorbent itself is restricted because of its high price and low hydrogen content. However, it can be used in low concentrations as an effective catalyst for (de)hydrogenation. Recently, several groups reported improved hydrogen physisorption properties for metal organic frameworks and ordered mesoporous carbons by deposition of Pd nanoparticles using incipient wetness impregnation. Kaskel *et al.* deposited 1 wt% Pd on MOF-5 using palladium acetylacetonate as precursor and chloroform as solvent [138]. Reduction under hydrogen flow and drying were performed at maximum temperatures of 200 °C. The deposition of Pd led to a significant decrease in surface area ($958\,m^2\,g^{-1}$) and pore volume ($0.39\,cm^3\,g^{-1}$) compared to pure MOF-5 ($2885\,m^2\,g^{-1}$, $1.18\,cm^3\,g^{-1}$), indicating also partial destruction of the MOF structure during synthesis. The Pd-doped MOF-5 showed improved hydrogen storage properties with a hydrogen uptake of 1.86 wt% at 1 bar and 77 K compared to 1.15 wt% found for the MOF-5 reference. Latroche *et al.* studied the hydrogen sorption properties of Pd and a PdNi alloy deposited on ordered mesoporous carbons replicated from SBA-15 on [165, 166]. To yield 10 wt% of 2 nm Pd nanoparticles on carbon, the support was impregnated with an acid tetrachloropalladinic (H_2PdCl_4) solution in acetone, followed by drying in air at 60 °C and reduction at 300 °C under Ar/H_2 flow. Samples with 16 wt% of 5 nm $Pd_{0.6}Ni_{0.4}$ nanoparticles were synthesized by impregnation with the same solution to which nickel nitrate had been added, followed by reduction under Ar/H_2 flow at 400 °C. Hydrogenation measurements showed that the addition of Pd and PdNi did not increase the hydrogen storage capacity at 77 K, however, the hydrogen uptake at room temperature was strongly improved. In the case of Pd on carbon, the capacity at RT was found to be eight times higher than the carbon reference (0.08 wt% instead of 0.01 wt% at 0.5 MPa), PdNi on carbon yielded a threefold increase (0.01 wt% instead of 0.027 wt% at 0.5MPa).

In all studies (MOF and carbon as support), the authors suggest hydrogen spillover as the mechanism for the improved capacity. Spillover is a phenomenon known from heterogeneous catalysis and describes the transport of an activated species sorbed on a first surface onto another surface that does not form that species under the same conditions [167]. In the described cases, hydrogen molecules first dissociate on the Pd nanoparticles to form hydrogen atoms that then spill over to the MOF/carbon support. The binding energy of the transferred hydrogen atoms is reportedly higher than for physisorbed hydrogen molecules, although the nature of the reversible bonding is not clear, thus explaining the higher hydrogen uptake at room temperature. Even though the hydrogen capacities are low compared to supported metal hydride systems, these studies show that hydrogen storage in classical physisorption systems can be considerably improved by using catalysts for the dissociation of hydrogen molecules.

10.7 Conclusions and Outlook

In this chapter we have discussed the impact of nanosizing and supporting on the hydrogen sorption properties of light metal (hydrides). It is well known from other fields of research that the stability and reactivity of small metal clusters are strongly influenced by their size, especially in the size regime below 10 nm. Much less information is available for small hydride particles. Nevertheless, some general conclusions on the stability can be drawn. For small clusters (up to ~5 nm) stability fluctuation with discrete cluster size is expected due to electronic and geometrical effects. Nanoparticles up to 10–30 nm are significantly less stable than the corresponding bulk materials due to the appreciable contribution of the surface energy to the total energy. In general, metals that form surface hydrides rather than bulk hydrides will have lower hydrogen equilibrium pressures for small hydride particles, while the opposite is true for the reverse situation. Often the rate of hydrogen desorption and absorption are limiting factors for application as a hydrogen storage material. Nanosizing is expected to have a strong beneficial effect on the kinetics, whether it is through the increased specific surface area (and hence hydrogen association and dissociation rates at the surface) or due to the faster diffusion (due to decreased effective solid-state diffusion distances).

It is relatively difficult to prepare nanoparticles and nanostructures of light metal (hydride)s due to their general sensitivity to oxidation, which is greatly enhanced by the large surface to volume ratio for these small structures. Gas-phase clusters of transition metals are interesting systems from a fundamental point of view. Furthermore, appreciable hydrogen storage capacities can be realized (up to several hydrogen atoms per metal atom being present in the cluster) even if the compounds only form surface hydrides. Palladium (hydride), although not relevant as a practical on-board hydrogen storage material, is an important model system to understand the impact of nanosize and surfaces on hydrogen sorption properties. The hydrogen sorption isotherms change dramatically when decreasing the particle size, which can be explained in terms of a significant part of the hydrogen sorption taking place in (sub)surface states. Experimental results also illustrate the importance of the boundary conditions (interface of particles with stabilizers, and complexing agents or supports for the thermodynamics of hydrogen sorption.

For the ionic light metal hydrides, most experimental results available concern magnesium (hydride). Nanoparticles and nanowires have been prepared, by a range of different preparation techniques. Invariably, a large improvement in the sorption kinetics, both in absorption and in desorption, is the result, as well as usually avoiding the need for "activation cycles". Hydrogen sorption data indicate that thermodynamic destabilization of the hydride is also possible, which would be important for application of the material. However, the results on the hydrogen sorption properties of these systems are rather preliminary. No successful preparation and hydrogen sorption measurements of complex hydrides in the form of nanoparticles or nanostructures has been reported so far.

The unsupported nanoparticles and nanostructures are less relevant for practical application, as it is unlikely that they would be stable upon repeated cycling. Metal particles smaller than 10 nm have a strong tendency to sinter, and nanowires (as has been shown experimentally for magnesium (hydride)) often fall apart due to mechanical stress upon cycling. From heterogeneous catalysis it is known that nanoparticles can effectively be stabilized by the presence of an inert support. Silica and carbon materials are standard supports in this field, and available at low cost in a variety of morphologies. Important parameters are the cost, purity, specific surface area and porosity of the support. Carbon has the advantage of being lightweight, thermally conducting, and, generally, less prone than silica to reactions with light metals. Supporting or confining the nanoparticles can not only stabilize the small particle size, but also affect the thermodynamics of hydrogen sorption due to the support's influence on particle shape, electronic interaction, and physical clamping or confinement. Furthermore, it can benefit the macroscopic system properties, such as mechanical stability and thermal conductivity. However, it does increase the total weight of the system.

The first experimental results on 3D-supported nanomaterials for hydrogen storage date from 2005, when a decrease in hydrogen desorption temperatures, and a different reaction pathway were reported for ammonia borane in mesoporous silica. Since then it has been shown, for alanates, magnesium hydride and boronhydrides, that dispersion of the active material on a support can dramatically improve the hydrogen release temperatures and reversibility of the systems. However, this research has just started, and understanding and tuning of these changes in properties is at an early stage of development.

In conclusion, we can say that nanosizing and supporting active hydrogen storage materials has been shown, both theoretically and experimentally, to greatly benefit hydrogen sorption properties such as release temperatures and pressures, loading and charging times, and reversibility. Furthermore, beneficial effects on macroscopic thermal and mechanical stability might be realized. However, this approach also has some drawbacks. It adds to the complexity of the preparation and handling due to the high oxidation sensitivity of nanosized materials, and increases costs. Furthermore, a support decreases the overall storage capacity of the system, and hence can only be used for systems that have high enough intrinsic capacities to compensate for this. On the other hand, thermodynamic, kinetic or reversibility of bulk metal hydrides are often unfavorable for practical application, and these issues might be relieved by nanosizing and supporting. Research on this approach of nanosizing and supporting has only started recently and the future will prove whether the beneficial effects on hydrogen sorption properties will outweigh the disadvantages.

References

1 Bogdanović, A.B. and Schwickardi, M. (1997) *J. Alloys Compd.*, **253–254**, 1–9.
2 Zaluska, A., Zaluski, L., and Ström-Olsen, J.O. (1999) *J. Alloys. Compd.*, **288**, 217–225.
3 Huot, J., Liang, G., and Schulz, R. (2001) *Appl. Phys. A.*, **72**, 187–195.
4 Vajo, J.J., Skeith, S.L., and Mertens, F. (2005) *J. Phys. Chem. B*, **109**, 3719–3722.
5 Barkhordarian, G., Klassen, T., and Bormann, R. (2006) Patent pending, WO2006/063627 A1.
6 Alapati, S.V., Johnson, J.K., and Sholl, D.S. (2006) *J. Phys. Chem. B*, **110**, 8769–8776.
7 Roduner, E. (2006) *Chem. Soc. Rev.*, **35**, 583–592.
8 Knight, W.D., Clemenger, K., de Heer, W.A., Saunders, W.A., Chou, M.Y., and Cohen, M.L. (1984) *Phys. Rev. Lett.*, **52**, 2141–2143.
9 Armentrout, P.B. (2001) *Annu. Rev. Phys. Chem.*, **52**, 423–461.
10 Martin, T.P., Bergmann, T., Göhlich, H., and Lange, T. (1991) *Chem.Phys. Lett.*, **176**, 343–347.
11 Martin, T.P. (1996) *Phys. Rep.*, **273**, 199–241.
12 Borel, J.P. (1981) *Surf. Sci.*, **106**, 1–9.
13 Tyson, W.R. and Miller, W.A. (1977) *Surf. Sc.*, **62**, 267–276.
14 Kumikov, V.K. and Khonokov, Kh.B. (1983) *J. Appl. Phys.*, **54**, 1346–1350.
15 Vitos, L., Ruban, A.V., Skriver, H.L., and Kollár, J. (1998) *Surf. Sc.*, **411**, 186–202.
16 Ge, Q. (2004) *J. Phys. Chem. A*, **108**, 8682–8690.
17 Vajeeston, P., Ravindran, P., and Fjellvåg, H. (2008) *Nanotechnology*, **19** (1–6), 275704.
18 Berube, V., Chen, G., and Dresselhaus, M.S. (2008) *Int. J. Hydrogen Energy*, **33**, 4122–4133.
19 Solliard, C. and Flueli, M. (1985) *Surf. Sci.*, **156**, 487–494.
20 Salomons, A., Griessen, R., de Groot, D.G., and Magerl, A. (1988) *Europhys. Lett.*, **5**, 449–454.
21 Wagemans, R.W.P., van Lenthe, J.H., de Jongh, P.E., van Dillen, A.J., and de Jong, K.P. (2005) *J. Am. Chem. Soc.*, **127**, 16675–16680.
22 Cheung, S., Deng, W.-Q., van Duin, A.C.T., and Goddard, W.A. III (2005) *J. Phys. Chem. A*, **109**, 851–859.
23 Liang, J.-j. (2005) *Appl. Phys. A*, **80**, 173–178.
24 Ojwang, J.G.O., van Santen, R., Kramer, G.J., van Duin, A.C.T., and Goddard, W.A. (2008) *J. Chem. Phys.*, **128**, 1647140.
25 Martin, M., Gommel, C., Borkhart, C., and Fromm, E. (1996) *J. Alloys Compd.*, **238**, 193–201.
26 Borgschulte, A., Westerwaal, R.J., Rector, J.H., Schreuders, H., Dam, B., and Griessen, R. (2006) *J. Catal.*, **239**, 263–271.
27 Zaluski, L., Zaluska, A., and Ström-Olsen, J.O. (1997) *J. Alloys Compd.*, **253–254**, 70–79.
28 Holleck, G.L. (1970) *J. Phys. Chem.*, **74**, 503–511.
29 Nishimura, C., Komaki, M., and Amano, M. (1999) *J. Alloys. Compd.*, **293**, 329–333.

30 Cermak, J. and Kral, L. (2008) *Acta Mater.*, **56**, 2677–2686.

31 Schimmel, H.G., Kearley, G.J., Huot, J., and Mulder, F.M. (2005) *J. Alloys. Compd.*, **404**, 235–237.

32 Sprunger, P.T. and Plummer, E.W. (1994) *Surf. Sci.*, **307–309**, 118–123.

33 Du, A.J., Smith, S.C., Yao, X.D., and Lu, G.Q. (2006) *Surf. Sci.*, **600**, 1854–1859.

34 Schmid, G. (1992) *Chem. Rev.*, **92**, 1709–1727.

35 Bönneman, H. and Richards, R.M. (2001) *Eur. J. Inorg. Chem.*, 2455–2480.

36 Ahmadi, T.S., Wang, Z.L., Green, T.C., Henglein, A., and El-Sayed, M.A. (1996) *Science*, **272**, 1924–1926.

37 Liu, K., Parks, E.K., Richtsmeier, S.C., Pobo, L.G., and Riley, S.J. (1985) *J. Chem. Phys*, **83**, 2882–2888.

38 Knickelbein, M.B. (1999) *Annu. Rev. Phys. Chem.*, **50**, 79–115.

39 Cox, D.M., Fayet, P., Brickman, R., Hahn, M.Y., and Kaldor, A. (1990) *Catal. Lett.*, **4**, 271–278.

40 Swart, I., de Groot, F.M.F., Weckhuysen, B.M., Gruene, P., Meijer, G., and Fielicke, A. (2008) *J. Phys. Chem. A.*, **112**, 1139–1149.

41 Yildirim, T. and Ciraci, S. (2005) *Phys. Rev. Lett.*, **94**, 175501.

42 Nützendadel, C., Züttel, A., Chartouni, D., Schmid, G., and Schlapbach, L. (2000) *Eur. Phys. J. D*, **8**, 245–250.

43 Knapton, A.G. (1977) *Plat. Met. Rev.*, **21**, 44.

44 Pundt, A. and Kirchheim, R. (2006) *Annu. Rev. Mater. Res.*, **36**, 555–608. and references therein.

45 Behm, R.J., Penka, V., Cattania, M.–G., Christmann, K., and Ertl, G. (1983) *J. Chem Phys.*, **78**, 7486–7490.

46 Tarascon, J.M. and Armand, M. (2001) *Nature*, **414**, 359–367.

47 Aurbach, D., Lu, Z., Schechter, A., Gofer, Y., Gizbar, H., Turgeman, R., Cohen, Y., Moshkovich, M., and Levi, E. (2000) *Nature*, **407**, 724–727.

48 Lu, Z., Schechter, A., Moshkovich, M., and Aurbach, D. (1999) *J. Electroanal. Chem.*, **466**, 203–217.

49 Aurbach, D., Weissman, I., Gofer, Y., and Levi, E. (2002) *Chem. Record*, **3**, 61–73.

50 van Tamelen, E.E. and Fechter, R.B. (1968) *J. Am. Chem. Soc.*, **90**, 6854.

51 Ashby, E.C., Liu, J.J., and Goel, A.B. (1978) *J. Org. Chem.*, **43**, 1557–1560.

52 Bogdanovíc, B. (1984) *Int. J. Hydrogen Energy*, **9**, 937–941.

53 Bogdanovíc, B. (1985) *Angew. Chem. Int. Ed. Engl.*, **24**, 262–273.

54 Zidan, R., Slattery, D.K., and Burns, J. (1991) *Int. J. Hydrogen Energy*, **16**, 821–827.

55 Combellas, C., Kanoufi, F., and Thiébault, A. (2001) *J. Electroanal. Chem.*, **499**, 144–151.

56 Imamura, H., Usui, Y., and Takashima, M. (1991) *J. Less-Common Metals*, **175**, 171–176.

57 Imamura, H., Kawahigashi, M., and Tsuchiya, S. (1983) *J. Less-Common Met.*, **95**, 157–160.

58 Haas, I. and Gedanken, A. (2008) *Chem. Commun.*, 1795–1797.

59 Aguey-Zinsou, K.-F. and Ares-Fernández, J.-R. (2008) *Chem. Mater.*, **20**, 376–378.

60 Oumellal, Y., Rougier, A., Nazri, G.A., Tarascon, J.-M., and Aymard, L. (2008) *Nature Mater.*, **7**, 916–921.

61 Klabunde, K.J., Efner, H.F., Satek, L., and Donley, W. (1974) *J. Organomet. Chem.*, **71**, 309–313.

62 Kimura, K. and Bandow, S. (1988) *Phys. Rev. B*, **37**, 4473–4481.

63 Kooi, B.J., Palasantzas, G., and de Hosson, J.Th. (2006) *Appl. Phys. Lett.*, **89** (1–3), 161914.

64 Friedrichs, O., Kolodziejczyk, L., Sánchez-López, J.C., Fernández, A., Lyubenova, L., Zander, D., Köster, U., Aguy-Zinsou, K.F., Klassen, T., and Bormann, R. (2008) *J. Alloys Compd.*, **463**, 539–545.

65 Li, W., Li, C., Ma, H., and Chen, J. (2007) *J. Am. Chem. Soc.*, **129**, 6710–6711.

66 Saita, I., Toshime, T., Tanda, S., and Akiyama, T. (2007) *J. Alloys. Compd.*, **446–447**, 80–83.

67 Zlotea, C., Sahlberg, M., Özbilen, S., Moretto, P., and Andersson, Y. (2008) *Acta Mater.*, **56**, 2421–2428.

68 Sietsma, J.R.A., Meeldijk, J.D., den Breejen, J.P., Versluijs-Helder, M., van Dillen, A.J., de Jongh, P.E., and de Jong,

K.P. (2007) *Angew. Chem. Int. Ed.*, **119**, 4631–4633.

69 Toebes, M.L., Zhang, Y., Hájek, J., Nijhuis, T.A., Bitter, J.H., van Dillen, J.A., Murzin, D.Y., and de Jong, K.P. (2004) *J. Catal.*, **226**, 215–225.

70 Sachtler, W.M.H. (1993) *Acc. Chem. Res.*, **26**, 383–387.

71 Luo, W. (2004) *J. Alloys Compd.*, **381**, 284–287.

72 Chen, P., Xiong, Z., Luo, J., Lin, J., and Tan, K.L. (2003) *J. Phys. Chem. B*, **107**, 10967–10970.

73 Bogdanović, B., Brand, R.A., Marjanović, A., Schwickardi, M., and Tölle, K. (2000) *J. Alloys Compd.*, **302**, 36–58.

74 Gutowska, A., Li, L.Y., Shin, Y.S., Wang, C.M.M., Li, X.H.S., Linehan, J.C., Smith, R.S., Kay, B.D., Schmid, B., Shaw, W., Gutowski, M., and Autrey, T. (2005) *Angew. Chem. Int. Ed.*, **44**, 3578–3582.

75 Goodman, D.W. (1995) *Chem. Rev.*, **95**, 523–536.

76 Gates, B.C. (1995) *Chem. Rev.*, **95**, 511–522.

77 Campbell, C.T. (1997) *Surf. Sci. Rep.*, **27**, 1–111.

78 Henry, C.R. (1998) *Surf. Sci. Rep.*, **31**, 231–325.

79 Stakheev, A.Y. and Kusthov, L.M. (1999) *Appl. Catal. A*, **188**, 3–35.

80 Schönberger, U., Andersen, O.K., and Methfessel, M. (1992) *Acta Metall. Mater.*, **40**, S1–S10.

81 Johnson, K.H. and Pepper, S.V. (1982) *J. Appl. Phys.*, **53**, 6634–6637.

82 Pacchioni, G. and Rösch, N. (1996) *J. Chem. Phys.*, **104**, 7329–7337.

83 Gunter, P.L.J., Niemantsverdriet, J.W., Ribiero, F.H., and Somorjai, G.A. (1997) *Catal. Rev.- Sci. Eng.*, **39**, 77–168.

84 Maiti, A. and Ricca, A. (2004) *Chem. Phys. Lett.*, **1–3**, 7–11.

85 Blick, K., Mitrellas, T.D., Hargreaves, J.S.J., Hutchings, G.L., Joyner, R.W., Kiely, C.J., and Wagner, F.E. (1998) *Catal. Lett.*, **50**, 211–218.

86 Attamny, F., Duff, D., and Baiker, A. (1995) *Catal. Lett.*, **34**, 305–311.

87 Larsen, G. and Haller, G.L. (1989) *Catal. lett.*, **3**, 103–110.

88 Sugimoto, M., Katsuno, H., Hayasaka, T., Ishikawa, N., and Hirasawa, K. (1993) *Appl. Catal. A*, **102**, 167–180.

89 Mojet, B.L., Miller, J.T., Ramaker, D.E., and Koningsberger, D.C. (1999) *J. Catal.*, **186**, 373–386.

90 Mojet, B.L., Hoogenraad, M.S., van Dillen, A.J., Geus, J.W., and Koningsberger, D.C. (1997) *J. Chem. Soc., Faraday Trans.*, **93**, 4371–4375.

91 Rodriguez-Reinoso, F. (1998) *Carbon*, **36**, 159–175.

92 Bartholomew, C.H. (1993) *Catalysis* (ed. J.J. Spivey), Royal Society of Chemistry, London, p. 41.

93 Pertsev, N.A., Zembilgotov, A.G., and Tagantsev, A.K. (1998) *Phys. Rev. Lett.*, **80**, 1988–1991.

94 Song, G., Remhof, A., theis-Bröhl, K., and Zabel, H. (1997) *Phys. Rev. Lett.*, **79**, 5062–5065.

95 Laudahn, U., Pundt, A., Bicker, M., van Hülsen, U., Geyer, U., Wagner, T., and Kirchheim, R. (1999) *J. Alloys Compd.*, **293–295**, 490–494.

96 Schwarz, R.B. and Khachaturyan, A.G. (2006) *Acta Mater.*, **54**, 313–323.

97 Wagner, S. and Pundt, A. (2008) *Appl. Phys. Lett.*, **92**, 051914.

98 Pivak, Y., Gremaud, R., Gross, K., Gonzalez-Silveira, M., Walton, A., Book, D., Schreuders, H., Dam, B., and Griessen, R. (2009) *Scr. Mater.*, **60**, 348–351.

99 Liang, J-j. and Kung, W.-C.P. (2005) *J. Phys. Chem. B*, **109**, 17837–17841.

100 Schlapbach, L. and Zuttel, A. (2001) *Nature*, **414**, 353–358.

101 Zhang, J.S., Fisher, T.S., Ramachandran, P.V., Gore, J.P., and Mudawar, I. (2005) *J. Heat Transfer-Trans Asme*, **127**, 1391–1399.

102 Levesque, S., Ciureanu, M., Roberge, R., and Motyka, T. (2000) *Int. J. Hydrogen Energy*, **25**, 1095–1105.

103 Sanchez, A.R., Klein, H.P., and Groll, M. (2003) *Int. J. Hydrogen Energy*, **28**, 515–527.

104 Nagel, M., Komazaki, Y., and Suda, S. (1986) *J. Less-Common Met.*, **120**, 35.

105 Kohlrausch, F. (1986) *Praktische Physik 3* (eds D. Hahn and S. Wagner), B.G. Teubner, Stuttgart.

106 Kim, K.J., Montoya, B., Razani, A., and Lee, K.H. (2001) *Int. J. Hydrogen Energy*, **26**, 609.

107 Ergun, S. (1952) *Chem. Eng. Progr.*, **48**, 88–94.

108 Marceau, E. (2008) *Handbook of Heterogeneous Catalysis*, vol. 1 (eds G. Ertl, H. Knözinger, F. Schüth, and J. Weitkamp), Wiley-VCH, p. 473.

109 Doesburg, E.B.M., de Jong, K.P., and van Hooff, J.H.C. (1999) *Catalysis: An integrated Approach* (eds R.A. van Santen, P.W.N.M. van Leeuwen, J.A. Moulijn, and B.A. Averill), Elsevier, Amsterdam.

110 Barton, T.J., Bull, L.M., Klemperer, W.G., Loy, D.A., McEnaney, B., Misono, M., Monson, P.A., Pez, G., Scherer, G.W., Vartuli, J.C., and Yaghi, O.M. (1999) *Chem. Mater.*, **11**, 2633–2656.

111 Hench, L.L. and Vasconcelos, W. (1990) *Annu. Rev. Mater. Sci.*, **20**, 269–298.

112 Fricke, J. (1986) Springer Proceedings in Physics 6.

113 Kresge, C.T., Leonowicz, M.E., Roth, W.J., Vartuli, J.C., and Beck, J.S. (1992) *Nature*, **359**, 710–712.

114 Zhao, D.Y., Feng, J.L., Huo, Q.S., Melosh, N., Fredrickson, G.H., Chmelka, B.F., and Stucky, G.D. (1998) *Science*, **279**, 548–552.

115 Ishizuka, N., Minakuchi, H., Nakanishi, K., Soga, N., and Tanaka, N. (1998) *J. Chromatogr. A*, **797**, 133–137.

116 de Jong, K.P., Glezer, J.H.E., Kuipers, H., Knoester, A., and Emeis, C.A. (1990) *J. Catal.*, **124**, 520–529.

117 Cabrera, K. (2004) *J. Sep. Sci.*, **27**, 843–852.

118 Zheng, S.Y., Fang, F., Zhou, G.Y., Chen, G.R., Ouyang, L.Z., Zhu, M., and Sun, D.L. (2008) *Chem. Mater.*, **20**, 3954–3958.

119 McNaught, A.D. and Wilkinson, A. (1997) *Compendium of Chemical Technology*, 2nd Edition, Blackwell Science.

120 de Jong, K.P. and Geus, J.W. (2000) *Cat. Rev. - Sci. Eng.*, **42**, 481–510.

121 Ebbesen, T.W. (1996) *Carbon Nanotubes: Preparation and Properties* (ed. T.W. Ebbesen), CRC Press LLC, p. 182.

122 Ryoo, R., Joo, S.H., and Jun, S. (1999) *J. Phys. Chem. B*, **103**, 7743–7746.

123 Meng, Y., Gu, D., Zhang, F.Q., Shi, Y.F., Cheng, L., Feng, D., Wu, Z.X., Chen, Z.X., Wan, Y., Stein, A., and Zhao, D.Y. (2006) *Chem. Mater.*, **18**, 4447–4464.

124 Adelhelm, P., Hu, Y.S., Chuenchom, L., Antonietti, M., Smarsly, B.M., and Maier, J. (2007) *Adv. Mater.*, **19**, 4012–4017.

125 Al-Muhtaseb, S.A. and Ritter, J.A. (2003) *Adv. Mater.*, **15**, 101–114.

126 Huesing, N. and Schubert, U. (1997) *Angew. Chem.*, **110**, 22–47.

127 Kyotani, T., Ma, Z.X., and Tomita, A. (2003) *Carbon*, **41**, 1451–1459.

128 Johnson, S.A., Brigham, E.S., Ollivier, P.J., and Mallouk, T.E. (1997) *Chem. Mater.*, **9**, 2448–2458.

129 Meyers, C.J., Shah, S.D., Patel, S.C., Sneeringer, R.M., Bessel, C.A., Dollahon, N.R., Leising, R.A., and Takeuchi, E.S. (2001) *J. Phys. Chem. B*, **105**, 2143–2152.

130 Ma, Z.X., Kyotani, T., and Tomita, A. (2002) *Carbon*, **40**, 2367–2374.

131 Chen, L., Singh, R.K., and Webley, P. (2007) *Micropor. Mesopor. Mater.*, **102**, 159–170.

132 Li, H., Eddaoudi, M., O'Keeffe, M., and Yaghi, O.M. (1999) *Nature*, **402**, 276–279.

133 Isaeva, V.I. and Kustov, L.M. (2007) *Russ. J. Gen. Chem.*, **77**, 721–739.

134 Rosi, N.L., Eckert, J., Eddaoudi, M., Vodak, D.T., Kim, J., O'Keeffe, M., and Yaghi, O.M. (2003) *Science*, **300**, 1127–1129.

135 Rowsell, J.L.C. and Yaghi, O.M. (2005) *Angew. Chem. Int. Ed.*, **44**, 4670–4679.

136 Hirscher, M. and Panella, B. (2007) *Scr. Mater.*, **56**, 809–812.

137 Hermes, S., Schroter, M.K., Schmid, R., Khodeir, L., Muhler, M., Tissler, A., Fischer, R.W., and Fischer, R.A. (2005) *Angew. Chem. Int. Ed.*, **44**, 6237–6241.

138 Sabo, M., Henschel, A., Fröde, H., Klemm, E., and Kaskel, S. (2007) *J. Mater. Chem.*, **17**, 3827–3832.

139 Geus, J.W. and van Veen, J.A.R. (1999) *Catalysis: An Integrated Approach* (eds R.A. van Santen, P.W.N.M. van Leeuwen, J.A. Moulijn, and B.A., Averill), Elsevier, Amsterdam.

140 de Jongh, P.E., Wagemans, R.W.P., Eggenhuisen, T.M., Dauvillier, B.S., Radstake, P.B., Meeldijk, J.D., Geus, J.W., and de Jong, K.P. (2007) *Chem. Mater.*, **19**, 6052–6057.

141 Lang, G. (2008/2009) Surface tension of liquid elements, *CRC Handbook of Chemistry and Physics*, 89th edn, CRC Press.

142 Dujardin, E., Ebbesen, T.W., Hiura, H., and Tanigaki, K. (1994) *Science*, **265**, 1850–1852.

143 Orimo, S.I., Nakamori, Y., Eliseo, J.R., Zuttel, A., and Jensen, C.M. (2007) *Chem. Rev.*, **107**, 4111–4132.

144 Lee, B.M., Jang, J.W., Shim, J.H., Cho, Y.W., and Lee, B.J. (2006) *J. Alloys Compd.*, **424**, 370–375.

145 Montone, A., Novakovic, J.G., Antisari, M.V., Bassetti, A., Bonetti, E., Fiorini, A.L., Pasquini, L., Mirenghi, L., and Rotolo, P. (2007) *Int. J. Hydrogen Energy*, **32**, 2926–2934.

146 Baitalow, F., Baumann, J., Wolf, G., Jaenicke-Rossler, K., and Leitner, G. (2002) *Thermochim. Acta*, **391**, 159–168.

147 Wolf, G., Baumann, J., Baitalow, F., and Hoffmann, F.P. (2000) *Thermochim. Acta*, **343**, 19–25.

148 Feaver, A., Sepehri, S., Shamberger, P., Stowe, A., Autrey, T., and Cao, G.Z. (2007) *J. Phys. Chem. B*, **111**, 7469–7472.

149 Claudy, P., Bonnetot, B., Chahine, G., and Letoffe, J.M. (1980) *Thermochim. Acta*, **38**, 75–88.

150 Gross, K.J., Thomas, G.J., and Jensen, C.M. (2002) *J. Alloys Compd.*, **330**, 683–690.

151 Schüth, F., Bogdanovic, B., and Akira, T. (2004) Materials encapsulated in porous matrices for the reversible storage of hydrogen, WO/2005/014469.

152 Baldé, C.P., Hereijgers, B.P.C., Bitter, J.H., and de Jong, K.P. (2006) *Angew. Chem. Int. Ed.*, **45**, 3501–3503.

153 Baldé, C.P., Hereijgers, B.P.C., Bitter, J.H., and de Jong, K.P. (2008) *J. Am. Chem. Soc.*, **130**, 6761–6765.

154 Sandrock, G., Gross, K., and Thomas, G. (2002) *J. Alloys Compd.*, **339**, 299–308.

155 Vegge, T. (2006) *Phys. Chem. Chem. Phys.*, **8**, 4853–4861.

156 Zheng, S.Y., Fang, F., Zhou, G.Y., Chen, G.R., Ouyang, L.Z., Zhu, M., and Sun, D.L. (2008) *Chem. Mater.*, **20**, 3954–3958.

157 Hsueh, H.S., Yang, C.T., Zink, J.I., and Huang, M.H. (2005) *J. Phys. Chem. B*, **109**, 4404–4409.

158 Bogerd, R., Adelhelm, P., Meeldijk, J.H., de Jong, K.P., and de Jongh, P.E. (2009) *Nanotechnology*, **20**, 204019.

159 Zaluska, A., Zaluski, L., and Strom-Olsen, J.O. (1999) *J. Alloys Compd.*, **289**, 197–206.

160 Ingason, A.S., Eriksson, A.K., and Olafsson, S. (2007) *J. Alloys Compd.*, **446**, 530–533.

161 Kostka, J., Lohstroh, W., Fichtner, M., and Hahn, H. (2007) *J. Phys. Chem. C*, **111**, 14026–14029.

162 Mauron, Ph., Buchter, F., Friedrichs, O., Remhof, A., Bielmann, M., Zwicky, C.N., and Züttel, A. (2008) *J. Phys. Chem. B*, **112**, 906–910.

163 Züttel, A., Rentsch, S., Fischer, P., Wenger, P., Sudan, P., Mauron, P., and Emmenegger, C. (2003) *J. Alloys Compd.*, **356**, 515–520.

164 Gross, A.F., Vajo, J.J., Van Atta, S.L., and Olson, G.L. (2008) *J. Phys. Chem. C*, **112**, 5651–5657.

165 Campesi, R., Cuevas, F., Gadiou, R., Leroy, E., Hirscher, M., Vix-Guterl, C., and Latroche, M. (2008) *Carbon*, **46**, 206–214.

166 Campesi, R., Cuevas, F., Leroy, E., Hirscher, M., Gadiou, R., Vix-Guterl, C., and Latroche, M. (2009) *Micropor. Mesopor. Mater.*, **117**, 511–514.

167 Conner, W.C. and Falconer, J.L. (1995) *Chem. Rev.*, **95**, 759–788.

Index

Handbook of Hydrogen Storage. Edited by Michael Hirscher

ISBN: 978-3-527-32273-2